Methods to Study Litter Decomposition

Felix Bärlocher • Mark O. Gessner
Manuel A. S. Graça
Editors

Methods to Study Litter Decomposition

A Practical Guide

Second Edition

Springer

Editors
Felix Bärlocher
Department of Biology
Mount Allison University
Sackville, NB, Canada

Manuel A. S. Graça
MARE – Marine and Environmental Sciences Centre & Department of Life Sciences
University of Coimbra
Coimbra, Portugal

Mark O. Gessner
Department of Experimental Limnology
Leibniz Institute of Freshwater Ecology and Inland Fisheries (IGB)
Stechlin, Germany

Department of Ecology
Berlin Institute of Technology (TU Berlin)
Berlin, Germany

ISBN 978-3-030-30517-8 ISBN 978-3-030-30515-4 (eBook)
https://doi.org/10.1007/978-3-030-30515-4

Cover Illustration: Núria Bonada.

This Springer imprint is published by the registered company Springer Nature Switzerland AG.
The registered company address is: Gewerbestrasse 11, 6330 Cham, Switzerland

Foreword

Terrestrial flora of vascular plants probably first appeared during the Middle Ordovician about 500 million years before present. But the development and spread of arborescent plants in the Middle and Late Devonian or perhaps the early Carboniferous, 360–385 million years before present, had major effects on the biosphere. These effects included increased rates of soil formation and weathering, major modifications in the global hydrologic cycle, reduced erosion and lower sediment yield, reduced atmospheric CO_2 concentrations, and global cooling.[1] Vascular plants were also a new resource for heterotrophic organisms, with the linkage between animal food webs and these plants probably being indirect, through detritivorous arthropods.[2] The remains of the structural components of vascular plants were highly resistant to decay relative to their algal ancestors and accumulated in soils and sediments of shallow aquatic ecosystems adjacent to these forests. Eventually, microorganisms and animals evolved to make use of this abundant organic resource. Thus vascular plant litter became a major component of both terrestrial and freshwater ecosystems.

As a result, modern freshwater ecosystems are supported not only by organic matter from autochthonous production by algae and macrophytes, they also receive allochthonous inputs of organic matter produced by photosynthesis elsewhere, typically by riparian trees. Across the world's freshwaters, the relative importance of these two pathways extends to both ends of the spectrum. As early as 1912, August Thienemann pointed out the importance of allochthonous matter in streams, but early limnology textbooks made almost no mention of terrestrial plant material. In his 1952 textbook, Franz Ruttner stated that the majority of stream animals are phytophagous, feeding on algae growing on stones, although Paul Welch's *Limnology*

---

[1] Algeo, T. J., & Scheckler, S. E. (1998). Terrestrial-marine teleconnections in the Devonian: Links between the evolution of land plants, weathering processes, and marine anoxic events. *Philosophical Transactions of the Royal Society of London B, 353*, 113–130.

[2] Behrensmeyer, A. K., Damuth, J. D., DiMichele, W. A., Potts, R., Sues, H., & Wing, S. L. (Eds.) (1992). *Terrestrial ecosystems through time: Evolutionary paleoecology of terrestrial plants and animals*. Chicago: University of Chicago Press.

published the same year included allochthonous detritus in an outline of sestonic materials. Also, Robert Coker (*Streams Lakes Ponds*, 1954) emphasized the importance of detritus to both lakes and streams: "A well-shaded stream may be cooler and have less capacity for the growth of green plants within the stream, but, where streams are well fed with organic matter through drainage from surrounding lands, what is called 'original production' by green plants within the stream is perhaps less significant than in still waters."

Research showing the importance of leaf litter to stream invertebrates includes the 1958 study by Donald Scott, who noted that the most abundant stream organisms in Georgia (USA) rivers fed on small organic particles and that a large part of this material might be derived from the adjacent land. Further support for the importance of forest litter to streams came from Herbert H. Ross in a 1963 publication where he reported a correlation between caddisfly distribution and terrestrial biome and from Egglishaw's 1964 study of the correlation between bottom fauna distribution and plant litter. In the first published book devoted to stream ecology, *The Ecology of Running Water*, Noel Hynes noted in 1970 that recognition of terrestrial forest litter as an important energy source to streams initially came from studies of aquatic invertebrate diets, including his own work on stoneflies, and from other studies showing that at least during autumn and winter most stream invertebrates were feeding on leaf litter. Ecosystem studies on energy flow in the 1950s, 1960s, and early 1970s by John Teal, Dan Nelson, W.C. Minkley, Wayne Minshall, and Stuart Fisher also clearly showed that many stream ecosystems were primarily dependent on allochthonous organic matter.

With these early studies, it became clear that knowledge of how this organic material is used by freshwater organisms is an important part of understanding ecosystem function in freshwaters. Streams and lakes are a critical part of our environment, providing drinking water, water for agriculture and industry, food, habitat for many organisms, and recreation. Understanding ecosystem function is critical to maintaining this important resource, and critical to this understanding are the field and laboratory methods to study litter decomposition presented in this book.

Methods for studying decomposition of vascular plant material were first developed for terrestrial systems. E. Melin's 1930 laboratory study of tree leaves and bracken stems is often cited as one of the first publications reporting decomposition rates, although in 1933 H.A. Lunt cited two 1890 field studies by E. Ramann assessing weight loss of oak and beech leaves. The litter bag method for measuring decomposition rate was first used by Bocock and Gilbert in a study published in 1957, motivated by the need to measure leaf litter decomposition in the field "to maintain the identity of the experimental material while allowing the full range of environmental factors to operate." They used nylon hairnets to measure leaf decomposition in a forest in the English Lake District. Prior to that, litter decomposition had been measured by confining leaves in wire mesh baskets or mesh-covered boxes. In perhaps the first intersite study of litter decomposition, Hans Jenny and his colleagues compared alfalfa decomposition in screen-covered cans placed at sites in Costa Rica, Colombia, and California, a study that was published in 1949. They were also the first to suggest a negative exponential model for litter mass loss. In another early methodological study, published in 1963, Martin Witkamp and Jerry Olson compared decomposition rates of oak leaves in mesh bags and tied on strings and discussed the advantages and disadvantages of each method.

Freshwater studies of litter decomposition go back to an investigation by Levanidov published in 1949 cited by Noel Hynes in his 1970 book. Other early studies published in 1969 and 1970 include the work by C.P. Mathews and A. Kowalczewski in the Thames River (England) and research by M. Witkamp and M. Frank and W.A. Thomas in forest, stream, and pond sites at Oak Ridge National Laboratory (Tennessee, USA). Also, in a study published in 1971, Narinder Kaushik and Noel Hynes used leaf discs in bags to compare decomposition rates in a polluted and a relatively unpolluted river in Ontario, Canada. In another classic study, published in 1974, instead of using mesh bags, Bob Petersen and Ken Cummins attached bunches of leaves to bricks in their very extensive study of leaf decomposition along a stream in Michigan (USA).

The earliest studies of vascular plant decomposition in lakes were those at Oak Ridge National Laboratory, followed by I.D. Hodkinson's measurements in 1975 of leaf litter and pine bark decomposition in mesh bags submerged in a beaver pond, and Jim Barnes and his colleagues published a study in 1978 that used leaf packs to measure tree leaf decomposition in a lake in Michigan (USA).

One of the most important reasons for measuring vascular plant litter decomposition is to enable comparisons of processes among ecosystems. This has sometimes been accomplished by planned studies using the same methods at many sites, such as the 2012 study by Guy Woodward and his colleagues. However, in other studies, meta-analysis is often difficult because of the variety of methods used in different studies, for example, the 2017 study by Jennifer Follstad-Shah and her colleagues. This book provides a compilation of protocols that provide opportunities to compare decomposition processes among freshwater ecosystems based on a wide range of approaches. Many of these methods will be equally useful for applications in other types of both aquatic and terrestrial ecosystems.

As I attempt to follow the growing literature on litter decomposition, I find that we know little about the actual mechanisms of decay. We know a lot about how fast litter decomposes and about the importance of different factors such as temperature and nutrient supply that are correlated with decomposition rates. We also know a lot about what invertebrate and microbial organisms are associated with decomposing litter. But what is actually happening during decomposition? What are these organisms doing? Microbial and ecosystem ecology are very different fields than they were 50 years ago when I learned about leaf decomposition from Martin Witkamp and Dac Crossley. The early studies of litter decomposition in streams had identified many of the invertebrates associated with the process, but it wasn't until 1973 when Ken Cummins synthesized information on the different feeding mechanisms of invertebrates that we began to understand more about how invertebrates actually participate in the decomposition process. Similarly, molecular techniques have provided great insight into the identity and diversity of fungi and bacteria associated with decaying litter, but with newer techniques, many of which are described in this book, we are beginning to achieve a better understanding of how these microorganisms actually participate in the litter decomposition process.

There has been a surge in studies of litter decomposition in freshwater in the last few decades. As these studies continue, many of the techniques detailed in this book will be useful to refine our knowledge of litter decomposition and integrate it into a more complete picture of freshwater ecosystem functioning.

Emeritus Professor of Ecology
Virginia Tech University,
Blacksburg, VA, USA

Jack Webster

Preface

Litter decomposition is a central ecosystem process in forests, grasslands and many other ecosystems, including marine and freshwaters situated at the interface where land meets water. Mangroves and marshes, lake margins and streams, which typically receive large inputs of plant litter from their riparian zones, are prominent examples. Indeed, most of the carbon and energy that flow through these ecosystems derive from vascular plant remains. This plant litter fuels 'brown' food webs characterized by heterotrophic microbes and detritivores that use and transfer dead organic matter to higher-level consumers, contrasting with 'green' food webs where carbon, energy and nutrients pass from live primary producers to higher trophic levels via herbivores.

Given the prevalence of plant litter as a resource, understanding the factors governing rates and routes of carbon, energy and nutrient flows during litter decomposition continues to be a key focus in ecosystem science. The field is dynamic, as evidenced by the number of publications on litter decomposition per year which has roughly doubled since the first edition of the book in 2005. This growth is driven by attempts to unravel fundamental relationships and mechanisms underlying the decomposition process, to understand natural fluctuations and large-scale patterns, and to assess the consequences of local and global environmental change on energy flow, carbon and nutrient cycling in ecosystems.

Reflecting these developments as well as new technologies that have emerged and several gaps identified in the first edition of the book, this second edition of "Methods to Study Litter decomposition – A Practical Guide" has been considerably expanded. It now comprises 63 chapters, a 50% increase over the first edition. Several of the new chapters address methodologies that are becoming increasingly relevant to the study of litter decomposition. All previous chapters were thoroughly revised and updated where needed. In addition, tables have been added to many chapters to provide guidelines on the range of values that can be expected when applying a particular method. Newly included are also links to various online documents that provide supplementary information, such as assigning invertebrate taxa to functional feeding groups, or spreadsheets and computer code to perform calculations presented in the corresponding chapters.

As one of the major additions, the new edition covers significant developments in the application of molecular methods to environmental samples. These methods open radically new opportunities to analyse patterns of biodiversity, abundance and activity of microorganisms associated with decomposing litter. Consequently, we introduced a new section on molecular microbial community analysis to accommodate two previous chapters on fingerprinting techniques (T-RFLP and DGGE) and five new chapters presenting methods to examine litter-associated microbial decomposers by meta-barcoding, quantitative real-time PCR, analyses of fungal precursor rRNA, transcriptomics and proteomics.

We also added a new chapter on polyunsaturated fatty acids (PUFAs), which have recently been found to play a role for plant litter consumers. The presented procedures to quantify critical PUFAs in leaf litter, invertebrates and other biological material will be useful for future studies to determine patterns of PUFAs in decomposing litter, characterize the main producers of PUFAs and assess how these contribute to the population dynamics of litter consumers.

An evident gap in the first edition of the book was the limited information on methods to examine litter-consuming invertebrates. The new edition redresses this imbalance by adding four chapters to a dedicated section on consumers. One of those chapters describes procedures to process aquatic invertebrates that colonize decomposing leaves, the others focus on determining invertebrate feeding and growth, energy budgets of individuals, and their importance in litter dynamics at the whole-stream scale. These methods will allow a more comprehensive appreciation of the trophic ecology of litter-feeding invertebrates and their roles in the decomposition process. That last objective will also be facilitated by a new chapter that was included to overcome, or mitigate, the difficulty of correctly identifying aquatic invertebrates as litter consumers. This is particularly important for tropical invertebrates, since the great majority of information available on litter consumers and their role in decomposition has originated from temperate zones.

The first edition of *Methods in Litter Decomposition* emphasized the overall dynamics and chemical and physical properties of leaf litter. Multiple chapters on microbial decomposers and enzymatic capabilities were also included. All these previous chapters were retained, but we added three new ones in the second edition to address previously neglected components of organic matter: fine particulate organic matter, wood (frequently the largest contributor to organic matter in forested streams), and aquatic macrophytes, which may be highly productive, accumulating large amounts of litter in ponds, littoral zones of lakes, freshwater wetlands and intertidal areas.

Early studies on litter decomposition were dominated by largely descriptive approaches, but hypotheses are best tested when system components can be adjusted and modified in a controlled way to assess responses. We therefore added an entirely new section focussing on litter manipulations. It introduces techniques to alter litter nutrient stoichiometry, label leaf litter with ^{15}N, and manipulate food quality. Furthermore, a chapter on techniques to inoculate leaf litter with aquatic hyphomycetes allows modifying fungal diversity and identity in experiments, for instance to explore relationships between biodiversity and ecosystem functioning in streams.

The short final section of the book on data analysis includes three new chapters. They complement a primer on statistics that had already been part of the first edition. The first of the three new chapters considers temperature normalization of litter decomposition rates, allowing comparisons across temperature regimes. A second chapter presents bioinformatic approaches and procedures needed to evaluate the large amounts of information generated by molecular approaches. Finally, the third chapter is an introduction to meta-analysis as a basis for combining and evaluating previously published data from various sources. It is meant to encourage syntheses of the notable body of information that has accumulated over five decades (streams) or even longer (forests).

Written primarily by stream ecologists, *Methods in Litter Decomposition* continues to have a strong bias towards running waters. This is particularly evident in the first section on organic matter dynamics and in the section on litter consumers. Nevertheless, most of the other methods presented in the book can be directly used or adapted with little modification to study litter decomposition and decomposers in other freshwater, marine or terrestrial ecosystems. This includes field procedures such as the study of decomposition by tagging plants instead of using litter bags, which would be appropriate to study decomposition in grasslands in general, as well as most of the chemical analytical procedures, litter manipulations, analyses of microbial decomposer communities, biomass and activities, and data analyses.

We would be remiss not to take the opportunity to thank once again the many authors for their enthusiasm when we approached them with the proposal for a new edition, for the fine revisions and new contributions we received, and for the diligence and patience during the editing process. We also extend our thanks to the publisher for approaching us in the first place and for granting us extensions for submitting the final manuscript. Preparing this revised edition was a more time-consuming endeavour than we had anticipated. We trust, however, that the extra effort and delay were worthwhile and produced a book that will be informative and a valuable hands-on aid both for beginning students and experienced scientists interested in litter decomposition.

Sackville, Canada — Felix Bärlocher
Stechlin, Germany — Mark O. Gessner
Coimbra, Portugal — Manuel A. S. Graça

Contents

Contributors

Manuela Abelho Instituto Politécnico de Coimbra, Escola Superior Agrária, Coimbra, Portugal
Centre for Functional Ecology, Department of Life Sciences, University of Coimbra, Coimbra, Portugal

Ricardo J. Albariño INstituto de Investigaciones en BIOdiversidad y MedioAmbiente (INIBIOMA), Laboratorio de Fotobiología, National University of Comahue – CONICET, Quintral, Argentina

Julio Arce-Funck MARE – Marine and Environmental Sciences Centre & Department of Life Sciences, University of Coimbra, Coimbra, Portugal

Maite Arroita Department of Plant Biology and Ecology, Faculty of Science and Technology, University of the Basque Country UPV/EHU, Bilbao, Spain

Mark O. Baerlocher Department of Vascular and Interventional Therapies, Royal Victoria Hospital, Barrie, ON, Canada

Petr Baldrian Laboratory of Environmental Microbiology, Institute of Microbiology of the Czech Academy of Sciences, Praha, Czech Republic

Felix Bärlocher Department of Biology, Mount Allison University, Sackville, NB, Canada

Christiane Baschien Leibniz-Institute DSMZ – German Collection of Microorganisms and Cell Cultures, Braunschweig, Germany

Severine Bienaimé Biogeochemistry of Forest Ecosystems, INRA, Champenoux, France

Elizabeth C. Bourne Department of Ecosystem Research, Leibniz Institute of Freshwater Ecology and Inland Fisheries (IGB), Berlin, Germany
Berlin Center for Genomics in Biodiversity Research, Berlin, Germany

Luz Boyero Department of Plant Biology and Ecology, University of the Basque Country, Leioa, Spain & Ikerbasque, Bilbao, Spain

Nanna Buesing Department of Aquatic Ecology, Eawag, Center for Ecology, Evolution, and Biogeochemistry (CEEB), Kastanienbaum, Switzerland

Marcos Callisto Instituto de Ciencias Biologicas, Federal University of Minas Gerais, Belo Horizonte, Brazil

Cristina Canhoto Centre for Functional Ecology, Department of Life Sciences, University of Coimbra, Coimbra, Portugal

Steffen C. Carl Leibniz-Institute DSMZ – German Collection of Microorganisms and Cell Cultures, Braunschweig, Germany

Umweltbundesamt, Berlin, Germany

Eric Chauvet EcoLab, Université de Toulouse, CNRS, Toulouse, France

Yoshimura Chihiro Department of Civil and Environmental Engineering, School of Environment and Society, Tokyo Institute of Technology, Tokyo, Japan

Francisco Correa-Araneda Unidad de Cambio Climático y Medio Ambiente, Instituto de Estudios de Hábitat (IEH), Facultad de Arquitectura y Construcción, Universidad Autónoma de Chile, Temuco, Chile

Etienne Dambrine CARRTEL, INRA, Le Bourget du Lac, France

Michael Danger University of Lorraine, Metz, France

Letitia Da Ros Department of Wood Science, University of British Columbia, Vancouver, BC, Canada

Enrique Descals Esporles, Mallorca, Spain

Michael Dobson APEM Ltd, Edinburgh Technopole, Penicuik, Midlothian, UK

Sofia Duarte Department of Biology, Centre of Molecular and Environmental Biology (CBMA), University of Minho, Braga, Portugal

Arturo Elosegi Department of Plant Biology and Ecology, Faculty of Science and Technology, University of the Basque Country, UPV/EHU, Bilbao, Spain

Andrea C. Encalada Colegio de Ciencias Biológicas y Ambientales, University of San Francisco de Quito, Quito, Ecuador

Vincent Felten University of Lorraine, Metz, France

Verónica Ferreira MARE – Marine and Environmental Sciences Centre & Department of Life Sciences, University of Coimbra, Coimbra, Portugal

Mogens R. Flindt Department of Biology, SDU – University of Southern Denmark, Odense, Denmark

Elisabeth Funke Department of Ecosystem Research, Leibniz Institute of Freshwater Ecology and Inland Fisheries (IGB), Berlin, Germany

Mark O. Gessner Department of Experimental Limnology, Leibniz Institute of Freshwater Ecology and Inland Fisheries (IGB), Stechlin, Germany

Department of Ecology, Berlin Institute of Technology (TU Berlin), Berlin, Germany

José M. González Departamento de Biología y Geología, Física y Química Inorgánica, Universidad Rey Juan Carlos, Madrid, Spain

Manuel A. S. Graça MARE – Marine and Environmental Sciences Centre & Department of Life Sciences, University of Coimbra, Coimbra, Portugal

Vladislav Gulis Department of Biology, Coastal Carolina University, Conway, SC, USA

Björn Hendel iFZ Research Centre for Biosystems, Land Use and Nutrition, Department of Animal Ecology and Systematics, Justus Liebig University, Giessen, Germany

Ellard R. Hunting School of Biological Sciences, University of Bristol, Bristol, UK

Paul R. Johnston Department of Ecosystem Research, Leibniz Institute of Freshwater Ecology and Inland Fisheries (IGB), Berlin, Germany

Berlin Center for Genomics in Biodiversity Research, Berlin, Germany

Institut für Biologie, Freie Universität Berlin, Berlin, Germany

Katharina M. Keiblinger Department of Forest and Soil Sciences, Institute of Soil Research, University of Natural Resources and Life Sciences Vienna, Vienna, Austria

Michiel H. S. Kraak Institute for Biodiversity and Ecosystem Dynamics (IBED-FAME), University of Amsterdam, Amsterdam, The Netherlands

Kevin A. Kuehn School of Biological, Environmental and Earth Sciences, The University of Southern Mississippi, Hattiesburg, MS, USA

Ana I. Lillebø Department of Biology & CESAM, University of Aveiro, Aveiro, Portugal

Shawn D. Mansfield Department of Wood Science, University of British Columbia, Vancouver, BC, Canada

Ludmila Marvanová Faculty of Science, Czech Collection of Microorganisms, Masaryk University, Brno, Czech Republic

Jürgen Marxsen iFZ Research Centre for Biosystems, Land Use and Nutrition, Department of Animal Ecology and Systematics, Justus Liebig University, Giessen, Germany

Frank Masese Department of Fisheries & Aquatic Sciences, University of Eldoret, Eldoret, Kenya

Jon Molinero Escuela de Gestión Ambiental, Pontificia Universidad Católica del Ecuador Sede Esmeraldas, Espejo y subida a Santa Cruz, Esmeraldas, Ecuador

Michael T. Monaghan Department of Ecosystem Research, Leibniz Institute of Freshwater Ecology and Inland Fisheries (IGB), Berlin, Germany
Berlin Center for Genomics in Biodiversity Research, Berlin, Germany

Marcelo S. Moretti Ecology of Aquatic Insects Lab, University of Vila Velha, Vila Velha, Brazil

Paul T. M. Neumann Friedrich-Kreusch-Weg 35, Langenfeld, Germany

Liliya G. Nikolcheva Mackenzie Orthodontics, Saint John, NB, Canada

Richard G. Pearson College of Science and Engineering, TropWater, James Cook University, Townsville, Australia

Frank Peeters Limnological Institute, University of Constance, Konstanz, Germany

Javier Pérez Department of Plant Biology and Ecology, Faculty of Science and Technology, University of the Basque Country (UPV/EHU), Bilbao, Spain

Jesús Pozo Department of Plant Biology and Ecology, Faculty of Science and Technology, University of the Basque Country, UPV/EHU, Bilbao, Spain

Alonso Ramírez Department of Biological Sciences, North Carolina State University, Raleigh, NC, USA
University of Puerto Rico, San Juan, PR, USA

John S. Richardson Department of Forest and Conservation Sciences, University of British Columbia, Vancouver, BC, Canada

Katharina Riedel Department of Microbial Physiology and Molecular Biology, Institute of Microbiology, University of Greifswald, Greifswald, Germany

Sahadevan Seena Department of Life Sciences, MARE – Marine and Environmental Sciences Centre, University of Coimbra, Coimbra, Portugal

Robert L. Sinsabaugh Department of Biology, University of New Mexico, Albuquerque, NM, USA

Libe Solagaistua Department of Plant Biology and Ecology, Faculty of Science and Technology, University of the Basque Country UPV/EHU, Bilbao, Spain

April E. Sparkman Department of Geography & Environmental Systems, University of Maryland, Baltimore, MD, USA

Daniel Steiner Department of Aquatic Ecology, Eawag, Centre for Ecology, Evolution, and Biogeochemistry (CEEB), Kastanienbaum, Switzerland

Martina Štursová Laboratory of Environmental Microbiology, Institute of Microbiology of the Czech Academy of Sciences, Praha, Czech Republic

Keller Suberkropp Department of Biological Sciences, University of Alabama, Tuscaloosa, AL, USA

Christopher M. Swan Department of Geography & Environmental Systems, University of Maryland, Baltimore, MD, USA

Faride Unda Department of Wood Science, University of British Columbia, Vancouver, BC, Canada

Gea H. Van Der Lee Institute for Biodiversity and Ecosystem Dynamics (IBED-FAME), University of Amsterdam, Amsterdam, The Netherlands

Eric von Elert Institute of Zoology, University of Cologne, Cologne, Germany

J. Arie Vonk Institute for Biodiversity and Ecosystem Dynamics (IBED-FAME), University of Amsterdam, Amsterdam, The Netherlands

Christian Wurzbacher Chair of Urban Water Systems Engineering, Technical University of Munich, Garching, Germany

Catherine M. Yule School of Science and Engineering, University of the Sunshine Coast, Queensland, Australia

Bernd Zeller Biogeochemistry of Forest Ecosystems, INRA, Champenoux, France

Martin Zimmer Leibniz Centre for Tropical Marine Research (ZMT), University of Bremen, Bremen, Germany

About the Editors

Felix Bärlocher is a Research Professor at Mt. Allison University, Sackville, New Brunswick, Canada. He retired from teaching in 2016. He has pioneered and published landmark studies on the ecology of aquatic hyphomycetes, a polyphyletic group of fungi that dominate the decomposition of riparian tree leaves in streams. His additional central research interests are the taxonomy and evolution of aquatic hyphomycetes, molecular approaches to characterize these fungi, and trophic interactions with litter-consuming invertebrates.

Mark O. Gessner is Department Head at the Leibniz Institute of Freshwater Ecology and Inland Fisheries (IGB) and Professor at the Berlin Institute of Technology (TU Berlin) in Germany. Trained as an aquatic ecosystem scientist, he is widely recognized for his work over 30 years on ecosystem processes, particularly many facets of the decomposition of plant litter in freshwaters, including the role of fungi, bacteria, and detritivorous invertebrates, as well as the importance of changing environmental conditions, biodiversity, and human activities in controlling decomposition rates.

Manuel A. S. Graça is a Researcher at MARE (Marine & Environmental Sciences Centre) and Professor at the University of Coimbra, Portugal. His extensive work for more than 30 years in Mediterranean and tropical climates has focused on shaded headwater streams, where he investigates particularly the trilogy of leaf litter – microbial decomposers – litter-consuming detritivores and their role in detrital food webs. Studies on anthropogenic impacts on streams and benthic stream invertebrates are another major long-standing research interest.

Part I
Litter Dynamics

Chapter 1
Litter Input

Arturo Elosegi and Jesús Pozo

Keywords Allochthonous resource · Coarse particulate organic matter · CPOM · Detritus · Fruits · Lateral litter inputs · Leaf litter · Litter dynamics · Litter fall · Litter transport · Litter traps · Streams · Vertical litter inputs · Wood

1 Introduction

Allochthonous plant litter is a major source of carbon and energy for stream organisms, especially in narrow reaches where riparian cover limits primary production (Vannote et al. 1980; Collins et al. 2016). Coarse particulate organic matter (CPOM) refers to organic particles larger than 1 mm, which primarily includes or originates from plant litter (Webster and Meyer 1997). Typically, several hundred grammes of litter dry mass per square metre of stream bed are received per year (Table 1.1). Even when macrophytes are abundant, the detrital pathway driven by allochthonous inputs is important for stream communities (Hill and Webster 1983). Litter inputs in forested sites mainly depend on forest composition and age, but other factors such as topography and wind also play a role (Bilby and Heffner 2016). Therefore, interannual variations in litter inputs are to be expected in managed forest through the harvest cycle (Santiago et al. 2011) and in all forests as a consequence of climatic variations (Sanpera-Calbet et al. 2016).

Litter inputs can be quantified for a whole stream or selected reaches (Cummins et al. 1983; Minshall 1996). It is important to realize the difference, because upstream import to individual reaches can be substantial, but is irrelevant when an entire stream system is considered. However, where reaches along a stream differ greatly in riparian vegetation or bank characteristics, whole-stream studies require extensive sampling to be meaningful.

A. Elosegi (✉) · J. Pozo
Department of Plant Biology and Ecology, Faculty of Science and Technology, University of the Basque Country, UPV/EHU, Bilbao, Spain
e-mail: arturo.elosegi@ehu.eus; jesus.pozo@ehu.eus

F. Bärlocher et al. (eds.), *Methods to Study Litter Decomposition*,
https://doi.org/10.1007/978-3-030-30515-4_1

Table 1.1 Vertical and lateral litter inputs to selected streams expressed as ash-free dry mass

Location	Vegetation	Stream order	Vertical input (g m^{-2} y^{-1})	Lateral input (g m^{-2} y^{-1})	References
Alaska, USA	Tundra	4	0	500	1
Québec, Canada	Taiga	1	417	344	2
Québec, Canada	Taiga	2	217	56	2
Oregon, USA	Temp. conifer forest	1	537	667	3
Oregon, USA	Temp. conifer forest	2–4	355[a]	57[a]	4
Oregon, USA	Temp. deciduous forest	2–4	454[a]	98[a]	4
Japan	Temp. deciduous forest	2	324[a]	–	5
Japan	Conifer plantation	2	338[a]	–	5
Japan	Clearcut	2	26[a]	–	5
Germany	Temp. mixed forest	1	700	–	3
UK	Temp. mixed forest	2	328–387	48–58	6
Spain	Temp. deciduous forest	1	611	104	7
Spain	Eucalyptus plantation	1	478	24	7
Spain	Mediterranean forest	3	520–680[b]	120–300[b]	8
Portugal	Temp. deciduous forests	1–2	261	–	9
Portugal	Eucalyptus plantation	1–2	204	–	9
Arizona, USA	Desert shrubs	5	17	3	10
Georgia, USA	Temp. mixed forest	6	843	3520	11
Panamá	Tropical rainforest	2	841–1022[a]	126–168[a]	12
Brazil	Tropical gallery forest	3	713	421	13
Brazil	Atlantic forest	2	627[a]	724[a]	14
Australia	Temp. eucalyptus forest	3	617	61	15

1, Harvey et al. (1997); 2, Naiman and Link (1997); 3, Benfield (1997); 4, Hart et al. (2013); 5, Inoue et al. (2012); 6, Pretty et al. (2005); 7, Pozo et al. (1997); 8, Acuña et al. (2007); 9, Abelho and Graça (1996); 10, Jones et al. (1997); 11, Meyer et al. (1997); 12, Colón-Gaud et al. (2008); 13, Afonso et al. (2000); 14, Gonçalves et al. (2014); 15, Campbell et al. (1992)

[a]Converted from DM to AFDM by multiplying by 0.90

[b]Converted from C to AFDM by multiplying by 2

A large part of the litter entering a stream channel consists of leaves from riparian vegetation, particularly in forested streams, although other more nutritious materials, such as fruits or insects, can be important for some stream organisms (Brett et al. 2017). Inputs include transport from upstream, direct (or vertical) inputs, and lateral inputs of material deposited on the forest floor and mobilized by wind or some other agent. Another significant contribution is made by wood (Díez et al. 2001). A large part of the wood inputs occur because of debris torrents, landslides, unusually severe storms, and similar extreme events (Harmon et al. 1986). Because of their sporadic occurrence, these inputs are not easily measured by routine procedures. Consequently, it is often useful to differentiate between inputs of coarse wood and other sorts of litter.

Of all pathways of litter input, transport from upstream is by far the most difficult to measure. It can be highly variable in response to discharge and is often impossible to estimate accurately during high flow. A large portion of litter deposited in the stream channel and on the stream banks is mobilized during spates (Webster et al. 1990), which tend to be unpredictable and hence difficult to sample. Therefore, measurements of long-term litter transport in streams tend to be gross underestimates, even when they are based on frequent sampling schedules (Golladay 1997). This shortcoming can be partly corrected for by plotting stream discharge versus the concentration of drifting litter from a long time series and extrapolating to the concentrations expected during floods (Webster et al. 1990). Nevertheless, discharge-concentration curves typically yield poor fits (Gurtz et al. 1980). The first flood after a long autumn base flow period will scour a much larger amount of litter than similar floods later in the season. Furthermore, the relation between concentration and discharge is characterized by hysteresis, with litter concentrations being typically lower during the falling limb of a storm hydrograph, because most of the litter deposited near the stream channel has already been scoured away, thus limiting further mobilization (Williams 1989).

In several studies, attempts have been made to sample continuously a portion of the flowing water or to build grid-like structures to retain all litter transported during extended periods (Likens and Bormann 1995; Eggert et al. 2012). However, these approaches require large effort and are seldom feasible, especially when streams are larger than first order. Here, we describe a less accurate but more readily applicable method for nonwoody CPOM inputs to small streams. Vertical inputs are collected with litter-fall baskets, blow-in inputs with lateral traps, and transport inputs with drift nets (Webster and Meyer 1997).

Fig. 1.1 (Left) A trap for determining vertical litter inputs. The mesh is fixed to a wooden or metallic frame hanging from nearby trees by four ropes. (Right) Three large traps set up in the riparian area and covering the total surface of the reach where inputs are estimated. The stream is further to the right, not visible on the photo

2 Site Selection and Equipment

2.1 Site Selection

Litter inputs can be measured in any stream, but they are most meaningful in small streams surrounded by riparian forests. Choosing a springbrook avoids having to measure inputs from upstream. Nevertheless, despite greater logistical difficulties, measurement uncertainties and hazards, even large rivers can be studied. Reach length depends on the objectives of the research and on the variability of the riparian areas. A 100-m reach without tributaries will suffice for most purposes. Streams with extensive floodplains are more difficult to study, as most inputs are likely to occur during floods.

2.2 Equipment and Material

- Litter-fall traps (Fig. 1.1) can be constructed from plastic laundry baskets or by sewing 1-mm mesh to any wooden or metallic frame. Although 0.25-m^2 traps appear to be most popular, traps ranging from 0.025 to 1 m^2 are described in the literature. More replicates are necessary when small traps are used. Alternatively, very large traps, covering the entire reach or a similar area in the riparian zone, can be used (Bañuelos et al. 2004; Fig. 1.1). Traps must allow rainwater to drain quickly, but ensure that no material larger than 1 mm is lost. Large traps are prone to losing material on windy days. To minimize this risk, build them deep enough and weight them down with stones. The number of traps necessary for reliable estimates depends on spatial variations of the riparian forest. For forest stream reaches with a fully closed canopy, 10 might be enough, although it is worth checking whether the estimated average inputs are precise enough. For

Fig. 1.2 A trap to determine lateral litter inputs to streams. The mesh is fixed to a wooden or metallic frame and secured to two stakes

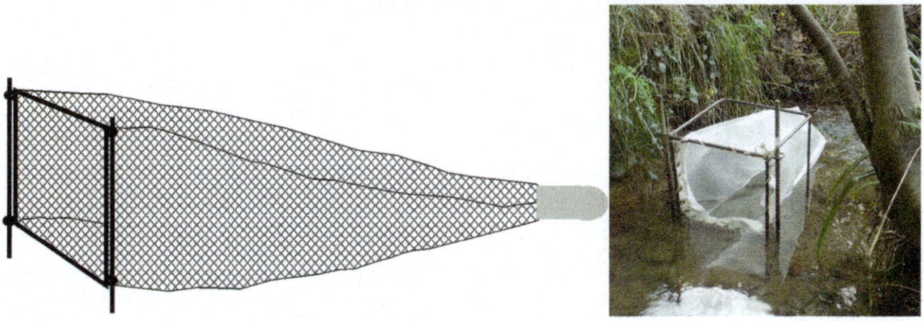

Fig. 1.3 (Left) Schematic of a typical drift net to sample CPOM in transport. The mesh is fixed to a metallic frame. Two metal rods can be used to secure the net in soft-bottom streams. A tube fixed at the end of the net makes sample collection easier. (Right) A modified version of the net shown in the drawing, deployed in a stream

instance, a preliminary sampling campaign could be performed to determine whether the traps yield standard deviations below 20% of the mean.

- Lateral-input traps (Fig. 1.2) can be constructed by tying a 1-mm mesh to a rectangular wooden or metallic frame. Traps 50 cm wide and 20 cm high are easy to transport and set level to the ground. As with the traps for measuring vertical inputs, ten lateral traps can be enough, but it is worth checking the precision of the estimated average input, as in the case of the vertical traps.
- Drift nets (Fig. 1.3) typically have a rectangular or square mouth and a long funnel (1-mm mesh size) to minimize flow resistance and delay clogging. Additional features can include rings in the frame to fix the net to the stream bottom with stakes and a plastic tube fixed at the end of the net to collect the retained CPOM. To sample during high flow, three to five nets should be used in all but the narrowest reaches.
- Aluminium trays
- Balance (±0.1 g precision)
- Crucibles

- Current meter
- Desiccator
- Drying oven
- Freezer
- Hammer
- Large ziplock plastic bags
- Measuring stick and ruler
- Measuring tape
- Muffle furnace
- Freeze-dryer (optional)
- Plastic trays
- Device to draw random numbers
- Ropes
- Tongs

3 Experimental Procedures

3.1 Reach Preparation

1. Suspend litter-fall baskets over the stream channel by tying them to nearby trees with ropes. Place the traps as low as possible above the water level but making sure they will not become submerged during the highest expected flood. Alternatively, in narrow streams under fully closed canopies, baskets can be placed on the banks, fixed to stakes. Distribute baskets randomly. Extend a measuring tape along the study reach. Draw a random number and go to the corresponding tape number. For instance, if the reach is 100 m long, randomly choose an integer between 0 and 100, with 1-m increments. Extend another measuring tape across the stream, and select the basket location with a second number. The precision of this number depends on channel width, but in general, the final location of the basket across the transect should be selected at random from at least five possible locations. Repeat for each basket. Random sampling should not be confounded with suspending baskets haphazardly or in "typical" places. Tag each basket with a number.
2. Place lateral-input traps randomly on the stream bank, perpendicular to the stream channel. Position the frame vertically to avoid direct inputs, and fix it tightly to stakes. The frame must be at ground level to allow free entry of litter from the forest floor. Tag each trap with a number.
3. Measure the surface area of the reach. Take 10–20 regularly spaced measurements of channel width, average them, and multiply by channel length. For more accurate estimates, prepare a detailed map of the reach, and measure surface area from the map. The wetted channel can vary greatly with discharge. Therefore, map the whole channel.

3.2 Sampling

1. Every other week, at least during peak leaf fall and at least once a month during the rest of the year, collect the material in litter-fall baskets and lateral traps, and enclose it in plastic bags. Mark each bag with the basket or trap number and the date. Discard any branches larger than 1 cm in diameter.
2. With at least the same frequency, sample inputs from upstream. If possible, locate a drift net at a narrow place where all stream water is funnelled into the net, at the upper end of the study reach. Keep the net in place for 4 h or as long as the material retained is not clogging the net. In the latter case, measure the time the net has been screening water. When the stream is too wide to be funnelled into a single net, use a stop net (mesh size $\leq$1 cm) covering the entire stream width. If a stop net with a sufficiently fine mesh is unavailable, several standard drift nets can be distributed across the channel. Measure the cross-sectional area intercepted by each net and the water velocity at their mouths, as soon as possible after setting the nets. Measure again the cross-sectional area and velocity of the intercepted water just before removing the nets. Enclose the collected material individually in plastic bags. Mark each bag.
3. Measure the cross-sectional area of the stream, and the water velocity, to calculate stream discharge.
4. CPOM transport during storms is particularly difficult to determine, but very important, because a large fraction of litter is transported during such events. However, nets can completely intercept high flow only in very small streams. Partial interception in larger streams must be taken into account to calculate total transport, as explained below. Furthermore, the concentration of drifting litter changes widely during a single storm, suggesting that repeated sampling during the rising and falling limb of the hydrograph is necessary for accurate estimates.
5. When collecting the net, pour away any water that has accumulated. Spin the sample dry if possible. Carry the collected material to the laboratory. If samples can be processed within the next few days, let them air-dry in open plastic trays. Mark each tray with the sample identification. Otherwise, freeze material immediately and dry later, preferably by freeze-drying.

3.3 Laboratory Procedures

1. Sort samples into leaves, fruits, bark, twigs, and other materials. Sort leaves by species. Put each of these categories in an aluminium tray. Mark all trays with the material and sample identifications.
2. Dry all samples at 50 °C to constant mass.
3. Cool the trays in a desiccator and weigh them to the nearest 0.1 g.

4. Transfer the material to pre-weighed crucibles. When the samples are too big for a crucible, shred by hand on an aluminium tray, mix thoroughly, and weigh a representative subsample.
5. Ash crucibles at 500 °C for 4 h.
6. Cool the crucibles in a desiccator and weigh them to the nearest 0.1 g.
7. If a subsample had to be ashed, correct the total sample mass for the determined ash concentration.

3.4 Calculations

1. Calculate ash-free dry mass (AFDM) of each category by subtracting ash mass from dry matter mass.
2. Sum up all categories in a sample to get total AFDM, and divide by the basket surface. Express results in g AFDM m^{-2} or in g AFDM m^{-2} d^{-1}.
3. To calculate the total amount of vertical inputs between sampling periods, multiply the above figure by the surface area of the reach.
4. Divide AFDM of lateral inputs by the trap length. Express results in g AFDM m^{-1} or in g AFDM m^{-1} d^{-1}.
5. To calculate the total amount of lateral inputs, multiply the above figure by the measured bank length. To calculate lateral inputs on an area basis, divide by the surface area of the reach.
6. To calculate the volume of water filtered by each drift net, multiply the cross-sectional area of the water funnelled into the net by the average water velocity measured immediately both after introducing and before removing the net.
7. To calculate the concentration of CPOM transported from upstream, divide the AFDM of transport inputs by the filtered water volume. Express results in g AFDM m^{-3}. When more than one net has been used simultaneously, calculate the average concentration of drifting litter during this period.
8. To calculate CPOM inputs from upstream, multiply the above concentration by stream discharge and by the time elapsed between samplings. It is worth exploring discharge-concentration relations. If the regression is significant and continuous discharge data are available, calculate CPOM inputs from this regression.
9. Divide CPOM inputs by the surface area of the reach to calculate the per-metre contribution of transport from upstream to total inputs.

4 Final Remarks

A relatively large number of replicates, both in time and space, are necessary to get reliable data, and the exact location of traps and nets can significantly affect results. To be ecologically meaningful, litter collections should at least encompass the main period of leaf fall (i.e. autumn in deciduous temperate forests) and preferably a whole year. Even then, caution is necessary when making long-term extrapolations, as inputs are far from constant among years (Cummins et al. 1983). Ideally, the measurements should begin before the onset of the main leaf-fall period, especially if annual data are of interest. An earlier onset of leaf fall due to unusual weather can strongly affect calculations.

References

Abelho, M., & Graça, M. A. S. (1996). Effects of eucalyptus afforestation on leaf litter dynamics and macroinvertebrate community structure of streams in central Portugal. *Hydrobiologia, 324*, 195–204.

Acuña, V., Giorgi, A., Muñoz, I., Sabater, F., & Sabater, S. (2007). Meteorological and riparian influences on organic matter dynamics in a forested Mediterranean stream. *Journal of the North American Benthological Society, 26*, 54–69.

Afonso, A. A. O., Henry, R., & Rodella, R. C. S. M. (2000). Allochthonous matter input in two different stretches of a headstream (Itatinga, São Paulo, Brazil). *British Archives of Biology and Technology, 43*, 335–343.

Bañuelos, R., Larrañaga, S., Elosegi, A., & Pozo, J. (2004). Effects of eucalyptus plantations on CPOM dynamics in headwater streams: A manipulative approach. *Archiv für Hydrobiologie, 159*, 211–228.

Benfield, E. F. (1997). Comparison of litterfall input to streams. *Journal of the North American Benthological Society, 16*, 104–108.

Bilby, R. E., & Heffner, J. T. (2016). Factors influencing litter delivery to streams. *Forest Ecology and Management, 369*, 29–27.

Brett, M. T., Bunn, S. E., Chandra, S., Galloway, A. W. E., Guo, F., Kainz, M. J., Kankaala, P., Lau, D. C. P., Moulton, T. P., Power, M. E., Rasmussen, J. B., Taipale, S. J., Thorp, J. H., & Wehr, J. D. (2017). How important are terrestrial organic carbon inputs for secondary production in freshwater ecosystems? *Freshwater Biology, 62*, 833–853.

Campbell, I. C., James, K. R., Hart, B. T., & Devereaux, A. (1992). Allochthonous coarse particulate organic material in forest and pasture reaches of two south-eastern Australian streams. I. Litter accession. *Freshwater Biology, 27*, 341–352.

Collins, S. M., Kohler, T. J., Thomas, S. A., Fetzer, W. W., & Flecker, A. S. (2016). The importance of terrestrial subsidies in stream food webs varies along a stream size gradient. *Oikos, 125*, 674–685.

Colón-Gaud, C., Peterson, S., Whiles, M. R., Kilham, S. S., Lips, K. R., & Pringle, C. M. (2008). Allochthonous litter inputs, organic matter standing stocks, and organic seston dynamics in Upland Panamanian streams: potential effects of larval amphibians on organic matter dynamics. *Hydrobiologia, 603*, 301–312.

Cummins, K. W., Sedell, J. R., Swanson, F. J., Minshall, G. W., Fisher, S. G., Cushing, C. E., Petersen, R. C., & Vannote, R. L. (1983). Organic matter budgets for stream ecosystems: Problems in their evaluation. In J. R. Barnes & G. W. Minshall (Eds.), *Stream ecology. Applications and testing of general ecological theory* (pp. 299–353). New York: Plenum Press.

Díez, J. R., Elosegi, A., & Pozo, J. (2001). Woody debris in north Iberian streams: Influence of geomorphology, vegetation and management. *Environmental Management, 28*, 687–698.

Eggert, S. L., Wallace, J. B., Meyer, J. L., & Webster, J. R. (2012). Storage and export of organic matter in a headwater stream: Responses to long-term detrital manipulations. *Ecosphere, 3*, 75.

Golladay, S. W. (1997). Suspended particulate organic matter concentration and export in streams. *Journal of the North American Benthological Society, 16*, 122–131.

Gonçalves, J. F., Jr., Rezende, R. S., Gregório, R. S., & Valentin, G. C. (2014). Relationship between dynamics of litterfall and riparian plant species in a tropical stream. *Limnologica, 44*, 40–48.

Gurtz, M. E., Webster, J. R., & Wallace, J. B. (1980). Seston dynamics in southern Appalachian streams: Effects of clear-cutting. *Canadian Journal of Fisheries and Aquatic Sciences, 37*, 624–631.

Harmon, M. E., Franklin, J. F., Swanson, F. J., Sollins, P., Gregory, S. V., Lattin, J. D., Anderson, N. H., Cline, S. P., Aumen, N. G., Sedell, J. R., Lienkaemper, G. W., Cromack, K., & Cummins, K. W. (1986). Ecology of coarse woody debris in temperate ecosystems. *Advances in Ecological Research, 15*, 133–302.

Hart, S. K., Hibbs, D. E., & Preakis, S. S. (2013). Riparian litter inputs to streams in the central Oregon Coast Range. *Freshwater Science, 32*, 343–358.

Harvey, C. J., Peterson, B. J., Bowden, W. B., Deegan, L. A., Finlay, J. C., Hershey, A. E., & Miller, M. C. (1997). Organic matter dynamics in the Kuparuk River, a tundra river in Alaska, USA. *Journal of the North American Benthological Society, 16*, 18–23.

Hill, B. H., & Webster, J. R. (1983). Aquatic macrophyte contribution to the New River organic matter budget. In T. Fontaine & S. Bartell (Eds.), *Dynamics of lotic ecosystems* (pp. 273–282). Ann Arbor: Ann Arbor Science.

Inoue, M., Shinotou, S., Maruo, Y., & Miyake, Y. (2012). Input, retention, and invertebrate colonization of allochthonous litter in streams bordered by deciduous broadleaved forest, a conifer plantation, and a clear-cut site in southwestern Japan. *Limnology, 13*, 207–219.

Jones, J. B., Schade, J. D., Fisher, S. G., & Grimm, N. B. (1997). Organic matter dynamics in Sycamore Creek, a desert stream in Arizona, USA. *Journal of the North American Benthological Society, 16*, 78–82.

Likens, G. E., & Bormann, F. H. (1995). *Biogeochemistry of a forested ecosystem* (2nd ed.). New York: Springer.

Meyer, J. L., Benke, A. C., Edwards, R. T., & Wallace, J. B. (1997). Organic matter dynamics in the Ogeechee River, a blackwater river in Georgia, USA. *Journal of the North American Benthological Society, 16*, 82–87.

Minshall, G. W. (1996). Organic matter budgets. In F. R. Hauer & G. A. Lamberti (Eds.), *Methods in stream ecology* (pp. 591–605). San Diego: Academic.

Naiman, R. J., & Link, G. L. (1997). Organic matter dynamics in 5 subarctic streams, Quebec, Canada. *Journal of the North American Benthological Society, 16*, 33–39.

Pozo, J., González, E., Díez, J. R., Molinero, J., & Elósegui, A. (1997). Inputs of particulate organic matter to streams with different riparian vegetation. *Journal of the North American Benthological Society, 16*, 602–611.

Pretty, J. L., Giberson, D. J., & Dobson, M. (2005). Resource dynamics and detritivore production in an acid stream. *Freshwater Biology, 50*, 578–591.

Sanpera-Calbet, I., Acuña, V., Butturini, A., Marcé, R., & Muñoz, I. (2016). El Niño southern oscillation and seasonal drought drive riparian input dynamics in a Mediterranean stream. *Limnology and Oceanography, 61*, 214–226.

Santiago, J., Molinero, J., & Pozo, J. (2011). Impact of timber harvesting on litterfall inputs and benthic coarse particulate organic matter (CPOM) storage in a small stream draining a eucalyptus plantation. *Forest Ecology and Management, 262*, 1146–1156.

Vannote, R. L., Minshall, G. W., Cummins, K. W., Sedell, J. R., & Cushing, C. E. (1980). The river continuum concept. *Canadian Journal of Fisheries and Aquatic Sciences, 37*, 130–137.

Webster, J. R., & Meyer, J. L. (Eds.). (1997). Stream organic matter budgets. *Journal of the North American Benthological Society, 16*, 3–161.

Webster, J. R., Golladay, S. W., Benfield, E. F., D'Angelo, D. J., & Peters, G. T. (1990). Effects of forest disturbance on particulate organic matter budgets of small streams. *Journal of the North American Benthological Society, 9*, 120–140.

Williams, G. P. (1989). Sediment concentration versus water discharge during single hydrologic events in rivers. *Journal of Hydrology, 111*, 89–106.

Chapter 2
Leaf Retention

Arturo Elosegi

Keywords Benthic organic matter · Coarse particulate organic matter · CPOM · Detritus · *Ginkgo biloba* leaves · Leaf litter · Litter storage · Litter transport · Streams · Stream retentiveness

1 Introduction

Allochthonous organic matter, especially leaf litter, is the main energy source of food webs in forested headwater streams (Vannote et al. 1980; Lohman et al. 1992; Cummins et al. 1989). Leaf litter enters streams mainly in a large burst during the period of leaf abscission (autumn in most temperate regions) and can either be trapped in the reach and thus become available for heterotrophs or transported downstream. Therefore, the capacity of a stream reach to retain materials (retentiveness) is important for the productivity and ecosystem efficiency of streams (Bilby and Likens 1980; Pozo et al. 1997).

Channel form is a key factor determining the capacity of streams to retain leaf litter, with small, rough-bottom streams being most retentive (Webster et al. 1994; Mathooko et al. 2001). Wood, especially when forming debris dams, enhances leaf retention and storage (Raikow et al. 1995; Díez et al. 2000; Koljonen et al. 2012; Fig. 2.1). Snagged (i.e. channels where "snags", or large wood pieces, have been removed) and channelized streams typically have poor retention capacity, which is partially recovered as a result of morphological restoration (Muotka and Syrjänen 2007; Scrimgeour et al. 2014). Changes in stream stage produce temporal variations in retention capacity, as higher discharge results in larger depth and width and higher hydraulic power, thus decreasing retentiveness (Ehrman and Lamberti 1992; Larrañaga et al. 2003). Because of the effects of changes in discharge, short-term

A. Elosegi (✉)
Department of Plant Biology and Ecology, Faculty of Science and Technology, University of the Basque Country, UPV/EHU, Bilbao, Spain
e-mail: arturo.elosegi@ehu.eus

F. Bärlocher et al. (eds.), *Methods to Study Litter Decomposition*,
https://doi.org/10.1007/978-3-030-30515-4_2

Fig. 2.1 This logjam in a north Spanish mountain stream retains large amounts of leaf litter

leaf retention can be a poor proxy for long-term benthic storage of organic matter (Elosegi et al. 2016).

Measuring leaf litter retention over short periods involves releasing leaves and estimating their downstream displacement. Four types of "leaves" may be used:

1. Natural leaves with some kind of mark that does not modify their short-term behaviour in water. Many paints affect leaf buoyancy and stiffness, so care must be taken to select a good dye; alternatively, a narrow line can be painted on both sides. The colours most easily recognized in streams are bright blue and blaze orange.
2. Leaves that do not occur naturally in the stream can sometimes be easily recognized. The bright yellow leaves of the exotic ginkgo tree (*Ginkgo biloba*), collected in autumn and stored dry between paper sheets, have often been used.
3. Artificial leaves of ornamental plastic plants, which can be painted in easily recognized colours. Although such artificial leaves may closely resemble natural ones, their floating behaviour can be different.
4. Any other material that is easily seen and behaves like leaves. Most commonly, strips (ca. 3 × 10 cm) from different types of plastic and in different colours are prepared.

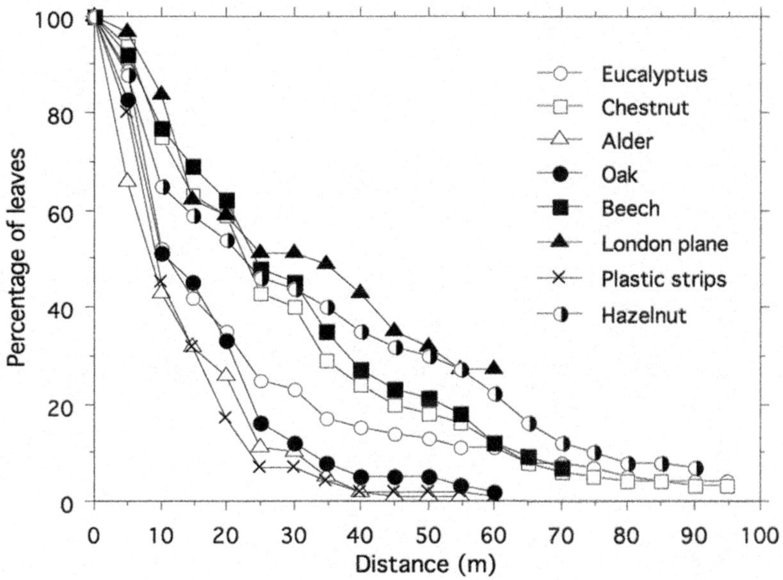

Fig. 2.2 Downstream decrease in the number of leaves transported in a third-order stream, expressed as a percentage of leaves released. Note that alder and plastic strips (3 × 10 cm) were most readily retained (average travel distance = 11.2 m), whereas London plane (Platanus × acerifolia) leaves travelled furthest (average travel distance = 50 m). (Data from Larrañaga et al. 2003)

Artificial materials are cheap and easily available throughout the year, whereas the use of natural leaves requires advanced planning, rather time-consuming collection, drying and storage. Furthermore, natural leaves tend to fragment easily and may therefore not be used repeatedly. However, released "leaf" material can be lost in the study reaches, especially when single-point collections are made (see below). Artificial or painted leaves should therefore only be used if one is confident that all leaves will be recovered. If this is not the case, it is preferable to use natural exotic leaves.

It is important to check that the materials used behave like natural leaves in a stream (see Fig. 2.2). If the goal is simply to compare the retention capacity of different reaches, any material can be used, but if the goal is to simulate the retention of real leaves, different kinds of materials need to be calibrated against the riparian leaf species most abundant in the study area. This is a point worth exploring in detail, as differences between materials can be substantial (Fig. 2.2; Young et al. 1978; Prochazka et al. 1991; Canhoto and Graça 1998) and the relationship between retention and leaf morphology is not straightforward (Larrañaga et al. 2003). Thus, a great deal of caution is necessary when comparing streams based on results obtained with different materials (Table 2.1).

This chapter describes a method to measure the capacity of stream reaches to retain leaf litter in the short term. This is done by monitoring the downstream displacement of leaves released at one point. The average travel distance of the leaves

Table 2.1 Average leaf travel distance (m) in selected publications

Site	Stream order	Leaf type	Travel distance (m)	References
Spain, control	1	Ginkgo	7.0 ± 5.5	1[a]
Spain, restored	1	Ginkgo	8.6 ± 8.3	1[a]
Spain, control	2	Ginkgo	38 ± 14	1[a]
Spain, restored	2	Ginkgo	14.1 ± 7.6	1[a]
Spain, control	3	Ginkgo	438 ± 297	1[a]
Spain, restored	3	Ginkgo	152.2 ± 72.8	1[a]
British Columbia, riffle	2	Red alder	226.2 ± 16.4	2[b]
British Columbia, pool	2	Red alder	94.7 ± 7.11	2[b]
Massachusetts, forested	NR	Acetate strips	5.6 ± 2.7	3[c]
Massachusetts, urban	NR	Acetate strips	8.9 ± 5.1	3[c]
Canada, reference	NR	Paper triangles	22.3 ± 5.8	4[d]
Canada,	CS	Paper triangles	58.1 ± 13.7	4[d]
Finland, high wood	AC	Plastic leaves	2 ± 0	5
Finland, channelized	AC	Plastic leaves	137 ± 87	5
Finland, boulder stream	AC	Plastic leaves	5 ± 2	5
Oregon, forested	2	Ginkgo	7.0 ± 4.0	6
Oregon, transition	2	Ginkgo	2.1 ± 0.6	6
Oregon, meadow	2	Ginkgo	5.7 ± 2.2	6

1, Elosegi et al. (2016); 2, Hoover et al. (2006); 3, Miller (2013); 4, Scrimgeour et al. (2014); 5, Koljonen et al. (2012); 6, Brookshire and Dwire (2003)
CS constructed streams in new landscapes after mining, *NR* not referred, *AC* artificial channel
[a]Average of multiple measurements under contrasting flows
[b]Measurement of single leaves in 1 day
[c]Each data is the average of seven sites
[d]Average and SD of values across several sites

is calculated by plotting the proportion of leaves in transport at a given point against the measured travel distance and fitting the data to an exponential decay model. This approach assumes that the number of leaves retained at any one point along the experimental reach is directly proportional to the number of leaves in transport. Two methods are given here.

The multiple-point collection method is best suited for small clear-water streams. All leaves are retrieved, and the distance travelled by each leaf is measured. A net is placed downstream of the reach as a safety device, to prevent the loss of leaves drifting past this point.

The single-point collection method may be slightly less accurate but is useful in larger reaches or turbid waters where many leaves are unlikely to be recovered after release. Instead of measuring the distance travelled by individual leaves, the proportion of leaves reaching a net placed downstream of an experimental reach is measured. If differentially marked leaves are released at various distances from the net, an exponential regression can be calculated as with the multiple-point collection method. Unlike in multiple-point collections, the distance between release point and net is critical. The experiment is not valid if more than 90% or less than 10% of leaves reach the net (Lamberti and Gregory 1996). Whereas both natural and artificial leaves can be used with multiple-point collection, only natural leaves should be used with single-point collection, to avoid polluting the reach.

Because retention distance varies with stream stage, inter-stream comparisons should be performed under similar hydrological conditions. This is most easily done during base-flow conditions, but distances so measured can grossly overestimate average retention efficiency, as leaves are more easily scoured during high flow. The relationship between travel distance and discharge can be studied for each reach by repeating retention experiments under different discharge conditions. If this is done, results from different reaches can be compared even if they do not correspond exactly to the same hydrological condition.

2 Site Selection and Equipment

2.1 Site Selection

Leaf retention can be measured in almost any stream or river, but measurements are easier in wadeable streams with clear water. In first- to second-order streams, most leaves are retained within a few metres during base-flow, but retention distance can increase to some tens of metres at higher discharge. Appropriate reach lengths are therefore normally 10–50 m. In larger streams and rivers, reaches 100–500 m long are recommended. As a rule of thumb, reach length should be ten times the wetted channel width (Lamberti and Gregory 1996), but especially with the single-point collection method, it is worth running preliminary experiments to determine the most appropriate length.

2.2 Equipment and Material

- Collect and air-dry recently fallen *Ginkgo biloba* leaves. Store groups of 100 leaves in a dark, well aerated place.
- Stop net (2–5 cm mesh size) per reach, wider than the stream channel
- Measuring tape
- Rope to tie the stop net to trees or other features
- Current meter
- Measuring stick and ruler

3 Experimental Procedures

3.1 Field Procedures

1. Before the experiment, soak the leaves overnight in water to give them neutral buoyancy.
2. Block the downstream end of the reach with the stop net.
3. Standing in the upstream end of the reach, release leaves one by one into the water, spreading them across the wetted channel. One hundred leaves are normally enough for first-order reaches, 500 for third-order reaches. Larger numbers are necessary with the single-point collection method.
4. Allow the stream to disperse leaves for 1 h.

Multiple-Point Collection

1. Extend measuring tape along the reach.
2. One hour after release, recover leaves that reached the stop net. Keep the net in place. Record the number of leaves.
3. Walking in a zigzag pattern upstream from the net, recover all leaves. Take care to monitor underneath logs, branches, boulders and overhanging banks, where many leaves can be retained. Record to the nearest metre (5 m in reaches >100 m) the distance travelled by each leaf.
4. For more exhaustive analysis, record the structure retaining each leaf (e.g. pool, riffle, channel margin, wood piece, roots, debris dam, boulder, gravel, sand).
5. After recovering all leaves, remove the stop net.

Single-Point Collection

1. One hour after release, recover and count the leaves that reached the net.
2. With either method, measure stream discharge after the retention experiment.
3. Additional information of interest may be channel gradient, average width and depth, bank slope, area covered by riffles and pools or area covered by different substrate categories (sand, gravel, etc.). Of particular significance is the abundance of woody debris, as it is one of the most retentive structures found in stream channels.

3.2 Calculations

1. The number of released leaves in transport is plotted against travel distance, and the data are fitted to the exponential decay model (Young et al. 1978):

$$L_d = L_0 \cdot \mathrm{e}^{-k \cdot d} \tag{2.1}$$

2. In the single-point collection method, L_d is the number of leaves recovered at the net, L_0 is the number of leaves released, d is the distance in metres between the release point and net, and k is the instantaneous retention rate, which is independent of reach length and number of released strips.
3. In the multiple-point collection, L_0 is the total number of leaves recovered (which should be close to the number released), and L_d is the number of leaves still in transport at distance d. This is calculated by subtracting the number of leaves retained between release point and distance d from the total number of leaves recovered in the experiment.
4. Calculations can be made with any standard statistical software or a calculator. Exponential regressions are calculated by first linearizing the data by ln-transformation and then calculating the linear regression. Alternatively, and more accurately, non-linear curve fitting may be used (see Chap. 6). The slope of the regression is the instantaneous retention rate.
5. Calculate the average travel distance as $1/k$ (Newbold et al. 1981).
6. Analysis of covariance (ANCOVA) or other approaches (see Chap. 6) can be used to test for statistically significant differences between slopes. When the multiple-point collection is chosen, the percentage of leaves retained by different channel structures can also be calculated.
7. Additionally, the relative retention efficiency of each substrate structure can be determined. To do this, simply divide the percentage of strips retained by a given structure by the percentage of wetted streambed area covered by the same structure.

References

Bilby, R. E., & Likens, G. E. (1980). Importance of organic debris dams in the structure and function of stream ecosystems. *Ecology, 61*, 1107–1113.

Brookshire, E. N. J., & Dwire, K. A. (2003). Controls on patterns of coarse organic particle retention in headwater streams. *Journal of the North American Benthological Society, 22*(1), 17–34.

Canhoto, C., & Graça, M. A. S. (1998). Leaf retention: A comparative study between two stream categories and leaf types. *Verhandlungen der internationalen Vereinigung für Limnologie, 26*, 990–993.

Cummins, K. W., Wilzbach, M. A., Gates, D. M., Perry, J. B., & Taliaferro, W. B. (1989). Shredders and riparian vegetation. *Bioscience, 39*, 24–30.

Díez, J. R., Larrañaga, S., Elosegi, A., & Pozo, J. (2000). Effect of removal of wood on streambed stability and retention of organic matter. *Journal of the North American Benthological Society, 19*, 621–632.

Ehrman, T. P., & Lamberti, G. A. (1992). Hydraulic and particulate matter retention in a 3rd-order Indiana stream. *Journal of the North American Benthological Society, 11*, 341–349.

Elosegi, A., Elorriaga, C., Flores, L., Martí, E., & Díez, J. R. (2016). Restoration of wood loading has mixed effects on water, nutrient and leaf retention in Basque mountain streams. *Freshwater Science, 35*, 41–54.

Hoover, T. M., Richardson, J. S., & Yonemitsu, N. (2006). Flow-substrate interactions create and mediate leaf litter resource patches in streams. *Freshwater Biology, 51*, 435–447.

Koljonen, S., Louhi, P., Mäki-Petäys, A., Huusko, A., & Muotka, T. (2012). Quantifying the effects of in-stream habitat structure and discharge on leaf retention: Implications for stream restoration. *Freshwater Science, 31*, 1121–1130.

Lamberti, G. A., & Gregory, S. V. (1996). Transport and retention of CPOM. In F. R. Hauer & G. A. Lamberti (Eds.), *Methods in stream ecology* (pp. 217–229). San Diego: Academic.

Larrañaga, S., Díez, J. R., Elosegi, A., & Pozo, J. (2003). Leaf retention in streams of the Agüera basin (northern Spain). *Aquatic Sciences, 65*, 158–166.

Lohman, K., Jones, J. R., & Perkins, B. D. (1992). Effects of nutrient enrichment and flood frequency on periphyton biomass in northern Ozark streams. *Canadian Journal of Fisheries and Aquatic Sciences, 49*, 1198–1205.

Mathooko, J. M., Morara, G. O., & Leichtfried, M. (2001). Leaf litter transport and retention in a tropical Rift Valley stream: An experimental approach. *Hydrobiologia, 443*, 9–18.

Miller, E. M. (2013). Local- and landscape-level controls on coarse particulate organic matter retention in urban and forested small streams of central Massachusetts. *Freshwater Science, 32*, 576–585.

Muotka, T., & Syrjänen, J. (2007). Changes in habitat structure, benthic invertebrate diversity, trout populations and ecosystem processes in restored forest streams: A boreal perspective. *Freshwater Biology, 52*, 724–737.

Newbold, J. D., Elwood, J. W., O'Neill, R. V., & VanWinkle, W. (1981). Measuring nutrient spiralling in streams. *Canadian Journal of Fisheries and Aquatic Sciences, 38*, 860–863.

Pozo, J., González, E., Díez, J. R., & Elosegi, A. (1997). Leaf litter budgets in two contrasting forested streams. *Limnetica, 13*, 77–84.

Prochazka, K., Stewart, B. A., & Davies, B. R. (1991). Leaf litter retention and its implications for shredder distribution in two headwater streams. *Archiv für Hydrobiologie, 120*, 315–325.

Raikow, D. F., Grubbs, S. A., & Cummins, K. W. (1995). Debris dam dynamics and coarse particulate organic matter retention in an Appalachian Mountain stream. *Journal of the North American Benthological Society, 14*, 535–546.

Scrimgeour, G. J., Tonn, W. M., & Jones, N. J. (2014). Quantifying effective restoration: Reassessing the productive capacity of a constructed stream 14 years after construction. *Canadian Journal of Fisheries and Aquatic Sciences, 71*, 589–601.

Vannote, R. L., Minshall, G. W., Cummins, K. W., Sedell, J. R., & Cushing, C. E. (1980). The river continuum concept. *Canadian Journal of Fisheries and Aquatic Sciences, 37*, 130–137.

Webster, J. R., Covich, A. P., Tank, J. L., & Crockett, T. V. (1994). Retention of coarse organic particles in the southern Appalachian Mountains. *Journal of the North American Benthological Society, 13*, 140–150.

Young, S. A., Kovalak, W. P., & Del Signore, K. A. (1978). Distances travelled by autumn leaves introduced into a woodland stream. *American Midland Naturalist, 100*, 217–222.

Chapter 3
Manipulating Litter Retention in Streams

Michael Dobson

Keywords Benthic organic matter · Coarse particulate organic matter · CPOM · Detritus · Leaf litter · Litter storage · Litter transport · Streams · Stream retentiveness

1 Introduction

Leaf litter is the dominant energy resource for food webs in low-order shaded streams (Wallace et al. 1999), but is only available to the vast majority of detritivores and microbial decomposers when retained on the streambed. Therefore, the retention capacity of the channel is crucial in determining the overall decomposition of litter in a stream reach or entire stream. Manipulating the retentive capacity of stream channels permits quantifying the importance of litter retention and testing hypotheses about the role of physical channel attributes and related parameters in leaf litter dynamics, including litter decomposition (Entrekin et al. 2008; Frainer et al. 2018), as well as the importance of litter for invertebrate and microbial communities (Dobson and Hildrew 1992; Tiegs et al. 2008). Manipulation of channel retentiveness can be in the form of enhancement or reduction. The procedures for these manipulations are straightforward, although some of the techniques that may be employed require major efforts or heavy machinery (Dobson and Hildrew 1992; Entrekin et al. 2008; Frainer et al. 2018).

The method presented here for enhancing litter retention has been adapted from Dobson et al. (1995). It consists of deploying on the stream bed a set of litter traps each made of two steel poles connected by a piece of plastic mesh screen. The method was originally designed to investigate the influence of increased retention in discrete patches on localized and overall numbers and biomass of detritivores and coarse benthic organic matter in low-order streams (Dobson and Hildrew 1992).

M. Dobson (✉)
APEM Ltd, Edinburgh Technopole, Penicuik, Midlothian, UK
e-mail: m.dobson@apemltd.co.uk

F. Bärlocher et al. (eds.), *Methods to Study Litter Decomposition*,
https://doi.org/10.1007/978-3-030-30515-4_3

A range of reach-level measurements (e.g. litter decomposition in mesh bags as described in Chap. 6) can be combined with such manipulations of stream retentiveness.

2 Equipment and Material

2.1 Equipment

- Surber-type sampler
- Drying oven (50 °C)
- Muffle furnace
- Top-loading balance

2.2 Materials

- Rigid plastic mesh (mesh size 8–10 mm is normally adequate), cut into 20 × 15 cm rectangles.
- Steel poles (e.g. rebars). These must be narrower in diameter than the mesh used. A range of lengths from 40 to 60 cm is useful, particularly if problems are envisaged in finding enough sites to sink them deeply into the stream bottom.
- Paper bags or aluminium pans for drying leaf material
- Materials to process invertebrate samples (see Chap. 49)

3 Experimental Procedures

3.1 Experimental Design

- Identify appropriate reference and experimental stretches, and sample all of them *before* manipulation.
- Determine the specific design, including frequency, size and relative placement of traps, depending on the aim of the project.
- Ensure that the density of traps is enough to achieve the aim. If the aim is to increase litter mass in the entire channel, then a high proportion of the stream bed needs to be covered by litter traps; for the trap type described above, at least 1 per m^2 is required. If the aim is simply to increase litter mass in discrete patches around the traps themselves, then a lower density can be used.
- Sample at appropriate spatial intervals. If stream reach outputs are being measured, then sample immediately downstream of each reach. For within-reach impacts, two alternatives are available. If the aim is to determine the influence of the manipulation on the entire channel, then sample points should be chosen *at*

random in both reference and experimental reaches, with no attempt to include or avoid litter traps in the latter reach. If the aim is to determine the localized influence of the litter traps, then the following is recommended for sampling: random points in the reference stretch, randomly chosen litter traps and random points in the experimental stretch *between* traps.
- Sampling a litter trap will remove its contents. Therefore, if multiple sampling dates are used, then sampling the same trap should not be resampled within a short time interval, unless an objective is to understand retention rates and litter colonization over such time periods.

3.2 Litter Trap Deployment

- Hammer steel poles into the river bed in pairs, 15–18 cm apart and orientated perpendicular to the flow relative to each other.
- Carefully thread the mesh rectangle onto the poles, and push down until its base is flush with the river bed (Fig. 3.1).
- Poles should protrude as little as possible above the mesh, and the entire trap is most efficient if it breaks the surface slightly at normal flow, thereby capturing litter floating at all depths.

Fig. 3.1 Newly placed litter traps in a stream. Photo: F. Peter

3.3 Sampling and Sample Processing

- Sample trapped litter and associated invertebrates with a standard Surber-type sampler. If sampling litter traps individually, the Surber sampler is carefully placed over the trap and its associated leaf pack. Then the leaf pack is removed and placed into a bag before sampling the exposed bed in the normal way. The plastic mesh of the trap can be removed and washed during this procedure and then replaced. It is normally instructive to retain the leaf pack in the trap and the bed sample around the trap as separate samples (Dobson and Hildrew 1992).
- Coarse benthic organic matter (CBOM) accumulated between litter traps may also be sampled with a Surber sampler, as may benthic invertebrates.
- Litter and other benthic organic matter can be processed, dried and ashed, and its mass determined as described in Chaps. 1, 4, and 6.

4 Final Remarks

Since traps or exclusion devices are likely to be left in place for several months or years, it is important to be aware of several risks: Protruding rebars can cause serious injury, so the river reach into which they are placed should not be frequented by people or animals. Furthermore, if stream discharge fluctuates greatly, and particularly if bed movement occurs during high flow events, the litter traps will act as sediment traps and will eventually fill in. This phenomenon needs to be closely monitored during long-term studies. In extreme cases, the traps will initiate development of a series of small islands. Conversely, litter traps may cause impacts to the river channel such as erosion or buildup of large amounts of debris followed by sudden release. They may also lead to risk of localized flooding. Thus, they should not be left unattended for long periods.

The small-scale manipulations described above have been run effectively for several years, with sampling intervals separated by months (Dobson et al. 1995). However, aggregation of leaf litter and animals can occur over a few days or even hours, so more frequent sampling is possible. The described traps are small enough to be completely enclosed by a Surber sampler, so traps can be sampled individually to examine small-scale patterns of retention. If, however, the aim is to determine output from an entire reach – for example, to study aquatic hyphomycete spore concentrations or nutrient concentrations in stream water – then using fewer but larger traps may be more practical. For example, the procedures can be scaled up by using logs that span a large proportion or the entire stream channel (e.g. Smock et al. 1989; Pretty and Dobson 2001; Entrekin et al. 2008; Frainer et al. 2018; Fig. 3.2).

Fig. 3.2 Use of wooden logs as litter traps, as used by Pretty and Dobson (2001). Note that each piece of wood is held in place by four steel poles, arranged in pairs upstream and downstream and each bent over the log at the top to stop it from being entrained by high water flows. Photo: M. Dobson

Reducing rather than enhancing retention would initially involve identification of the major retention structures in the channel and then their systematic removal (e.g. Wallace et al. 1999; Díez et al. 2000). If retention is mainly by woody debris, then this is straightforward clearance of wood from the channel. If it is cobbles or river bank features such as trailing roots, then removal can be difficult. However, natural retention is generally low in such streams. Depending upon the source of leaf litter, reduction is also possible by stop-netting upstream of the experimental area (e.g. Tiegs et al. 2008), although such nets need constant vigilance as a large mass of debris upstream can quickly build up and may cause them to break. For small-scale projects, small stop-nets can be placed to create localized patches of reduced retention (Fig. 3.3).

Fig. 3.3 Small exclusion nets, designed to reduce inputs of leaf litter from upstream transport to localized patches, in order to monitor the influence of benthic organic matter upon leaf decay rates in mesh bags. Note the mesh bag at (**a**), the unwanted buildup of leaf litter at (**b**) and the large piece of wood caught by the trap at (**c**). Photo: M. Dobson

References

Díez, J. R., Larrañaga, S., Elosegi, A., & Pozo, J. (2000). Effect of removal of wood on streambed stability and retention of organic matter. *Journal of the North American Benthological Society, 19*, 621–632.

Dobson, M., & Hildrew, A. G. (1992). A test of resource limitation among shredding detritivores in low order streams in southern England. *Journal of Animal Ecology, 61*, 69–77.

Dobson, M., Hildrew, A. G., Orton, S., & Ormerod, S. J. (1995). Increasing litter retention in moorland streams: Ecological and management aspects of a field experiment. *Freshwater Biology, 33*, 325–337.

Entrekin, S. A., Tank, J. L., Rosi-Marshall, E. J., Hoellein, T. J., & Lamberti, G. A. (2008). Responses in organic matter accumulation and processing to an experimental wood addition in three headwater streams. *Freshwater Biology, 53*, 1642–1657.

Frainer, A., Polvi, E., Jansson, R., & McKie, B. G. (2018). Enhanced ecosystem functioning following stream restoration: The roles of habitat heterogeneity and invertebrate species traits. *Journal of Applied Ecology, 55*, 377–385.

Pretty, J. L., & Dobson, M. (2001). A possible strategy for enhancing the stream biota in areas of intensive commercial forestry. *Verhandlungen der Internationalen Vereinigung für Theoretische und Angewandte Limnologie, 27*, 1075–1078.

Smock, L. A., Metzler, G. W., & Gladden, J. E. (1989). Role of debris dams in the structure and functioning of low gradient headwater systems. *Ecology, 70*, 764–775.

Tiegs, S. D., Peter, F. D., Robinson, C. T., Uehlinger, U., & Gessner, M. O. (2008). Leaf decomposition and invertebrate colonization responses to manipulated litter quantity in streams. *Journal of the North American Benthological Society, 27*, 321–331.

Wallace, J. B., Eggert, S. L., Meyer, J. L., & Webster, J. R. (1999). Effects of resource limitation on a detrital-based ecosystem. *Ecological Monographs, 69*, 409–442.

Chapter 4
Coarse Benthic Organic Matter

Jesús Pozo and Arturo Elosegi

Keywords CBOM · Coarse benthic organic matter · Coarse particulate organic matter · Detritus · Litter fall · Leaf litter · Riparian vegetation · Streams

1 Introduction

Coarse benhic organic matter (CBOM) is the primary energetic basis of communities in forest streams (Hall et al. 2000). Riparian forests in particular provide streams with substantial amounts of this material (e.g. Cummins et al. 1983; Webster et al. 1995; Abelho 2001), although primary production within the stream can be an additional source of energy. The major components of CBOM are wood, leaves and leaf fragments, fruits and flowers, with leaves generally dominating in terms of both absolute amount and regularity of input (Fisher and Likens 1972; Pozo et al. 1997; Abelho 2001). Once in the stream, CBOM is often efficiently retained under baseflow conditions (Quinn et al. 2007), although there is substantial spatial and temporal variation depending on current velocity, flow regime, channel form, substrate composition, the presence of debris dams and other factors (Larrañaga et al. 2003). Thus, together with the structure of the riparian canopy and hence litter inputs, these factors affecting retention efficiency control the accumulation of CBOM in stream beds (Smock 1990).

Temporal variability of CBOM can also be high as a consequence of leaf-fall phenology and hydrologic regime (e.g. Molinero and Pozo 2004). In temperate deciduous forest streams, leaf fall peaks in autumn and early winter, and this pattern is typically reflected in the amounts of CBOM in the stream at this time (Iversen et al. 1982; Bärlocher 1983), whereas lowest values are often found during spring and summer. Although there is less information from tropical streams, seasonal variations both in litter inputs and storage are also found, mostly related to dry periods, when

---

Electronic Supplementary Material The online version of this chapter (https://doi.org/10.1007/978-3-030-30515-4_4) contains supplementary material, which is available to authorized users.

J. Pozo (✉) · A. Elosegi
Department of Plant Biology and Ecology, Faculty of Science and Technology, University of the Basque Country, UPV/EHU, Bilbao, Spain
e-mail: jesus.pozo@ehu.es; arturo.elosegi@ehu.eus

F. Bärlocher et al. (eds.), *Methods to Study Litter Decomposition*,
https://doi.org/10.1007/978-3-030-30515-4_4

most leaf litter falls (Bambi et al. 2017; Tonin et al. 2017). Often, however, the input of leaves correlates poorly with the dynamics of CBOM under changing discharge conditions. Only when high inputs coincide with periods of low flow are increases in CBOM in the stream expected (Molinero and Pozo 2004). Clearly, studies on CBOM in streams need to consider variability across both spatial and temporal scales.

Disturbances of the riparian vegetation such as clear-cutting and plantations of exotic species alter the quantity, quality and temporal and spatial distribution of inputs and storage in streams (Webster et al. 1990; Graça et al. 2002; Santiago et al. 2011). Conversely, introduction of large pieces of wood in stream channels, whether naturally or as a restoration measure, generally enhances CBOM storage and temporal stability (Flores et al. 2011). In addition to stream characteristics, diverging sampling methods and size fractionation in different studies also partly contribute to high variability of CBOM storage reported in the literature (Abelho 2001). This fact needs to be considered when comparing data on CBOM storage in streams, such as those compiled in Table 4.1.

CBOM estimates have been based on a variety of methods: random sampling of the wetted channel (González and Pozo 1996), sampling of transects either at random (Golladay et al. 1989) or at regular intervals along the stream (Wallace et al. 1995) and stratified random sampling (Mulholland 1997). Samples are usually taken with a Surber-type sampler or a power-vacuum assisted, cylindrical corer. This chapter describes a method to estimate the amounts of CBOM stored in small streams with a Surber-type sampler. Potential applications of this method include assessments of differences in CBOM among similar-sized streams experiencing different degrees of disturbance of the riparian vegetation, assessment of in-channel restoration works or studies of trophic networks. In addition, relationships of CBOM storage with the retention capacity of streams (see Chap. 2), their flow regime, the phenology of allochthonous inputs or other temporal changes, may be explored.

2 Equipment and Materials

- Aluminium trays
- Balance
- Bucket
- Crucibles
- Desiccator
- Drying oven
- Freezer
- Labelled plastic bags
- Surber-type sampler (mesh size of 1 mm)
- Muffle furnace
- Set of nested sieves (1 mm and 1 cm mesh sizes)
- Plastic trays
- Random number table or cell phone app
- Small shovel
- Tape measure
- Tongs

Table 4.1 Coarse benthic organic matter (excluding large wood) from selected streams

Location	Latitude	Riparian vegetation	Stream order	CBOM (g AFDM m^{-2})	Reference
Alaska, USA	65°N	Taiga	1	3	1
			2	8	1
Denmark	56°N	Deciduous forest	1	135*	2
Quebec, Canada	50°N	Boreal forest	1	968	3
			2	317	3
			5	456	3
Switzerland	47°N	Deciduous forest	3	27**	4
Oregon, USA	45°N	Coniferous forest	1	1012–5117	5
			3	388	5
			5	61	5
New Hampshire, USA	44°N	Deciduous forest	2	509***	6
Spain	43°N	Deciduous forest	1	60	7
			3	20	7
		Eucalyptus plantation	1	15–200	8
Pennsylvania, USA	40°N	Agricultural and deciduous forest	3	118	9
Virginia, USA	37°N	Deciduous forest	1	739	10
North Carolina, USA	35°N	Deciduous forest	1	391	11
		Logged deciduous forest	2	286	11
Arizona, USA	33°N	Desert scrub	5	5	12
Panama	8°N	Rainforest	1–2	48–113	13
Ecuador	0°S	Tropical Andean forest	1	12–125	14
Federal District, Brazil	15.5°S	Cerrado savannah	3	247**	15
Minas Gerais, Brazil	20°S	Atlantic forest	2	327**	16
South Africa	33°S	Fynbos	2	19–32**	17
Victoria, Australia	37°S	Eucalyptus forest	4	105	18

*Leaves only. **Converted from dry mass (DM), assuming that AFDM = 0.9 × DM. ***Converted from kcal, assuming 10 kcal = 1 g C = 2 g AFDM. 1, Irons and Oswood (1997); 2, Iversen et al. (1982); 3, Naiman and Link (1997); 4, Bärlocher (1983); 5, Webster and Meyer (1997); 6, Fisher and Likens (1972); 7, González and Pozo (1996); 8, Santiago et al. (2011); 9, Newbold et al. (1997); 10, Smock (1990); 11, Webster et al. (1990); 12, Jones et al. (1997); 13, Colón-Gaud et al. (2008); 14, Ríos et al. (2009); 15, Bambi et al. (2017); 16, França et al. (2009); 17, King et al. (1987); 18, Treadwell et al. (1997)

3 Experimental Procedures

3.1 Sampling

1. Choose a suitable and accessible stream segment, typically 50 m long or 10 times longer than the active channel width, for example, in a small (stream order 1 or 2) forested headwater stream.
2. Use a random number table or cell phone app and tape measure to choose five points along the selected stream reach to establish a 0.5-m-wide transect across the stream from bank to bank, including any dry parts of the channel.
3. Note the width of the channel in each transect.
4. If the transect includes a dry section, use the following procedures:

   (a) On the dry section, collect the substrate with a small shovel to a depth of 5 cm when possible.
   (b) Eliminate large mineral substrates before putting the collected material in a set of nested sieves (sieve of 1 cm mesh size on top of a sieve of 1 mm mesh size).
   (c) Rinse the sample with stream water and eliminate, as much as possible, all inorganic materials and wood pieces >1 cm in diameter that are retained by the 1 cm sieve.
   (d) Transfer the rest to a labelled plastic bag.
   (e) Transfer the material retained by the 1 mm sieve to the same plastic bag.

5. In the submerged section, use the following procedures:

   (a) Collect the CBOM with a Surber-type sampler, making sure to disturb the substrate to a defined depth (at least 5 cm) in a standardized fashion.
   (b) Transfer the material retained by the net of the Surber-type sampler to the nested sieves.
   (c) Discard large pieces of mineral substrate and wood (>1 cm in diameter) before putting the rest of the sample in a labelled plastic bag.

6. Proceed in the same way with the other transects of the stream segment.
7. Carry the collected CBOM to the laboratory, and freeze it upon arrival, unless samples can be processed within the next few days.
8. In temporally extended studies, collect samples twice a month during periods of heavy litter fall or sudden changes in discharge, and monthly thereafter, while avoiding that the same transect is sampled repeatedly.
9. Obtain flow measurements or discharge records from the nearest gauging station as a critical piece of information to interpret temporal changes in CBOM.

3.2 Laboratory Procedures

1. Remove attached sand and silt from the CBOM collected in the nested sieves by rinsing with tap water, and then transfer all the organic matter to a plastic tray.
2. Remove any remaining branches >1 cm in diameter.
3. Sort the remaining CBOM into the following categories: leaves, twigs, bark, fruits and flowers and unidentifiable fragments >1 mm.
4. Sort leaves by species, and put the material in each CBOM category in a separate pre-weighed aluminium tray.
5. Dry at 50 °C to constant mass (generally 48 h).
6. Let the samples cool in a desiccator and weigh.
7. Transfer the material of each tray to a pre-weighed crucible, or, if a sample is large, use a weighed subsample.
8. Put the crucibles in the muffle furnace, and ash the organic matter at 500 °C for 4 h.
9. Let the crucibles cool in a desiccator and weigh.

3.3 Calculations

1. To calculate the area (m^2) of each transect, multiply the respective channel width (m) by the width of the sampled transect (i.e. 0.5 m).
2. Calculate the ash-free dry mass (AFDM) of each CBOM category by subtracting the ash mass from the dry mass.
3. Divide the results by the area of the respective transect to express the data in terms of g AFDM m^{-2}.

4 Final Remarks

Mosses or other macrophytes, when present, can easily be removed from the remaining CBOM, allowing estimates of their respective contributions. Since a considerable fraction of the total CBOM can be found below the streambed surface (Cummins et al. 1983), it is critical to take this buried material into account.

If the main objective of the assessment is to determine the availability of food to aquatic invertebrates, sampling can be restricted to the wetted channel. However, if the objective is to construct an organic matter budget (see Chap. 10), the whole channel has to be sampled, including both dry and wet areas.

References

Abelho, M. (2001). From litterfall to breakdown in streams: A review. *The ScientificWorld, 1*, 656–680.

Bambi, P., Rezende, R. S., Feio, M. J., Leite, G. F. M., Alvin, E., Quintao, J. M. B., Araújo, F., & Goncalves, J. G. (2017). Temporal and spatial patterns in inputs and stock of organic matter in Savannah streams of Central Brazil. *Ecosystems, 20*, 757–768.

Bärlocher, F. (1983). Seasonal variation of standing crop and digestibility of CPOM in a Swiss Jura stream. *Ecology, 64*, 1266–1272.

Colón-Gaud, C., Peterson, S., Whiles, M. R., Kilham, S. S., Lips, K. R., & Pringle, C. M. (2008). Allochthonous litter inputs, organic matter standing stocks, and organic seston dynamics in Upland Panamanian streams: Potential effects of larval amphibians on organic matter dynamics. *Hydrobiologia, 603*, 301–312.

Cummins, K. W., Sedell, J. R., Swanson, F. J., Minshall, F. W., Fisher, S. G., Cushing, C. E., Peterson, R. C., & Vannote, R. L. (1983). Organic matter budgets for stream ecosystems: Problems in their evaluation. In J. R. Barnes & G. W. Minshall (Eds.), *Stream Ecology. Application and testing of general ecological theory* (pp. 299–353). New York: Plenum Press.

Fisher, S. G., & Likens, G. E. (1972). Stream ecosystem: Organic energy budget. *BioScience, 22*, 33–35.

Flores, L., Larrañaga, A., Díez, J., & Elosegi, A. (2011). Experimental wood addition in streams: effects on organic storage and breakdown. *Freshwater Biology, 56*, 2156–2167.

França, J. S., Gregório, R. S., de Paula, J. A., Gonçalves, J. F., Jr., Ferreira, F. A., & Callisto, M. (2009). Composition and dynamics of allochthonous organic matter inputs and benthic stock in a Brazilian stream. *Marine and Freshwater Research, 60*, 990–998.

Golladay, S. W., Webster, J. R., & Benfield, E. F. (1989). Changes in stream benthic organic matter following watershed disturbance. *Holarctic Ecology, 12*, 96–105.

González, E., & Pozo, J. (1996). Longitudinal and temporal patterns of benthic coarse particulate organic matter in the Agüera stream (northern Spain). *Aquatic Sciences, 58*, 355–366.

Graça, M. A. S., Pozo, J., Canhoto, C., & Elosegi, A. (2002). Effects of *Eucalyptus* plantations on detritus, decomposers, and detritivores in streams. *The ScientificWorld, 2*, 1173–1185.

Hall, R. O., Jr., Wallace, J. B., & Eggert, S. L. (2000). Organic matter flow in stream food webs with reduced detrital resource base. *Ecology, 81*, 3445–3463.

Irons, J. G., & Oswood, M. W. (1997). Organic matter dynamics in 3 subarctic streams of interior Alaska, USA. *Journal of the North American Benthological Society, 16*, 23–28.

Iversen, T., Thorup, J., & Skriver, J. (1982). Inputs and transformation of allochthonous particulate organic matter in a headwater stream. *Holarctic Ecology, 5*, 10–19.

Jones, J. B., Schade, J. D., Fisher, S. G., & Grimm, N. B. (1997). Organic matter dynamics in Sycamore Creek, a desert stream in Arizona, USA. *Journal of the North American Benthological Society, 16*, 78–82.

King, J. M., Day, J. A., Davies, B. R., & Henshall-Howard, M. P. (1987). Particulate organic matter in a mountain stream in the South-western Cape, South Africa. *Hydrobiologia, 154*, 165–187.

Larrañaga, S., Díez, J. R., Elosegi, A., & Pozo, J. (2003). Leaf retention in streams of the Agüera basin (northern Spain). *Aquatic Sciences, 65*, 158–166.

Molinero, J., & Pozo, J. (2004). Impact of a eucalyptus (*Eucalyptus globulus* Labill.) plantation on the nutrient content and dynamics of coarse particulate organic matter (CPOM) in a small stream. *Hydrobiologia, 528*, 143–165.

Mulholland, P. J. (1997). Organic matter dynamics in the West Fork of Walker Branch, Tennessee, USA. *Journal of the North American Benthological Society, 16*, 61–67.

Naiman, R. J., & Link, G. L. (1997). Organic matter dynamics in 5 subarctic streams, Quebec, Canada. *Journal of the North American Benthological Society, 16*, 33–39.

Newbold, J. D., Bott, T. L., Kaplan, L. A., Sweeney, B. W., & Vannote, R. L. (1997). Organic matter dynamics in White Clay Creek, Pennsylvania, USA. *Journal of the North American Benthological Society, 16*, 46–50.

Pozo, J., González, E., Díez, J. R., & Elosegi, A. (1997). Leaf-litter budgets in two contrasting forested streams. *Limnetica, 13*, 77–84.

Quinn, J. M., Phillips, N. R., & Parkyn, S. M. (2007). Factors influencing retention of coarse particulate organic matter in streams. *Earth Surface Processes and Landforms, 32*, 1186–1203.

Rios, B., Encalada, A. C., & Prat, N. (2009). Leaf litter dynamics and its use by invertebrates in a high-altitude tropical Andean stream. *International Review of Hydrobiology, 94*, 357–371.

Santiago, J., Molinero, J., & Pozo, J. (2011). Impact of timber harvesting on litterfall inputs and benthic coarse particulate organic matter (CPOM) storage in a small stream draining a eucalyptus plantation. *Forest Ecology and Management, 262*, 1146–1156.

Smock, L. A. (1990). Spatial and temporal variation in organic matter storage in low-gradient, headwater streams. *Archiv für Hydrobiologie, 118*, 169–184.

Tonin, A. M., Goncalves, J. F., Jr., Bambi, P., Couceiro, S. R. M., Feitoza, L. A. M., Fontana, L. E., Hamada, N., Hepp, L. U., Lezan-Kowalczuk, V. G., Leite, G. F. M., Lemes-Silva, A. L., Lisboa, L. K., Loureiro, R. C., Martins, R. T., Medeiros, A. O., Morais, P. B., Moretto, Y., Oliveria, P. C. A., Pereira, E. B., Ferreira, L. P., Perez, J., Petrucio, M. M., Reis, D. F., Rezende, R. S., Roque, N., Santos, L. E. P., Siegloch, A. E., Tonello, G., & Boyero, L. (2017). Plant litter dynamics in the forest-stream interface: Precipitation is a major control across tropical biomes. *Scientific Reports, 7*, 10799.

Treadwell, S. A., Campbell, I. C., & Edwards, R. T. (1997). Organic matter dynamics in Keppel Creek, Southeastern Australia. *Journal of the North American Benthological Society, 16*, 58–61.

Wallace, J. B., Whiles, M. R., Eggert, S., Cuffney, T. F., Lugthart, G. J., & Chung, K. (1995). Long-term dynamics of coarse particulate organic matter in three Appalachian Mountain streams. *Journal of the North American Benthological Society, 14*, 217–232.

Webster, J. R., & Meyer, J. L. (1997). Stream organic matter budgets-introduction. *Journal of the North American Benthological Society, 16*, 5–13.

Webster, J. R., Golladay, S. W., Benfield, E. F., D'Angelo, D. J., & Peters, G. T. (1990). Effects of forest disturbance on particulate organic matter budgets of small streams. *Journal of the North American Benthological Society, 9*, 120–140.

Webster, J. R., Wallace, J. B., & Benfield, E. F. (1995). Organic processes in streams of the Eastern United States. In C. E. Cushing, K. W. Cummins, & G. W. Minshall (Eds.), *Ecosystems of the World 22: River and stream ecosystems* (pp. 117–187). Amsterdam: Elsevier.

Chapter 5
Leaching

Felix Bärlocher

Keywords Analysis of leaching data · Dissolved organic matter · Dry leaves · Fresh leaves · Leaching in laboratory conditions · Leaching in streams · Leaf mass loss

1 Introduction

The decomposition of autumn-shed leaves has traditionally been subdivided into three more or less distinct phases: leaching, microbial colonization, and invertebrate feeding (Petersen and Cummins 1974; Gessner et al. 1999). Leaching is defined as the abiotic removal of soluble substances, among them phenolics, carbohydrates, and amino acids (for analyses of these compounds, see Chaps. 13, 15, and 18). It is largely completed within the first 24–48 h after immersion in water and results in a loss of up to 30% of the original mass, depending on leaf species. Gessner and Schwoerbel (1989) showed that no such rapid leaching loss can be observed when fresh, rather than pre-dried, alder and willow leaves are used. Fungal colonization proceeded more slowly on fresh than on pre-dried alder and willow leaves (Bärlocher 1991; Chergui and Pattee 1992), dynamics of chemical leaf constituents differed between fresh and pre-dried leaves during subsequent decomposition (Gessner 1991), but no effects on invertebrate colonization have been observed (Chergui and Pattee 1993; Gessner and Dobson 1993). In a survey of 27 leaf species, drying significantly changed the magnitude of leaching in a majority of cases (Taylor and Bärlocher 1996), although the direction of change was variable among species with

Electronic Supplementary Material The online version of this chapter (https://doi.org/10.1007/978-3-030-30515-4_5) contains supplementary material, which is available to authorized users.

F. Bärlocher (✉)
Department of Biology, Mount Allison University, Sackville, NB, Canada
e-mail: fbaerloc@mta.ca

F. Bärlocher et al. (eds.), *Methods to Study Litter Decomposition*,
https://doi.org/10.1007/978-3-030-30515-4_5

Table 5.1 Percentages of mass losses of fresh and dried leaves over 48 h in distilled water

	Leaf mass loss (% dry mass)		
Leaf species	Fresh		Dried
Acer saccharum	15.2 ± 7.9	<	21.4 ± 7.6
A. negundo	14.7 ± 3.3	<	30.5 ± 2.1
A. circinatum	6.3 ± 5.2	<	23.7 ± 1.5
A. rubrum	16.6 ± 9.2	=	24.5 ± 8.2
Fagus grandifolia	5.1 ± 7.0	=	7.4 ± 6.6
A. macrophyllum	10.2 ± 6.7	>	5.9 ± 0.9
Betula papyrifera	15.1 ± 2.4	>	11.7 ± 1.3

Data from Taylor and Bärlocher (1996)
Mean ± SD. <, loss significantly greater in dried leaves; =, no significant difference; >, loss significantly greater from fresh leaves

drying actually decreasing leaching in several cases. Some representative data are listed in Table 5.1.

Changes in types and amounts of compounds retained by leaves may affect their breakdown rate by selectively stimulating or inhibiting colonization by aquatic microorganisms (Bengtsson 1983, 1992) and by modifying palatability to leaf-eating invertebrates (review in Bärlocher 1997). In addition, they will influence the dynamics of the dissolved organic matter (DOM) pool in the water column, its flocculation into solid particles, and its entrapment and processing at liquid-solid interfaces (Bärlocher et al. 1989; Armstrong and Bärlocher 1989a,b; Meyer et al. 1998; Allan and Castillo 2007; Brett et al. 2017; Findlay and Parr 2017).

In some areas, the yearly leaf fall may overlap with the first night frosts. Freezing living or senescent leaves can have a similar effect as drying them: it may damage cell membranes, which generally accelerates leaching (Bärlocher 1992). In other areas, leaf senescence may coincide with hot, dry weather, and leaves may dry on the tree.

The method described here allows assessing how drying leaves influences leaching. Freshly collected, non-dried leaves (fresh leaves) and leaves that are dried after collection (dried leaves) are exposed in fine-mesh bags (to prevent access by macroinvertebrates) in a stream. After 2 days, the remaining mass is measured. During this early period of decomposition, leaching generally predominates. If drying significantly increases leaching, we expect higher losses in dried leaves. If desired, identically treated leaves can be examined for colonization by aquatic hyphomycetes. To study the temporal course of leaching losses in greater detail, leaf bags should be prepared to allow daily samples for an extended period of 7 days. Or, leaves can be submerged in distilled or stream water in the laboratory, and daily samples can be taken (Gessner and Schwoerbel 1989; Taylor and Bärlocher 1996).

2 Equipment, Chemicals, and Solutions

2.1 Equipment and Material

- Oven (50 °C)
- Leaves of *Alnus glutinosa* or other species
- Litter bags (10 × 10 cm, mesh size 0.5 or 1 mm)
- Plastic labels
- Balance (±1 mg)

3 Experimental Procedures

3.1 Sample Preparation

1. Dry leaves: collect leaves from a single tree by gently shaking branches and collecting fallen leaves. Dry for 2 days at 50 °C to constant mass. Randomly select two to three leaves and weigh to the nearest mg. Moisten leaves to avoid breakage by placing the leaves in a small tray and spraying them with water (avoid highly chlorinated tap water), and place them in a litter bag (see Chap. 6). Label the bag. Prepare a total of 20 bags.
2. Fresh leaves: harvest leaves from the same tree. Return them to laboratory in a cool, closed container. Randomly select two to three leaves, weigh them, and place them in a litter bag. Label the bag. Prepare a total of 20 bags. To determine wet mass/dry mass ratio of fresh leaves, individually weigh 20 fresh leaves, dry them, and weigh them again. Calculate an average correction factor, D = (oven-dry mass)/(fresh mass). Apply to fresh leaves to calculate their original dry mass (see Chap. 6).

$$M_d = M_f \cdot D \tag{5.1}$$

3.2 Experiment

1. Expose all bags in a stream.
2. Recover all bags after 2 days. Leaching in more recalcitrant leaves or conifer needles is generally delayed, and 4–7 days of immersion may be more appropriate. However, this increases the probability that some of the mass loss is due to microbial decomposition.

3. Rinse leaves under running tap water, dry for 48 h at 50 °C to constant mass, and weigh them.
4. Express mass loss as percentage of original leaf mass.

3.3 *Leaching Under Controlled Conditions*

The described method can be modified for the laboratory. Leaf material is prepared and weighed as above. Individual replicates of 3–6 g fresh leaves (or 1–2 g of dried leaves) are placed in 1 l of deionized or filtered stream water in glass bottles and incubated in the dark at a low temperature to minimize microbial attack (e.g., 4 °C; Taylor and Bärlocher 1996). Gessner and Schwoerbel (1989) recommend gentle agitation on a shaker to provide oxygenation and prevent buildup of gradients; Taylor and Bärlocher (1996) incubated leaves under static conditions to avoid stimulation of microbial activity. There is no study comparing the effects of turbulence and incubation temperature on the relative impact of microbial decomposition.

3.4 *Statistical Analysis*

Mass losses of fresh and dried leaves can be compared with a t-test or a permutation test (see Chap. 58). If >2 species are compared, a two-way ANOVA is recommended. Since some values are likely to be below 20%, normal distribution cannot be assumed, and arcsine transformation of proportion p is advisable before applying a standard t-test ($p' = \arcsin \sqrt{p}$).

For the permutation test, we assume that the values for fresh and dried leaves belong to the same population (H_0, null hypothesis). We therefore pool all values. Next, we randomly divide the 40 values into two groups of 20. We determine the difference between mean mass losses of the two groups (or we can use another test statistic, see Edgington and Onghena 2007). We do this 10,000 times and plot the distribution of the differences. Next, we determine the actual difference between the original data from fresh and dried leaves. How "extreme" is it? If it is at least as extreme as 5% of the population of differences based on the permutated data (this corresponds to $p \leq 0.05$), we reject the null hypothesis. For details of the procedure, see 5_Suppl1_final (XLSX 4217 kb).

References

Allan, J. D., & Castillo, M. M. (2007). *Stream ecology. Structure and function of running water* (2nd ed.). Dordrecht: Springer.

Armstrong, S. M., & Bärlocher, F. (1989a). Adsorption and release of amino acids from epilithic biofilms in streams. *Freshwater Biology, 22*, 153–159.

Armstrong, S. M., & Bärlocher, F. (1989b). Adsorption of three amino acids to biofilms on glass-beads. *Archiv für Hydrobiologie, 115*, 391–399.
Bärlocher, F. (1991). Fungal colonization of fresh and dried leaves in the River Teign (Devon, England). *Nova Hedwigia, 52*, 349–357.
Bärlocher, F. (1992). Effects of drying and freezing autumn leaves on leaching and colonization by aquatic hyphomycetes. *Freshwater Biology, 28*, 1–7.
Bärlocher, F. (1997). Litter breakdown in rivers and streams. *Limnetica, 13*, 1–11.
Bärlocher, F., Tibbo, P. G., & Christie, S. H. (1989). Formation of phenol-protein complexes and their use by two stream invertebrates. *Hydrobiologia, 173*, 243–249.
Bengtsson, G. (1983). Habitat selection in two species of aquatic hyphomycetes. *Microbial Ecology, 9*, 15–26.
Bengtsson, G. (1992). Interactions between fungi, bacteria and beech leaves in a stream microcosm. *Oecologia, 89*, 542–549.
Brett, M. D., Bunn, S. E., Chandra, S., Galloway, A. W. E., Guo, F., Kainz, M. J., Kankaala, P., Lau, D. C. P., Moulton, T. P., Power, M. E., Rasmussen, J. P., Thorp, J. H., & Wehr, J. D. (2017). How important are terrestrial organic carbon inputs for secondary production in freshwater ecosystems? *Freshwater Biology, 62*, 833–853.
Chergui, H., & Pattee, E. (1992). Processing of fresh and dry *Salix* leaves in a Moroccan river system. *Acta Oecologica, 13*, 291–298.
Chergui, H., & Pattee, E. (1993). Fungal and invertebrate colonization of *Salix* fresh and dry leaves in a Moroccan river system. *Archiv für Hydrobiologie, 127*, 57–72.
Edgington, E. S., & Onghena, P. (2007). *Randomization tests* (4th ed.). Boca Raton: Chapman & Hall/CRC.
Findlay, S., & Parr, T. (2017). Dissolved organic matter. In R. F. Hauer & G. A. Lamberti (Eds.), *Methods in stream ecology* (Vol. 2, 3rd ed.). London: Academic. in press.
Gessner, M. O. (1991). Differences in processing dynamics of fresh and dried leaf litter in a stream ecosystem. *Freshwater Biology, 26*, 387–398.
Gessner, M. O., & Dobson, M. (1993). Colonization of fresh and dried leaf-litter by lotic macroinvertebrates. *Archiv für Hydrobiologie, 127*, 141–149.
Gessner, M. O., & Schwoerbel, J. (1989). Leaching kinetics of fresh leaf-litter with implications for the current concept of leaf-processing in streams. *Archiv für Hydrobiologie, 115*, 81–90.
Gessner, M. O., Chauvet, E., & Dobson, M. (1999). A perspective on leaf litter breakdown in streams. *Oikos, 85*, 377–384.
Meyer, J. L., Wallace, J. B., & Eggert, S. L. (1998). Leaf litter as a source of dissolved organic carbon in streams. *Ecosystems, 1*, 240–249.
Petersen, R. C., & Cummins, K. W. (1974). Leaf processing in a woodland stream. *Freshwater Biology, 4*, 343–368.
Taylor, B. R., & Bärlocher, F. (1996). Variable effects of air-drying on leaching losses from tree leaf litter. *Hydrobiologia, 325*, 173–182.

Chapter 6
Leaf Mass Loss Estimated by the Litter Bag Technique

Felix Bärlocher

Keywords Analysis of mass loss data · ANCOVA · Coarse-mesh litter bags · Comparison of slopes · Fine-mesh litter bags · Mass loss models · Regression analysis · Two-phase exponential decay model · Litter breakdown · Litter decomposition

1 Introduction

Leaf litter is a dominant component of coarse particulate organic matter (CPOM) in streams, and its decomposition has received considerable attention (Webster and Benfield 1986; Allan 1995; Gessner et al. 1999). In view of its central place in ecosystem functioning, Gessner and Chauvet (2002) proposed using leaf litter breakdown to evaluate functional stream integrity. In order to increase the sensitivity and robustness of the assay, "noise" due to nonstandardized procedures has to be minimized. Many studies have used predried leaves or leaf disks enclosed in litter bags. Several aspects of this approach have been criticized as introducing artificial modifications of the natural process (Petersen and Cummins 1974; Wieder and Lang 1982; Boulton and Boon 1991; Bärlocher 1997). Mass loss in litter bags (1 mm mesh size) resembles that of loose, naturally entrained leaves in depositional zones, while mass loss in litter packs (leaves tied together and tethered in streams) is close to that of loose leaves in riffle areas (Cummins et al. 1980). The use of litter bags with different mesh sizes allows size-selective exclusion of macro-consumers.

Electronic Supplementary Material The online version of this chapter (https://doi.org/10.1007/978-3-030-30515-4_6) contains supplementary material, which is available to authorized users.

F. Bärlocher (✉)
Department of Biology, Mount Allison University, Sackville, NB, Canada
e-mail: fbaerloc@mta.ca

F. Bärlocher et al. (eds.), *Methods to Study Litter Decomposition*,
https://doi.org/10.1007/978-3-030-30515-4_6

Table 6.1 Daily decay rates, k and number of days to reach 50% mass loss (T_{50}) of selected leaf species

Category	Species	k	T_{50}
Fast	*Fraxinus americana*	0.0120	58
	Tilia americana	0.0175	40
Medium	*Carya glabra*	0.0089	78
	Salix lucida	0.0078	89
Slow	*Fagus grandifolia*	0.0025	277
	Quercus alba	0.0022	315

Data from Petersen and Cummins (1974), based on leaf packs

The mass loss of leaf litter, including conifer needles, as a function of time is most often approximated by an exponential decay model:

$$M_t = M_0 \cdot e^{-kt} \tag{6.1}$$

where M_t is mass at time t; M_0, mass at time 0; k, exponential decay coefficient; and t, time in days. Based on daily decay coefficients in a stream in Michigan, leaves have been classified as "fast" ($k > 0.01$), "medium" ($k = 0.005 - 0.001$) and "slow" ($k < 0.005$) (Petersen and Cummins 1974). However, the decomposition rate of a given leaf species can vary greatly among streams (Lecerf and Chauvet 2008; Benfield et al. 2017), suggesting that Petersen and Cummins' (1974) classification has limitations when applied uncritically. A few typical values of k, with number of days required for 50% mass loss, are listed in Table 6.1.

The exponential decay equation is typically converted to a linear form before regression is performed.

$$\ln[M_t] = \ln\left[M_0 \cdot e^{-kt}\right] = \ln[M_0] - kt \tag{6.2}$$

rewritten as

$$Y = a + bX \tag{6.3}$$

Y is the dependent variable, corresponding to M_t. The independent variable X equals time in days. The linear regression procedure, which minimizes the sum of squares, determines the slope b (equals decay coefficient k) and the intercept a (calculated mass at time 0). The intercept should be close to 100%. If it is <100%, rapid initial leaching of soluble compounds may have occurred. If it exceeds 100%, an initial slow phase of decomposition may be succeeded by an accelerated phase (e.g. Maria et al. 2006).

Most computer programs will calculate R^2 of linear regressions. This indicates how much of the variance among the data is due to the linear relationship between the X and Y values. For example, a value of 0.95 corresponds to 95%.

Linear regression calculations are only valid when the experimental uncertainty of replicate *Y* values is not related to the values of *X* or *Y* (Zar 2010). This is not usually the case after data transformation, which tends to enhance errors associated with small *Y* values. These points will be emphasized by linear regressions, and points with large *Y* values will be relatively ignored. Thus, linearizing transformations is not an ideal procedure because it distorts experimental errors (Motulsky and Rasnsnas 1987; Motulsky and Christopoulos 2003).

An alternative is nonlinear regression analysis. This is defined as fitting data to any selected equation. As with linear regression, nonlinear regression procedures determine values of the parameters that minimize the sum of the squares of the distances of the data points to the curve. The approach is only appropriate when the experimental uncertainty is normally distributed and not related to the values of *X* or *Y*. Nonlinear regression (or curve fitting) must be solved iteratively, rather than analytically, and an initial estimate of each parameter must usually be provided. During the fitting procedure, these values are modified to increasingly improve the fit (lower the sum of squared deviations) of the curve to the data. These iterations are continued until additional improvements are negligible.

Often two sets of data are fitted to the same model, and the question is whether the two sets of data differ significantly. For example, do eucalypt and alder leaves decay at significantly different rates? A good introduction to this topic can be found in Motulsky and Christopoulos (2003) and at http://www.graphpad.com/curvefit. The recommended approach is to repeat the experiment several times and compare the resulting estimates of the parameter *k* with a *t*-test, which compares a difference with the standard error of that difference. This method is labour-intensive and statistically conservative; the calculated *p* value may be too high. If the experiment has only been done once, the best-fit value of two groups can still be compared with a *t*-test by using the standard error reported by the curve-fitting program. This again assumes normal error distribution, which is approximately true for the exponential decay equation.

The more commonly used approach is to analyse the two data sets separately as well as simultaneously. This method, known as comparison of slopes by analysis of covariance or ANCOVA, is again strictly valid only for a linear relationship between *X* and *Y*. The details of this test can be found in many statistics textbooks (e.g. Motulsky 1995; Quinn and Keough 2002; Zar 2010; MacDonald 2014). The main purpose of ANCOVA is to compare a Y variable (e.g. remaining mass) among groups (e.g. different leaf species or same species in different streams) while statistically controlling for variation in Y caused by variation in the X variable (X corresponds to time in days). It tests two hypotheses: (1) homogeneity of slopes (i.e. are decay rates the same?); (2) if slopes are identical, are the Y intercepts the same (or are adjusted means identical)? In the context of decomposition studies, the first hypothesis is of primary interest. It tests whether there is a significant interaction between the Y values of different treatments (e.g. leaf species) and the covariate time. If this is confirmed, the second hypothesis is generally irrelevant (though it is possible to test the null hypothesis that the regression lines have the same Y value for a particular X value (Tsutakawa and Hewett 1978; MacDonald 2014). Selected examples of ANCOVA are presented in Suppl. 6.1.

Another approach of comparing decomposition models is based on Akaike's information criterion (AIC), which answers the following questions: Which model is more likely to have generated the data? How much more likely? The theory behind AIC is quite difficult; it combines maximum likelihood, information theory and the concept of information entropy. Fortunately, the computations and interpretation of the results are straightforward (Motulsky and Christopoulos 2003).

Two or more data sets can also be compared by a paired t-test or repeated-measures ANOVA, respectively, where data from a given sampling date are "paired". The result informs about whether the values for the remaining mass of one species are consistently different from those of a second species. This approach is particularly useful when two sets of data do not follow the same decay model (e.g. one is best fitted by a single and the second by a double exponential model). Alternatively, one can compare time in days to lose a specified percentage of the original mass (e.g. 50%).

Instead of parametric tests, permutation or randomization procedures can be used (see Chap. 59). The first step is to define a test statistic *S*, such as the difference between several estimates of the *k* values of alder and eucalypt. The value of *S* is then calculated for the original data set. Next, all values within a sampling date are pooled and randomly assigned to alder or eucalypt. The difference is calculated. This is repeated at least 1000 times, giving the distribution of all possible values of *S*. The final question is: How extreme is the *S* of the original data compared to all possible values? If it is more extreme than 5% (or 1%) of the population, the null hypothesis (H_0) is rejected. If we do not want to make the assumption of linear decay (or exponential decay after log transformation of the mass loss data), we can choose another test statistic. For example, we can add up all data (% remaining mass on all sampling dates) separately for the two leaf species and determine the difference. The null hypothesis is that low and high values are randomly distributed between alder and eucalypt. We again pool the data, randomly distribute them between alder and eucalypt and calculate the new value of *S*. This is repeated at least 1000 times; we then determine how extreme the original *S* is.

Sometimes the single exponential model is clearly inappropriate (Minderman 1968; Wieder and Lang 1982). When the leaf consists of two clearly defined components decaying at different rates, a double exponential equation is more realistic (e.g. Lousier and Parkinson 1976). Or, when decomposition does not appear to proceed beyond a certain point, an asymptotic model provides a better fit (Sridhar et al. 2002).

How does one decide which model gives a better fit? (Zen Koan of Statistics: the person with one watch knows what time it is; the person with two watches is never sure.) A simple comparison of sum of squares (or R^2) values is inappropriate, since a curve with more parameters nearly always has a lower sum of squares because it has more inflection points (Kvålseth 1985). The question is whether this decrease is worth the "cost" of the additional variables, whose inclusion in the model results in a loss of degrees of freedom. Two approaches are commonly used: an *F* test (extra sum-of-squares test) or Akaike's information criterion. The appropriate considerations for making the decision are described in Motulsky and Christopoulos (2003).

The chapter describes the basic procedures to estimate mass loss with the litter bag technique. Other approaches, e.g. the use of leaf packs, are described in Benfield et al. (2017). When there is concern about substantial mineral deposits (silt) on the leaves, all data should be converted to ash-free dry mass (Chap. 4).

2 Equipment, Chemicals and Solutions

2.1 Equipment and Material

- Autumn-shed leaves
- Litter bags (10 × 10 cm, 1 mm mesh size; alternatively, 0.5 mm and 10 mm mesh size)
- Plastic labels (DYMO or laser-printed numbers on transparencies)
- Cool box and plastic bags (for transporting litter bags to and from stream)
- Bricks and/or steel pegs or other devices (e.g. rebars) to secure litter bags on the stream bed. Alternatively, litter bags can be attached with ropes to roots or tree trunks. To avoid excessive movement of the bags, cover ropes with rocks.
- Drying oven (50 °C)
- Balance (±1 mg precision)
- Statistics program
- Optional: muffle furnace and crucibles

3 Experimental Procedures

3.1 Sample Preparation

1. Collect leaf litter are and prepare as in Chap. 5. Depending on objectives, mass loss rates of different leaf species, leaves in different mesh size bags or leaves placed in different rivers can be compared.
2. Place 2–4 preweighed leaves (equivalent of 5 g) in each labelled litter bag.
3. Prepare a sufficient number of bags to allow 4–6 replicates per sampling date plus two extra sets, one set to convert air-dry mass to oven-dry mass and another set to determine handling losses (see below). Optional: determine ash-free dry mass by exposing additional samples to 500 °C for 12 h (Chap. 4).

3.2 Exposure of Litter Bags

1. Place litter bags in a cool box and transport to stream.
2. Anchor leaf bags to the stream bed with bricks, steel pegs, etc. Depending on objectives, all bags may be placed in riffle or pool areas. Care must be taken not to place too many bags close to each other, because this may drastically change current patterns and affect colonization by microorganisms and invertebrates. Alternatively, leaf bags may be attached to rebars anchored to the stream bed.

3.3 Recovery of Bags and Analysis

3. Depending on pretreatment, leaves may have to be corrected for water content (for an example, see Suppl. 6.1). Weigh a control set of leaves before (air-dry) and after drying (oven-dry) to constant mass (e.g. 48 h at 50 °C). Calculate an average correction factor, D = (oven-dry mass)/(air-dry mass). This correction factor only needs to be applied to the initial leaf mass, since recovered leaves are dried to constant mass (see step 6).
4. The initial oven-dry mass of each leaf pack brought to the streams is estimated by multiplying the measured air-dry mass by the average correction factor, D.
5. A second set of bags should be recovered immediately upon exposure in the stream. This allows an estimate of losses due to initial handling. Calculate a handling correction factor, H = (leaf mass before handling)/(leaf mass after handling).
6. The corrected leaf mass in the bags is calculated as $M_{corr} = M_{initial} * D * H$.
7. Subsequent samples are taken according to a pre-planned schedule, for example, after 3, 7, 14, 21, 28 and 35 days. With slowly decomposing leaves, an extended sampling schedule of 3 months or longer may be necessary.
8. Rinse the recovered leaves under running tap water, dry to constant mass and weigh.

3.4 Statistical Analysis

1. Express mass loss as percentage of the original mass after correction for (1) water content and (2) handling (100% = mass after corrections). Do not include values at time 0, since for all treatments, mean remaining mass is 100% by definition (Wieder and Lang 1982; Fox 2017).
2. Run a regression analysis. Do not force the model through mass at time 0 (see Final Remarks). (a) Transform [% mass remaining] to ln[% mass remaining], and run a linear regression (time = independent variable, ln[% mass remaining] = dependent variable). Alternatively, run (b) a nonlinear curve-fitting

program (exponential decay; in most programs, you will have to type in the equation and provide initial estimates of the parameters that are to be determined; in the exponential decay model, provide an estimate of *a* and *k*).

3. If variation among replicate samples is small, the differences in slope estimates between the two models are also small, but the nonlinear curve-fit often provides the best estimate for the intercept (which, in theory, should be 100%) and slope.
4. If data from more than one series have been collected (e.g. two or more leaf species), an analysis of covariance using time as covariate can be run (for examples, see Suppl. 6). This procedure is part of most comprehensive computer programs. Provided the estimated initial leaf masses of the two series are similar, the decay coefficients are significantly different if the *p* value for the *interaction* between time and series is <0.05. Alternatively, an appropriate test statistic may be formulated and a corresponding permutation test performed (e.g. with Resampling Stats, commercial software; Statistics101, giftware; R package "resample", freeware; see also Chap. 59).
5. If a pronounced steep decline during the early phase of decomposition is observed (which can be due to strong leaching), try fitting the data to a more complex model (e.g. double exponential decay) or calculate post-leaching decomposition rates. Motulsky and Christopoulos (2003) can be a useful guide to decide which model is more appropriate.

4 Final Remarks

"Leaf mass loss" is a broad term that includes leaching, biological and mechanical fragmentation, chemical conversion, etc., all of which are influenced by physico-chemical conditions of the environment and may change repeatedly and unpredictably during the course of decomposition.

Since mineral deposits are often difficult to remove from recovered leaves, dry mass is often converted to ash-free dry mass. This can be achieved by combusting organic matter at 500–550 °C (see Chap. 4). The remaining material, mineral ash, is then subtracted from the initial dry mass.

Both drying conditions and length of the drying period may affect the final dry mass of leaf litter. Freeze-drying and exposure at various temperatures of up to 105 °C has no notable effect on the outcome, provided sufficient time is given (M.O. Gessner, pers. comm.). Forced-air ovens are recommended. Avoid placing large quantities of wet samples in an oven where leaves have already been dried. To preclude the resulting problem of remoistening, two ovens may be used in sequence: one for wet samples and one for predried samples.

Drying also changes the physico-chemical properties of plant litter. This will affect the dynamics of leaching (Gessner and Schwoerbel 1989; Chap. 5), leaf chemical composition (Gessner 1991), colonization by aquatic hyphomycetes and invertebrates (Bärlocher 1991; Legssyer et al. 2003) and precipitation of proteins by tannins (Hagerman et al. 1998). The latter is temperature-sensitive (Hagerman et al. 1998);

if leaves are used to investigate the decomposition process, drying at low temperatures or the use of fresh leaves is recommended.

There is no such thing as all-purpose experiment or model; designs can be chosen for realism, generality or precision, but these three qualities cannot be maximized simultaneously (Levins 1968; Boulton and Boon 1991). It is therefore essential to decide beforehand which particular aspect is of paramount interest in any given investigation.

Similarly, all mathematical models have been said to be wrong, though some are more economical (Occam's razor) and therefore more useful than others (Box 1976). The exponential decay model provides a simple and intuitive characterization of mass loss rates, but does not capture the true dynamics and complexities of the underlying processes.

When fitting a regression through mass loss data, it is tempting to force the line through the origin (i.e. [Time = 0; $Mass_{remaining}$ = 100%]). Fox (2017) summarized arguments against this approach (https://tinyurl.com/yadtur2x). In his view, it would only be justified if we knew for certain what the true relationship between the two variables was (which we never do in ecology).

Acknowledgements Tim Moulton, Rio de Janeiro State University, Brazil, was instrumental in pointing out crucial issues and misunderstandings in the statistical analysis of litter mass loss data.

References

Allan, J. D. (1995). *Stream ecology. Structure and function of running waters*. London: Chapman & Hall.

Bärlocher, F. (1991). Fungal colonization of fresh and dried leaves in the River Teign (Devon, England). *Nova Hedwigia, 52*, 349–357.

Bärlocher, F. (1997). Pitfalls of traditional techniques when studying decomposition of vascular plant remains in aquatic habitats. *Limnética, 13*, 1–11.

Benfield, E. F., Fritz, K. M., & Tiegs, S. D. (2017). Leaf-litter breakdown. In G. A. Lamberti & F. R. Hauer (Eds.), *Methods in stream ecology. Vol. 2: Ecosystem function*. London: Academic/Elsevier.

Boulton, A. J., & Boon, P. I. (1991). A review of the methodology used to measure leaf litter decomposition in lotic environments: Time to turn over an old leaf? *Australian Journal of Marine and Freshwater Research, 42*, 1–43.

Box, G. E. P. (1976). Science and statistics. *Journal of the American Statistical Association, 71*, 791–799.

Cummins, K. W., Spengler, G. L., Ward, G. M., Speaker, R. M., Ovink, R. W., Mahan, D. C., & Mattingly, R. L. (1980). Processing of confined and naturally entrained leaf litter in a woodland stream ecosystem. *Limnology and Oceanography, 25*, 952–957.

Fox, J. (2017). *Don't force your regression through zero just because you know the true intercept has to be zero*. https://tinyurl.com/yadtur2x.

Gessner, M. O. (1991). Differences in processing dynamics of fresh and dried leaf litter in a stream ecosystem. *Freshwater Biology, 26*, 387–398.

Gessner, M. O., & Chauvet, E. (2002). A case for using litter breakdown to assess functional stream integrity. *Ecological Applications, 12*, 498–510.

Gessner, M. O., & Schwoerbel, J. (1989). Leaching kinetics of fresh leaf-litter with implications for the current concept of leaf-processing in streams. *Archiv für Hydrobiologie, 115*, 81–90.
Gessner, M. O., Chauvet, E., & Dobson, M. (1999). A perspective on leaf litter breakdown in streams. *Oikos, 85*, 377–384.
Hagerman, A. E., Rice, M. E., & Ritchard, N. T. (1998). Mechanisms of protein precipitation for two tannins, pentagalloyl glucose and epicatechin$_{16}$ (4→8) catechin (procyanidin). *Journal of Agricultural and Food Chemistry, 46*, 2590–2595.
Kvålseth, T. O. (1985). Cautionary note about R^2. *The American Statistician, 39*, 279–285.
Lecerf, A., & Chauvet, E. (2008). Intraspecific variability in leaf traits strongly affects alder leaf decomposition in a stream. *Basic and Applied Ecology, 9*, 598–605.
Legssyer, B., Chergui, H., & Maamri, A. (2003). Invertebrate dynamics during the decomposition of dry and fresh willow leaves in Oued Zegzel (Eastern Morocco). *Annales de Limnologie – International Journal of Limnology, 39*, 27–33.
Levins, R. (1968). *Evolution in changing environments: Some theoretical explorations*. Princeton: Princeton University Press.
Lousier, J. D., & Parkinson, D. (1976). Litter decomposition in a cool temperate deciduous forest. *Canadian Journal of Botany, 54*, 419–436.
MacDonald, J. H. (2014). *Handbook of biological statistics*. Baltimore: Sparky House Publishing.
Maria, G. L., Sridhar, K. R., & Bärlocher, F. (2006). Decomposition of dead twigs of *Avicennia officinalis* and *Rhizophora mucronata* in a mangrove in southwestern India. *Botanica Marina, 49*, 450–455.
Minderman, G. (1968). Addition, decomposition and accumulation of organic matter in forests. *Journal of Ecology, 56*, 355–362.
Motulsky, H. J. (1995). *Intuitive biostatistics*. Oxford: Oxford University Press.
Motulsky, H. J., & Christopoulos, A. (2003). *Fitting models to biological data using linear and nonlinear regression. A practical guide to curve fitting*. San Diego: GraphPad Software, Inc. www.graphpad.com.
Motulsky, H. J., & Rasnsnas, L. A. (1987). Fitting curves to data using nonlinear regression: A practical and non-mathematical review. *Journal of the Federation of American Societies for Experimental Biology, 1*, 365–374.
Petersen, R. C., & Cummins, K. W. (1974). Leaf processing in a woodland stream. *Freshwater Biology, 4*, 343–368.
Quinn, G. P., & Keough, M. J. (2002). *Experimental design and data analysis for biologists*. Cambridge: Cambridge University Press.
Sridhar, K. R., Krauss, G., Bärlocher, F., Raviraja, N. S., Wennrich, R., & Baumbach, R. (2002). Decomposition of alder leaves in two heavy metal polluted streams in Central Germany. *Aquatic Microbial Ecology, 26*, 73–80.
Tsutakawa, R. K., & Hewett, J. E. (1978). Comparison of two regression lines over a finite interval. *Biometrics, 34*, 391–398.
Webster, J. R., & Benfield, E. F. (1986). Vascular plant breakdown in freshwater ecosystems. *Annual Review of Ecology and Systematics, 17*(1): 567–594.
Wieder, R. K., & Lang, G. E. (1982). A critique of the analytical methods used in examining decomposition data obtained from litter bags. *Ecology, 63*, 1636–1642.
Zar, J. H. (2010). *Biostatistical analysis* (5th ed.). New York: Prentice-Hall.

Chapter 7
Determining Litter Mass Loss by the Plant Tagging Approach

Kevin A. Kuehn and Mark O. Gessner

Keywords Carbon cycling · Emergent macrophytes · Floating-leaved macrophytes · Grassland · Litter decomposition · Litter breakdown · Marshes · Nutrient cycling · Standing-dead litter · Wetlands

1 Introduction

The spatial and temporal context in which plant litter decomposes is a critical consideration when selecting methods to examine plant. For most questions typically addressed in these studies, experimental conditions need to be chosen in a way that does not fundamentally alter the natural sequence and environmental conditions of decay (Bärlocher 1997). The litter-bag approach (Chap. 6) generally serves this purpose well when plant parts such as leaves become naturally detached from their parent plant before decomposition begins. This is typically the case when trees and shrubs shed leaves following senescence at the end of the growing season. However, in other plants, including in the grass, sedge and rush families, leaves do not typically abscise, and a large fraction of plant biomass remains attached to shoots for extended periods of time, so that litter decomposition begins in an upright position (Newell 1993; Kuehn 2016). Likewise, leaves of floating-leaved macrophytes, such as water lilies (Nymphaeaceae), will remain

K. A. Kuehn (✉)
School of Biological, Environmental and Earth Sciences, The University of Southern Mississippi, Hattiesburg, MS, USA
e-mail: kevin.kuehn@usm.edu

M. O. Gessner
Department of Experimental Limnology, Leibniz Institute of Freshwater Ecology and Inland Fisheries (IGB), Stechlin, Germany

Department of Ecology, Berlin Institute of Technology (TU Berlin), Berlin, Germany
e-mail: gessner@igb-berlin.de

F. Bärlocher et al. (eds.), *Methods to Study Litter Decomposition*,
https://doi.org/10.1007/978-3-030-30515-4_7

floating on the water surface, while they senesce and start to decompose (Klok and van der Velde 2017).

Current understanding of macrophyte decomposition is mostly based on studies where plant material is harvested, enclosed in and placed on sediments or suspended in the overlying water (Polunin 1984; Webster and Benfield 1986). However, when applied to emergent and floating-leaved macrophytes, this approach bears a serious risk of introducing artefacts when natural decay conditions are not adequately reflected (Newell 1993; Kuehn 2016). Results from such studies are particularly unlikely to represent naturally occurring processes when the plant material is collected in a living green state or an early state of senescence. The reason is that prematurely disconnecting the collected plant organs (e.g. leaves) from their parent plant disrupts physiological plant processes. In particular, the normal orderly translocation of carbon and nutrients to below-ground plant parts during shoot senescence is prevented (Kuehn et al. 1999; Kuehn et al. 2011; Su et al. 2015), thus circumventing an important pathway of mass loss and changes in plant tissue chemistry. As a consequence, concentrations of labile carbon compounds, nitrogen and in the collected plant material are likely to be unnaturally high and could influence the colonization and activities of and involved in decomposition. Furthermore, given vastly different environmental conditions and microbial source communities, the colonization dynamics and activities on shoots differ fundamentally from those occurring at the sediments or in surface waters (Newell and Fallon 1989; Newell et al. 1995; Kuehn et al. 2000).

An alternative approach to examine decomposition of standing-dead plants is to follow mass loss as well as nutrient and microbial dynamics on shoots that have been tagged in a natural position (Newell and Fallon 1989). Several studies have used this technique to examine senescence (Kuehn and Suberkropp 1998) and mass loss of emergent macrophytes, as well as nutrient and microbial dynamics of naturally positioned standing shoots (Table 7.1). Similarly, a non-destructive leaf-tagging method has been applied to examine the senescence and turnover of floating-leaved macrophytes (Kok et al. 1990; Klok and van der Velde 2017). The most common method involves the use of brightly coloured electrical cable ties or monofilament line that can be used in combination with other markers (adhesive tape or Dymo tape) to number and follow individual shoots or leaves during the course of senescence and decomposition. Tagged shoots are periodically collected and mass loss of specific plant organs (e.g. leaf blades, leaf sheaths and stems of tall grasses) are determined based on declines in either area-specific mass (e.g. Gessner 2001) or other morphometric measures used to estimate the initial dry mass of the tagged plant parts (Kuehn et al. 1999).

Table 7.1 Mass loss (%) and exponential decay rate (k) observed for standing-dead emergent macrophyte species

Species	Plant organ	Loss of ash-free dry mass (%)	Exposure time (day)	$-k$ (day^{-1})	Reference
Carex walteriana	Leaf blades	46	~240	0.0010	1
Erianthus giganteus	Leaf blades	32	82	n.d.	2
	Stems	25	190	n.d.	2
Lythrum salicaria	Leaf blades	41	84	n.d.	3
Phragmites australis	Leaf blades	28	28	n.d.	4
	Leaf blades	31	80	0.0055	5
	Leaf sheaths	28	195	0.0016	5
	Stems	25	195	0.0013	5
Spartina alterniflora	Leaf blades	60	83	0.0110	6
Spartina alterniflora	Leaf blades	70	137	n.d.	7
Spartina alterniflora	Leaf blades	60	90	0.0110	8
Typha latifolia	Leaf blades	55	210	n.d.	3
Typha angustifolia	Leaf blades	55[a]	358	0.0021	9
Typha domingensis	Leaf blades	37[a]	340	0.0012	10

1, Newell et al. (1995); 2, Kuehn et al. (1999); 4, Gessner (2001); 5, Kuehn and Gessner (unpublished data); 6, Newell et al. (1989); 7, Newell and Fallon (1989); 8, Samiaji and Bärlocher (1996); 3, Bärlocher and Biddiscombe (1996); 9, Kuehn et al. (2011); 10, Su et al. (2015)

n.d. not determined

[a]Based on the assumption that percent litter carbon loss equalled ash-free dry mass loss

2 Equipment, Chemicals and Solutions

2.1 Equipment and Material

- Electrical cable ties or monofilament line
- Plastic labels (e.g. Dymo)
- Freeze-dryer or drying oven
- Balance (±1 to 10 mg precision, depending on the sample dry mass)
- Computerized scanner, scanning photocopier or leaf area meter (e.g. LiCor-3100)
- Image analysis programme, such as ImageJ (Schneider et al. 2012) or Easy Leaf Area (Easlon and Bloom 2014), both of which are free, open-source software (https://imagej.nih.gov/ij; https://github.com/heaslon/Easy-Leaf-Area)
- Measuring tape, ruler and calipers if morphometric measurements are made

3 Experimental Procedures

3.1 Field Procedures

1. Establish replicate random plots or transects in the field where the plant shoots or leaves of interest will be tagged and followed.
2. Tag shoots or leaves when they are still living and at peak biomass, to determine mass loss patterns during both senescence and decomposition.
3. Tag randomly selected leaves or shoots with electrical cable ties or monofilament line that has numbered markers (adhesive tape or Dymo tape) to be able to follow individual shoots or leaves over time (Fig. 7.1). Cable ties with numbered flags are particularly well suited to tag emergent macrophytes; Dymo plastic labels have also proved useful for tagging leaf blades of floating-leaved macrophytes (Klok and van der Velde 2017). Tags can be placed at various sampling heights in the canopy of emergent macrophytes to examine spatial variation in decay of particular plant organs (Gessner 2001). Care must be taken not to damage shoots or leaves during tagging.
4. Tag a sufficient number of leaves or shoots in each plot to enable the collection of at least three replicates per plot at each envisaged sampling date.
5. Randomly collect replicate tagged leaves or shoots from each plot on the day of tagging in order to determine the initial area, mass and area-mass relationships.

Fig. 7.1 Senescing shoot of the emergent macrophyte *Phragmites australis* (left) and a floating leaf of the water lily *Nymphaea alba* tagged with cable ties and a numbered label. Photos. M. Abelho

For leaf blades, area-mass relationships work well. However, other plant organs may require establishing different types of relationships (see below).

6. Take subsequent samples according to a predetermined schedule reflecting the anticipated decay rate, for example, after 3, 7, 14, 21, 28, 42, 56 and 70 days. Depending on the plant species of interest and environmental conditions (e.g. moisture availability), longer periods may be necessary. For example, with slowly decomposing stems of tall emergent macrophytes, a sampling schedule exceeding a year can be necessary.

3.2 Laboratory Procedures

7. Determine the surface area (or other appropriate measures such as stem section length and diameter) of the tagged plant material by scanning, photocopying or measuring various dimensions of the collected plant parts. This generally works best for leaves when they are wet or moist, but care needs to be taken not to lose any soluble plant matter when wetting the samples.
8. Care is also needed to reproduce the original proportions of the plant material. This is ensured by using scanners, including scanning photocopy machines, which literally scan objects. In contrast, devices such as many photocopiers taking a shot of the objects distort the proportions on the photocopy, which leads to notable inaccuracies in the estimation of surface area.
9. Determine the surface area of the scanned or photocopied images using image analysis software, such as the freeware ImageJ or Easy Leaf Area. Alternatively, surface areas of collected leaf blades can be rapidly determined with a leaf area meter (e.g. LiCor-3100).
10. In some instances, surface area may not be the most suitable measure to establish relationships with litter mass (e.g. stems of emergent macrophytes). In this situation, various alternative morphometric measurements, such as total length and diameter, could be recorded. Multiple linear regression analysis can then generate models that predict the initial mass of the tagged litter collected over the course of an experiment (Kuehn et al. 1999).
11. Following area or morphometric measurement, dry or freeze-dry the plant material to constant weight, and weigh to the nearest 10 mg or less, depending on the sample mass.
12. Either determine the area-specific mass of the tagged plant material (e.g. leaf blades) or estimate the initial mass of the tagged plant material based on morphometric measurements other than surface area (e.g. stems).
13. Express the mass remaining as percentage of the original estimated dry mass or ash-free dry mass as described in Chap. 6.
14. Determine and statistically compare mass loss rates, chemical and physical properties of the remaining litter and microbial community structure and biomass as described in other chapters.

4 Final Remarks

The main strength of the litter-tagging method is that it provides accurate mass loss data in many situations where the litter-bag approach falls short of mimicking the natural sequence of the decomposition process in the field. The tagging approach has been particularly valuable in the case of emergent and floating-leaved macrophytes (Newell et al. 1989; Klok and van der Velde 2017). However, its application range could also include leaves in the canopy of some tree or terrestrial grass species that tend to retain a significant portion or their foliage for extended periods after senescence and death. In some instances, tagging of individuals leaves or shoots may not be necessary. If plants collectively senescence and die as a single cohort, then individual non-tagged leaves or shoots could be randomly collected through time and their mass loss determined (Table 7.1; Kuehn et al. 2011; Su et al. 2015).

The litter-tagging method is generally less precise than the litter-bag approach, since the initial litter mass must be indirectly estimated by establishing some sort of an allometric relationship, generally a curvilinear surface area-dry mass relationship. However, when the method is carefully employed, its precision is sufficient to obtain reliable mass loss data (Table 7.1). A critical condition, however, is that enough replicate samples of plant material be initially collected at the time of tagging (e.g. 25–50 plant shoots) to generate predictive relationships that ensure reliable estimates of the initial plant dry mass.

Although the tagging method overcomes limitation of the litter-bag approach, it is not a silver bullet that is easily applicable in all situations. For example, while divers may tag submerged macrophytes in a similar way as emergent macrophytes are tagged in wetlands or on land, tagging does not offer a straightforward solution to examining the natural decomposition of below-ground plant parts such as roots and rhizomes.

References

Bärlocher, F. (1997). Pitfalls of traditional techniques when studying decomposition of vascular plant remains in aquatic habitats. *Limnetica, 13*, 1–11.

Bärlocher, F., & Biddiscombe, N. R. (1996). Geratology and decomposition of *Typha latifolia* and *Lythrum salicaria* in a freshwater marsh. *Archiv für Hydrobiologie, 136*, 309–325.

Easlon, H. M., & Bloom, A. J. (2014). Easy leaf area: Automated digital image analysis for rapid and accurate measurement of leaf area. *Applications in Plant Sciences, 2*, 1400033.

Gessner, M. O. (2001). Mass loss, fungal colonization and nutrient dynamics of *Phragmites australis* leaves during senescence and early aerial decay. *Aquatic Botany, 69*, 325–339.

Klok, P. F., & van der Velde, G. (2017). Plant traits and environment: Floating leaf blade production and turnover of waterlilies. *PeerJ, 5*, e3212.

Kok, C. J., van der Velde, G., & Landsbergen, K. M. (1990). Production, nutrient dynamics and initial leaf decomposition of floating leaves of *Nymphaea alba* L. and *Nuphar lutea* (L.) Sm. (Nymphaeaceae) in alkaline and acid waters. *Biogeochemistry, 11*, 235–250.

Kuehn, K. A. (2016). Lentic and lotic habitats as templets for fungal communities: Traits, adaptations and their significance to litter decomposition within freshwater ecosystems. *Fungal Ecology, 19*, 135–154.

Kuehn, K. A., & Suberkropp, K. (1998). Decomposition of standing litter of the freshwater emergent macrophyte *Juncus effusus*. *Freshwater Biology, 40*, 717–727.

Kuehn, K. A., Gessner, M. O., Wetzel, R. G., & Suberkropp, K. (1999). Standing litter decomposition of the emergent macrophyte *Erianthus giganteus*. *Microbial Ecology, 38*, 50–57.

Kuehn, K. A., Lemke, M. J., Suberkropp, K., & Wetzel, R. G. (2000). Microbial biomass and production associated with decaying leaf litter of the emergent macrophyte *Juncus effusus*. *Limnology and Oceanography, 45*, 862–870.

Kuehn, K. A., Ohsowski, B. M., Francoeur, S. N., & Neely, R. K. (2011). Contribution of decomposer fungi to standing-dead leaf litter decomposition of the emergent macrophyte *Typha angustifolia*. *Limnology and Oceanography, 56*, 529–539.

Newell, S. Y. (1993). Decomposition of shoots of a saltmarsh grass: Methodology and dynamics of microbial assemblages. *Advances in Microbial Ecology, 13*, 301–326.

Newell, S. Y., & Fallon, R. D. (1989). Litterbags, leaf tags and decay of nonabscised intertidal leaves. *Canadian Journal of Botany, 67*, 2324–2327.

Newell, S. Y., Fallon, R. D., & Miller, J. D. (1989). Decomposition and microbial dynamics for standing, naturally positioned leaves of the salt-marsh grass *Spartina alterniflora*. *Marine Biology, 101*, 471–481.

Newell, S. Y., Moran, M. A., Wicks, R., & Hodson, R. E. (1995). Productivities of microbial decomposers during early stages of decomposition of leaves of a freshwater sedge. *Freshwater Biology, 34*, 135–148.

Polunin, N. V. C. (1984). The decomposition of emergent macrophytes in fresh water. *Advances in Ecological Research, 14*, 115–166.

Samiaji, J., & Bärlocher, F. (1996). Geratology and decomposition of *Spartina alterniflora* Loisel. In a New Brunswick saltmarsh. *Journal of Experimental Marine Biology and Ecology, 201*, 233–252.

Schneider, C. A., Rasband, W. S., & Eliceiri, K. W. (2012). NIH image to ImageJ: 25 years of image analysis. *Nature Methods, 9*, 671–675.

Su, R., Kuehn, K. A., & Phipps, S. W. (2015). Carbon and nutrient flow into decomposer fungi during standing-dead *Typha domingensis* decomposition in a subtropical freshwater marsh. *Freshwater Biology, 60*, 2100–2112.

Webster, J. R., & Benfield, E. F. (1986). Vascular plant breakdown in freshwater ecosystems. *Annual Review of Ecology and Systematics, 17*, 567–594.

Chapter 8
Wood Decomposition

Arturo Elosegi, Maite Arroita, and Libe Solagaistua

Keywords Carbon cycling · Ice-cream stick · Logs · Lignin · Snags · Streams · Tongue depressor · Wood breakdown

1 Introduction

Wood is an extremely abundant component of detrital biomass in many ecosystems around the world (Pregitzer and Euskirchen 2004). As trees die or are uprooted by wind, snowfall, or other forces, they fall to the ground and into nearby freshwaters, where they can accumulate in densities of up to thousands of m^3 per ha of riverbed (Harmon et al. 1986). Wood density in streams can sometimes be higher than in nearby forests. Wood, primarily composed of cellulose, hemicellulose, and lignin, is very resistant to decomposition and therefore plays an important physical role in freshwaters, enhancing habitat diversity, slowing down water current velocity, and promoting the retention of sediments and organic matter (Wohl 2017). Submersed wood can be an important substratum for fungal growth (Golladay and Sinsabaugh 1991); microbial respiration per unit surface area is larger on wood than on leaf litter (Fuss and Smock 1996). Additionally, because of its high abundance in some streams and despite its low nutrient concentration and slow decomposition, wood can make a significant contribution to the total flux of carbon and nutrients, even exceeding that of leaf litter (Elosegi et al. 2007).

Wood breakdown involves leaching of soluble materials, physical abrasion and fragmentation, biological shredding and ingestion of wood particles by animals, and mineralization (respiration) by both animals and microbes (Bilby 2003). The relative contributions of these processes are not clear (Maser and Sedell 1994). Wood leaching has usually been considered of minor importance (Triska and Cromack

A. Elosegi (✉) · M. Arroita · L. Solagaistua
Department of Plant Biology and Ecology, Faculty of Science and Technology, University of the Basque Country, UPV/EHU, Bilbao, Spain
e-mail: arturo.elosegi@ehu.eus; maite.arroita@ehu.eus; libe.solagaistua@ehu.eus

F. Bärlocher et al. (eds.), *Methods to Study Litter Decomposition*,
https://doi.org/10.1007/978-3-030-30515-4_8

1980), but its contribution cannot be neglected (France et al. 1997). Physical fragmentation can be caused either directly by the force of the flowing water or secondarily as suspended solids abrade the wood surface (Aumen 1985). Water-saturated wood decomposition seems mainly driven by microbes, primarily by single-celled bacteria and actinomycetes growing in superficial biofilms (Aumen et al. 1983). However, some invertebrate taxa ingest and/or burrow in wood tissues previously colonized and softened by microbes, thereby enhancing the breakdown of wood in streams (Hoffman and Hering 2000).

In freshwaters, unlike in oceans, there are few or no wood-boring organisms, and thus, decomposition occurs fundamentally at the surface of wood. Spänhoff and Meyer (2004) showed that most of the variability in wood decomposition rates observed across the world (e.g., Table 8.1) was linearly related to the wood surface-to-volume ratio. Wood pieces can range from huge trees to small twigs, with their

Table 8.1 Ranges of decomposition rates of wood across world rivers

Substrate	Diameter (cm)	Latitude	Location	k (y^{-1})	References
Natural	Entire logs	43°N	British Columbia, Canada	0.010	1[a]
	Entire logs	40°N	New Mexico, USA	0.003–0.065	2[b]
	2.5–7.8	51–53°N	Germany	0.065–0.086	3
	1.3–3.6	35°N	North Carolina, USA	0.062–0.376	4
	3	43°N	Spain	0.016–0.271	5
	1–3	35°N	North Carolina, USA	0.107–0.281	6
	0.5–1.5	35°N	North Carolina, USA	0.113–0.157	7
	0.06–1	31°N	Alabama, USA	0.256–0.584	8
Artificial	Tongue depressors	43°N	Spain	0.036–0.986	9
	Tongue depressors	36–43°N	Spain	0.124–6.01	10
	Tongue depressors and coffee stirrers	41–42°S	New Zealand	0.040–0.358	11
	Coffee stirrers	37°S	New Zealand	0.986	12
	Veneer	35–36°N	North Carolina, USA	0.548–2.77	13
	Veneers	35°N	North Carolina, USA	0.533–3.50	14
	0.6 dowels	47°N	Minnesota, USA	0.13–0.22	15
	Chips	74°N	Quebec, Canada	0.040–1.20	16

1, Chen et al. (2005); 2, Ellis et al. (1999); 3, Spänhoff and Meyer (2004); 4, Webster et al. (1999); 5, Diez et al. (2002); 6, Golladay and Webster (1988); 7, Eggert and Wallace (2003); 8, Fritz et al. (2006); 9, Estevez et al. (2017); 10, Aristi et al. (2012); 11, Niyogi et al. (2013); 12, Collier (2014); 13, McTammany et al. (2008); 14, Tank and Webster (1998); 15, Wold and Hershey (1999); 16, Melillo et al. (1983)

[a]Breakdown rate calculated by indirect methods

[b]Breakdown in riparian area

surface-to-volume ratio (and, thus, their decomposition rates) differing by several orders of magnitude. Underwater decomposition of large wood pieces is extremely slow and can virtually stop for millennia in anoxic sediments (Blanchette et al. 1991; Hyatt and Naiman 2001). Decomposition is much faster if the wood is subject to the frequent drying and rewetting (Noetzli et al. 2008) of parafluvial areas. Apart from that, wood decomposition tends to be affected by the same environmental factors as leaf litter decomposition, such as nutrient content (Diez et al. 2002), oxygen concentration (Zeikus 1980), water temperature, and pollution (Ferreira et al. 2016), with abrasion and physical fragmentation also being important, especially in logs transported by rivers (Wohl 2017). However, wood decomposition seems to be more sensitive to nutrient content than leaf litter decomposition (Ferreira et al. 2015). Wood decomposition rates differ greatly among tree species and tissue types, which is mainly related to their composition. Gymnosperms generally contain more lignin and lower nutrient concentrations and therefore decompose more slowly than angiosperms (Harmon et al. 1986). Regarding tissue types, bark is richer in nutrients that the inner tissues. Besides, in some tree species the older, inner tissues of the logs form heartwood, a darker and more resilient wood than the outer bark or sapwood, which can be a confounding factor in breakdown rates. The proportions of different wood tissues are related to piece size, with smaller pieces typically having a relatively high proportion of bark and sapwood and little heartwood.

There are several methods to assess the degree of decay of wood pieces, based on characteristics such as the presence or absence of bark or the softness of outer wood (e.g., Triska and Cromack 1980; Spänhoff et al. 2001). On the other hand, breakdown rates are calculated gravimetrically following a procedure similar to that of leaf litter. Special attention must be paid to the size and surface-to-volume ratio of wood used when comparing data from different authors. Wood decomposition experiments usually are performed either with naturally shaped fragments, such as entire logs or pieces cut from branches, or with manufactured pieces, such as ice-cream sticks, tongue depressors, wood veneers, or wood chips (Fig. 8.1). The former can be used to calculate the total mass of wood decomposed in a known reach (Elosegi et al. 2007), whereas the later can be used for monitoring ecosystem functioning at

Fig. 8.1 Tongue depressors deployed (left) and recovered (right) from a river after 3 months of incubation. The shape has not changed, but they have suffered considerable mass loss

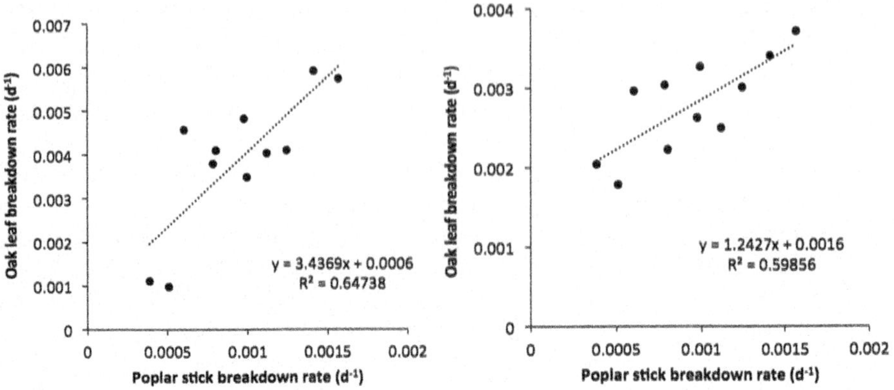

Fig. 8.2 Comparison between the breakdown rates of poplar sticks and oak leaves in coarse (left) and fine (right) mesh bags measured simultaneously at 11 streams across the Ebro River basin (Spain). (Oak breakdown data taken from Monroy et al. 2016)

multiple sites (Aristi et al. 2012). Manufactured pieces such as ice-cream sticks are cheap and homogenous, can be easily mailed over large distances, can be deployed and treated much more easily than leaf-litter bags (Arroita et al. 2012), and respond to environmental factors much in the same manner as leaves (Fig. 8.2).

Here we propose two methods, one for estimating wood decomposition in a reach and the other to compare the decomposition potential of multiple reaches for monitoring purposes.

2 Equipment

2.1 Natural Wood Pieces

- Measuring tape
- Saw
- Balance (±0.01 g)
- 1-cm mesh bags
- Plastic or metal tag
- Very strong fishing line
- Metal bars (optional)
- Blotting paper envelops
- Cooler
- Crucibles
- Drying oven (50 °C)
- Muffle furnace
- Desiccator

2.2 *Standard Substrates*

- Standard substrates, e.g., tongue depressors, ice-cream sticks, or coffee stirrers
- Punch
- Balance (±0.01 g)
- Plastic or metal tag
- Weights (optional)
- Fishing line
- Metal bars (optional)
- Blotting paper envelops
- Cooler
- Crucibles
- Drying oven
- Muffle furnace
- Desiccator

3 Experimental Procedures

3.1 *Sample Preparation*

Natural Wood Pieces

1. Cut fresh branches from the tree species to be studied into ~10-cm-long and 1–10-cm-diameter pieces. Cut a sufficient number of pieces to allow at least nine replicates that cover in a balanced way the whole diameter range per tree species and site (at least three sizes and three replicates per size, species, and site).
2. Air-dry pieces at room temperature to constant weight.
3. Weigh pieces to obtain the initial mass (M_{initial}).
4. Calculate the surface-to-volume (*S/V*) ratio for each piece.
5. Enclose each piece in a 1-cm mesh bag identified with a plastic or metal tag.
6. Estimate the initial % ash-free dry mass in the piece. For this purpose:

    (a) Identify and weigh another subset of ten pieces per species to obtain the initial mass (M_0).
    (b) Dry at 50 °C to constant mass (generally 48 h).
    (c) Cool samples in a desiccator and weigh to obtain dry mass (DM_0).
    (d) Transfer to pre-weighed crucibles.
    (e) Incinerate in a muffle furnace at 500 °C for 12 h.
    (f) Weigh after cooling in a desiccator to obtain ash mass (AM_0).
    (g) Calculate the ash-free dry mass (AFDM_0) of each piece by subtracting the AM_0 from DM_0:

$$\text{AFDM}_0 = DM_0 - AM_0 \tag{8.1}$$

(h) Determine the initial % ash-free dry mass (% $\text{AFDM}_{\text{initial}}$) in the pieces:

$$\%\text{ADFM}_{\text{initial}} = \frac{\text{AFDM}_0}{M_0} \times 100 \tag{8.2}$$

Standard Substrates

1. Select a standard substrate.
2. Air-dry substrates at room temperature to constant weight.
3. Punch substrates so that fishing line can pass through.
4. Weight substrates to obtain the initial mass (M_{initial}).
5. Add a plastic or metal label using fishing line.
6. Optionally, include weights to submerge substrates.
7. Prepare a sufficient number of substrates to allow four to six replicates per site.
8. Determine the initial % ash-free dry mass of substrates following the procedure described in Sect. 3.1.6.

3.2 *Field Procedures*

1. Tie samples (bags containing natural wood pieces or standard substrates) to metal bars or roots in the stream channels using fishing line and anchor to the stream bed using weights or boulders.
2. Retrieve samples from streams after at least 1 year (natural pieces) or after 3–6 months (manufactured substrates).
3. Transport samples to the laboratory in individual blotting paper envelops. If samples cannot be processed within the next few hours, air-dry and store in a dark, aerated place.

3.3 *Laboratory Procedures*

1. Remove attached invertebrates and mineral particles from substrates by rinsing with tap water.
2. Dry at 50 °C to constant weight. For standard substrates 48 h will usually suffice, thick natural pieces can need up to several weeks.
3. Cool in a desiccator and weigh to obtain final dry mass (DM_{final}).
4. Transfer to pre-weighed crucibles.
5. Incinerate in a muffle furnace at 500 °C for 12 h.
6. After cooling in a desiccator, weigh to obtain ash mass (AM_{final}).

3.4 Calculations

1. Correct the initial mass (M_{initial}) of each substrate with the mean initial % ash-free dry mass (% $\text{AFDM}_{\text{initial}}$), to obtain the initial ash-free dry mass ($\text{AFDM}_{\text{initial}}$).

$$\text{AFDM}_{\text{initial}} = M_{\text{initial}} \times \frac{\%\text{ADFM}_{\text{initial}}}{100} \tag{8.3}$$

2. Calculate the final ash-free dry mass ($\text{AFDM}_{\text{final}}$) of each substrate by subtracting the ash mass (AM_{final}) from dry mass (DM_{final}).

$$\text{AFDM}_{\text{final}} = DM_{\text{final}} - AM_{\text{final}} \tag{8.4}$$

3. Convert $\text{AFDM}_{\text{final}}$ for each substrate to % AFDM remaining:

$$\%\text{AFDM remaining} = \frac{\text{AFDM}_{\text{final}}}{\text{AFDM}_{\text{initial}}} \times 100 \tag{8.5}$$

4. If the duration of the decomposition experiments is exactly the same for all substrates, % AFDM remaining can be used for comparisons. If the incubation time differs, calculate breakdown rates (k) as the slope of the regression of the natural log (ln) % AFDM over incubation time (t, dependent variable), following Petersen and Cummins (1974):

$$k = -\frac{\ln(\%\text{AFDM remaining}) - \ln(100)}{t} \tag{8.6}$$

5. In the case of natural wood pieces, calculate regressions between breakdown rates (k) in relation to the surface-to-volume (S/V) ratio of sticks for each site, following Spänhoff and Meyer (2004).

4 Final Remarks

Multiple collections can also be made (Chap. 6), but it makes the whole procedure more complex, and little is gained. An initial characterization of substrates (e.g., nutrient and lignin content) could provide important information, especially if different tree species or substrate types will be used. Interesting ancillary variables to monitor during decomposition include discharge, water temperature, and dissolved nutrient concentration. This method can easily be adapted for use in riffles, pools, dry surfaces, or even within the sediments.

References

Aristi, I., Ramon Diez, J., Larranaga, A., Navarro-Ortega, A., Barcelo, D., & Elosegi, A. (2012). Assessing the effects of multiple stressors on the functioning of Mediterranean rivers using poplar wood breakdown. *Science of the Total Environment, 440*, 272–279.

Arroita, M., Aristi, I., Flores, L., Larranaga, A., Diez, J., Mora, J., Romaní, A. M., & Elosegi, A. (2012). The use of wooden sticks to assess stream ecosystem functioning: Comparison with leaf breakdown rates. *Science of the Total Environment, 440*, 115–122.

Aumen, N.G. (1985). *Characterization of lignocellulose decomposition in stream wood samples using [14] C and ^{15}N techniques*. Ph.D. dissertation. Oregon State University.

Aumen, N. G., Bottomley, P. J., Ward, G. M., & Gregory, S. V. (1983). Microbial decomposition of wood in streams: Distribution of microflora and factors affecting [^{14}C] lignocellulose mineralization. *Applied and Environmental Microbiology, 46*, 1409–1416.

Bilby, R. E. (2003). Decomposition and nutrient dynamics of wood in streams and rivers. In S. V. Gregory, K. L. Boyer, & A. M. Gurnell (Eds.), *Ecology and management of wood in world rivers* (pp. 135–147). Bethesda: American Fisheries Society.

Blanchette, R. A., Iiyama, K., Abad, A. R., & Cease, K. R. (1991). Ultrastructure of ancient buried wood from Japan. *Holzforschung, 45*, 161–168.

Chen, X. Y., Wei, X. H., & Scherer, R. (2005). Influence of wildfire and harvest on biomass, carbon pool, and decomposition of large woody debris in forested streams of southern interior British Columbia. *Forest Ecology and Management, 208*, 101–114.

Collier, K. J. (2014). Wood decay rates and macroinvertebrate community structure along contrasting human pressure gradients (Waikato, New Zealand). *New Zealand Journal of Marine and Freshwater Research, 48*, 97–111.

Diez, J., Elosegi, A., Chauvet, E., & Pozo, J. (2002). Breakdown of wood in the Aguera stream. *Freshwater Biology, 47*, 2205–2215.

Eggert, S. L., & Wallace, J. B. (2003). Litter breakdown and invertebrate detritivores in a resource-depleted Appalachian stream. *Archiv für Hydrobiologie, 156*, 315–338.

Ellis, L. M., Molles, M. C., & Crawford, C. S. (1999). Influence of experimental flooding on litter dynamics in a Rio Grande riparian forest, New Mexico. *Restoration Ecology, 7*, 193–204.

Elosegi, A., Díez, J., & Pozo, J. (2007). Contribution of dead wood to the carbon flux in forested streams. *Earth Surface Processes and Landforms, 32*, 1219–1228.

Estevez, E., Rodriguez-Castillo, T., Alvarez-Cabria, M., Penas, F. J., González-Ferreras, A. M., Lezcano, M., & Barquín, J. (2017). Analysis of structural and functional indicators for assessing the health state of mountain streams. *Ecological Indicators, 72*, 553–564.

Ferreira, V., Castagneyrol, B., Koricheva, J., Gulis, V., Chauvet, E., & Graça, M. A. S. (2015). A meta-analysis of the effects of nutrient enrichment on litter decomposition in streams. *Biological Reviews, 90*, 669–688.

Ferreira, V., Koricheva, J., Duarte, S., Niyogi, D. K., & Guerold, F. (2016). Effects of anthropogenic heavy metal contamination on litter decomposition in streams – A meta-analysis. *Environmental Pollution, 210*, 261–270.

France, R., Culbert, H., Freeborough, C., & Peters, R. (1997). Leaching and early mass loss of boreal leaves and wood in oligotrophic water. *Hydrobiologia, 345*, 209–214.

Fritz, K. M., Feminella, J. W., Colson, C., Lockaby, B. G., Governo, R., & Rummer, R. B. (2006). Biomass and decay rates of roots and detritus in sediments of intermittent coastal plain streams. *Hydrobiologia, 556*, 265–277.

Fuss, C. L., & Smock, L. A. (1996). Spatial and temporal variation of microbial respiration rates in a blackwater stream. *Freshwater Biology, 36*, 339–349.

Golladay, S. W., & Sinsabaugh, R. L. (1991). Biofilm development on leaf and wood surfaces in a Boreal river. *Freshwater Biology, 25*, 437–450.

Golladay, S. W., & Webster, J. R. (1988). Effects of clear-cut logging on wood breakdown in Appalachian mountain streams. *American Midland Naturalist, 119*, 143–155.

Harmon, M. E., Franklin, J. F., Swanson, F. J., Sollins, P., Gregory, S. V., Lattin, J. D., Anderson, N. H., Cline, S. P., Aumen, N. G., Sedell, J. R., Lienkamper, G. W., Cromack, K., Jr., & Cummins, K. W. (1986). Ecology of coarse woody debris in temperate ecosystems. *Advances in Ecological Research, 15*, 133–302.

Hoffman, A., & Hering, D. (2000). Wood-associated macroinvertebrate fauna in Central European streams. *International Review of Hydrobiology, 85*, 25–48.

Hyatt, T. L., & Naiman, R. J. (2001). The residence time of large woody debris in the Queets River, Washington, USA. *Ecological Applications, 11*, 191–202.

Maser, C., & Sedell, J. R. (1994). *From the forest to the sea. The ecology of wood in streams, rivers, estuaries, and oceans*. Delray Beach: St Lucie.

McTammany, M. E., Benfield, E. F., & Webster, J. R. (2008). Effects of agriculture on wood breakdown and microbial biofilm respiration in southern Appalachian streams. *Freshwater Biology, 53*, 842–854.

Melillo, J. M., Naiman, R. J., Aber, J. D., & Eshleman, K. N. (1983). The influence of substrate quality and stream size on wood decomposition dynamics. *Oecologia, 58*, 281–285.

Monroy, S., Menéndez, M., Basaguren, A., Pérez, J., Elosegi, A., & Pozo, J. (2016). Drought and detritivores determine leaf litter decomposition in calcareous streams of the Ebro catchment (Spain). *Science of the Total Environment, 573*, 1450–1459.

Niyogi, D. K., Harding, J. S., & Simon, K. S. (2013). Organic matter breakdown as a measure of stream health in New Zealand streams affected by acid mine drainage. *Ecological Indicators, 24*, 510–517.

Noetzli, K., Boell, A., Graf, F., Sieber, T. N., & Holdenrieder, O. (2008). Influence of decay fungi, construction characteristics, and environmental conditions on the quality of wooden check-dams. *Forests Products Journal, 58*, 72–79.

Petersen, R. C., & Cummins, K. W. (1974). Leaf processing in a woodland stream. *Freshwater Biology*, 4, 343-368.

Pregitzer, K. S., & Euskirchen, E. S. (2004). Carbon cycling and storage in world forests: Biome patterns related to forest age. *Global Change Biology, 10*, 2052–2077.

Spänhoff, B., & Meyer, E. I. (2004). Breakdown rates of wood in streams. *Journal of the North American Benthological Society, 23*, 189–197.

Spänhoff, B., Alecke, C., & Meyer, E. I. (2001). Simple method for rating the decay stages of submerged woody debris. *Journal of the North American Benthological Society, 20*, 385–394.

Tank, J. L., & Webster, J. R. (1998). Interaction of substrate and nutrient availability on wood biofilm processes in streams. *Ecology, 79*, 2168–2179.

Triska, F. J., & Cromack, K. C. (1980). The role of wood debris in forests and streams. In R. H. Waring (Ed.), *Forest: Fresh perspectives from ecosystem analysis* (pp. 171–1900). Corvallis: O.S.U. Press.

Webster, J. R., Benfield, E. F., Ehrman, T. P., Schaeffer, M. A., Tank, J. L., Hutchens, J. J., & D'Angelo, D. J. (1999). What happens to allochthonous material that falls into streams? A synthesis of new and published information from Coweeta. *Freshwater Biology, 41*, 687–705.

Wohl, E. (2017). Geomorphology bridging the gaps: An overview of wood across time and space in diverse rivers. *Geomorphology, 279*, 3–26.

Wold, A. K. F., & Hershey, A. E. (1999). Effects of salmon carcass decomposition on biofilm growth and wood decomposition. *Canadian Journal of Fisheries and Aquatic Sciences, 56*, 767–773.

Zeikus, J. G. (1980). Fate of lignin and related aromatic substrates in anaerobic environments. In T. H. Kirk, T. Higuchi, & H. M. Chang (Eds.), *Lignin, Biodegradation: Microbiology, Chemistry and potential Applications* (pp. 101–109). Boca Raton: C.R.C. Press.

Chapter 9
Decomposition of Fine Particulate Organic Matter

Chihiro Yoshimura

Keywords Decay model · Decomposition rate · FPOM · *Gammarus* · Microbes · Shredders · Streams

1 Introduction

Fine particulate organic matter (FPOM) is defined as the bulk fraction of organic particles that is smaller than coarse particulate organic matter (CPOM, >1 mm) and larger than dissolved organic matter (DOM, <0.45 μm) (Wotton 1994). In streams, this fraction is highly mobile and, depending on hydraulic conditions, found suspended in flowing water and deposited on and in the stream bed. FPOM is derived from various sources such as leaf litter, algae, microbes, and DOM via physical, chemical, and biological processes. The dynamics of FPOM in streams have been investigated in situ from various angles, ranging from its importance in organic matter budgets to assessments of instream processing and longitudinal transport (Webster and Meyer 1997; Newbold et al. 2005; Sakamaki and Richardson 2011; Rowland et al. 2017). Meybeck (1981) estimated that FPOM accounts for up to 40% of the total annual organic carbon transported by large rivers to the oceans.

FPOM plays an important role as a food source for aquatic fauna, substrate for heterotrophic microorganisms, and carrier of nutrients, metals, and other chemicals. Aspects studied in relation to FPOM dynamics in streams include decomposition rate (Mattingly 1986; Jackson and Vallaire 2007; Yoshimura et al. 2008), biogeochemical properties (Peters et al. 1989; Vignati and Dominik 2003; Yoshimura et al. 2010), associated microbes and enzyme activities (Sinsabaugh et al. 1992; Bonin et al. 2000; Jackson and Vallaire 2007; Wurzbacher et al. 2016), microbial metabolism (Sinsabaugh and Findlay 1995; Tank et al. 2010; Tant et al. 2015), and faunal activities (Wotton and Malmqvist 2001; Joyce et al. 2007; Joyce and Wotton 2008).

C. Yoshimura (✉)
Department of Civil and Environmental Engineering, School of Environment and Society, Tokyo Institute of Technology, Tokyo, Japan
e-mail: yoshimura.c.aa@m.titech.ac.jp

F. Bärlocher et al. (eds.), *Methods to Study Litter Decomposition*,
https://doi.org/10.1007/978-3-030-30515-4_9

This body of literature indicates that FPOM in streams is highly variable in terms of biogeochemical properties and interactions with microbes, despite its rather narrow size range. Other largely unexplored and yet important themes are the role of FPOM in ecosystem functioning (Kominoski et al. 2010), anthropogenic impacts on FPOM dynamics, and the role of FPOM in mediating ecotoxicological effects (Bundschuh and McKie 2016).

Our understanding of FPOM dynamics in streams is still very incomplete, hindering a comprehensive view of its dynamics and ecological roles. This lack contrasts with the wealth of information available on CPOM dynamics. FPOM decomposition has received particularly scant attention, and the few published studies to date indicate complex interaction with microbes that depends on particle size. The microbial decomposition of FPOM in the size range of 100–500 μm has been reported to be slower than that of leaf litter (Yoshimura et al. 2008). In another study (Jackson and Vallaire 2007), the fraction of 63–250 μm decomposed faster than the larger FPOM particles (250–1000 μm) (Table 9.1), although the chemical quality of smaller particles suggests greater recalcitrance (Peters et al. 1989). Differences in

Table 9.1 Daily decay rate coefficients (k) and number of days to reach 50% mass loss (T_{50}) of FPOM in streams

Source (pretreatment)	Diameter (μm)	k (day^{-1})[a]	T_{50} (day)	References
Alnus rubra leaf (ground and leached in distilled water)	53–125	0.0053	131	Mattingly (1986)
	125–250	0.0063	110	
	250–500	0.0126	55	
	500–1000	0.0088	79	
Acer macrophyllum leaf (ground and leached in distilled water)	53–125	0.0047	147	Mattingly (1986)
	125–250	0.0053	131	
	250–500	0.0048	144	
	500–1000	0.0067	103	
Polystichum munitum leaf (ground and leached in distilled water)	53–125	0.0028	248	Mattingly (1986)
	125–250	0.0051	136	
	250–500	0.0043	161	
	500–1000	0.0023	301	
Fraxinus excelsior leaf (conditioned in a stream and processed by *Gammarus* spp.)	100–500	0.0015	462	Yoshimura et al. (2008)
Quercus robur leaf (conditioned in a stream and processed by *Gammarus* spp.)	100–500	0.0013	533	Yoshimura et al. (2008)
Sediment in a cypress-tupelo swamp (air-dried)	63–250[b]	0.0027	207[c]	Jackson and Vallaire (2007)
	63–250[b]	0.0029	167[c]	
	250–1000[b]	0.0018	303[c]	
	250–1000[b]	0.0018	326[c]	

Rate coefficients (k) for *Alnus rubra*, *Acer macrophyllum*, and *Polystichum munitum* were calculated from the data reported by Mattingly (1986)

[a]Coefficients (k) are first-order rate constants in Mattingly (1986) and Yoshimura et al. (2008) and zero-order rate constants in Jackson and Vallaire (2007)

[b]Results from two different sites

[c]T_{50} determined from the graphs (time-series in mass remaining) reported by Jackson and Vallaire (2007)

the origin of the two size fractions is a possible reason for this unexpected outcome, suggesting that careful attention to both size and source is essential for understanding FPOM decomposition.

This chapter describes a method to determine decomposition rates of FPOM in streams. The FPOM used in such studies can be obtained by (1) collecting FPOM in the field, (2) instream conditioning and subsequent processing of CPOM by macroinvertebrate shredders in controlled laboratory conditions, or (3) grinding CPOM. Since FPOM and CPOM differ greatly in chemical quality, option 3 results in a poor proxy of natural FPOM. Therefore, this chapter focuses on procedures for options 1 and 2.

2 Site Selection and Equipment

2.1 Site Selection

Choose a suitable study site. FPOM decomposition is readily investigated in small streams or shallow parts of rivers and floodplains. In a third-order stream, a reach of 20 m normally provides sufficient space.

2.2 Equipment and Material

- Plankton net to collect natural FPOM (≤100 μm mesh size)
- Particulate organic matter (POM, e.g. leaf litter) as FPOM source
- Shredders to produce FPOM (gammarids, limnephilid caddisflies, etc.)
- Litter bags (1 cm mesh size) to condition CPOM
- Aquaria (e.g. 25 cm × 40 cm × 25 cm; number depends on types of CPOM and required mass of FPOM)
- Rectangular frames slightly smaller than the aquarium, equipped with 5 cm plastic walls, a mesh screen bottom (500 μm mesh, e.g. polyester or nylon), and 5 cm plastic legs at every corner to keep the mesh screen 5 cm above the bottom of the aquarium
- Air pumps
- Balance (at least ±0.1 mg precision)
- Porcelain or aluminium dishes
- Freeze-dryer or drying oven
- Desiccator
- Refrigerator and freezer
- Tubes (e.g. 15 cm length, 1.5 cm diameter) made of mesh screen finer than the particle size of the FPOM (e.g. 7 μm)
- Forceps

- Sieves (two mesh sizes, e.g. 7 μm and 100 μm)
- Plastic clips
- Small metal cages (e.g. 30 cm × 15 cm × 10 cm)
- String
- Ice boxes
- Small shovels
- Rebars
- Plastic tape as label (optional)
- Temperature logger (optional)
- Mortar or mill (optional)

3 Experimental Procedures

3.1 Sample Preparation

1. Select appropriate types of FPOM according to the goals of the study, either particles produced naturally (steps 2–3 and 17–23) or particles derived from a known single source of leaf litter or other organic matter (steps 4–16 and 17–23).
2. Use a suitable net (e.g. plankton net) to collect sufficient amounts of naturally produced FPOM (min. 80 g wet mass). A mesh size of 100 μm is convenient under typical stream flow conditions, but note that this does not collect the finer particles that account for the majority of FPOM in streams.
3. Place a net facing upstream. To obtain samples with a relatively high organic matter content, preferably collect only suspended matter from the water column, where the organic matter content tends to be higher than for material deposited on the stream bottom. One could disturb the stream bed in front of the net to collect fine organic particles in large amounts with little effort. However, in most cases such samples would include a higher fraction of mineral particles. This is undesirable and would in all cases require determining ash-free dry mass (AFDM) as explained in step 3.3.7.
4. To derive FPOM from a single source of CPOM, collect and freeze-dry or air-dry sufficient amounts of leaf litter or small pieces of wood or other types of CPOM (min. 1 kg air-dry mass), preferably from the stream where the decomposition experiment will be conducted. Place the collected leaf litter in 1 cm mesh size bags and condition (e.g. for 10 days). Condition other sources of CPOM in an analogous manner.
5. Set up aquaria in a temperature-controlled chamber or outdoors under a roof.
6. Insert in each aquarium a rectangular frame with a mesh screen (500 μm mesh size) to partition the aquaria into a bottom chamber (~5 cm height) and a top chamber.

7. Fill each aquarium with stream water passed through a mesh screen (≤100 µm). The water level should be above the mesh screen but below the top of the frame to keep all shredders on the screen.
8. Aerate with aquarium pumps.
9. Collect a sufficient number of shredders, ideally from the stream where the decomposition experiment will be conducted.
10. Transfer the collected shredders to the top chamber (≥100 individuals per aquarium). *Gammarus* spp. are well suited as they often occur in large numbers, do not emerge, and shred leaves efficiently (Kelly et al. 2002; Dangles et al. 2004), but limnephilid caddisflies and other shredders can also be used.
11. Add conditioned CPOM (equivalent to 10 g dry mass) to each aquarium every week to allow shredders to feed *ad libitum.*
12. Inspect the aquaria frequently (at least every 3 days), and replace any dead individuals to avoid their consumption by the survivors.
13. Empty the lower aquarium chamber after 3 days, and discard the accumulated FPOM to avoid contamination by FPOM originating from unknown food consumed in the stream before collecting the shredders.
14. Next, collect FPOM settled in the lower aquarium chamber by passing the water and particles through a 100 µm mesh screen to obtain FPOM in the size range of 100–500 µm.
15. Continue this process until the required amount of FPOM has been produced (e.g. a total of 8 g dry mass).
16. Determine water content of each sample by weighing, drying to constant mass (50 °C, 48 h), and reweighing.
17. Store the FPOM slurry in a fridge. If samples cannot be processed within a few days, freeze them.
18. For longer storage, samples may be dried, but this is not recommended because drying and rewetting changes particle properties.
19. A day before starting the field experiment, set up all mesh tubes and metal cages needed to expose FPOM *in situ*.
20. Homogenize the wet FPOM, and insert a known amount (equivalent of 200–400 mg dry mass) to each tube while one of the tube ends is closed by a plastic clip. Dialysis tubes may be applied as an alternative, but they tend to be too fragile to be used in streams.
21. Prepare and expose replicate empty control tubes to determine particles entering the tubes during the experiment.
22. Tightly close each tube at both ends with plastic clips, and fix the tubes in small metal cages (or permeable boxes) using durable strings (Fig. 9.1). Allocate samples so that sets of samples and negative controls can be randomly retrieved on every sampling occasion.
23. Store the boxes in a cool place during transport to the experimental site (e.g. in an ice box).

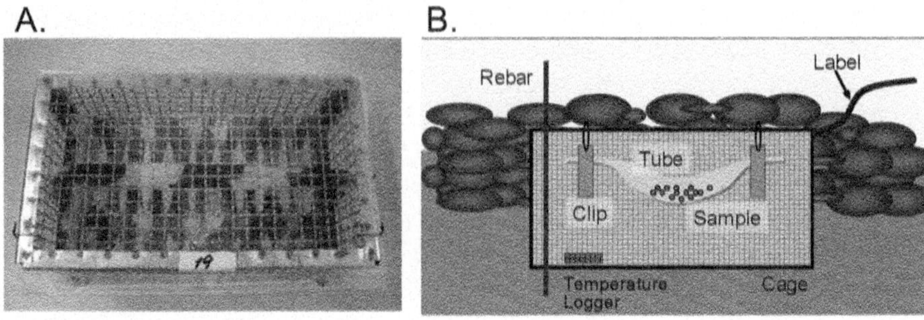

Fig. 9.1 Example of sample tubes fixed in a metal cage (**a**) and its configuration when installed in a stream (**b**)

3.2 *Field Procedures*

1. Secure metal cages on the stream bed with rebars (or other means), and create shaded condition to prevent algal growth on the tube surfaces. Alternatively, bury the cages in the stream bed (Fig. 9.1).
2. Optionally install a temperature logger within or near by a cage.
3. Immediately after exposure, retrieve the first set of randomly chosen FPOM samples and negative controls to determine FPOM recovery before decomposition starts (i.e. initial mass).
4. On every subsequent sampling occasion, retrieve an appropriate set of cages, including negative controls. Choose the time intervals of sample collection based on expected decomposition rates (Table 9.1). At least five time points for the sample collection are recommended.
5. Place all retrieved cages in ice boxes and immediately transport to the laboratory.

3.3 *Sample Treatment and Calculations*

1. Remove tubes from the cages, retrieve FPOM from the tubes, and collect the remaining FPOM on sieves equipped with the same mesh size as the tubes.
2. If chemical and biological characteristics of FPOM are to be determined, suspend the retrieved FPOM homogenously in a flask containing a known volume of filtered stream water or possibly deionized water.
3. Refer to Chaps. 11 through 48 for determining chemical and biological characteristics of FPOM.
4. Take appropriate aliquots of the FPOM suspension, pass through a washed and tared glass fibre filter, and dry and weigh the filter.

5. Determine the total mass of FPOM that remained in the tubes on the basis of the ratio of the aliquot to the total sample volume.
6. Determine the remaining FPOM dry mass in porcelain or aluminium dishes, placed in a freeze-dryer or drying oven (50 °C for 48 h) and then in a desiccator.
7. Remove any confounding effect of external particles that entered the tubes by subtracting mass in the negative controls from the FPOM remaining for each sampling occasion. If the initial FPOM sample or the negative control contain non-negligible amount of inorganic particles (e.g. clay and silt), the mass change needs to be expressed in terms of AFDM.
8. Fit the obtained time series of dry-mass data to a first-order decay model to determine decomposition rate (*k* as in Table 9.1), following the same procedure as for CPOM (see Chaps. 6 and 60).

4 Final Remarks

A limitation of the described method is that decomposition is assessed under constrained conditions. This might cause an underestimate of decomposition rates (Sinsabaugh et al. 1994). Furthermore, the method has not been extensively tested in different settings, and the smallest size fractions below 100 µm are only partly covered by the present method. It might hence be necessary to consider appropriate modifications of the described procedures, according to the specific goals of a given study.

References

Bonin, H. L., Griffiths, R. P., & Caldwell, B. A. (2000). Nutrient and microbiological characteristics of fine benthic organic matter in mountain streams. *Journal of the North American Benthological Society, 19*, 235–249.

Bundschuh, M., & McKie, B. G. (2016). An ecological and ecotoxicological perspective on fine particulate organic matter in streams. *Freshwater Biology, 61*, 2063–2074.

Dangles, O., Gessner, M. O., Guérold, F., & Chauvet, E. (2004). Impacts of stream acidification on litter breakdown: Implications for assessing ecosystem functioning. *Journal of Applied Ecology, 41*, 365–378.

Jackson, C. R., & Vallaire, S. C. (2007). Microbial activity and decomposition of fine particulate organic matter in a Louisiana cypress swamp. *Journal of the North American Benthological Society, 26*, 742–752.

Joyce, P., & Wotton, R. S. (2008). Shredder fecal pellets as stores of allochthonous organic matter in streams. *Journal of the North American Benthological Society, 27*, 521–528.

Joyce, P., Warren, J. L., & Wotton, R. S. (2007). Faecal pellets in streams: Their binding, breakdown and utilization. *Freshwater Biology, 52*, 1868–1880.

Kelly, D. W., Dick, J. T. A., & Montgomery, W. I. (2002). The functional role of *Gammarus* (Crustacea, Amphipoda): Shredders, predators, or both? *Hydrobiologia, 485*, 199–203.

Kominoski, J. S., Hoellein, T. J., Leroy, C. J., Pringle, C. M., & Swan, C. M. (2010). Beyond species richness: Expanding biodiversity–ecosystem functioning theory in detritus-based streams. *River Research and Applications, 26*, 67–75.

Mattingly, R. L. (1986). Mass loss and qualitative change in stream-incubated fine particulate organic matter derived from leaves differing in rate of processing. *Hydrobiologia, 135*, 207–214.

Meybeck, M. (1981). River transport of organic carbon to the ocean. In Committee on flux of organic carbon the ocean (Ed.), *Carbon dioxide effects research and assessment program: Flux of organic carbon by rivers to the oceans* (pp. 219–269). Washington, DC: US Department of Energy.

Newbold, J. D., Thomas, S. A., Minshall, G. W., Cushing, C. E., & Georgian, T. (2005). Deposition, benthic residence, and resuspension of fine organic particles in a mountain stream. *Limnology and Oceanography, 50*, 1571–1580.

Peters, G. T., Benfield, E. F., & Webster, J. R. (1989). Chemical composition and microbial activity of seston in a southern Appalachian headwater stream. *Journal of the North American Benthological Society, 8*, 74–84.

Rowland, R., Inamdar, S., & Parr, T. (2017). Evolution of particulate organic matter (POM) along a headwater drainage: Role of sources, particle size class, and storm magnitude. *Biogeochemistry, 133*, 181–200.

Sakamaki, T., & Richardson, J. S. (2011). Biogeochemical properties of fine particulate organic matter as an indicator of local and catchment impacts on forested streams. *Journal of Applied Ecology, 48*, 1462–1471.

Sinsabaugh, R. L., & Findlay, S. (1995). Microbial production, enzyme activity, and carbon turnover in surface sediments of the Hudson River estuary. *Microbial Ecology, 30*, 127–141.

Sinsabaugh, R. L., Weiland, T., & Linkins, A. E. (1992). Enzymatic and molecular analysis of microbial communities associated with lotic particulate organic matter. *Freshwater Biology, 28*, 393–404.

Sinsabaugh, R. L., Osgood, M. P., & Findlay, S. (1994). Enzymatic models for estimating decomposition rates of particulate detritus. *Journal of the North American Benthological Society, 13*, 160–169.

Tank, J. L., Rosi-Marshall, E. J., Griffiths, N. A., Entrekin, S. A., & Stephen, M. L. (2010). A review of allochthonous organic matter dynamics and metabolism in streams. *Journal of the North American Benthological Society, 29*, 118–146.

Tant, C. J., Rosemond, A. D., Mehring, A. S., Kuehn, K. A., & Davis, J. M. (2015). The role of aquatic fungi in transformations of organic matter mediated by nutrients. *Freshwater Biology, 60*, 1354–1363.

Vignati, D., & Dominik, J. (2003). The role of coarse colloids as a carrier phase for trace metals in riverine systems. *Aquatic Science, 65*, 129–142.

Webster, J. R., & Meyer, J. L. (1997). Steam organic matter budgets. *Journal of the North American Benthological Society, 16*, 3–4.

Wotton, R. S. (1994). *The biology of particles in aquatic systems* (2nd ed.). Boca Raton: CRC Press.

Wotton, R. S., & Malmqvist, B. (2001). Feces in aquatic ecosystems. *BioScience, 51*, 537–544.

Wurzbacher, C., Wannicke, N., Grimmett, I. J., & Bärlocher, F. (2016). Effects of FPOM size and quality on aquatic heterotrophic bacteria. *Limnologica, 59*, 109–115.

Yoshimura, C., Gessner, M. O., Tockner, K., & Furumai, H. (2008). Chemical properties, microbial respiration, and decomposition of coarse and fine particulate organic matter. *Journal of the North American Benthological Society, 27*, 664–673.

Yoshimura, C., Fujii, M., Omura, T., & Tockner, K. (2010). Instream release of dissolved organic matter from coarse and fine particulate organic matter of different origins. *Biogeochemistry, 100*, 151–165.

Chapter 10
Coarse Particulate Organic Matter Budgets

Jesús Pozo and Jon Molinero

Keywords Carbon budget · CPOM · ecosystem efficiency · litter inputs · litter outputs · litter storage · organic matter outputs · riparian forest disturbance · riparian vegetation · streams

1 Introduction

During the last four decades, great efforts have been dedicated to the study of terrestrial-aquatic linkages, in particular to the riparian origin and fate in streams of coarse particulate organic matter (CPOM) (see Abelho 2001; Tank et al. 2010 for reviews). A considerable amount of litter entering streams is efficiently retained within the channel. Consequently, the amounts and the composition of benthic CPOM are closely related to the structure of the local riparian vegetation, underlining the strong influence of the terrestrial surroundings on the energy basis of low-order forest streams (Wallace et al. 1999). In addition, the discharge regime of streams directly affects the retention capacity for CPOM (Larrañaga et al. 2003; Chap. 2) and thus the availability of organic matter to stream consumers and decomposers. When peaks of litter input coincide with high discharge, downstream

Electronic Supplementary Material The online version of this chapter (https://doi.org/10.1007/978-3-030-30515-4_10) contains supplementary material, which is available to authorized users.

J. Pozo (✉)
Department of Plant Biology and Ecology, Faculty of Science and Technology, University of the Basque Country, UPV/EHU, Bilbao, Spain
e-mail: jesus.pozo@ehu.es

J. Molinero
Escuela de Gestión Ambiental, Pontificia Universidad Católica del Ecuador Sede Esmeraldas, Espejo y subida a Santa Cruz, Esmeraldas, Ecuador

F. Bärlocher et al. (eds.), *Methods to Study Litter Decomposition*,
https://doi.org/10.1007/978-3-030-30515-4_10

displacement of CPOM is favoured, whereas CPOM tends to accumulate in the streambed when litter inputs coincide with low flow (Molinero and Pozo 2004) or even no flow (Datry et al. 2018).

An organic matter budget is a systematic account of the organic matter inputs (*I*) and outputs (*O*) in a given ecosystem or parts thereof. The general approach for constructing a CPOM budget consists of determining the various inputs (Chap. 1) and ascertaining outputs (Chap. 6) and storage (= standing stock, *S*; Chap. 4) of this material (Fig. 10.1). Inputs to a stream (vertical, lateral and from upstream), outputs (downstream transport or export and biological and physical breakdown) and changes in storage should all be measured independently. However, this approach is very time-consuming, and often one or more variables are obtained by addition or subtraction of the measured components, taken from the literature, or ignored (Cummins et al. 1983; Minshall 1996).

The general mass balance for CPOM in a stream reach is given by:

$$O = I + \Delta S \tag{10.1}$$

where the total output in a given time period (*O*) equals the total input (*I*) plus the change in the standing stock (ΔS) in that time period. In a more detailed form, the standing stock of CPOM at time *t* (S_t) is given by the standing stock at time *0* (S_0) plus the total inputs (*I*) from time *0* to time *t*, minus the losses resulting from downstream transport (O_T) and biological and physical breakdown (O_B) in the same time period:

$$S_t = S_0 + I - O_T - O_B \tag{10.2}$$

The determination of complete budgets is rare. Frequent omissions include simple but important aspects such as accurate measurements of streambed area. Knowing the streambed area makes it possible to express exports in terms of mass per streambed surface per unit time, facilitating comparisons among streams. However, very often (see Webster and Meyer 1997) CPOM exports are expressed as kg year^{-1}, making comparisons of streams differing in size difficult. Similarly, some areas of the stream channel or the stream itself may dry up during the annual hydrological cycle modifying the temporal and spatial patterns of CPOM accumulation (Datry et al. 2018), and this fact is often ignored despite its importance for accurate budget calculations. Furthermore, the validity of the assumption that natural streams are in a steady state depends on both the temporal and spatial scales considered (Cummins et al. 1983). For periods from days to a few weeks, changes in storage can be negligible, unless stream discharge fluctuates widely (Minshall 1996). Although budgets should ideally consider entire streams (e.g. Fisher and Likens 1973) and be based on long-term data (Cummins et al. 1983), most studies have been restricted to short reaches and periods of 1 year or less (Webster and Meyer 1997). Budgets that include the stream channel and the riparian corridor are rare (e.g. Ruffing et al. 2016), although they provide valuable information on the feedbacks between the aquatic and terrestrial environments.

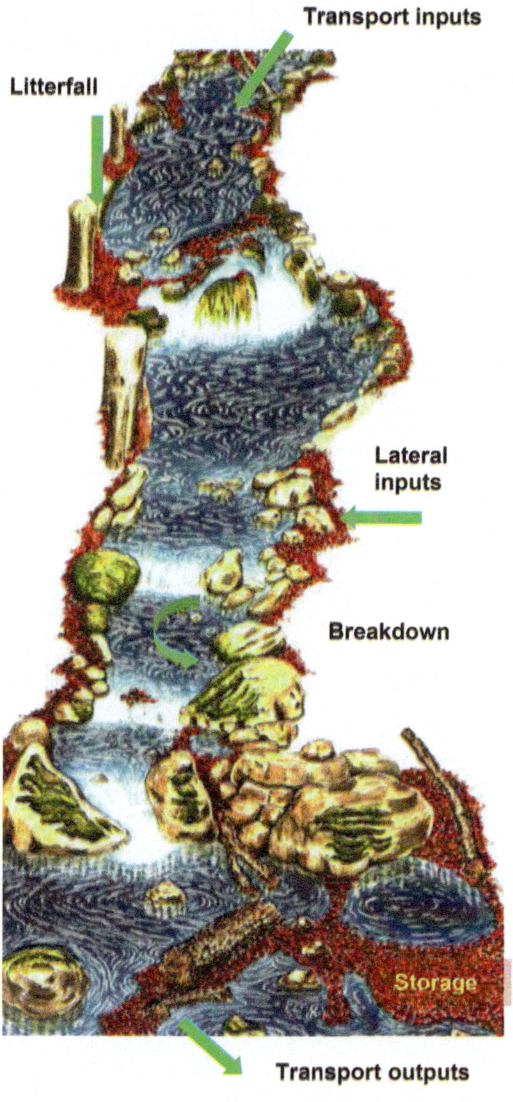

Fig. 10.1 Components of a coarse particulate organic matter budget for a stream reach

Organic matter budgets quantify storage and fluxes in ecosystems, thus providing estimates of organic matter utilization efficiencies and assessments of changes resulting from disturbances. Furthermore, it is beneficial to incorporate the considerable effort invested in measuring leaf breakdown rates during the last few decades into calculations of CPOM budgets (e.g. Pozo et al. 1997; Molinero and Pozo 2006), which synthesize information covered in the earlier chapters of this section.

Inputs (amount and timing), storage and outputs (transport and breakdown) depend on the type of material and, therefore, on the riparian vegetation. Some studies have demonstrated that forest disturbance modifies export and turnover of benthic CPOM (e.g. Webster et al. 1990; Pozo et al. 1997) and that these changes can impact stream communities (Graça et al. 2002). Although notable spatial and temporal variability can make accurate measurements of the individual components of organic matter budgets difficult, these budgets are useful to assess effects of human-induced disturbances on stream ecosystem functioning (e.g. Epstein et al. 2016). The restoration of headwater streams is not complete unless their energy base is restored by reconnecting lost aquatic-terrestrial linkages (Wallace et al. 2015). This chapter describes procedures to construct CPOM budgets in streams. For comparison, Table 10.1 shows selected CPOM turnover rates using data of terrestrial CPOM inputs and instream storage from the literature.

Table 10.1. Turnover rates (i.e. terrestrial inputs/storage) of coarse particulate organic matter (excluding large wood) calculated from selected streams

Location	Latitude	Vegetation	Stream order	Turnover (year^{-1})	References
Alaska, USA	65 °N	Taiga	1 2	13.8 7.6	Irons and Oswood (1997)
Denmark	56 °N	Beech forest	1	3.8	Iversen et al. (1982)
Quebec, Canada	50 °N	Spruce and deciduous forest	1 2 5	0.8 0.9 0.1	Naiman and Link (1997)
Oregon, USA	45 °N	Coniferous forest	1 3 5	0.3–0.7 1.9 12.0	Webster and Meyer (1997)
New Hampshire, USA	44 °N	Deciduous forest	2	1.0	Fisher and Likens (1973)
Spain	43 °N	Deciduous forest *Eucalyptus* plantation	1 1	16.9[a] 5.3[a]	Molinero and Pozo (2006)
Pennsylvania, USA	40 °N	Agricultural land and woodland	3	2.4	Newbold et al. (1997)
Virginia, USA	37 °N	Mixed forest	1	0.3	Smock (1990)
North Carolina, USA	35 °N	Deciduous and *Rhododendron* forest	1	3.9	Wallace et al. (1999)
North Carolina, USA	35 °N	Deciduous forest Logged deciduous forest	1 2	1.6 1.5	Webster et al. (1990)
Arizona, USA	33 °N	Desert scrub	5	3.8	Jones et al. (1997)
Minas Gerais, Brazil	19° S	Atlantic forest	2	0.5	Gonçalves et al. (2014)
South Africa	33 °S	Fynbos	2	7.6–13.9	King et al. (1987)
Victoria, Australia	37 °S	Eucalyptus forest	4	6.0	Treadwell et al. (1997)
Victoria, Australia	37 °S	Urban (*Eucalyptus, Acacia, Melaleuca*)	1-2	0.1–1.4	Imberger et al. (2011)

[a]Leaves only

2 Site Selection and Equipment

2.1 Site Selection

- Select an entire small wadeable stream or, which is often sufficient, a 100-m stream reach. A CPOM budget is most easily constructed for narrow, forested headwater streams.

2.2 Equipment and Material

- Basket traps to determine vertical inputs by litter fall, placed randomly either across the stream (see Chap. 1) or, if the canopy cover above the channel and in the riparian forest is similar, on the forest floor
- Lateral traps to determine horizontal CPOM inputs (see Chap. 1)
- Drift nets (1-mm mesh size; see Chap. 1)
- Equipment and materials presented in Chap. 4 for measuring CPOM storage
- Materials to determine local leaf breakdown rates (see Chap. 6) to estimate losses by biological decomposition and physical processes, unless suitable literature values of breakdown rates are available
- Equipment and materials to measure respiration rates of CPOM (see Chap. 33) unless appropriate literature values are available

3 Experimental Procedures

3.1 Sampling

1. Measure length of the stream reach and, at several locations, channel width to calculate the average width and surface area of the stream segment.
2. Measure widths and depths at the upper and lower ends of the stream reach to obtain cross-sections profiles and determine discharge (Chap. 1).
3. To determine terrestrial inputs, collect materials in litter baskets and in lateral traps as described in Chap. 1.
4. To determine inputs and outputs by transport, collect material retained by drift nets located at both the upper and lower ends of the stream reach and spanning the entire width of the channel as described in Chap. 1.
5. To determine storage, collect benthic CPOM as described in Chap. 4, taking care to sample entire channel transects (i.e. including dry parts). Begin sampling downstream and move upstream to avoid extra CPOM transport caused by the disturbance of the streambed.

6. To determine breakdown and respiration rates or both, follow the instructions in Chaps. 6 and 33, respectively.

3.2 Laboratory Procedures

1. Separate CPOM from inorganic materials in all samples with the aid of nested sieves and under tap water, if necessary (see Chaps. 1 and 4).
2. Determine dry mass and ash-free dry mass (AFDM) of the collected material (see Chaps. 1 or 4).

3.3 Calculations

1. Calculate ash-free dry mass from laboratory determinations of total CPOM and its components as described in Chaps. 1 and 4.
2. Express direct litter inputs (I_D) in terms of g AFDM m^{-2} d^{-1} by dividing the amount of collected material by the surface area of the basket and by the number of days elapsed between successive sampling dates.
3. Express lateral inputs (I_L) in terms of g AFDM m^{-2} d^{-1} by dividing twice the amount of material collected in a given lateral trap ($2M_L$) by the length of the trap in metres (l), the mean channel width in metres (W) and the elapsed time in days (t):

$$I_L = \frac{2 \times M_L}{(l \times W \times t)} \tag{10.3}$$

4. Alternatively, divide M_L by l, multiply by the total length of banks (2 × channel length) and divide by the area of the stream segment (channel length × mean channel width) and by the number of days (t).
5. Express CPOM transport (inputs, I_T, and outputs, O_T) in terms of g AFDM m^{-2} d^{-1} according to Chap. 1.
6. Calculate CPOM storage (S) in terms of g AFDM m^{-2} according to Chap. 4. To express changes in storage in terms of g AFDM m^{-2} d^{-1}, calculate the difference in standing stocks between sampling dates and divide by the number of days.
7. If breakdown rates (k) for particular leaf species are used, calculate them from leaf mass loss measured according to Chap. 6 or use published breakdown rates of the dominant leaf species in the stream.
8. If respiration rates are used (the most frequent measures refer to benthic respiration measurements without further differentiation (Webster et al. 1995)), express them in terms of g AFDM m^{-2} d^{-1}. Alternatively, use published rates with appropriate conversion factors.

9. Assuming that the stream is in a steady state with respect to inputs and outputs, calculate the turnover of the standing stock (T; d^{-1}) as the ratio of total litter input (I), in g m^{-2} d^{-1}, to the litter standing stock (S), in g m^{-2}:

$$T = \frac{I}{S} \tag{10.4}$$

10. Litter inputs are either from the riparian vegetation (i.e. direct plus lateral inputs, $I_D + I_L$; g m^{-2} d^{-1}) or by transport from upstream (I_T; g m^{-2} d^{-1}). Similarly, litter outputs occur either by downstream transport (O_T) or by breakdown (O_B). Therefore, turnover can be also determined as the sum of the loss rates by transport (k_T; d^{-1}) and breakdown (k; d^{-1}). Thus:

$$\frac{\left(I_D + I_L + I_T\right)}{S} = \frac{\left(O_T + O_B\right)}{S} = k_T + k \tag{10.5}$$

11. Estimate outputs by downstream transport (O_T; g m^{-2} d^{-1}) and by breakdown (O_B; g m^{-2} d^{-1}) as:

$$O_T = S \times k_T \tag{10.6}$$

$$O_B = S \times k \tag{10.7}$$

12. If information on the contribution of various processes to leaf litter breakdown exists (e.g. Baldy et al. 2007), outputs can be further assigned to instream processes. If breakdown rates have been estimated in coarse (k_c, d^{-1}) and fine bags (k_f, d^{-1}, see Chap. 6), outputs can be assigned to fragmentation (O_{BFRA}; g m^{-2} d^{-1}), which includes macro-invertebrate-mediated breakdown and physical abrasion, and other processes such as microbial-mediated breakdown and leaching (O_{BNFRA}; g m^{-2} d^{-1}):

$$O_{BFRA} = S \times \overline{\lambda_F} \tag{10.8}$$

$$O_{BNFRA} = O_B - O_{BFRA} \tag{10.9}$$

where $\overline{\lambda_F}$ is the average fragmentation rate calculated following Lecerf (2017):

$$\overline{\lambda_F} = k_c - \frac{k_f - k_c}{\ln\left(k_f\right) - \ln\left(k_c\right)} \tag{10.10}$$

13. Similar to using the quotient of respiration to inputs as an indicator of ecosystem efficiency (e.g. Fisher and Likens 1973; Webster and Meyer 1997), a measure of stream ecosystem efficiency has been proposed based on breakdown (Stream Breakdown Index, SBI). It can be obtained as:

$$SBI = \frac{O_B}{I} \tag{10.11}$$

14. Other ecologically meaningful ratios (i.e. quotients between budget components) can be calculated as well.
15. Compare whether calculated values of a given variable match the results of its measurement (i.e. validation).
16. If all budget components have been measured independently, use the ratio between total inputs (I) and total outputs (O) to check whether the steady-state assumption is correct. An estimate of the unaccounted organic matter as a percentage of the total input (I) can also be used as a measure of the accuracy (A) of the budget (Waletzko and Mitsch 2013):

$$A = \frac{100 \cdot (O - I)}{I} \tag{10.12}$$

4 Final Remarks

A CPOM budget can be constructed regardless of the time interval considered (e.g. a week, month or year). However, if an annual budget is desired, samples should be collected over at least 1 year, twice a month during heavy litter fall and at least monthly during the rest of the year. In addition, special care needs to be taken to sample CPOM transport adequately during storms, as a large fraction of the organic carbon exported from catchments occurs during high-discharge events (e.g. Webster and Meyer 1997).

Budgets can be refined by sorting the collected CPOM into categories. Leaves can be sorted into species, and twigs (less than 1 cm in diameter), bark, fruits and flowers and debris (unidentified fragments) can also be distinguished. In a wider context, organic matter budgets can benefit from the incorporation of stable carbon isotope analysis and models to assess the fate of organic carbon in streams (Ford and Fox 2015).

References

Abelho, M. (2001). From litterfall to breakdown in streams: A review. *The ScientificWorld, 1*, 656–680.

Baldy, V., Gobert, V., Guerold, F., Chauvet, E., Lambrigot, D., & Charcosset, J.-Y. (2007). Leaf litter breakdown budgets in streams of various trophic status: Effects of dissolved inorganic nutrients on microorganisms and invertebrates. *Freshwater Biology, 7*, 1322–1335.

Cummins, K. W., Sedell, J. R., Swanson, F. J., Minshall, G.W., Fisher, S. G., Cushing, C. E., Petersen, R. C., & Vannote, R. L. (1983). Organic matter budgets for stream ecosystems:

problems in their evaluation. In J. R. Barnes & G. W. Minshall (Eds.), *Stream ecology. Application and testing of general ecological theory* (pp. 299–353). New York: Plenum Press.

Datry, T., Foulquier, A., Corti, R., von Schiller, D., Tockner, K., Mendoza-Lera, C., Clément, J. C., Gessner, M. O., Moleón, M., Stubbington, R., Gücker, B., Albariño, R., Allen, D. C., Altermatt, F., Arce, M. I., Arnon, S., Banas, D., Banegas-Medina, A., Beller, E., Blanchette, M. L., Blanco-Libreros, J. F., Blessing, J. J., Boëchat, I. G., Boersma, K. S., Bogan, M. T., Bonada, N., Bond, N. R., Brintup Barría, K. C., Bruder, A., Burrows, R. M., Cancellario, T., Canhoto, C., Carlson, S. M., Cauvy-Fraunié, S., Cid, N., Danger, M., de Freitas Terra, B., De Girolamo, A. M., de la Barra, E., del Campo, R., Diaz-Villanueva, V. D., Dyer, F., Elosegi, A., Faye, E., Febria, C., Four, B., Gafny, S., Ghate, S. D., Gómez, R., Gómez-Gener, L., Graça, M. A. S., Guareschi, S., Hoppeler, F., Hwan, J. L., Jones, J. I., Kubheka, S., Laini, A., Langhans, S. D., Leigh, C., Little, C. J., Lorenz, S., Marshall, J. C., Martín, E., McIntosh, A. R., Meyer, E. I., Milisa, M., Mlambo, M. C., Morais, M., Moya, N., Negus, P. M., Niyogi, D. K., Papatheodoulou, A., Pardo, I., Paril, P., Pauls, S. U., Pesic, V., Polásec, M., Robinson, C. T., Rodríguez-Lozano, P., Rolls, R. J., Sánchez-Montoya, M. M., Savic, A., Shumilova, O., Sridhar, K. R., Steward, A. L., Storey, R., Taleb, A., Uzan, A., Vander Vorste, R., Waltham, N. J., Woelfle-Erskine, C., Zak, D., Zarfl, C., & Zoppini, A. (2018). A global analysis of terrestrial plant litter dynamics in non-perennial waterways. *Nature Geoscience, 11*, 497–503.

Epstein, D. M., Kelso, J. E., & Baker, M. A. (2016). Beyond the urban stream syndrome: organic matter budget for diagnostics and restoration of an impaired urban river. *Urban Ecosystems, 19*, 1623–1643.

Fisher, S. G., & Likens, G. E. (1973). Energy flow in Bear Brook, New Hampshire: an integrative approach to stream ecosystem metabolism. *Ecological Monographs, 43*, 421–439.

Ford, W. I., & Fox, J. F. (2015). Isotope-based fluvial organic carbon (ISOFLOC) Model: Model formulation, sensitivity, and evaluation. *Water Resources Research, 51*, 4046–4064.

Gonçalves, J. F., Jr., Rezende, R. S., Gregório, R. S., & Valentin, G. C. (2014). Relationships between dynamics of litterfall and riparian plant species in a tropical stream. *Limnologica, 44*, 40–48.

Graça, M. A. S., Pozo, J., Canhoto, C., & Elosegi, A. (2002). Effects of *Eucalyptus* plantations on detritus, decomposers, and detritivores in streams. *The ScientificWorld, 2*, 1173–1185.

Imberger, S. J., Thompson, R. M., & Grace, M. R. (2011). Urban catchment hydrology overwhelms reach scale effects of riparian vegetation on organic matter dynamics. *Freshwater Biology, 56*, 1370–1389.

Irons, J. G., & Oswood, M. W. (1997). Organic matter dynamics in 3 subarctic streams of interior Alaska, USA. *Journal of the North American Benthological Society, 16*, 23–28.

Iversen, T., Thorup, J., & Skriver, J. (1982). Inputs and transformation of allochthonous particulate organic matter in a headwater stream. *Holarctic Ecology, 5*, 10–19.

Jones, J. B., Schade, J. D., Fisher, S. G., & Grimm, N. B. (1997). Organic matter dynamics in Sycamore Creek, a desert stream in Arizona, USA. *Journal of the North American Benthological Society, 16*, 78–82.

King, J. M., Day, J. A., Davies, B. R., & Hensall-Howard, M. P. (1987). Particulate organic matter in a mountain stream in the South-western Cape, South Africa. *Hydrobiologia, 154*, 165–187.

Larrañaga, S., Díez, J. R., Elosegi, A., & Pozo, J. (2003). Leaf retention in streams of the Agüera basin (northern Spain). *Aquatic Sciences, 65*, 158–166.

Lecerf, A. (2017). Methods for estimating the effect of litterbag mesh size on decomposition. *Ecological Modelling, 362*, 65–68.

Minshall, G. W. (1996). Organic matter budgets. In F. R. Hauer & G. A. Lamberty (Eds.), *Methods in stream ecology* (pp. 591–605). San Diego: Academic.

Molinero, J., & Pozo, J. (2004). Impact of a eucalyptus (*Eucalyptus globulus* Labill.) plantation on the nutrient content and dynamics of coarse particulate organic matter (CPOM) in a small stream. *Hydrobiologia, 528*, 143–165.

Molinero, J., & Pozo, J. (2006). Organic matter, nitrogen and phosphorus fluxes associated with leaf litter in two small streams with different riparian vegetation: a budget aproach. *Archiv für Hydrobiologie, 166*, 363–385.
Naiman, R. J., & Link, G. L. (1997). Organic matter dynamics in 5 subarctic streams, Quebec, Canada. *Journal of the North American Benthological Society, 16*, 33–39.
Newbold, J. D., Bott, T. L., Kaplan, L. A., Sweeney, B. W., & Vannote, R. L. (1997). Organic matter dynamics in White Clay Creek, Pennsylvania, USA. *Journal of the North American Benthological Society, 16*, 46–50.
Pozo, J., González, E., Díez, J. R., & Elosegi, A. (1997). Leaf-litter budgets in two contrasting forested streams. *Limnetica, 13*, 77–84.
Ruffing, C. M., Dwire, K. A., & Daniels, M. D. (2016). Carbon pools in stream-riparian corridors: legacy of disturbance along mountain streams of south eastern Wyoming. *Earth Surface Processes and Landforms, 41*, 208–223.
Smock, L. A. (1990). Spatial and temporal variation in organic matter storage in low-gradient, headwater streams. *Archiv für Hydrobiologie, 118*, 169–184.
Tank, J. L., Rosi-Marshall, E. J., Griffiths, N. A., Entrekin, S. A., & Stephen, M. L. (2010). A review of allochthonous organic matter dynamics and metabolism in streams. *Journal of the North American Benthological Society, 29*, 118–146.
Treadwell, S. A., Campbell, I. C., & Edwards, R. T. (1997). Organic matter dynamics in Keppel Creek, southeastern Australia. *Journal of the North American Benthological Society, 16*, 58–61.
Waletzko, E. J., & Mitsch, W. J. (2013). The carbon balance of two riverine wetlands fifteen years after their creation. *Wetlands, 33*, 989–999.
Wallace, J. B., Eggert, S. L., Meyer, J. L., & Webster, J. R. (1999). Effects of resource limitation on a detrital-based ecosystem. *Ecological Monographs, 69*, 409–442.
Wallace, J. B., Eggert, S. L., Meyer, J. L., & Webster, J. R. (2015). Stream invertebrate productivity linked to forest subsidies: 37 stream-years of reference and experimental data. *Ecology, 96*, 1213–1228.
Webster, J. R., & Meyer, J. L. (Eds.). (1997). Stream organic matter budgets. *Journal of the North American Benthological Society, 16*, 3–161.
Webster, J. R., Golladay, S. W., Benfield, E. F., D'Angelo, D. J., & Peters, G. T. (1990). Effects of forest disturbance on particulate organic matter budgets of small streams. *Journal of the North American Benthological Society, 9*, 120–140.
Webster, J. R., Wallace, J. B., & Benfield, E. F. (1995). Organic processes in streams of the Eastern United States. In C. E. Cushing, K. W. Cummins, & G. W. Minshall (Eds.), *Ecosystems of the world 22: River and stream ecosystems* (pp. 117–187). Amsterdam: Elsevier.

Part II
Chemical and Physical Leaf Properties

Chapter 11
Total Phosphorus, Nitrogen and Carbon in Leaf Litter

Mogens R. Flindt, Ana I. Lillebø, Javier Pérez, and Verónica Ferreira

Keywords Acid digestion · Alkaline digestion · Elemental analyser · Kjeldahl · Leaf conditioning · Litter quality · Loss on ignition · Nutrient immobilization · Spectrophotometry · Wet chemical analysis

1 Introduction

Plant litter decomposition is a fundamental ecosystem process in forests that sustains soil food webs (Gessner et al. 2010; Garcia-Palacios et al. 2016). In many freshwater, coastal and marine environments, plant litter of various origins is also abundant and an important source of energy and nutrients for aquatic food webs (e.g. Wolanski and Elliott 2015; Garcia-Palacios et al. 2016). Nitrogen (N) and phosphorus (P) are important nutrients determining the quality and decomposability of plant litter (Enriquez et al. 1993), although other litter characteristics also play important roles in controlling rates of litter decomposition (Webster and Benfield 1986; Ostrofsky 1997; Aerts 1997; Abelho 2001). Most P is used by biota for the

M. R. Flindt (✉)
Department of Biology, SDU – University of Southern Denmark, Odense, Denmark
e-mail: mrf@biology.sdu.dk

A. I. Lillebø
Department of Biology & CESAM, University of Aveiro, Aveiro, Portugal
e-mail: lillebo@ua.pt

J. Pérez
Department of Plant Biology and Ecology, Faculty of Science and Technology, University of the Basque Country (UPV/EHU), Bilbao, Spain
e-mail: javier.perezv@ehu.eus

V. Ferreira
MARE – Marine and Environmental Sciences Centre & Department of Life Sciences, University of Coimbra, Coimbra, Portugal
e-mail: veronica@ci.uc.pt

F. Bärlocher et al. (eds.), *Methods to Study Litter Decomposition*,
https://doi.org/10.1007/978-3-030-30515-4_11

Table 11.1 Summary of nitrogen and phosphorus concentrations as well as C:N and C:P molar ratios in plant litter of a broad range of species. Except for macroalgae, plant litter refers to leaves

Plant material	N (%)	C:N	P (%)	C:P	Reference
2 macroalgae	2.50–3.92	12.6–18.7	0.19–0.36	224–544	1
2 seagrass species	2.64–2.76	15.6–24.3	0.50–0.55	191–258	1
4 freshwater angiosperms	1.37–3.44	12.7–27.8	0.43–0.85	140–249	1
6 *Carex* spp.	0.18–1.07	46.1–315	0.016–0.150	729–7847	1
2 *Pinus* spp.	0.40–1.51		0.017–0.131		1
5 *Acer* spp.	0.73–1.71	24.1–60.9	0.112–0.411		2
2 *Salix* spp.	0.83–2.24	21.0–56.5	0.121–0.281		2
3 *Populus* spp.	0.92–2.38	19.7–52.4	0.083–0.092		2
Alnus glutinosa	1.38–2.98		0.02–0.15		3
Quercus robur	0.56–2.13		0.01–0.35		3
Fraxinus excelsior	1.1		0.073		4
Fagus sylvatica	0.68		0.032		4
Prunus avium	0.52		0.048		4
Corylus avellana	1.05		0.072		4
Platanus hybrida	0.52		0.017		4
67 temperate riparian tree species	0.16–3.20		0.010–0.300		5
101 tropical riparian tree species	0.18–3.84		0.010–0.230		5

1 = Enriquez et al. (1993); 2 = Ostrofsky (1997); 3 = Graça and Poquet (2014); 4 = Gessner and Chauvet (1994); 5 = Boyero et al. (2017); 6 = Pérez-Harguindeguy et al. (2000)

synthesis of RNA, DNA, ATP and phospholipids, whereas N is mainly associated with chitin and proteins but also an important component of nucleic acids and some secondary plant metabolites (Sterner and Elser 2002). Litter characterized by low ratios of C:N and C:P (i.e. high N and P concentrations) and low concentrations of structural plant polymers such as lignin tend to decompose fast, whereas litter containing nutrients at low concentrations tend to be more recalcitrant (Flindt et al. 1999; Kadlec and Wallace 2009). Typical concentrations of N and P in plant litter from diverse environments are summarized in Table 11.1, based on a comprehensive compilation of nutrient concentrations in more than 250 photosynthetic organisms ranging from phytoplankton to trees (Enríquez et al. 1993), 48 deciduous tree leaf species (Ostrofsky 1997), riparian leaf litter analysed in a global-scale study (Boyero et al. 2017) and a meta-analysis of C:N and C:P ratios of leaf litter in aquatic environments from 44 primary studies (Kennedy and El-Sabaawi 2017).

A large fraction of the P is rapidly leached from dead leaf tissue during the initial phase of decomposition, especially when autumn-shed leaves dry before they enter the aquatic environment (Gessner 1991; Flindt et al. 1999). N is not generally leached upon senescence and death, although initially decreasing concentrations have occasionally been observed (Meyer and Johnson 1983). Subsequently, N and P concentrations of leaves usually increase during decomposition (Webster and Benfield 1986; Gessner 1991). This increase is attributed to microbial colonization (Gulis and Suberkropp 2003; Gulis et al. 2006; Ferreira et al. 2006, 2015), which enhances the nutritional quality of leaf litter for detritivores (Webster and Benfield 1986; Bärlocher and Sridhar 2014).

Total N in litter can be determined by the standard Kjeldahl method for plant samples or a modification thereof (Ferskvandsbiologisk Laboratorium University of Copenhagen 1992a). Leaf-litter samples are dried, homogenized and digested in concentrated sulphuric acid to reduce all nitrogen species to ammonium (NH_4^+). After neutralizing the resulting solution, it is filtered and analysed spectrophotometrically. Alternatively, the total N concentration of litter can be quantified by means of a CHN elemental analyser. Advantages are that the analysis is accurate and easy and that total carbon is determined simultaneously, enabling the calculation of C:N ratios based on analyses of both elements in the same sample. However, the equipment is costly and not readily available in all laboratories.

Total P can also be analysed spectrophotometrically after digestion. Dried and homogenized litter samples are digested by autoclaving with sodium hydroxide and sodium persulphate or the samples are first combusted and the ash digested in concentrated hydrochloric acid (Ferskvandsbiologisk Laboratorium University of Copenhagen 1992b). The resulting solutions are filtered and total P is quantified as orthophosphate (PO_4^{3-}) by using the ascorbic acid method. Both procedures are presented here. The first method is preferable when individual plant samples are large. The second method is advantageous when many samples are to be processed or the amount of plant material is low, because the method is fast and handling of small amounts of ash is avoided.

A combined method for total N and total P determination can be performed by the total N procedure. However, the pH during neutralization of samples must never exceed 6, otherwise dissolved phosphorus will precipitate. The samples are then filtered and analysed spectrophotometrically. As N and P are rapidly recycled by biological processes, samples should in all cases be analysed as soon as possible after the digestion step. When this is not possible, samples should be preserved by acidification, refrigeration, or freezing, depending on the method.

2 Phosphorus Using an Acid Digestion Method

2.1 Equipment and Material (Acid-Washed Glassware)

- Drying oven (50 °C)
- Mill or mortar and pestle
- Spectrophotometer
- Cuvettes
- Hot plate
- pH meter
- Analytical balance (± 0.1 mg)
- Muffle furnace

- Porcelain crucibles
- Desiccator
- Vortex
- Glass fibre filters (Whatman GF/C or equivalent)
- Syringes (20 ml) with adapter for ∅ 25 mm glass fibre filters
- Pipettes (200, 1000, 5000 and 10,000 μl)
- Erlenmeyer flasks (50–100 ml)
- Glass tubes (> 15 ml)
- Cooling bulbs
- Gloves

2.2 Chemicals (Analytical Grade)

- Deionized water
- HCl (37%) – concentrated hydrochloric acid
- KH_2PO_4 – potassium dihydrogen phosphate
- 2-[$(CH_3)_2CH$]C_6H_3–5-(CH_3)OH – thymol (crystalline)
- $(NH_4)_6Mo_7O_{24} \bullet 4\ H_2O$ – ammonium heptamolybdate tetrahydrate
- H_2SO_4 (98%) – sulphuric acid concentrate
- $K(SbO)C_4H_4O_6 \bullet 5\ H_2O$ – potassium antimony (III) oxide tartrate pentahydrate, extra pure
- $C_6H_8O_6$ – L(+)ascorbic acid (vitamin C)

2.3 Solutions

- Solution 1 – PO_4^{3-} stock solution (40 mg P l^{-1}): dissolve 175.75 mg KH_2PO_4 in 1000 ml of deionized water, add one crystal of thymol, and store the stock solution for up to 3 months at ambient temperature.
- Solution 2 – PO_4^{3-} working solution (10 mg P l^{-1}) for preparing a standard curve: dilute 4 ml of the stock solution in 1000 ml of deionized water; this working solution must be freshly prepared every day.
- Solution 3 – stock solution of reagent; to 12 g of $(NH_4)_6Mo_7O_{24} \bullet 4\ H_2O$ in 500 ml of deionized water, add very carefully 140 ml of concentrated H_2SO_4. After mixing and cooling, add 275 mg of $K(SbO)C_4H_4O_6 \bullet 5\ H_2O$ and let it dissolve, then adjust the volume to 1000 ml with deionized water, and store the solution for up to 3 months at ambient temperature.
- Solution 4 – working solution of reagent: immediately before use add 1.06 g of ascorbic acid to 100 ml of the stock reagent solution.

2.4 Sample Preparation

1. Dry and grind plant material, and then put it into porcelain crucibles.
2. Place the samples in a muffle furnace for 4 hours at 500 °C.
3. Determine dry and ash-free dry mass of the samples as indicated in the C determination method (see Sect. 6.2).
4. Weigh portions of approximately 5 mg ash to the nearest 0.1 mg.
5. Place the sample in a 50 or 100 ml Erlenmeyer flask with 25 ml of deionized water.
6. Add 1 ml of concentrated HCl.
7. Place the Erlenmeyer flask on a heating plate to evaporate the water.
8. Place a cooling bulb on top of the flask and wait for a few minutes until the solution starts tingling gently.
9. Continue heating the samples (temperature < 110 °C, because the boiling point of HCl is 110 °C) until the solution turns yellow and translucid.
10. Rotate the flask from time to time to remove ash particles from the walls (wear gloves!).
11. If the water evaporates before the solution turns yellow and translucid, add more HCl in 1 ml aliquots.
12. Transfer the solution to a 100 ml flask and adjust the sample volume to 100 ml with deionized water.
13. Filter the samples through a glass fibre filter connected to a 20 ml syringe.
14. Analyse the filtered samples immediately after extraction or, if not possible, store them frozen at −20 °C.

2.5 Spectrophotometric Analysis

1. Run a standard curve using concentrations of 40, 80, 100, 200, 400 and 800 μg P l^{-1} (Table 11.2).
2. Dilute samples if higher P concentrations are expected, because the standard curve is linear only up to 1000 μg P l^{-1}.
3. Transfer 10 ml of sample into a glass tube.

Table 11.2 Preparation of the standard curve to calculate PO_4^{3-} concentrations

PO_4^{3-} working solution (ml)	Deionized H_2O (ml)	Final volume (ml)	Final P concentration(μg l^{-1})
0.8	99.2	100	80
1.0	99.0	100	100
2.0	98.0	100	200
4.0	96.0	100	400
8.0	92.0	100	800

4. Add 1 ml of the working reagent and vortex.
5. Wait for 15 min for the reaction to complete
6. Measure absorbance at 882 nm, using deionized water as reference.
7. Use as a blank 10 ml of deionized water treated like the samples.
8. Calculate the P concentration of the samples based on the standard curve, taking into account the absorbance reading, blank values and moisture and ash contents of the litter samples.

3 Phosphorus Using an Alkaline Digestion Method

3.1 Equipment and Material (Acid-Washed Glassware)

- Drying oven
- Mill or mortar and pestle
- Porcelain crucibles
- Desiccator
- Analytical balance (±0.1 mg)
- Spatula
- Pipettes (1000, 5000 and 10,000 µl)
- Extraction glass tubes with cap (> 30 ml; e.g. Pyrex tubes)
- Autoclave
- Fridge
- Syringes (20 ml) with adapter for ∅ 25 mm glass fibre filters
- Syringe filters (Whatman GF/C or equivalent)
- Glass tubes (> 15 ml)
- pH paper
- Spectrophotometer
- Cuvettes

3.2 Chemicals

- Ethanol
- Deionized water
- NaOH – sodium hydroxide
- $Na_2S_2O_8$ – sodium persulphate
- H_2SO_4 (96–98%) – concentrated sulphuric acid
- $K(SbO)C_4H_4O_6 \cdot 0.5H_2O$ – potassium antimonyl tartrate hemihydrate
- $(NH_4)_6Mo_7O_{24} \cdot 4H_2O$ – ammonium molybdate tetrahydrate
- $C_6H_8O_6$ – ascorbic acid (vitamin C)
- KH_2PO_4 – potassium dihydrogen phosphate

3.3 Solutions

- Solution 1–1 M sodium hydroxide: dissolve 3.9997 g NaOH in 100 ml of deionized water, and keep indefinitely in the dark at room temperature.
- Solution 2–3 M sodium hydroxide: dissolve 12 g NaOH in 100 ml of deionized water, and keep indefinitely in the dark at room temperature.
- Solution 3–0.525 M sodium persulphate: dissolve 12.5 $Na_2S_2O_8$ g in 100 ml of deionized water, and keep at 4 °C for up to several months.
- Solution 4–15% sulphuric acid: add 15 ml concentrated H_2SO_4 to 75 ml of deionized water, and keep at 4 °C for up to 2 months.
- Solution 5 – potassium antimonyl tartrate (2.8 g l^{-1}): dissolve 0.14 g K(SbO) $C_4H_4O_6 \cdot 0.5\ H_2O$ in 50 ml of deionized water, and keep at 4 °C for up to 1 month.
- Solution 6 – ammonium molybdate (40 g l^{-1}): dissolve 2 g $(NH_4)_6Mo_7O_{24} \cdot 4H_2O$ in 50 ml of deionized water, and keep at 4 °C in the dark for up to 2–3 weeks.
- Solution 7 – combined reagent: mix 50 ml of 15% sulphuric acid +5 ml of potassium antimonyl tartrate +15 ml of ammonium molybdate +30 ml of deionized water, and keep at 4 °C in the dark for up to 2–3 weeks.
- Solution 8 – ascorbic acid (20 g l^{-1}): always freshly prepare solution by dissolving 1 g vitamin C in 50 ml of deionized water.
- Solution 9 – concentrated potassium dihydrogen phosphate (50 mg P l^{-1}): dissolve 21.97 mg KH_2PO_4 in 100 ml of deionized water, and keep at 4 °C for up to 2 months.
- Solution 10 – diluted potassium dihydrogen phosphate (1 mg P l^{-1}): mix 1 ml of the concentrated KH_2PO_4 with deionized water, and make up to 50 ml.
- Solution 11 – diluted KH_2PO_4, (2 mg P l^{-1}): if high P concentrations are expected in the samples, mix 2 ml of concentrated KH_2PO_4 with deionized water, and make up to 50 ml.

3.4 Sample Preparation

1. Grind samples and determine dry mass and ash-free dry mass as indicated in Sect. 6.2.
2. Weigh 2–4 mg (± 0.1 mg) of ground plant material and transfer to labelled glass tubes.
3. Add 1 ml of 1 M NaOH, which results in a bright yellow translucid coloration.
4. Add 2.38 ml of 0.525 M sodium persulphate.
5. Add 6.62 ml of deionized water and make up to a total volume of 10 ml.
6. Loosely close the tubes and place them inside the autoclave, noting the tube positions in the racks and in the autoclave as the labelling may be erased.
7. Autoclave at 120 °C for 2 h.

8. Let the samples cool until the solution's yellow coloration disappears and the litter turns grey or whitish.
9. Check the pH with pH paper to ensure it is between 2 and 7
10. If outside this range, adjust the pH by adding 15% H_2SO_4 or 3 M NaOH (e.g. 4 mg alder litter results in a pH of ~1.5, which can be raised to 3 by adding 0.5 ml of NaOH).
11. Adjust the volume in the tube to 25 ml with deionized water.
12. Filter 10 ml of the sample into a glass tube.

3.5 Spectrophotometric Analysis

1. Add consecutively the volumes indicated in Table 11.3 of Solution 10 (KH_2PO_4), Solution 1 (1 M NaOH), Solution 3 (0.525 M $Na_2S_2O_8$) and deionized water, including the last two concentrations in the table only if high P concentrations are expected.
2. Dilute samples when the final P concentrations are high, because the standard curve is linear only up to 1000 μg P l^{-1} (Table 11.3).
3. Add 2 ml of combined reagent (Solution 7) to the tubes, including to those used to establish the standard curve.
4. Add 0.5 ml of ascorbic acid solution.
5. Wait for 30 min for the reaction to complete.
6. Determine absorbance in the spectrophotometer at 880 nm, using deionized water as reference.
7. Use as a blank 10 ml of deionized water treated like a sample.

Table 11.3 Preparation of solutions to establish a standard curve for the determination of P concentrations in leaf litter

KH_2PO_4 solution (mg l^{-1})	KH_2PO_4 solution (ml)	1 M NaOH (ml)	0.525 M $Na_2S_2O_8$ (ml)	Deionized H_2O (ml)	Final volume (ml)	Final P concentration (μg l^{-1})
1	0	1	2.38	6.62	10	0
1	0.05	1	2.38	6.57	10	5
1	0.10	1	2.38	6.52	10	10
1	0.25	1	2.38	6.37	10	25
1	1.5	1	2.38	5.12	10	150
1	2.5	1	2.38	4.12	10	250
1	4.0	1	2.38	2.62	10	400
1	5.0	1	2.38	1.62	10	500
2	5.0	1	2.38	1.62	10	1000

8. Calculate the P concentration of the samples based on the standard curve, taking into account the absorbance reading, blank values and moisture and ash contents of the litter samples.

4 Nitrogen by a Modified Kjeldahl Method

4.1 Equipment and Material

- Drying oven (50 °C)
- Mill or mortar and pestle
- Spectrophotometer
- Hot plate
- pH meter
- Analytical balance (± 0.1 mg)
- Porcelain crucibles
- Desiccator
- Vortex
- Glass fibre filters (Whatman GF/C or equivalent)
- Syringes (20 ml) with adapter for ⌀ 25 mm glass fibre filters
- Pipettes (200, 1000 and 5000 μl)
- Erlenmeyer flasks (50–100 ml)
- Glass tubes (> 15 ml)
- Cooling bulbs
- Gloves

4.2 Chemicals (Analytical Grade)

- Deionized water
- $CuSO_4 \cdot 5\,H_2O$ – copper sulphate pentahydrate
- H_2SO_4 (96–98%) – concentrated sulphuric acid
- $(NH_4)_2SO_4$ – ammonium sulphate
- 2-[$(CH_3)_2CH$]C_6H_3–5-(CH_3)OH – thymol (crystalline)
- C_6H_5OH – phenol (crystalline)
- $Na_2[Fe(CN)_5NO] \cdot 2\,H_2O$ – sodium nitroprusside dihydrate
- NaOH – sodium hydroxide
- NaOCl (15%) – 15% sodium hypochlorite solution

4.3 Solutions

- Solution 1 – copper sulphate: dissolve 10 g $CuSO_4 \bullet 5\ H_2O$ in 100 ml of deionized water.
- Solution 2 – ammonium stock solution (100 mg N l^{-1}): dissolve 0.472 g $(NH_4)_2SO_4$ in 1000 ml of deionized water, add one crystal of thymol, and keep the stock solution for up to 3 months at ambient temperature.
- Solution 3 – ammonium working solution (10 mg N l^{-1}) for preparing a standard curve: dilute 10 ml of the stock solution in 1000 ml of deionized water; this working solution must be freshly prepared every day.
- Solution 4 – reagent A: dissolve 50 g phenol and 0.25 g sodium nitroprusside dihydrate in 1000 ml of deionized water.
- Solution 5 – reagent B: dissolve 25 g NaOH and 20 ml 15% NaOCl in 1000 ml of deionized water.

4.4 Sample Preparation

1. Determine fresh and dry mass of litter samples (see Sect. 6.2).
2. Weigh portions of dried and ground litter samples ($\leq$2 mg) to the nearest 0.1 mg.
3. Place the sample in an Erlenmeyer flask with 25 ml of deionized water.
4. Add 0.2 ml of Solution 1 and 1 ml of H_2SO_4.
5. Place the Erlenmeyer flasks on a hot plate and boil off the water (wear gloves!).
6. When the water has evaporated and a light smoke appears, place a cooling bulb on top of the flasks, which will result in a gentle tingling after a few minutes.
7. Observe the start of the organic matter digestion (temperature ~150 °C; the boiling point of H_2SO_4 is 290 °C), which is noticeable by the appearance of a dark coloured tar that is subsequently mineralized to CO_2 and H_2O, as indicated by white smoke appearing inside the flask.
8. Rotate the flask from time to time to remove residues from the walls.
9. If necessary, add more H_2SO_4 in 1 ml aliquots until the digestion is complete, as indicated by the acid solution becoming light yellow and translucid.
10. Continue the digestion for another 30 min.
11. After cooling, transfer the solution to a beaker and rinse the flask walls with deionized water, ignoring that the solution temperature will exceed 30 °C.
12. Adjust the samples with 1 M NaOH to the point where $Cu(OH)_2$ flocculates (pH ~3), ignoring that the solution temperature may increase to >50 °C.
13. Use a dilute (0.1 M) NaOH solution to raise the pH to 5–8, being most careful because pH shifts rapidly from 3 to 10 and ensuring that it is similar among all samples.
14. Adjust the sample volume to 100 ml with deionized water.
15. Filter the samples through a glass fibre filter connected to a 20 ml syringe.

Table 11.4 Preparation of the standard curve to calculate NH_4^+ concentrations

NH_4^+ working solution (10 mg N l^{-1})	Deionized H_2O (ml)	Final volume (ml)	Final N concentration (μg l^{-1})
0.5 ml	99.5	100	50
1.0 ml	99	100	100
2.0 ml	98	100	200
4.0 ml	96	100	400
8.0 ml	92	100	800

16. Analyse samples immediately or, if not feasible, store acid samples at 4 °C for up to 2 months or neutralized and filtered samples at −20 °C for up to 6 months.

4.5 *Spectrophotometric Analysis*

1. Run a standard curve with concentrations of 50, 100, 200, 400 and 800 μg N l^{-1} as shown in Table 11.4.
2. Dilute samples if higher N concentrations are expected, because the standard curve is linear only up to 1000 μg N l^{-1}.
3. Transfer 10 ml of sample into a glass tube.
4. Add 1 ml of reagent A and vortex.
5. Wait for about 1 min, and then add 1 ml of reagent B and mix again; if many samples are processed simultaneously, pipetting can be staggered such that the delay between the addition of reagents A and B is about 1 min.
6. Protect the developing blue colour from sunlight.
7. Wait for a defined time, but at least 1 h for the reaction to complete.
8. Measure absorbance at 630 nm, using deionized water as reference.
9. Use as a blank 10 ml of deionized water treated like the samples.
10. Calculate the N concentration of the sample based on the standard curve, taking into account the absorbance reading, blank values and moisture content of the litter sample.

5 Total Nitrogen and Carbon with an Elemental Analyser

5.1 *Equipment and Material*

- Drying oven (50 °C)
- Desiccator
- Analytical balance (± 0.1 mg)
- Mill or mortar and pestle
- CHN Elemental Analyser
- Metal cups (tin, aluminium or silver) suitable for the CHN analyser

5.2 Procedure

1. Calibrate the CHN analyser according to the instructions of the manufacturer.
2. Grind the dry plant material in a mill or with mortar and pestle (powder <1 mm particle diameter), and carefully clean the equipment between grinding successive samples.
3. Weigh 2–5 mg dry mass of ground litter in suitable metal cups for the CHN analyser.
4. Determine the C and N contents of the litter samples and appropriate standards according to the instructions of the manufacturer.
5. Calculate the N and C contents of the litter samples by taking into account the measured values and the moisture and ash contents of the samples.
6. Calculate the molar or atomic C:N ratio of the litter samples by taking into account the measured C and N contents as well as the atomic masses of C and N.

6 Total Carbon by a Combustion Method

6.1 Equipment and Material

- Drying oven (50 °C)
- Mill or mortar and pestle
- Analytical balance (± 0.1 mg)
- Muffle furnace (500 °C)
- Porcelain crucibles
- Desiccator

6.2 Procedure

1. Place pre-weighed fresh plant material in the oven and dry to constant weight following a standard protocol (e.g. 50 °C for at least 48 h).
2. Grind the dry plant material in a mill or with mortar and pestle (powder <1 mm particle diameter); carefully clean the equipment between grinding successive samples.
3. Place the ground litter in pre-weighed porcelain crucibles and dry at 50 °C for 48 h.
4. Cool the samples to room temperature in a desiccator and re-weigh to calculate dry mass.
5. Ignite the samples in the muffle furnace for at least 3 h at 500 °C.
6. Cool the ashes to room temperature in a desiccator and weigh.
7. Calculate litter ash-free dry mass.
8. Estimate carbon content by assuming 50% in litter AFDM, which is a reasonably good approximation for a wide variety of leaf litter types.

7 Final Remarks

Leaf litter samples collected in the field must be handled carefully to minimize external or between-sample contamination. Always use very clean material and latex gloves. Gently clean litter samples to remove as much sediment as possible, without scraping off the microbial biofilm. Use stream water for this purpose that has been passed through a 100 μm mesh screen.

Results can be expressed as %N and %P of leaf litter dry mass, if samples are not contaminated by mineral particles. Otherwise, it is preferable to measure the ash content in a representative subsample and express the data in terms of %N or %P ash-free dry mass. This can be particularly important for samples from calcareous streams where travertine precipitates can incrust decomposing litter (Martínez et al. 2015). Alternatively, nutrient content in litter can be expressed as elemental ratios (C:N, C:P and N:P), as is common in analyses of ecological stoichiometry (Sterner and Elser 2002). The molar or atomic ratio represents the proportion between elements in the number of atoms; therefore, the molar mass of each element must be considered in the calculations. For example, a sample containing 50% C and 2.92% N, the ratio is calculated as follows: molar C:N = (50/12)/(2.92/14), where 12 and 14 refer to the atomic mass of C and N, respectively.

When assessing litter nutrient dynamics during the decomposition process, it is often essential to ensure an initially high temporal resolution to capture the leaching losses, particularly of P, within the first 24–48 h of litter submergence in streams or other aquatic environments (e.g. Gessner 1991).

Ebina et al. (1983) have devised an alternative method to determine total N and P concentrations simultaneously in a single procedure. This approach greatly reduces the total analysis time. The combined analysis is facilitated by inducing a pH shift during the sample digestion, resulting in a sequential digestion of N and P species. A downside is that the method is less robust than the standard methods described in this chapter, requiring careful analyses by skilled analysts and care not to exceed the total N contents in the analysed samples (100 μg N with the described procedure). However, the method has been successfully applied to leaf litter (e.g. Gessner et al. 1998).

References

Abelho, M. (2001). From litterfall to breakdown in streams: A review. *TheScientific World, 1*, 656–680.

Aerts, R. (1997). Climate, leaf litter chemistry and leaf litter decomposition in terrestrial ecosystems: A triangular relationship. *Oikos, 79*, 439–449.

Bärlocher, F., & Sridhar, K. R. (2014). Association of animals and fungi in leaf decomposition. In G. Jones, K. Hyde, & K.-L. Pang (Eds.), *Freshwater fungi and fungus-like Organisms* (Marine and freshwater Botany) (pp. 413–441). Berlin: De Gruyter.

Boyero, L., Graça, M. A. S., Tonin, A. M., Pérez, J., Swafford, A. J., Ferreira, V., Landeira-Dabarca, A., Alexandrou, M. A., Gessner, M. O., McKie, B. G., Albariño, R. J., Barmuta,

L. A., Callisto, M., Chara, J., Chauvet, E., Colon-Gaud, C., Dudgeon, D., Encalada, A. C., Figueroa, R., Flecker, A. S., Fleituch, T., Frainer, A., Gonçalves, J. F., Jr., Helson, J. E., Iwata, T., Mathooko, J., M'Erimba, C., Pringle, C. M., Ramirez, A., Swan, C. M., Yule, C. M., & Pearson, R. G. (2017). Riparian plant litter quality increases with latitude. *Scientific Reports, 7*, 10562.

Ebina, J., Tsutsui, T., & Shirai, T. (1983). Simultaneous determination of total nitrogen and total phosphorus in water using peroxodisulfate oxidation. *Water Research, 17*, 1721–1726.

Enríquez, S., Duarte, M., & Sand-Jensen, K. (1993). Patterns in decomposition rates among photosynthetic organisms: The importance of detritus C:N:P content. *Oecologia, 94*, 457–471.

Ferreira, V., Chauvet, E., & Canhoto, C. (2015). Effects of experimental warming, litter species, and presence of macroinvertebrates on litter decomposition and associated decomposers in a temperate mountain stream. *Canadian Journal of Fisheries and Aquatic Sciences, 72*, 206–216.

Ferreira, V., Gulis, V., & Graça, M. A. S. (2006). Whole-stream nitrate addition affects litter decomposition and associated fungi but not invertebrates. *Oecologia, 149*, 718–729.

Ferskvandsbiologisk Laboratorium University of Copenhagen (Ed.). (1992a). *Limnologisk Metodik*. Copenhagen: Akademisk Forlag. (Nitrogen method pp. 28–29 and 36–37).

Ferskvandsbiologisk Laboratorium University of Copenhagen (Ed.). (1992b). *Limnologisk Metodik*. Copenhagen: Akademisk Forlag. (Phosphorus method pp. 40–41 and 106–107).

Flindt, M. R., Pardal, M. A., Lillebø, A. I., Martins, I., & Marques, J. C. (1999). Nutrient cycling and plant dynamics in estuaries: A brief review. *Acta Oecologica, 20*, 237–248.

Garcia-Palacios, P., McKie, B. G., Handa, I. T., Frainer, A., & Hättenschwiler, S. (2016). The importance of litter traits and decomposers for litter decomposition: A comparison of aquatic and terrestrial ecosystems within and across biomes. *Functional Ecology, 30*, 819–829.

Gessner, M. O. (1991). Differences in processing dynamics of fresh and dried leaf litter in a stream ecosystem. *Freshwater Biology, 26*, 387–398.

Gessner, M. O., & Chauvet, E. (1994). Importance of stream microfungi in controlling breakdown rates of leaf litter. *Ecology, 75*, 1807–1817.

Gessner, M. O., Robinson, C. T., & Ward, J. V. (1998). Leaf breakdown in streams of an alpine glacial floodplain: Dynamics of fungi and nutrients. *Journal of the North American Benthological Society, 17*, 403–419.

Gessner, M. O., Swan, C. M., Dang, C. K., McKie, B. G., Bardgett, R. D., Wall, D. H., & Hättenschwiler, S. (2010). Diversity meets decomposition. *Trends in Ecology & Evolution, 25*, 372–380.

Graça, M. A. S., & Poquet, J. M. (2014). Do climate and soil influence phenotypic variability in leaf litter, microbial decomposition and shredder consumption? *Oecologia, 174*, 1021–1032.

Gulis, V., & Suberkropp, K. (2003). Leaf litter decomposition and microbial activity in nutrient-enriched and unaltered reaches of a headwater stream. *Freshwater Biology, 48*, 123–134.

Gulis, V., Ferreira, V., & Graça, M. A. S. (2006). Stimulation of leaf litter decomposition and associated fungi and invertebrates by moderate eutrophication: Implications for stream assessment. *Freshwater Biology, 51*, 1655–1669.

Kadlec, R. H., & Wallace, S. D. (2009). *Treatment wetlands* (2nd ed.). Boca Raton: CRC Press. 1016 pp.

Kennedy, K., & El-Sabaawi, R. W. (2017). A global meta-analysis of exotic versus native leaf decay in stream ecosystems. *Freshwater Biology, 6*, 977–989.

Martínez, A., Pérez, J., Molinero, J., Sagarduy, M., & Pozo, J. (2015). Effects of flow scarcity on leaf-litter processing under oceanic climate conditions in calcareous streams. *Science of the Total Environment, 503*, 251–257.

Meyer, J. L., & Johnson, C. (1983). The influence of elevated nitrate concentration on rate of leaf decomposition in a stream. *Freshwater Biology, 13*, 177–184.

Ostrofsky, M. L. (1997). Relationship between chemical characteristics of autumn-shed leaves and aquatic processing rates. *Journal of the North American Benthological Society, 16*, 750–759.

Pérez-Harguindeguy, N., Díaz, S., Cornelissen, J. H. C., Vendramini, F., Cabido, M., & Castellanos, A. (2000). Chemistry and toughness predict leaf litter decomposition rates over a wide spectrum of functional types and taxa in Central Argentina. *Plant and Soil, 218*, 21–30.

Sterner, R. W., & Elser, J. J. (2002). *Ecological stoichiometry: The biology of elements from molecules to the biosphere*. Princeton: Princeton University Press.

Webster, J. R., & Benfield, E. F. (1986). Vascular plant breakdown in freshwater systems. *Annual Review of Ecology and Systematics, 17*, 567–594.

Wolanski, E., & Elliott, M. (2015). *Estuarine Ecohydrology* (2nd ed.). Amsterdam: Elsevier. 322 pp.

Chapter 12
Total Protein

Mark O. Baerlocher

Keywords Bicinchoninic acid assey · Biuret reaction · Coomassie Blue · Leaf conditioning · Litter chemistry · Litter quality · Lowry method · Pierce BCA assay · Protein extraction · Spectrophotometry

1 Introduction

Detrital plant matter, which supplies most of the energy and nutritional needs for stream communities (Allan and Castillo 2007), consists primarily of structural plant polysaccharides (Chaps. 14 and 21). These are generally inaccessible to invertebrates without microbial assistance, termed 'conditioning' (Chaps. 27 and 32). An important aspect of microbial conditioning is the enrichment of the substrate with nitrogen, especially in the form of protein, derived from microbial cells and excretions (Kaushik and Hynes 1971; Bärlocher et al. 1989). During decomposition, nitrogen and protein levels tend to decline somewhat in the first few days and then often rise to a level that may remain constant for several weeks (Fig. 12.1).

Factors that influence protein concentrations in decomposing leaves include leaf species, concentrations of dissolved nutrients in the stream (primarily N and P), and length of exposure in the stream. There are numerous analytical methods to estimate protein content, which have varying degrees of accuracy. A common approach in forage analysis is to multiply the nitrogen concentration (normally estimated by the Kjeldahl procedure) with an empirical factor of 6.25 (AOAC 1990). When this method is applied to leaves or other plant structures decomposing in streams, it yields protein concentrations higher than those estimated with other assays (Kaushik and Hynes 1971; Gessner 1991; Fiset et al. 2017). This is due to the fact that nitrogen is not restricted to protein; it occurs, for example, in chitin (a constituent of

M. O. Baerlocher (✉)
Department of Vascular and Interventional Therapies, Royal Victoria Hospital, Barrie, ON, Canada
e-mail: mark.baerlocher@alumni.utoronto.ca

F. Bärlocher et al. (eds.), *Methods to Study Litter Decomposition*,
https://doi.org/10.1007/978-3-030-30515-4_12

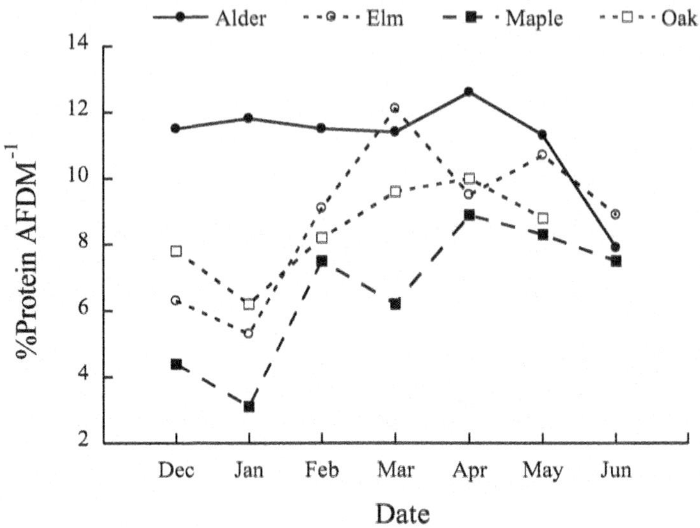

Fig. 12.1 Protein contents (as % of ash-free dry mass) of leaves decaying in the Speed River (Kaushik and Hynes 1971). The leaves were exposed on December 7 and recovered at monthly intervals. Alder, *Alnus glutinosa*; Elm, *Ulmus americana*; Maple, *Acer saccharum*; Oak, *Quercus alba*

fungal cell walls) or in 'artefact lignins', complexes that may form between leaf phenolics and compounds released by microbes during decomposition (Suberkropp et al. 1976; Odum et al. 1979).

An alternative to deriving protein concentrations from N measurements is to hydrolyze proteins to amino acids which can then be quantified by HPLC or other assays (Craig et al. 1989, Chap. 13). Similarly, to estimate amounts of protein available (i.e. digestible) to detritivores, the sample can be exposed to invertebrate enzyme extracts and the released amino acids measured (Bärlocher 1983; Bärlocher and Howatt 1986; Craig et al. 1989). These methods are relatively elaborate and not needed where the objective is simply to estimate total protein concentrations.

The first step in protein analysis is extraction from the substrate (Scopes 1982). Protein solubility is pH dependent: in the presence of tannins, complexes tend to form that are most stable near neutrality and dissociate more readily with increasing alkalinity (Swain 1979) or in the presence of surfactants (Bärlocher et al. 1989). One of the more aggressive solvents is NaOH (0.1 or 0.5 N; Kaushik and Hynes 1971; Gessner 1991). However, this also extracts phenolics bound through esterification to cell walls, which may interfere with various protein assays.

The protein-containing extract is then mixed with a chemical that reacts stoichiometrically to form a coloured product measured in the visible region of the spectrum. The Biuret method is based on the formation of a violet complex between cupric (Cu^+) ions and two or more peptide bonds. The intensity of the colour is proportional to the number of peptide bonds (Bailey 1962). In its original form, it is relatively insensitive. Lowry et al. (1951) combined the copper reaction in Biuret's approach with Folin's reagent to produce a strong dark blue colour with the phenolic amino acid tyrosine found in proteins (cf. Chap. 18).

Unfortunately, many substances interfere with this assay, not least among them non-proteinaceous phenolics that may be present in high concentrations in leaf litter. To circumvent this, proteins can be precipitated from the extract with TCA (trichloroacetic acid), which does not react with other phenolics. The pellets can then be redissolved in 0.1 N NaOH, and the reaction is performed with this new solution (Kaushik and Hynes 1971).

Another very popular and more sensitive method is based on the non-specific binding of Coomassie Blue (Brilliant Blue) to almost all proteins in an approximately stoichiometric manner (Bradford 1976). The binding results in a shift in the absorption maximum of the dye from 465 to 595 nm. This assay is generally less susceptible to interference by other substances than the Lowry or Biuret method, but strongly alkaline conditions and high concentrations of detergents can distort the result.

A final option is the bicinchoninic acid (BCA) protein assay. As noted above, Cu^{+2} is reduced to Cu^{+1} in an alkaline medium via the biuret reaction. When mixed with BCA, 2 molecules of BCA chelate with 1 cuprous ion (Cu^{+1}) to form a purple-coloured product that exhibits strong absorption at 562 nm that is linear over a large concentration range. There is no final colour with this method, i.e. the colour continues to develop over time. However, the rate of colour development/change is slow enough that many samples can be assayed together (https://tools.thermofisher.com/content/sfs/manuals/MAN0011430_Pierce_BCA_Protein_Asy_UG.pdf). The colour formation is strongly influenced by four amino acid residues (cysteine, cystine, tyrosine, tryptophan). Unlike the Coomassie dye-binding methods, the universal peptide backbone also contributes to colour formation, which minimizes variability due to differences in amino acid sequences (Wiechelman et al. 1988).

The results from samples are usually compared to a calibration curve with a defined protein. For greatest accuracy, the calibration protein should be one that dominates in the sample. This has rarely been attempted in ecological studies. For convenience, bovine serum albumin (BSA) has most often been used. Since leaf litter contains a mix of proteins, which changes during senescence and decomposition (Woo et al. 2013), any estimates of total protein based on a single standard are approximations.

This chapter introduces the Pierce BCA protein assay to follow changes in protein concentrations during leaf senescence and decomposition. The protocol has been modified from Smith et al. (1985) and Baerlocher et al. (2004). It involves extraction of total protein (i.e. both hydrophobic and hydrophilic proteins) from both leaf material and microorganisms such as fungi and bacteria, should the leaf be colonized.

2 Equipment, Chemicals and Solutions

2.1 Equipment and Material

- Leaves of various stages
- Thermos to carry liquid nitrogen (optional)
- Freezer at −80 °C (optional)
- Analytical balance
- Drying oven
- Mortar and pestle
- Sterile sand
- Microcentrifuge, capable of 12,000 *g*
- Eppendorf centrifuge tubes (1.5 ml)
- Ultrasonication probe (e.g. Sonic 300 Dismembrator, Artek Systems, 300 W)
- Spectrophotometer
- Cuvettes (1 cm)
- Liquid nitrogen (optional)
- Crushed ice
- Bowl
- Hotplate and boiling water bath
- Waterbath at 37 °C
- Parafilm
- Brown bottle
- Test tubes or small bottles with capacity of at least 3 ml
- 200 μl pipettor
- Graduated cylinder, or 1000 μl pipettor
- Scanner (optional)
- Glass fibre filters (optional)

2.2 Chemicals (Analytical Grade)

- Trizma® base
- Sucrose
- Ethylenediaminetetraacetic acid (EDTA)
- Sodium dodecyl sulphate (SDS)
- Double distilled water (ddH_2O)
- Bovine serum albumin (BSA)
- Aqueous 0.1 N NaOH
- Sodium tartrate ($C_4H_4O_6Na_2$)
- Sodium carbonate monohydrate (Na_2CO_3-H_20)
- Sodium bicarbonate ($NaHCO_3$)
- Copper sulphate pentahydrate ($CuSO_3$-$5H_2O$)

2.3 Solutions

- Solubilization buffer: dissolve 100 nM Trizma base, 160 nM sucrose, 1 nM EDTA, 1% SDS (w:v) in ddH_2O; store at 4 °C. Add 0.5% SDS immediately before use.
- Standard solution of bovine serum albumin (BSA):
  1. 20 mg BSA per 10 ml of the same diluents for samples (solubilization buffer)
  2. Dilute to 200, 400, 600, 800, 1000, and 1200 μg/ml
- Assay reagent A:
  1. Aqueous solution of 1% BCA, disodium salt ($BCA\text{-}Na_2$)
  2. 2% $Na_2CO_3\text{-}H_20$
  3. 0.16% $Na_2CO_3\text{-}H_20$
  4. 0.4% NaOH
  5. 0.95% $NaHCO_3$ with pH adjusted to 11.25 (using aqueous NaOH)
- Assay reagent B:
  1. 4% $CuSO_3\text{-}5H_2O$
- Standard working reagent (S-WR):
  1. Mix 50 vol of reagent A with 1 vol of reagent B

Reagents A and B are stable indefinitely at room temperature and are commercially available. Standard Working Reagent (S-WR) is prepared weekly or as needed.

3 Experimental Procedures

3.1 Leaf Protein Extraction

1. Collect leaves in the field. With senescent leaves, it is best to place them immediately in liquid nitrogen and in the laboratory, store at −80 °C to minimize protease activity.
2. When ready for extraction, a sample is gently cleaned and dried with paper towels.
3. With scissors or a cork borer, cut out between 0.5 and 2 cm^2 of leaf tissue, and determine its fresh mass to the nearest 0.1 or 0.01 mg. The sample is flattened under a glass plate in a scanner and scanned to determine leaf area.
4. Dry at 50 °C to constant mass (48 h) and reweigh to determine dry mass.
5. Grind each sample for ca. 10 min into a fine powder using a clean mortar and pestle, liquid nitrogen (optional) and 50 mg sand.

6. If living or senescent leaves are used, the samples should not be allowed to thaw at any time from this point on, as this may activate proteases. Additional liquid nitrogen must therefore periodically be added to the sample as it is being ground. If the sample strongly adheres to the pestle, a longer drying period may be required.
7. Transfer powdered sample to a 1.5 ml Eppendorf centrifuge tube.
8. Clean mortar and pestle by adding another 50 mg of sand, grind for an additional 2 min and transfer powder to the same Eppendorf tube. Repeat a third time. Periodically add more liquid nitrogen. Keep the Eppendorf tubes on ice at all times.
9. Add 250 μl of solubilization buffer to each Eppendorf microcentrifuge tube and vortex.
10. Add liquid nitrogen to a mortar, and then dip the Eppendorf microcentrifuge tube into the nitrogen freezing the contents. Sonicate the tube until the contents have thawed.
11. Repeat the freeze/thaw step.
12. Incubate tubes in a water bath at 90 °C for 5 min. Be careful that the tubes do not burst open, spraying some of their contents. Boil-proof tubes are available. Alternatively, secure the tube caps with parafilm or labelling tape. Centrifuge tubes for 5 min at 12,000 *g*.
13. Apportion the supernatant with the protein into two tubes, and store in a freezer (preferably −80 °C) until use.

3.2 Spectrophotometric Analysis

1. Pipet duplicate 0.1 ml aliquots of the BSA standard solution, or of sample extracts, into 5 ml test tubes.
2. Pipet 2 ml of S-WR into each test tube with 0.1 ml of standard. Make sure solutions are well mixed.
3. Place tubes in 37 °C water bath. Incubate for 30 min.
4. Adjust spectrophotometer absorbance to 0 in a cuvette with S-WR solution at 562 nm. The absorbance reading of the blank is subtracted from each standard.
5. Calculate protein concentration as BSA equivalents from absorbance readings of sample and standard curve.

4 Final Remarks

The Pierce BCA assay only gives a linear standard curve within a relatively narrow range. For highest accuracy and precision, sample concentrations should be adjusted to fall within that range.

The protein extraction procedure described here breaks disulphide bonds. This allows separating and analysing the extracted mix of proteins based on their sizes without additional treatment. For example, the mix can be analysed by sodium dodecyl sulphide polyacrylamide gel electrophoresis (SDS-PAGE).

There are several alternatives for extracting and measuring proteins. Pierce's BCA is believed to generally give the most consistent results (A. Cockshutt, Mt. Allison University, pers. comm.), though there is no published comparison of the various approaches as applied to decaying leaves. Interference by other leaf components is always a possibility, for example, by copper-chelating compounds. Strategies to minimize these effects are listed on the company's website (https://tools.thermofisher.com/content/sfs/manuals/MAN0011430_Pierce_BCA_Protein_Asy_UG.pdf).

References

Allan, J. D., & Castillo, M. (2007). *Stream ecology. Structure and function of running waters* (2nd ed.). Dordrecht: Springer.

AOAC. (1990). Protein (crude) determination of animal feed: Copper catalyst Kjeldahl method (984.13). In *Official methods of analysis of the Association of Official Analytical Chemists* (15th ed.). Gaithersburg: AOAC International.

Baerlocher, M. O., Campbell, D. A., & Ireland, R. J. (2004). Developmental progression of photosystem II electron transport and CO_2 uptake in *Spartina alterniflora*, a facultative halophyte, in a northern salt marsh. *Canadian Journal of Botany, 82*, 365–375.

Bailey, J. L. (1962). *Techniques in protein chemistry*. New York: Elsevier.

Bärlocher, F. (1983). Seasonal availability and digestibility of CPOM in a stream. *Ecology, 64*, 1266–1272.

Bärlocher, F., & Howatt, S. L. (1986). Digestion of carbohydrate and protein by *Gammarus mucronatus* say (Amphipoda). *Journal of Experimental Marine Biology and Ecology, 104*, 229–237.

Bärlocher, F., Tibbo, P. G., & Christie, S. H. (1989). Formation of phenol-protein complexes and their use by two stream invertebrates. *Hydrobiologia, 173*, 243–249.

Bradford, M. M. (1976). A rapid and sensitive method for the quantitation of microgram quantities of protein using the method of protein dye bonding. *Analytical Biochemistry, 72*, 248–254.

Craig, D., Ireland, R. J., & Bärlocher, F. (1989). Seasonal variation in the organic composition of seafoam. *Journal of Experimental Marine Biology and Ecology, 130*, 71–80.

Fiset, C., Liefer, J., Irwin, A. J., & Finkel, Z. V. (2017). Methodological biases in estimates of macroalgal macromolecular composition. *Limnology and Oceanography. Methods, 15*, 618–630.

Gessner, M. O. (1991). Differences in processing dynamics of fresh and dried leaf litter in a stream ecosystem. *Freshwater Biology, 26*, 387–398.

Kaushik, N. K., & Hynes, H. B. N. (1971). The fate of the dead leaves that fall into streams. *Archiv für Hydrobiologie, 68*, 465–515.

Lowry, O. H., Rosebrough, N. J., Farr, A. L., & Randall, R. J. (1951). Protein measurement with the Folin phenol reagent. *Journal of Biological Chemistry, 193*, 265–275.

Odum, W. E., Kirk, P. W., & Zieman, J. C. (1979). Non-protein nitrogen compounds associated with particles of vascular plant detritus. *Oikos, 32*, 363–367.

Scopes, R. (1982). *Protein purification*. New York: Springer.

Smith, P. K., Krohn, R. I., Hermanson, A. K., Mallia, F. H., Gartner, F. H., Provenzano, E. K., Fujimoto, N. M., Goeke, B. J., & Klenk, D. C. (1985). Measurement of protein using bicinchoninic acid. *Analytical Biochemistry, 150*, 76–85.

Suberkropp, K., Godshalk, G. L., & Klug, M. J. (1976). Changes in the chemical composition of leaves during processing in a woodland stream. *Ecology, 57*, 720–727.

Swain, T. (1979). Tannins and lignins. In G. A. Rosenthal & D. H. Janzen (Eds.), *Herbivores: Their interactions with secondary plant metabolites* (pp. 657–682). New York: Academic Press.

Wiechelman, K., Braun, R. D., & Fitzpatrick, J. D. (1988). Investigation of the bicinchoninic acid protein assay: Identification of the groups responsible for color formation. *Analytical Biochemistry, 175*, 231–237.

Woo, H. R., Kim, H. J., Nam, H. G., & Lim, P. O. (2013). Plant leaf senescence and death – regulation by multiple layers of control and implications for aging in general. *Journal of Cell Science, 126*, 4823–4833.

Chapter 13
Free Amino Acids

Shawn D. Mansfield and Mark O. Baerlocher

Keywords Amino acid extraction · Derivatization · HPLC · Leaching · Litter chemistry · Litter quality · Ninhydrin · Nitrophenylisothiocyanate

1 Introduction

Amino acids are one of the components that may contribute to the leaching effect most commonly observed in dried leaves (Gessner 1991). Amino acids, along with soluble carbohydrates (Chap. 15), represent a readily assimilated and digested source of nutrients for most organisms. This is in contrast to low molecular weight phenolics, which are also liberated during leaching, but are not easily metabolized by invertebrates or microorganisms. Several amino acids are essential to many animals, i.e. they are needed for protein synthesis but cannot be made de novo, and thus have to be provided in the diet. The number and type of essential amino acids varies among taxa and also depends on other nutrients supplied in the diet (Croset et al. 2016; Hill et al. 2016).

The total concentration of free amino acids in decaying leaves represents a dynamic equilibrium between losses (e.g. by leaching) and abiotic or biotic uptake. Positively charged amino acids are generally adsorbed more readily to surfaces (Armstrong and Bärlocher 1989a), and this adsorption is affected by the concentration of Ca^{2+} in the stream water, which selectively influences the behaviour of different amino acids (Armstrong and Bärlocher 1989b). Uptake of amino acids from stream water by aquatic hyphomycetes may affect the protein level of decaying leaves, as well as the fungal species that successfully colonize newly immersed leaves (Bengtsson 1988).

S. D. Mansfield (✉)
Department of Wood Science, University of British Columbia, Vancouver, BC, Canada
e-mail: shawn.mansfield@ubc.ca

M. O. Baerlocher
Department of Vascular and Interventional Therapies, Royal Victoria Hospital, Barrie, ON, Canada
e-mail: mark.baerlocher@alumni.utoronto.ca

F. Bärlocher et al. (eds.), *Methods to Study Litter Decomposition*,
https://doi.org/10.1007/978-3-030-30515-4_13

There are many approaches to identify and quantify amino acids in sample extracts. A quick and simple estimate of total amino acid concentration is based on the reaction of reduced ninhydrin with the amino group (Rosen 1957). A more accurate technique uses pre-column derivatization and high performance liquid chromatography (HPLC), to facilitate the identification of individual amino acids (Craig et al. 1989; Cohen 1990; Hodisan et al. 1998). The derivatized amino acids can be separated and quantified by fluorescence (Hodisan et al. 1998) or UV detection (Cohen 1990).

The details of the HPLC protocol used depend on the type and availability of HPLC systems. The following protocol is intended to serve as a general guide and may have to be adapted to the particular equipment available in the laboratory. It should also be noted that polypeptides and proteins can be hydrolysed into their individual amino acids (boiling in 6 M HCl under vacuum for 20–24 h; Craig et al. 1989). This protocol results in the conversion of tryptophan, glutamine and asparagine to their respective acids, and these amino acids are therefore not measured. However, the hydrolysis step is not included in the current protocol, as the intention here is to determine the identity and quantity of free amino acids, in contrast to amino acids present in peptides or proteins which are referred to as bound amino acids. Therefore, the protocol presented here includes a step which removes polypeptides and proteins to prevent contamination of the HPLC column or co-elution. The method has been adopted from Hodisan et al. (1998).

2 Equipment, Chemicals and Solutions

2.1 *Equipment and Material*

- Latex gloves
- Analytical balance
- Desiccator (containing phosphorus pentoxide or another desiccant)
- Micropipettors
- Scissors or cork borer
- Vortex
- Drying oven (50 °C)
- Mortar and pestle
- Sterile sand
- Centrifuge (12,000 *g*)
- Thermos to carry liquid nitrogen (optional)
- Liquid nitrogen (optional)
- Freezer at −80 °C (optional)
- Scanner (optional)
- HPLC system to run gradient elution profiles and equipped with a UV detector

- C-18 reversed-phase column, i.e. Lichrospher ODS 100 (5 μm particle diameter; 250 × 4.6 mm)
- C-18 guard column
- pH meter,
- HPLC vials and caps.
- HPLC filters (0.45 μm pore size)
- Glassware such as test tubes, centrifuge tubes, graduated cylinders, acid-washed (e.g. soaked in 10% nitric acid overnight and then rinsed thoroughly with deionized distilled water)
- Organic matter samples (i.e. leaves harvested from trees or recovered from a stream or soil)

2.2 *Chemicals and Solutions*

- Double distilled water (ddH$_2$O)
- HCl (1%)
- Acetonitrile (HPLC grade)
- 4-Nitrophenylisothiocyanate in acetonitrile (NPITC), 50 mM stock solution prepared daily and stored in a dark bottle at 4 °C
- 10% triethylamine (TEA) in acetonitrile
- Derivatization reagent (working solution): 490 μl of the NPITC stock solution, 50 μl of the 10% TEA stock solution, and 50 μl of ddH$_2$O water
- Toluene (HPLC grade)
- Glacial acetic acid
- Sodium acetate (0.14 M)
- Amino acid standards (individual amino acid standards or, more efficiently, a commercially available standard mix with known amounts of several amino acids, diluted in ddH$_2$O to a known concentration of between 50 and 250 μM; 100 μM is suggested)

3 Experimental Procedure

3.1 *Extraction of Free Amino Acids*

1. Collect leaves in the field. When using living or senescent leaves, place them immediately in liquid nitrogen, and store them at −80 °C in the laboratory to minimize protease activity.
2. When ready for extraction, gently clean sample and dry with paper towels.
3. With scissors or a cork borer, cut out between 0.5 and 2 cm^2 of leaf tissue, and determine its fresh mass to the nearest 0.1 or 0.01 mg.

4. Flatten sample under a glass plate in a scanner and scan to determine leaf area.
5. Dry at 50 °C to constant weight and reweigh to determine dry mass.
6. Grind sample for ca. 10 min into a fine powder using a clean mortar and pestle, liquid nitrogen, and 50 mg sand.
7. If living or senescent leaves are employed, ensure samples are not thawed at any time from this point on, as this may activate proteases. Additional liquid nitrogen must therefore periodically be added to the sample as it is being ground. If the sample strongly adheres to the pestle, a longer drying period may be required.
8. Transfer powdered sample to a 10 ml centrifuge tube.
9. Add 5 ml of 0.1% HCl and vortex.
10. Centrifuge tube for 5 min at 12,000 *g*.
11. The supernatant is unlikely to contain large amounts of proteins. Nevertheless, preferably use an HPLC guard column to prevent clogging the column.
12. Store the supernatant in a freezer (preferably at −80 °C) until use.

3.2 *Derivatization of Amino Acids*

1. Freeze-dry samples and standards and stored overnight in a desiccator before derivatization.
2. Record sample weight from freeze-dried sample stock to the nearest 0.01 mg.
3. Derivatize dried samples and standards in Pyrex tubes by adding 25 µl of the derivatization working solution.
4. Vortex for 1 min and allow to react for 10 min.
5. After exactly 10 min, add 100 µl of ddH_2O.
6. Extract reaction mixture with 125 µl of toluene with gentle mixing.
7. Allow phases to separate, collect aqueous layer and discard organic layer in appropriate container.
8. Filter aqueous sample through 0.45 µm filter into an HPLC vial and cap.

3.3 *HPLC Analysis and Amino Acid Determination*

1. Degas eluants (i.e. solvents used as mobile phase) before HPLC analysis.
2. Set HPLC to the conditions suggested in Table 13.1.
3. Run standards on HPLC.
4. Determine retention time and peak area for each amino acid in the standard solution.
5. Correct for dilutions if any is required, and use this information to determine the concentration of each individual amino acid in the standard.

Table 13.1 HPLC conditions for amino acids determination

Parameter	Condition
Eluent A	94% 0.14 M sodium acetate, 0.05% TEA (v:v) in 6% acetonitrile, pH adjusted to 6.4 with glacial acetic acid
Eluent B	60% acetonitrile in deionized distilled dd H_2O
Column equilibration before injection	5 min with 85% eluant A and 15% eluant B
Gradient	Linear gradient from 85% eluant A and 15% eluant B to 40% eluant A and 60% eluant B over 30 min; wash column with 100% eluant B for 5 min, and re-equilibrate for 10 min at 85% eluant A and 15% eluant B
Flow rate	1 ml min^{-1}
Column temperature	45 °C
Detection	UV detection at 340 nm (alternatively use 254 nm)
Injection volume	20 μl

6. Prepare standard curves by plotting areas under the peaks versus concentrations of each amino acid in the standards mixture.
7. Run each sample, always using the same volumes as for the calibration. If the sample peaks are off scale, dilute with ddH_2O water.
8. Determine which peak represents which amino acid (based on retention times), and subsequently determine sample concentrations (μmol μl^{-1}) based on the standard curves for each amino acid.
9. Adjust the total concentration of each individual amino acid in the sample for the total sample volume and divide by the total weight of the sample material used (mg).
10. The individual amino acids are then recorded as μg mg^{-1} of the original sample. Addition of all individual concentrations gives the total concentration of free amino acids in the sample.

4 Final Remarks

Many alternative approaches and procedures exist to analyse amino acids by HPLC (Hodisan et al. 1998). For example, if both primary and secondary amines are to be detected, ninhydrin is a better choice than the 4-nitrophenylisothiocyanate (NPITC) used in the protocol presented here to derivatize amino acids. However, the various methods typically measure most but not all amino acids.

The risk of contamination is considerable. To minimize this risk, it is critical to use high-quality chemicals and very clean, acid-washed glassware. Latex gloves should be worn at all times.

References

Armstrong, S. M., & Bärlocher, F. (1989a). Adsorption and release of amino acids from epilithic biofilms in streams. *Freshwater Biology, 22*, 153–159.

Armstrong, S. M., & Bärlocher, F. (1989b). Adsorption of three amino acids to biofilms on glass-beads. *Archiv für Hydrobiologie, 115*, 391–399.

Bengtsson, G. (1988). The impact of dissolved amino acids on protein and cellulose degradation in stream waters. *Hydrobiologia, 164*, 97–102.

Cohen, S. A. (1990). Analysis of amino acids by liquid chromatography after pre-column derivatization with 4-nitrophenylisocyanate. *Journal of Chromatography, 512*, 283–290.

Craig, D., Ireland, R. J., & Bärlocher, F. (1989). Seasonal variation in the organic composition of seafoam. *Journal of Experimental Marine Biology and Ecology, 130*, 71–80.

Croset, V., Schleyer, M., Arguello, J. R., Gerber, B., & Benton, R. (2016). A molecular and neuronal basis for amino acid sensing in the *Drosophila* larva. *Scientific Reports, 6*, 34871.

Gessner, M. O. (1991). Differences in processing dynamics of fresh and dried leaf litter in a stream ecosystem. *Freshwater Biology, 26*, 387–398.

Hill, R. W., Wyse, G. A., & Anderson, M. (2016). *Animal physiology. Sinauer associates*. Oxford: Oxford University Press.

Hodisan, T., Culcea, M., Cimpoiu, C., & Cot, A. (1998). Separation, identification and quantitative determination of free amino acids from plant extracts. *Journal of Pharmaceutical and Biomedical Analysis, 18*, 319–323.

Rosen, H. (1957). A modified ninhydrin colorimetric analysis for amino acids. *Archives of Biochemistry and Biophysics, 67*, 10–15.

Chapter 14
Determination of Total Carbohydrates

Shawn D. Mansfield

Keywords Acid hydrolysis · Anion exchange chromatography · Carbohydrates · Cellulose · Gas chromatography · GC · Hemicellulose · High-performance liquid chromatography · HPLC · Lignocellulose · Litter chemistry · Litter quality · Plant cell walls · Sugar monomer quantification

1 Introduction

The plant cell wall is a composite of cellulose microfibrils intricately associated with other polysaccharides, commonly referred to as hemicelluloses, and embedded in an amorphous matrix of lignin (Bidwell 1974). Many other non-structural constituents, organic and inorganic, are present in the cell wall, most notably starch and proteins, which may also be linked to carbohydrates. The proportions of all these components vary among species, age, geographical location, and growing conditions (Groover et al. 2010) and will change during decomposition (Skyba et al. 2013; Skyba et al. 2016) due to differential attack by various degradative enzymes (Chap. 41). The intricate association of various macromolecules in plant cell walls makes analysing the individual components a difficult task.

The main component of plant cell walls is cellulose, a long, unbranched, homopolymer of D-glucose units linked by β-1,4-glycosidic bonds. The linear polymers form chains varying in size that can exceed 10,000 glucose residues and associate together to form larger macromolecules (Hon and Shiraishi 1991; Mansfield et al. 1999; Joshi and Mansfield 2007). Like cellulose, the amylose component of starch consists of 1,4-glycosidic-linked glucopyranose units. However, in starch these units are α- rather than β-anomers. This stereochemical characteristic gives starch its unique physical and chemical behaviour. Amylose generally occurs as a helix in the solid state and sometimes in solution. A second

S. D. Mansfield (✉)
Department of Wood Science, University of British Columbia, Vancouver, BC, Canada
e-mail: shawn.mansfield@ubc.ca

F. Bärlocher et al. (eds.), *Methods to Study Litter Decomposition*,
https://doi.org/10.1007/978-3-030-30515-4_14

major starch component in plants, amylopectin, is 1,4-α-D-glucan that is highly branched and often associated with pectin (Sjöström 1993).

Hemicelluloses are heteropolymeric polysaccharides of low molecular weight. Associated with cellulose and lignin, they play an integral role in defining cell wall characteristics. Most function as supporting material, but they can also significantly influence interactions between adjacent cells. In contrast to cellulose and starch, hemicelluloses are constructed from a number of different residues, the most common of which are D-xylose, D-mannose, D-galactose, D-glucose, L-arabinose, D-rhamnose, D-galacturonic acid, D-glucuronic acid and 4-O-methyl-D-glucuonic acid (Fengel and Wegener 1983; Sjöström 1993). The complexity and chemical nature of the hemicelluloses vary among both cell types and species. Hemicelluloses fall into four classes: unbranched chains, such as 1,4-linked xylans or mannans; helical chains, such as 1,3-linked xylans; branched chains, such as 1,4-linked galactoglucomannans; and pectic substances, such as polyrhamnogalacturonic acid. Many hemicelluloses also show substantial degrees of acetylation (Sjöström 1993; Johnson et al. 2017).

In evaluating cell wall carbohydrates, two approaches can be taken; one involves determining total available carbohydrates and ignores the composition and origin of the material, following a modified method of White and Kennedy (1986). For details of this method, follow the orcinol total sugar method described in Chap. 15, using the hydrolysed sample as an unknown. The second approach is more specific and quantifies the individual carbohydrates to calculate the total amount of polysaccharides in a substrate (Mansfield et al. 1997; Porth et al. 2013). However, both methods require that lignocellulose first be degraded from its macromolecular structure to its individual monomeric constituents.

2 Equipment and Chemicals to Degrade Lignocellulose

2.1 Equipment and Material

- Analytical balance
- Autoclave
- Desiccator (containing anhydrous calcium sulphate)
- Drying oven (at 105 °C)
- Grinding mill (i.e. Wiley Mill) or mortar and pestle
- Water bath (at 20 °C)
- Ice bath
- Test tubes

- Test tube rack
- Glass rods
- Septum-sealed serum bottles (150 ml)
- Medium coarseness sintered-glass crucibles (Gooch crucibles)
- Dried leaves at various stages of decay

2.2 Chemicals

- Standard sugars (e.g. glucose, galactose, arabinose, mannose, xylose, rhamnose)
- 72% sulphuric acid (665 ml conc. H_2SO_4 + 335 ml distilled H_2O made to 1 l in a volumetric flask)
- Deionized water
- Liquid nitrogen

3 Experimental Procedures to Degrade Lignocellulose

3.1 General

1. Grind a leaf sample to pass through a 0.5 mm mesh using a Wiley or similar mill. If a mill is unavailable, freeze-dry the sample and use a mortar and pestle to pulverize it in liquid nitrogen.
2. Dry the sample overnight at 105 °C.
3. If samples are not immediately used, store in a desiccator containing anhydrous calcium sulphate.
4. Weigh out 200 mg sample into a test tube and record weight to the nearest 0.1 mg.
5. Place the test tube in an ice bath.
6. Add exactly 3 ml of 72% (w/w) H_2SO_4. If multiple samples are to be hydrolysed, start time should be staggered appropriately (e.g. 2–3 min between samples) to ensure appropriate mixing and constant time for hydrolysis.
7. Mix and macerate the sample with a glass rod.
8. Immediately transfer the sample to a 20 °C water bath for exactly 2 h, with continuous mixing every 10 min using a glass rod.
9. After 2 h, transfer the content of the lignocellulose digestion to a serum bottle. Rinse out the reaction flask with exactly 112 ml of deionized H_2O and transfer washings to the serum bottle (total volume in bottle should be 115 ml; acid is now diluted to 3%), then seal with septum.

3.2 Sugar Standards

1. Prepare sugar standards as indicated in Table 14.1 to control for loss by decomposition during autoclaving and to generate standard curve for quantification. This step is only required if HPLC analysis is used (see below). Sugar type and concentration will vary depending on the type of sample (i.e. leaf material, wood, etc.)
2. Autoclave samples and sugar standards for 1 h at 121 °C and allow flasks to cool.
3. Vacuum-filter hydrolysates through dry, pre-weighed, sintered-glass crucibles (Gooch crucibles). Ensure all solids (acid-insoluble lignin) are recovered from the serum bottles by washing with small volumes (i.e. 2× 10 ml) of hydrolysate while filtering through crucible.
4. Quantify carbohydrates in hydrolysates by any of the methods described below.

4 Equipment and Chemicals to Determine Monomeric Carbohydrates by High Performance Anion Exchange Liquid Chromatography

4.1 Equipment

- HPLC with either electrochemical detection by pulsed amperometry, or refractive index detection
- HPLC filters (0.45 μm pore size)

Table 14.1 Standard sugar solutions in 50 ml deionized H_2O

Standard	Composition
Sugar stock solutions	Arabinose:200 mg l^{-1}
	Galactose:200 mg l^{-1}
	Glucose:4.0 g l^{-1}
	Xylose:1.2 g l^{-1}
	Mannose:1.2 g l^{-1}
High standard	30 ml sugar stock solution
	82 ml deionized H_2O
	3 ml 72% H_2SO_4
Medium standard	10 ml sugar stock solution
	102 ml deionized H_2O
	3 ml 72% H_2SO_4
Low standard	5 ml sugar stock solution
	107 ml deionized H_2O
	3 ml 72% H_2SO_4

- HPLC vials and caps
- Disposable syringes
- Micropipettors

4.2 *Chemicals*

- Deionized distilled water (degassed)
- 1 M NaOH solution (degassed)
- 200 mM NaOH solution (degassed)
- Internal standard: Fucose (5 mg ml^{-1}) to control for injection volume

5 Experimental Procedures to Determine Monomeric Carbohydrate by High Performance Anion Exchange Liquid Chromatography

5.1 *Sample and HPLC Preparation*

1. In an Eppendorf tube, add 950 μl of sample or standard and 50 μl of fucose internal standard, and then mix well with a vortex.
2. Use a disposable syringe to remove sample from Eppendorf tube, filter through 0.45 μm filter into HPLC vial, and cap.
3. Set HPLC for electrochemical detection using pulsed amperometry or refractive index detection (Table 14.2).

5.2 *Carbohydrate Determination*

1. Run standards and samples on HPLC (Fig. 14.1).
2. Normalize peak areas for internal standards (controls for injection variation).
3. Prepare standard curves by plotting area under the peak versus concentration for each sugar in the standards mixture.
4. Determine sample concentrations from area using linear regression for unknown sugars; results will be in mg ml^{-1} of each monomeric sugar.
5. Correct these values for the conversion from polymeric to monomeric constituents, since the standard curves are generated from anhydrous sugar standards. That is, account for the incorporation of a water molecule for each bond degraded during the hydrolysis of polymeric carbohydrates to their corresponding monomeric moiety. Correct the concentration of all hexoses

Table 14.2 HPLC conditions for monomeric carbohydrate determination

Conditions	Electrochemical detection using pulsed amperometry (gold electrode)	Refractive index detection
Column wash	15 min, 250 mM NaOH (degassed)	–
Column equilibration (prior to injection)	10 min of deionized H_2O	
Mobile phase	Deionized H_2O (degassed)	Deionized H_2O (degassed)
Flow rate	1 ml min^{-1}	0.4 ml min^{-1}
Column	Dionex PA-1	Biorad HPX-87P
Post-column mobile phase	200 mM NaOH (at 0.5 ml min^{-1})	–
Injection volume	20 μl	20 μl

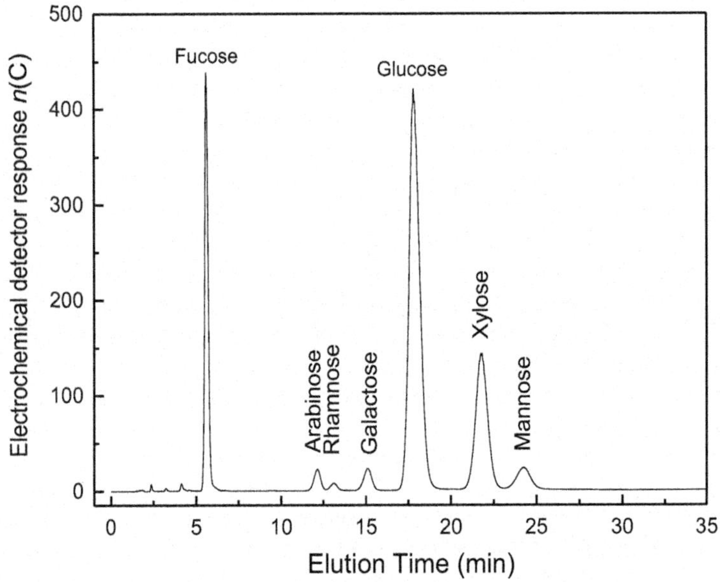

Fig. 14.1 Chromatograph of neutral wood carbohydrates determined by ion exchange chromatography with electrochemical detection (pulsed amperometry)

(glucose, mannose, galactose, etc.) and pentoses (xylose, arabinose, etc.) by multiplying the determined concentration (mg ml^{-1}) by 0.90 and 0.88, respectively.

6. Adjust the total concentration in the sample for the total volume of hydrolysate (i.e. multiply by 115 ml) and then divided by the total sample mass (mg). The individual sugars are recorded as mg mg^{-1} of original sample. Addition of all individual monomers will result in the total carbohydrate content of a sample.

6 Equipment and Chemicals to Determine Monomeric Carbohydrate by Gas Chromatography

6.1 Equipment and Material

- Gas chromatograph (GC)
- Disposable filters (0.45 μm pore size)
- GC vials and caps
- Disposable syringes
- Separatory funnel
- 10 ml reaction vials with Teflon caps
- Micropipettors
- Beaker

6.2 Chemicals

- Deionized distilled water
- 3 M NH_4OH
- 2.8 M KBH_4 dissolved in 3 M NH_4OH
- Glacial acetic acid
- 1-Methylimidazole
- Acetic anhydride
- Dichloromethane
- Internal standard: inositol (5 mg ml^{-1})

7 Experimental Procedures to Determine Monomeric Carbohydrate by Gas Chromatography

7.1 Sample and GC Preparation

1. In a 10 ml reaction vial, add 200 μl of hydrolysed sample or standard and 50 μl of inositol internal standard.
2. Add 40 μl of 3 M NH_4OH and 100 μl of 2.8 M KBH_4 solutions, and incubate for 90 min at 40 °C.
3. Stop reaction after 90 min with 100 μl of glacial acetic acid.
4. Add 500 μl of 1-methylimidazole and 2 ml of acetic anhydride to complete the acetylation of the sugar alcohols.
5. Vortex and allow reacting for 30 min at room temperature.
6. Add 5 ml of distilled deionized water to stop the reaction.

Table 14.3 Conditions for GC analysis with flame ionization detection

Parameter	Condition
Carrier gas	Helium
Flow rate	30 cm s^{-1}
Split ratio	1:25
Column	DB-225 capillary column (30 m × 0.25 mm i.d.)
Column temperature	220 °C
Detection	Flame ionization detection (FID)
Detector temperature	240 °C
Injector temperature	240 °C
Injection volume	2 μl
Run time	30 min

7. Dispense reaction mixture to a separatory funnel, where 2 ml of dichloromethane are added and mixed vigorously.
8. Allow the phases in the mixture to separate, and remove approximately 1.5 ml of the dichloromethane phase.
9. Using a disposable syringe, filter the sample through a 0.45 μm filter into a GC vial, and cap vial.
10. Set the GC analyser to the conditions indicated in Table 14.3.

7.2 Carbohydrate Determination

1. Run standards and samples on GC.
2. Follow steps 2–6 in Sect. 5.2.

References

Bidwell, R. G. S. (1974). *Plant physiology*. New York: MacMillian.

Fengel, D., & Wegener, G. (1983). *Wood: Chemistry, ultrastructure, reactions*. New York: Walter de Gruyter.

Groover, A. T., Nieminen, K., Helariutta, Y., & Mansfield, S. D. (2010). Wood formation in poplar. In S. Jansson, R. P. Bhalerao, & A. T. Groover (Eds.), *Plant genetics/genomics vol. 8: Genetics and genomics of Populus* (pp. 201–224). New York: Springer.

Hon, D. N.-S., & Shiraishi, N. (1991). *Wood and cellulosic chemistry*. New York: Marcel Dekker.

Johnson, A. M., Kim, H., Ralph, J., & Mansfield, S. D. (2017). Natural acetylation impacts carbohydrate recovery during deconstruction of *Populus trichocarpa* wood. *Biotechnology for Biofuels, 10*, 48.

Joshi, C. P., & Mansfield, S. D. (2007). The cellulose paradox – Simple molecule, complex biosynthesis. *Current Opinion in Plant Biology, 10*, 220–226.

Mansfield, S. D., de Jong, E., & Saddler, J. N. (1997). Cellobiose dehydrogenase, an active agent in cellulose depolymerization. *Applied and Environmental Microbiology, 63*, 3804–3809.

Mansfield, S. D., Mooney, C., & Saddler, J. N. (1999). Substrate and enzyme characteristics that limit cellulose hydrolysis. *Biotechnology Progress, 15*, 804–816.

Porth, I., Klápště, J., Skyba, O., Lai, B. S. K., Geraldes, A., Muchero, W., Tuskan, G. A., Douglas, C. J., El-Kassaby, Y. A., & Mansfield, S. D. (2013). *Populus trichocarpa* cell wall chemistry and ultrastructure trait variation, genetic control, and genetic correlations. *New Phytologist, 197*, 777–790.

Sjöström, E. (1993). *Wood chemistry: Fundamentals and applications* (2nd ed.). London: Academic.

Skyba, O., Douglas, C. J., & Mansfield, S. D. (2013). Syringyl-rich lignin renders poplars more resistant to degradation by wood decay fungi. *Applied and Environmental Microbiology, 79*, 2560–2571.

Skyba, O., Cullen, D., Douglas, C. J., & Mansfield, S. D. (2016). Gene expression patterns of wood decay fungi *Postia placenta* and *Phanerochaete chrysosporium* are influenced by wood substrate composition during degradation. *Applied and Environmetal Microbiology, 82*, 4387–4400.

White, C. A., & Kennedy, J. F. (1986). Oligosaccharides. In M. F. Chaplin & J. F. Kennedy (Eds.), *Carbohydrate analysis: A practical approach* (pp. 37–54). Oxford: IRL Press.

Chapter 15
Determination of Soluble Carbohydrates

Letitia Da Ros, Faride Unda, and Shawn D. Mansfield

Keywords Carbohydrate extraction · High performance liquid chromatography · HPLC · Leaching · Litter chemistry · Litter quality · Monomeric carbohydrates · Oligomeric carbohydrates · Oligosaccharides · Orcinol reagent · Spectrophotometry · Sugar monomer quantification

1 Introduction

Leaf litter and other plant material consist primarily of structural polysaccharides and lignin, neither of which is readily accessible to stream invertebrates (Bärlocher and Sridhar 2014). However, more easily digestible soluble carbohydrates, such as sucrose and other tri-, di- and monosaccharides (i.e. raffinose, fructose, glucose) are also present in notable concentrations. Initially, these compounds may account for up to 16% of total dry mass (e.g. in hickory leaves; Suberkropp et al. 1976), but leaching from dead leaves can reduce this value by ≥80% within a few days (Gessner 1991). However, the extent of leaching may be significantly influenced by treatment of the leaves before immersion in water (Chaps. 5, 6; Gessner 1991; Bärlocher 1997).

During decomposition, microbial enzymes attack plant cell wall polymers, releasing a mixture of oligomeric and monomeric carbohydrates, which again are more accessible to invertebrates than the original macromolecular compounds (Bärlocher and Porter 1986). In addition, fungi colonizing leaves (which can account for 10–20% of litter dry mass at intermediate stages of decay; Gessner 1997; Kuehn 2016) contain soluble carbohydrates in their mycelium.

L. Da Ros · F. Unda · S. D. Mansfield (✉)
Department of Wood Science, University of British Columbia, Vancouver, BC, Canada
e-mail: shawn.mansfield@ubc.ca

F. Bärlocher et al. (eds.), *Methods to Study Litter Decomposition*,
https://doi.org/10.1007/978-3-030-30515-4_15

Analysis of total soluble carbohydrates facilitates the quantification of nutritionally valuable carbon fractions of leaf material. Identification of the individual compounds, combined with analyses of hydrolysable polysaccharides (i.e. remaining polymeric carbohydrates), facilitates a detailed characterization of the sequence of enzymatic breakdown of leaf constituents. The same methods can be modified to assess activities of selected degradative enzymes present in microorganisms or invertebrates (Chap. 41).

Two approaches can be taken to analyse soluble carbohydrates. The first involves determining all soluble carbohydrates while ignoring their composition. The procedure presented here for this bulk analysis follows a modified method of White and Kennedy (1986). The second approach specifically quantifies individual monosaccharides, which can then be used to calculate the total amount of soluble carbohydrates present in a sample (Cataldi et al. 2000). Both methods require that the soluble sugars first be extracted from the lignocellulosic material; the extraction procedure described below follows a modified protocol of Coleman et al. (2006). This method has recently been shown to compare well with several alternatives that have been historically used to quantify soluble carbohydrates in plant biomass (Quentin et al. 2015).

2 Equipment, Chemical, and Solutions

2.1 Equipment and Material

- Freeze-drier
- Analytical balance
- Spectrophotometer
- Vacuum concentrator
- Desiccators (containing anhydrous calcium sulphate or similar desiccant)
- Mortar and pestle
- Acid-washed glass test tubes (10 ml; wash with 10% nitric acid overnight, then rinse thoroughly with distilled water)
- Polypropylene conical centrifuge tubes (15 and 50 ml)
- Microcentrifuge tubes (2 ml)
- Test tube rack
- Hot water bath or heated test tube reactor
- Thermometer
- Ice water bath
- Vortex
- Fridge (4 °C)
- Micropipettors
- Cuvettes (disposable ones are suitable)
- Spectrophotometer (set at 540 nm)

- Laboratory timer or stop watch
- Aluminium foil
- High-performance liquid chromatography (HPLC) with electrochemical detector using pulsed amperometry
- HPLC filters (0.45 μm pore size)
- HPLC vials and caps
- Disposable syringes

2.2 *Chemicals*

- Selected sugars or sugar alcohols to prepare standards: glucose, sucrose, fructose, galactose, raffinose, stachyose, myoinositol, glucuronic acid, galacturonic acid
- Liquid nitrogen
- Methanol/chloroform/water (12:5:3; v:v:v) solution
- Distilled water (degassed)
- 0.2 M sodium hydroxide (NaOH; prepared in a volumetric flask and degassed)
- 10 mM sodium acetate (NaOAc; prepared in a volumetric flask and degassed)
- Internal standard: galactitol (10 mg ml^{-1})
- 10% (w:w) nitric acid
- Sugar standard: glucose or other sugar dissolved in water or appropriate buffer (e.g. 50 mM sodium phosphate or sodium acetate), with concentrations ranging from 10 to 120 μg ml^{-1}
- Freshly prepared 0.2% orcinol reagent (2 g l^{-1} orcinol dissolved in concentrated sulphuric acid), which can be stored for up to 1 week at 4 °C

3 Experimental Procedures

3.1 *Sample Preparation*

1. Wash leaf litter or similar sample thoroughly with distilled water.
2. Freeze-dry sample overnight.
3. Using a mortar and pestle, grind freeze-dried material in liquid nitrogen into a powder.
4. Store ground material in sample vials in a desiccator (with anhydrous calcium sulphate) until extraction.
5. Weigh out 50 mg of pulverized sample into a 15 ml polypropylene tube, and record weight to nearest 0.1 mg. This should be done in at least duplicate for each sample.

6. To each 15 ml tube containing ground sample, add 50 μl of galactitol as internal standard.
7. Add 4 ml of methanol/chloroform/water (12:5:3) solution to each tube, mix, close the lid, and let stand overnight at 4 °C.
8. Remove samples from fridge and mix well with a vortex.
9. Centrifuge samples (5000 *g*) for 10 min at 4 °C, and then pipet supernatant into a 50 ml polypropylene tube.
10. Add 4 ml of methanol/chloroform/water (12:5:3) solution to the pellet, vortex and centrifuge for 10 min. Remove supernatant and pool with sample in 50 ml tube. Repeat a second time and pool.
11. Add 5 ml of distilled water to the pooled supernatant, cap, and mix thoroughly.
12. Centrifuge samples (5000 *g*) for 4 min at 4 °C to allow for phase separation.
13. Collect 2 ml of the upper polar phase into a microcentrifuge tube. Collect the remaining upper polar phase into a 15 ml tube and store at 4 °C.
14. Evaporate all solvent from the microcentrifuge tube in a vacuum centrifuge at 30 °C (16–18 h).
15. Resuspend dried sample in 1 ml of distilled water.
16. Remove sample from microcentrifuge tube with a disposable syringe.
17. When using the HPLC method for quantification, pass sample through a 0.45 μm filter into an HPLC vial and cap.

3.2 Analysis of Total Soluble Carbohydrate

1. Add 0.5 ml each of the carbohydrate standard, water (blank) and/or sample to separate 10 ml test tubes with a pipettor.
2. Cover each tube with aluminium foil.
3. Immerse test tubes in ice bath (below ~4 °C) for 15 min.
4. Add 2 ml of orcinol reagent (start timer with first sample and proceed with each subsequent sample at 1 min intervals).
5. Vortex reaction mixtures vigorously and incubate in an 80 °C water bath for exactly 15 min.
6. Terminate the reaction by rapid cooling in an ice bath (below ~4 °C) for 5 min.
7. Equilibrate tubes to room temperature.
8. Measure absorbance of the reaction mixture with a spectrophotometer at 540 nm, using the water blank for zeroing the spectrophotometer.
9. Determine total carbohydrates in the samples by reference to an appropriate standard curve generated from a standard glucose solution.

Table 15.1 HPLC conditions for soluble carbohydrate determination

Parameter	Condition
Mobile phase	16 mM NaOH and 2 mM NaOAc for 30 min; 200 mM NaOH from 30–40 min; 16 mM NaOH and 2 mM NaOAc from 40–60 min
Flow rate	0.8 ml min^{-1}
Column	Dionex PA1 (4 x 250 mm)
Column temperature	30 °C
Detection	Pulsed amperometry (using gold electrode)
Injection volume	15 μl

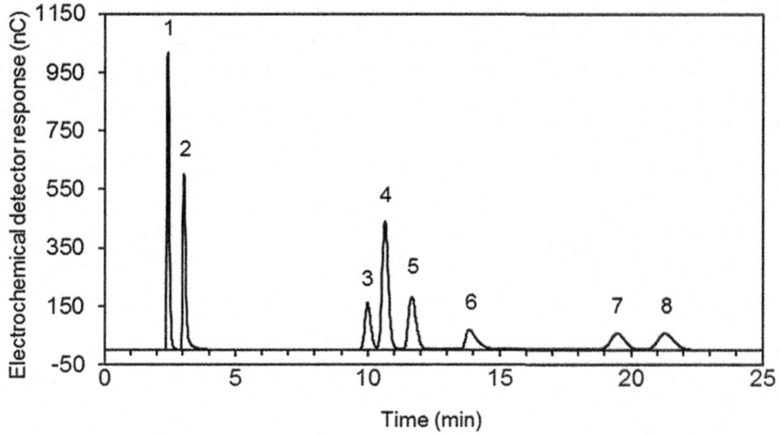

Fig. 15.1 Chromatogram of soluble carbohydrate standards obtained with a Dionex HPLC system equipped with an electrochemical detector (pulsed amperometry). Eluted sugars are (1) myoinositol, (2) galactitol, (3) galactose, (4) glucose, (5) sucrose, (6) fructose, (7) raffinose, (8) stachyose

3.3 Analysis by High Performance Liquid Chromatography (HPLC)

1. Set HPLC to the conditions indicated in Table 15.1.
2. Prepare sugar standards (glucose, sucrose, fructose, etc.) at concentrations ranging from 0.1 to 2 mg ml^{-1}. A range of sugars, including sugar alcohols, can be used as standards, depending on the material being analysed or specific carbohydrates of interest.
3. Run standards and samples on the HPLC (Fig. 15.1).
4. Normalize peaks for internal standards.

5. Prepare standard curves by plotting areas under the peaks versus concentrations for each sugar in the standards mixture.
6. Calculate the concentration of each monomeric sugar in sample dry mass. The sum of all individual monomers gives the total soluble carbohydrate concentration.

4 Final Remarks

Samples analysed by either method may require dilution if the unknown samples give absorbance or detection readings greater than the highest value of the standard curve. Should this occur, dilute samples, record dilution volume, and repeat the analysis.

The spectrophotometric assay to determine total soluble carbohydrates is subject to interference from particles or air bubbles in sample or reaction solutions. Different sugars give different quantitative responses (e.g. glucose ≠ xylose). Therefore, choice of the standards will depend on the carbohydrate composition of the sample. Sensitivity of the method is approximately 5–10 μg ml^{-1} carbohydrate (1 ml sample required); sensitivity of the HPLC method is ~1 μg ml^{-1} carbohydrate.

Orcinol is a harmful substance and special care needs to be taken when handling, especially when it is made up as a solution in concentrated sulphuric acid. Therefore, laboratory coats, protective eyewear, and gloves are required at all stages of the analysis.

References

Bärlocher, F. (1997). Pitfalls of traditional techniques when studying decomposition in aquatic habitats. *Limnetica, 13*, 1–11.

Bärlocher, F., & Porter, C. W. (1986). Digestive enzymes and feeding strategies of three stream invertebrates. *Journal of the North American Benthological Society, 5*, 58–66.

Bärlocher, F., & Sridhar, K. R. (2014). Association of animals and fungi in leaf decomposition. In E. B. Gareth Jones, K. D. Hyde, & K.-L. Pang (Eds.), *Freshwater fungi and fungal-like organisms* (pp. 412–441). Berlin: De Gruyter.

Cataldi, T. R., Margiotta, G., Iasi, L., Di Chio, B., Xiloyannis, C., & Bufo, S. (2000). Determination of sugar compounds in olive plant extracts by anion-exchange chromatography with pulsed amperometric detection. *Analytical Chemistry, 72*, 3902–3907.

Coleman, H. D., Ellis, D., Gilbert, M., & Mansfield, S. D. (2006). Up-regulation of sucrose synthase and UDP-glucose pyrophosphorylase impacts plant growth and metabolism. *Plant Biotechnology Journal, 4*, 87–101.

Gessner, M. O. (1991). Differences in processing dynamics of fresh and air-dried leaf litter in a stream ecosystem. *Freshwater Biology, 26*, 387–398.

Gessner, M. O. (1997). Fungal biomass, production and sporulation associated with particulate organic matter in streams. *Limnetica, 13*, 33–44.

Kuehn, K. A. (2016). Lentic and lotic habitats as templets for fungal communities: Traits, adaptation, and their significance to litter decomposition within freshwater ecosystems. *Fungal Ecology, 19*, 135–154.

Quentin, A. G., Pinkard, E. A., Ryan, M. G., Tissue, D. T., Baggett, L. S., Adams, H. D., Maillard, P., Marchand, J., Landhäusser, S. M., Lacointe, A., Gibon, Y., Anderegg, W. R. L., Asao, S., Atkin, O. K., Bonhomme, M., Claye, C., Chow, P. S., Clément-Vidal, A., Davies, N. W., Dickman, L. T., Dumbur, R., Ellsworth, D. S., Falk, K., Galiano, L., Grünzweig, J. M., Hartmann, H., Hoch, G., Hood, S., Jones, J. E., Koike, T., Kuhlmann, I., Lloret, F., Maestro, M., Mansfield, S. D., Martínez-Vilalta, J., Maucourt, M., McDowell, N. G., Moing, A., Muller, B., Nebauer, S. G., Niinemets, Ü., Palacio, S., Piper, F., Raveh, E., Richter, A., Rolland, G., Rosas, T., Joanis, B., Sala, A., Smith, R. A., Sterck, F., Stinziano, J. R., Tobias, M., Unda, F., Watanabe, M., Way, D. A., Weerasinghe, L. K., Wild, B., Wiley, E., & Woodruff, D. R. (2015). Non-structural carbohydrates in woody plants compared among laboratories. *Tree Physiology, 35*, 1146–1165.

Suberkropp, K., Godshalk, G. L., & Klug, M. J. (1976). Changes in the chemical composition of leaves during processing in a woodland stream. *Ecology, 57*, 720–727.

White, C. A., & Kennedy, J. F. (1986). Oligosaccharides. In M. F. Chaplin & J. F. Kennedy (Eds.), *Carbohydrate analysis: A practical approach* (pp. 37–54). Oxford: IRL Press Ltd.

Chapter 16
Total Lipids

Mark O. Gessner and Paul T. M. Neumann

Keywords Food quality · Invertebrate lipids · Litter quality · Lipid extraction · Litter chemistry · Spectrophotometry · Sulphophosphovanillin · Vanillin assay

1 Introduction

Lipids are a major class of chemical compounds in plant tissues that have rarely been considered in litter decomposition studies and assessments of litter nutritional quality for aquatic detritivores. This neglect partly reflects the fact that emphasis in the chemical characterization of decomposing litter has been placed on inorganic nutrients, such as nitrogen and phosphorus (Enriquez et al. 1993), as well as refractory litter constituents such as lignin (e.g. Gessner and Chauvet 1994; Palm and Rowland 1997). However, the few data currently available on litter lipid contents show that lipids can be a sizeable fraction of decomposing leaves (Table 16.1), suggesting that information on total lipid content may be useful in efforts to model decomposition rate as a function of litter chemistry (e.g. Moorhead et al. 1999). Since the ^{13}C signature of lipids tends to show a depletion relative to that of bulk carbon in organisms, lipid analyses may also prove useful to improve assessments of trophic relationships in litter-based food webs (Post et al. 2007; Logan et al. 2008).

More important, there is evidence that lipids can provide critical cues to detritivore feeding (Anderson and Cargill 1987; Chap. 18). For example, the sequence of food preference of two detritivores (*Gammarus tigrinus* and *Pycnopsyche guttifer*) for certain combinations of fungal species grown on leaves could be reproduced by applying lipid extracts of the fungi to uncolonized leaves (Rong et al. 1995). For holometabolic insects with limited food acquisition during

M. O. Gessner (✉)
Department of Experimental Limnology, Leibniz Institute of Freshwater Ecology and Inland Fisheries (IGB), Stechlin, Germany

Department of Ecology, Berlin Institute of Technology (TU Berlin), Berlin, Germany
e-mail: gessner@igb-berlin.de

P. T. M. Neumann
Friedrich-Kreusch-Weg 35, Langenfeld, Germany

F. Bärlocher et al. (eds.), *Methods to Study Litter Decomposition*,
https://doi.org/10.1007/978-3-030-30515-4_16

Table 16.1 Total lipid content (mean ± SD, where available) of leaf litter of various tree species. AFDM = ash-free dry mass. Leaf litter is undecomposed unless noted otherwise

Leaf species	Common name	Lipids (% AFDM)	Reference
Acacia melanoxylon	Blackwood acacia	1.9[a]	1
Carya glabra	Hickory	5.2	2
Eucalyptus viminalis	Manna eucalyptus	12[a]	1
Liquidambar styraciflua	Sweetgum	1.7[a, b]	3
Pomaderris aspera	Hazel pomaderris	3.8[a]	1
Quercus alba	White oak	4.9	2
Quercus nigra	Water oak	1.0[a, b]	3
Quercus robur	English oak	2.9 ± 0.5[a]	4
Quercus robur	English oak	3.5 ± 0.2[a, c]	4
Quercus robur	English oak	3.0 ± 0.1[a, d]	4

1 = Campbell et al. (1992); 2 = Suberkropp et al. (1976); 3 = Mills et al. (2001); 4 Mas-Martí et al. (2015)

[a]AFDM assumed to be 90% of dry mass

[b]Based on fatty acid content of saponified methanol extracts

[c]Litter conditioned in a stream for 3 weeks at about 12 °C

[d]Litter conditioned in a stream for 3 weeks d at about 15 °C

the adult stage, lipid content may be a particularly important litter-quality attribute, because late instars of these invertebrates benefit from consuming food with a high energy content to build up energy reserves before emergence. Support for this hypothesis has come from litter-consuming caddisflies (Hanson et al. 1983; Cargill et al. 1985a,b; Mas-Martí et al. 2015).

Bulk lipid analyses have classically adopted a gravimetric approach (Suberkropp et al. 1976). Lipids are extracted from the tissue with an apolar solvent, which is then evaporated before weighing the dried residue. The main limitation of this approach is that relatively large sample sizes are needed for accurate analyses. An attractive alternative is a spectrophotometric assay that allows analysis of small samples. More than 50 years ago, Zöllner and Kirsch (1962) described such a method for analysing blood lipids. This method, known as the sulphophosphovanillin (SPV) assay, was later applied to estimate lipid contents of algae (Ahlgren and Merino 1991; Mishra et al. 2014), aquatic invertebrates (Barnes and Blackstock 1973; Meyer and Walther 1989; Mas-Martí et al. 2015), fine-particulate organic matter (Neumann 1995), stream biofilms and seston (Sanpera-Calbet et al. 2017), and leaf leachates (Canhoto and Laranjeira 2007). The analysis is based on the reaction of lipid degradation products with aromatic aldehydes, which results in a red coloration that can be quantified at 528 nm (Zöllner and Kirsch 1962). With particulate organic matter, it is essential to extract the lipids from the bulk sample before performing the sulphophosphovanillin assay, since interference by nonlipidic compounds such as carbohydrates results in high nonspecific absorbance after heating the sample in sulphuric acid and addition of the vanillin reagent (Ahlgren and Merino 1991; Neumann 1995).

The protocol described here has been adopted from Neumann (1995). Although developed for fine-particulate organic matter (FPOM), it has also proved reliable for decomposing leaves from streams (M.O. Gessner, unpubl. data; Mas-Martí et al. 2015).

2 Equipment, Chemicals, and Solutions

2.1 Equipment and Material

- Freeze-dryer
- Mill
- Analytical balance (±0.1 mg precision)
- Glass centrifuge tubes (12 ml, preferably pressure-resistant, with Teflon-lined screw-caps)
- Centrifuge tubes (4 ml)
- Pipettes allowing precise pipetting of solvents with low viscosity (e.g. Eppendorf Multipette® or Varipette®)
- Standard laboratory centrifuge
- Evaporating centrifuge (e.g. SpeedVac concentrator SPD131DDA, Thermo Electron Corp., Woburn, MA, USA)
- Vials (4 ml, with Teflon-lined screw-caps)
- Vortex mixer
- Pear-shaped glass bulbs or marbles (only needed if centrifuge tubes do not resist high pressure)
- Water or dry baths (20 and 100 °C)
- Timer
- Spectrophotometer (set at 528 nm)
- Volumetric flasks

2.2 Chemicals

- Chloroform ($CHCl_3$), for residual analysis
- Methanol (CH_3OH), for residual analysis
- Deionized water (e.g. Nanopure®)
- NaCl, reagent grade
- Concentrated sulphuric acid (H_2SO_4, 95–97%), reagent grade
- Concentrated phosphoric acid (H_3PO_4, 85%), reagent grade
- Vanillin
- Cholesterol (5-cholesten-3β-ol)

2.3 Solutions

- 0.9% (w:v) NaCl solution
- Phosphoric acid-vanillin reagent: 20 ml 0.6% (w:v) vanillin solution and 85% H_3PO_4 in a total volume of 100 ml
- Standard solutions: Cholesterol standards in chloroform at concentrations ranging from 10 to 100 mg ml^{-1}

3 Experimental Procedures

3.1 Lipid Extraction

1. Freeze-dry leaves and grind to powder that passes through a 0.5 mm mesh screen.
2. Weigh out 25 mg of sample to the nearest 0.1 mg in a 12-ml screw-cap glass centrifuge tube with Teflon-lined caps.
3. Add 7 ml chloroform/methanol (2:1, v:v).
4. Shake for 1 min, then let stand for 2 h, with shaking for 1 min every 30 min.
5. Centrifuge for 1 min at about 3000 *g*, rinse tube walls to suspend any adhering particles, then centrifuge for another 10 min to separate particles from the lipid extract.
6. Transfer 5 ml of the lipid extract to a clean tube containing 1 ml of 0.9% NaCl solution.
7. Shake for 1 min, then centrifuge for 10 min at about 3000 *g* to separate phases.
8. Remove and discard the upper aqueous phase.
9. Rinse inner walls of the tube twice with 1 ml chloroform:methanol:water (3:48:47, v:v:v).
10. Remove rinsing solution.
11. Evaporate sample to dryness in a SpeedVac concentrator at about 45 °C.
12. Transfer residue with 2 ml chloroform to a clean 4-ml screw-cap vial with Teflon-lined cap.
13. Evaporate sample again to dryness in the SpeedVac concentrator.
14. Redissolve residue in 1 ml chloroform, close vial tightly, and run spectrophotometric assay or store at −20 °C until analysis.
15. Run control without sample material in the same way.

3.2 Spectrophotometric Analysis

1. Place 100 μl of the lipid extract in a 12-ml test tube.
2. Evaporate solvent in SpeedVac concentrator at about 45 °C.

3. Add 200 µl of conc. H_2SO_4 and vortex.
4. Close tube tightly or, if not pressure-resistant, cover it with a pear-shaped glass bulb (or marble), and heat for 10 min to 100 °C in a water or dry bath.
5. Let cool for 5 min in a water bath at 20 °C.
6. Add 2.5 ml of H_3PO_4-vanillin reagent and vortex.
7. Measure absorbance after 60–65 min at 528 nm.
8. Run cholesterol standards in the same way.
9. Calculate lipid concentration as cholesterol equivalents from absorbance reading of sample and standard curve.

4 Final Remarks

The time course of colour development depends strongly on the ratio of sulphuric acid to the H_3PO_4-vanillin solution (Neumann 1995). It is essential, therefore, to keep this ratio strictly constant in a given sample series (e.g. at 1:12.5 as in the protocol described above). Furthermore, at a 1:12.5 ratio of sulphuric acid to the H_3PO_4-vanillin reagent, absorbance readings must be taken 60–65 min after addition of reagent to the sample. Earlier or later readings result in an underestimation of lipid concentrations.

Since the assay is sensitive to fatty acid composition (Higgins et al. 2014) and lipids are a highly heterogeneous class of molecules, choice of an appropriate standard is critical to facilitate accurate quantitative estimates. Neumann (1995) found that cholesterol and a lipid extract from fine-particulate organic matter (FPOM) collected in a stream gave identical responses with the protocol described here. However, use of a specific standard (i.e. a lipid solution from a representative sample, with the lipid content determined gravimetrically) may be preferable when precise information about the absolute magnitude of lipid concentrations is required.

For finely ground FPOM, the efficiency of lipid extraction in a single step as described above was 93% compared to three successive extraction steps (Neumann 1995).

Reducing the sample and reagent volume of the method presented here enables using a microplate reader for the final spectrophotometric analysis (Higgins et al. 2014). This would significantly increase sample throughput.

References

Ahlgren, G., & Merino, L. (1991). Lipid analysis of freshwater microalgae: A method study. *Archiv für Hydrobiologie, 121*, 295–306.

Anderson, N. H., & Cargill, A. S. (1987). Nutritional ecology of aquatic detritivorous insects. In F. Slansky Jr. & J. G. Rodriguez (Eds.), *Nutritional ecology of insects, mites, spiders, and related invertebrates* (pp. 903–925). New York: Wiley.

Barnes, H., & Blackstock, J. (1973). Estimation of lipids in marine animals and tissues: Detailed investigation of the sulphophosphovanillin method for 'total' lipids. *Journal of Experimental Marine Biology and Ecology, 12*, 103–118.

Campbell, I. C., James, K. R., Hart, B. T., & Devereaux, A. (1992). Allochthonous coarse particulate organic material in forest and pasture reaches of two south-eastern Australian streams. 2. Litter processing. *Freshwater Biology, 27*, 353–365.

Canhoto, C., & Laranjeira, C. (2007). Leachates of *Eucalyptus globulus* in intermittent streams affect water parameters and invertebrates. *International Review of Hydrobiology, 92*, 173–182.

Cargill, A. S., II, Cummins, K. W., Hanson, B. J., & Lowry, R. R. (1985a). The role of lipids as feeding stimulants for shredding aquatic insects. *Freshwater Biology, 15*, 455–464.

Cargill, A. S., II, Cummins, K. W., Hanson, B. J., & Lowry, R. R. (1985b). The role of lipids, fungi, and temperature in the nutrition of a shredder caddisfly, *Clistoronia magnifica*. *Freshwater Invertebrate Biology, 4*, 64–78.

Enríquez, S., Duarte, C. M., & Sand-Jensen, K. (1993). Patterns in decomposition rates among photosynthetic organisms: The importance of detritus C:N:P content. *Oecologia, 94*, 457–471.

Gessner, M. O., & Chauvet, E. (1994). Importance of stream microfungi in controlling breakdown rates of leaf litter. *Ecology, 75*, 1807–1817.

Hanson, B. J., Cummins, K. W., Cargill, A. S., II, & Lowry, R. R. (1983). Dietary effects on lipid and fatty acid composition of *Clistoronia magnifica* (Trichoptera, Limnephilidae). *Freshwater Invertebrate Biology, 2*, 2–15.

Higgins, B. T., Thornton-Dunwoody, A., Labavitch, J. M., & VanderGheynst, J. S. (2014). Microplate assay for quantitation of neutral lipids in extracts from microalgae. *Analytical Biochemistry, 465*, 81–89.

Logan, J. M., Jardine, T. D., Miller, T. J., Bunn, S. E., Cunjak, R. A., & Lutcavage, M. E. (2008). Lipid corrections in carbon and nitrogen stable isotope analyses: Comparison of chemical extraction and modelling methods. *Journal of Animal Ecology, 77*, 838–846.

Mas-Martí, E., Muñoz, I., Oliva, F., & Canhoto, C. (2015). Effects of increased water temperature on leaf litter quality and detritivore performance: A whole-reach manipulative experiment. *Freshwater Biology, 60*, 184–197.

Meyer, E., & Walther, A. (1989). Methods for the estimation of protein, lipid, carbohydrate and chitin levels in fresh water invertebrates. *Archiv für Hydrobiologie, 113*, 161–177.

Mills, G. L., McArthur, J. V., Wolfe, C., Aho, J. M., & Rader, R. B. (2001). Changes in fatty acid and hydrocarbon composition of leaves during decomposition in a southeastern blackwater stream. *Archiv für Hydrobiologie, 152*, 315–328.

Mishra, S. K., Suh, W. I., Farooq, W., Moon, M., Shrivastav, A., Park, M. S., & Yang, J. W. (2014). Rapid quantification of microalgal lipids in aqueous medium by a simple colorimetric method. *Bioresource Technology, 155*, 330–333.

Moorhead, D. L., Currie, W. S., Rastetter, E. P., Parton, W. J., & Harmon, M. E. (1999). Climate and litter quality controls of decomposition: An analysis of modelling approaches. *Global Biogeochemical Cycles, 13*, 575–589.

Neumann, P. (1995). *Untersuchungen zur Nahrungsqualität von benthischem feinpartikulärem Detritus für Feinpartikelsammler, unter dem Aspekt seiner biochemischen Zusammensetzung im Breitenbach*. Doctoral thesis, University of Marburg, Germany, 135 pp.

Palm, C. A., & Rowland, A. P. (1997). A minimum data set for characterization of plant quality for decomposition. In G. Cadish & K. E. Giller (Eds.), *Driven by nature: Plant litter quality and decomposition* (pp. 379–391). Wallingford: CAB International.

Post, D. M., Layman, C. A., Arrington, D. A., Takimoto, G., Quattrochi, J., & Montaña, C. G. (2007). Getting to the fat of the matter: Models, methods and assumptions for dealing with lipids in stable isotope analyses. *Oecologia, 152*, 179–189.

Rong, Q., Sridhar, K. R., & Bärlocher, F. (1995). Food selection in three leaf-shredding stream invertebrates. *Hydrobiologia, 316*, 173–181.

Sanpera-Calbet, I., Ylla, I., Romaní, A. M., Sabater, S., & Muñoz, I. (2017). Biochemical quality of basal resources in a forested stream: Effects of nutrient enrichment. *Aquatic Sciences, 79*, 99–112.

Suberkropp, K., Godshalk, G. L., & Klug, M. J. (1976). Changes in the chemical composition of leaves during processing in a woodland stream. *Ecology, 57*, 720–727.

Zöllner, N., & Kirsch, K. (1962). Über die quantitative Bestimmung von Lipoiden (Mikromethode) mittels der vielen natürlichen Lipoiden (allen bekannten Plasmalipoiden) gemeinsamen Sulphophosphovanillin-Reaktion. *Zeitschrift für die gesamte experimentelle Medizin, 135*, 545–561.

Chapter 17
Polyunsaturated Fatty Acids in Decomposing Leaf Litter

Eric von Elert

Keywords Aquatic hyphomycetes · Eicosapentaenoic acid · EPA · Food quality · GC · Invertebrates · Lipid extraction · Litter quality · PUFA · Shredders

1 Introduction

Polyunsaturated fatty acids (PUFAs) are a subclass of fatty acids characterized by more than one double bond arranged in *cis*-configuration and separated by a single $–CH_2–$ group. PUFAs are divided into two major classes distinguished by the position of the first double bond when counted from the terminal methyl group. In n-3 PUFAs, this double bond is at the third position from the terminal carbon atom, and in n-6 PUFAs it is at the sixth position. As this difference reflects distinct biosynthetic pathways, PUFAs can be converted within but not between these two classes. One function of PUFAs is that they serve as precursors for hormones. For example, long-chain PUFAs such as arachidonic acid (ARA; C20:4n-6) and eicosapentaenoic acid (EPA; C20:5n-3) serve as precursors of prostaglandins, which have several hormonal functions in arthropod reproduction (Stanley 2000; Schlotz et al. 2012). Furthermore, PUFAs are an integral constituent of cell membranes, where they mainly occur in phospholipids.

Arthropods are unable to synthesize long-chain PUFAs *de novo* and hence require a dietary source to satisfy their physiological needs (Harrison 1990). This makes PUFAs an essential class of lipids and suggests constraints on consumers when diets have low PUFA contents. Evidence for such a limitation in freshwater arthropods includes strong correlations between PUFA contents of natural phytoplankton and the growth rate of *Daphnia* (Müller-Navarra 1995; Müller-Navarra et al. 2000; Wacker and von Elert 2001; Hartwich et al. 2012), with EPA and

E. von Elert (✉)
Institute of Zoology, University of Cologne, Cologne, Germany
e-mail: evelert@uni-koeln.de

F. Bärlocher et al. (eds.), *Methods to Study Litter Decomposition*,
https://doi.org/10.1007/978-3-030-30515-4_17

α-linolenic acid (α-LA, C18:3n-3) producing the strongest relationship. Supplementation experiments have confirmed that low concentrations of EPA or α-LA in food algae can indeed limit not only the somatic growth of *Daphnia* (von Elert 2002; Becker and Boersma 2005) but also parthenogenetic egg production (Wacker and Martin-Creuzburg 2007; Martin-Creuzburg and von Elert 2009), and similar effects have also been demonstrated for the zebra mussel, *Dreissena polymorpha* (Wacker and von Elert 2003a, b).

PUFA deficiency in one food item can be mitigated by another that is rich in PUFAs (Marzetz et al. 2017). Similarly, temporal fluctuations in dietary PUFA content can be buffered (Koussoroplis et al. 2017), because consumers can store PUFAs and thus profit from ingesting a PUFA-rich diet slightly earlier or later than a PUFA-deficient diet (Koussoroplis et al. 2017). Such mitigating effects can mask the potentially limiting nature of low dietary PUFA contents. At low temperatures, PUFA limitation becomes more severe (Sperfeld and Wacker 2012). This effect has been attributed to reduced membrane fluidity (Hazel 1995), which poikilotherms try to counter by increasing the content of unsaturated fatty acids in their membranes (Farkas 1979; Hazel 1995; Hazel and Williams 1990). Thus, PUFA-deficient diets can constrain consumer fitness and limit the use of cold habitats (Brzezinski and von Elert 2015).

Given the importance of PUFAs as a determinant of food quality and huge amounts accumulating in some benthic invertebrates (*Hydropsyche* spp., *Ephemerella* spp., isopods, oligochaetes; Torres-Ruiz et al. 2007; gammaridae; Makhutova et al. 2016), these lipids could also play an important role for stream detritivores. Indeed, one hypothesis to explain the observation that microbial colonization and partial decomposition of litter improve the palatability to, and promote the growth of, litter consumers in streams (Graça 2001) is that microbial lipids enhance litter quality as food (Cargill et al. 1985). This hypothesis is supported by the high relative content of the PUFAs C18:2n-6 and C18:3n-3 in aquatic hyphomycetes (Arce Funck et al. 2015). Poor growth of *Gammarus fossarum* feeding on undecomposed leaf litter from alder (*Alnus glutinosa*) was indeed significantly increased by supplementing the food with a diatom (*Nitzschia palea*) rich in PUFAs, although supplementation with aquatic hyphomycetes affected neither consumption (Aßmann et al. 2011) nor growth (Crenier et al. 2017). The aquatic hyphomycetes were rich in C18:2n-6, whereas the diatom was rich in C20:5n-3, suggesting a limitation of *Gammarus* growth on leaf litter by C20:5n-3 or possibly other n-3 PUFAs. Notably, however, PUFA contents in submerged leaves colonized by microbes were found to be very low (Torres-Ruiz and Wehr 2010), and *G. roeselii* tested in food-choice experiments showed no clear preference for leaf litter of alder (*Alnus glutinosa*) coated with lipid extracts of fungi and an oomycete (Aßmann and von Elert 2009; see also Rong et al. 1995). In view of these mixed results and the fact that the

Table 17.1 Fatty acid content of abscissed maple leaves incubated in tap water for 1 week (*Acer platanoides*), freshly fallen leaf litter of common hornbeam (*Carpinus betulus*) (1) and a leaf-litter mixture of *Acer rubrum*, *Quercus rubra*, and *Platanus occidentalis*

	Fatty acid content (μg g^{-1} dry mass)		
Fatty acid	*A. platanoides* (1)	*C. betulus* (1)	Leaf-litter mixture (2)
C 14:0	439 (26)	50 (5)	
C 14:1 n-9	279 (13)	208 (32)	
C 16:0	4466 (128)	1467 (55)	
C 16:1 n-9	388 (23)	20 (1)	
C 18:0	747 (1)	502 (36)	
C 18:1 n-9 c	126 (8)	0 (0)	
C 18:2 n-6 c	2160 (70)	525 (50)	670 (50)
C 18:3 n-6	17 (1)	29 (1)	
C 18:3 n-3	20,777 (762)	2112 (144)	2050 (220)
C 22:0	62 (1)	105 (2)	
C 24:0	442 (1)	238 (12)	
Total	29,902 (750)	5258 (340)	

Values are means (SE) of three replicates
1 = von Elert (unpublished data), 2 = Torres-Ruiz and Wehr (2010)

role of PUFAs is still poorly investigated in detritus-based food webs, general conclusions about the importance of PUFAs in these systems are currently premature.

Some data are available on the composition of fatty acids in leaf litter (e.g., Guo et al. 2018). Information on the fatty acid content is scarce, however; the total fatty acid content in freshly fallen leaves ranges from 5 to 30 mg g^{-1} dry mass in the few available studies to date (Table 17.1). In maple leaves (*A. platanoides*) and leaf litter of common hornbeam (*C. betulus*), the most abundant PUFAs were α-LA (C18:3n-3) and C18:2n-6, whereas C_{20} PUFAs were undetectable (unpublished data). This result is in accordance with Torres-Ruiz and Wehr (2010) and suggests that C_{20} PUFAs are scarce in microbially conditioned leaf litter. Choice experiments have indicated that α-LA (C18:3n-3) can serve as an indicator of food palatability (Vonk et al. 2016). Although this finding points to a role of PUFAs in determining detritivore food preference, supplementation experiments with specific PUFAs are needed to assess the potential importance for growth and reproduction of litter consumers. The high PUFA contents of some benthic invertebrates mentioned above (Torres-Ruiz et al. 2007) might even point to PUFA sources other than leaf litter to mitigate PUFA deficiency for litter-consuming detritivores (Guo et al. 2018).

The method presented here details procedures to quantify individual PUFAs in leaf litter, invertebrates, and other types of samples, complementing a method to determine the total lipid content in leaf litter as described in Chap. 16.

2 Equipment, Chemicals, and Solutions

2.1 *Equipment and Materials*

- Freezer (−20 or −80 °C)
- Refrigerator (4 °C)
- Stream of nitrogen gas under a fume hood
- Heating block or water bath (40 and 70 °C)
- Vortex
- Ultrasonic bath
- Centrifuge suited for glass reagent tubes (e.g., 16 mm × 100 mm; 3000 *g*)
- Glass reagent tubes (e.g., 16 mm × 100 mm) with screw caps
- Sample vials with micro-inserts (200 μl) and septum-lined caps
- Pasteur pipettes
- Gas chromatograph (GC) with a splitless injector and a flame ionization detector (FID), equipped with a capillary GC column suited for the analysis of fatty acid methyl esters (FAMEs), e.g., an Agilent J&W DB-225 column (length 30 m, ID 0.25 mm, film thickness 0.25 μm; von Elert 2002).

2.2 *Chemicals*

- Liquid nitrogen
- Dichloromethane, HPLC grade
- Methanol, HPLC gradient grade
- Isohexane, gas chromatography FID grade
- HCl in methanol (3 M), CAS number 7647-01-0
- Reference compounds
  - Heptadecanoid acid methyl ester (C17:0 ME), CAS number 1731-92-6
  - Nonadecanoid acid methyl ester (C19:0 ME), CAS number 1731-94-8
  - Tricosanoic acid methyl ester (C23:0 ME), CAS number 2433-97-8
  - Commercially available mixture of fatty acid methyl esters (FAME), such as the Sigma™ 37 Component FAME Mix, menhaden oil, or the bacterial acid methyl ester (BAME) mix

2.3 *Solutions*

- Extraction solvent: dichloromethane/methanol (2:1, v:v)
- C17:0 ME, C19:0 ME and/or C23:0 ME (each 200 μg ml^{-1} isohexane)

3 Experimental Procedures

3.1 Sample Preparation

1. Collect leaf litter or animal sample in the field or from a laboratory experiment and protect from light and elevated temperatures throughout the analysis to avoid oxidation and auto-degradation of polyunsaturated lipids.
2. Blot-dry samples when water adheres to the surface (e.g., on Kim-Wipes™) and extract lipids from these fresh samples (the solvent efficiently extracts lipids from living tissue, so that freeze-drying of samples is not critical).
3. If results are to be related to litter dry mass, best split samples and use one portion for dry mass determination, the other for fatty acid analysis.
4. If a sample is used for both dry mass and fatty acid analyses, freeze the sample in liquid N_2, and store at −20 or −80 °C if immediate processing is not possible.
5. Freeze-dry and homogenize samples using pestle and mortar. Minimize storage at this stage (even at − 20 or −80 °C) to avoid a possible slow oxidation of unsaturated fatty acids; however, freeze-dried and homogenized samples immersed immediately in solvent for lipid extraction can be stored at −20 °C for several weeks.
6. Determine the dry mass of samples prior to adding solvent for lipid extraction.

3.2 Lipid Extraction

1. Transfer leaf litter or animal sample in screw-cap reagent tube and add 5 ml of extraction solvent with a Pasteur pipette.
2. Add internal standards (IS), with the type and amounts depending on the sample (e.g., 100 μl C17:0 ME, C19:0 ME, and/or C23:0 ME per 2–4 mg of litter dry mass or per 200–600 μg of animal dry mass).
3. Incubate overnight at 4–8 °C.
4. Perform all subsequent steps at low light and especially avoid exposure to direct sunlight.
5. Vortex the sample tubes and place them in an ultrasonic bath for 1 min.
6. Centrifuge for 5 min at 3000 *g* (without brake!)
7. Use a Pasteur pipette to transfer the extract into a clean screw-cap reagent tube, add another 3 ml of extraction solvent to the original tube, close the tube, and repeat the sample extraction by ultrasonication for 1 min and subsequent centrifugation.
8. Use a Pasteur pipette to combine the second extract with the first.
9. Spin down any debris in the combined extract by centrifuging the reagent tubes at 3000 *g* for 5 min (without brake!).

10. Carefully (sensitive pellet!) transfer the supernatant with a Pasteur pipette to a clean screw-cap reagent tube.
11. Resuspend the pellet with another 3 ml of extraction solvent, vortex, and then centrifuge for 5 min at 3000 *g* (without brake!) and combine the supernatants.
12. Place the reagent tube containing the combined supernatants into a heating block or water bath (max. temperature 40 °C) and evaporate the solvent under a stream of nitrogen gas, immediately removing the tube from the heating block when dry.

3.3 Transesterification

1. Proceed immediately with transesterification by adding 5 ml of 3 M methanolic HCl to the reagent tube, closing the screw cap, and incubating the sample for 20 min at 70 °C in a heating block or water bath.
2. Let the sample cool down to room temperature or lower (e.g., by placing the tube at 4–8 °C in a refrigerator).
3. Add 2 ml of isohexane with a Pasteur pipette, and vortex three times for several seconds, allowing the phases to separate each time between mixings.
4. Transfer the upper hexane phase with a Pasteur pipette into a new screw-cap reagent tube.
5. Repeat this isohexane extraction two more times, ensuring that no solvent from the lower phase is transferred by leaving a few microliters of the upper layer in the sample.
6. Evaporate the combined isohexane extracts under a stream of nitrogen gas at max. 40 °C in a heating block or water bath.
7. Dissolve the dry deposit in the reagent tube in 100 μl isohexane and transfer the solution to a microliter-insert in a sample vial.
8. Repeat this step to obtain a total volume of 200 μl in the sample vial.
9. Gently evaporate the solvent again under a stream of nitrogen and dissolve the dry deposit in the microliter-insert of a sample in a final volume of 100 μl isohexane.
10. Close the vial with a septum cap and store at −20 °C until measurement by gas chromatography.

3.4 Gas Chromatography

1. Set the flow of the carrier gas (e.g., 35 cm s^{-1}, helium).
2. Set FID detector and injector to 220 °C.
3. Set the temperature program for the GC oven, e.g., 60 °C (hold for 1 min) and then to 180 °C at 120 °C min^{-1}, next to 200 °C at 50 °C min^{-1} (hold for 10.5 min), and then to 220 °C at 120 °C min^{-1} (hold for 7.5 min), according to von Elert (2002), resulting in a total run time per sample of 20.6 min.

4. Inject a 1 µl aliquot of each sample in splitless mode.
5. Inject reference compounds to identify peaks in a given sample as FAMEs with retention times identical to those of the reference compounds.
6. For quantification by means of internal standards (IS), establish dose-response curves for each of the FAMEs of interest, based on different ratios of IS with the FAME of interest; this is most conveniently achieved by preparing a mixture of all FAMEs of interest with known concentrations and mixing increasing aliquots with known amounts of IS.
7. Subsequently, derive calibration curves for each FAME of interest from splitless injections of 1 µl aliquots of these solutions.

4 Final Remarks

Several types of reference compounds can be used to identify FAMEs by comparing retention times. For routine analyses of fatty acids with an even number of carbon atoms, these include a commercially available mixture containing 37 components (Sigma™ 37 Component FAME mix), menhaden oil, and specifically prepared PUFA mixes. Bacterial acid methyl ester (BAME) mixtures serve well for fatty acids with an odd number of carbon atoms (e.g., bacterial fatty acids). Petroselinic acid (C18:1n-12) and oleic acid (C18:1n-9) cannot be distinguished.

The presented GC method quantifies absolute amounts of fatty acids, which can be related to sample dry mass or organic carbon. The detection limit is 10 ng of FAME mg^{-1} dry mass. Quantification requires the systematic addition of an internal standard (IS) to the samples. If the fatty acid profiles of samples are unknown, it is good practice first to perform a qualitative analysis without IS to check which FAMEs (C17:0 ME, C19:0 ME, C23:0 ME) are best suited as IS. Unless precluded by other constraints, C17:0 ME and C23:0 ME are best used simultaneously to relate FAMEs with a low retention time to C17:0 ME and those with a high retention time to C23:0 ME. Separate calibration curves with different ratios of IS need to be established for all FAMEs of interest. The amount of IS may have to be adjusted to the fatty acid content of the sample, to ensure that the ratios of IS to each fatty acid are covered by the calibration curve. A dry mass of 3–4 mg litter and 200–600 µg animal tissue are good starting points.

References

Arce Funck, J., Bec, A., Perrière, F., Felten, V., & Danger, M. (2015). Aquatic hyphomycetes: A potential source of polyunsaturated fatty acids in detritus-based stream food webs. *Fungal Ecology, 13*, 205–210.

Aßmann, C., & von Elert, E. (2009). The impact of fungal extracts on leaf litter on the food preference of *Gammarus roeselii*. *International Review of Hydrobiology, 94*, 484–496.

Aßmann, C., Rinke, K., Nechwatal, J. A. N., & von Elert, E. (2011). Consequences of the colonisation of leaves by fungi and oomycetes for leaf consumption by a gammarid shredder. *Freshwater Biology, 56*, 839–852.

Becker, C., & Boersma, M. (2005). Differential effects of phosphorus and fatty acids on *Daphnia magna* growth and reproduction. *Limnology and Oceanography, 50*, 388–397.

Brzezinski, T., & von Elert, E. (2015). Predator evasion in zooplankton is suppressed by polyunsaturated fatty acid limitation. *Oecologia, 179*, 687–697.

Cargill, A. S., II, Cummins, K. W., Hanson, B. J., & Lowry, R. R. (1985). The role of lipids, fungi and temperature in the nutrition of a shredder caddisfly, *Clistoronia magnifica*. *Freshwater Invertebrate Biology, 4*, 64–78.

Crenier, C., Arce-Funck, J., Bec, A., Billoir, E., Perrière, F., Leflaive, J., Guérold, F., Felten, V., & Danger, M. (2017). Minor food sources can play a major role in secondary production in detritus-based ecosystems. *Freshwater Biology, 62*, 1155–1167.

Farkas, T. (1979). Adaptation of fatty-acid compositions to temperature - study on planktonic crustaceans. *Comparative Biochemistry and Physiology B – Biochemistry and Molecular Biology, 64*, 71–76.

Graça, M. A. S. (2001). The role of invertebrates on leaf litter decomposition in streams – A review. *International Review of Hydrobiology, 86*, 383–393.

Guo, F., Bunn, S. E., Brett, M. T., Fry, B., Hager, H., Ouyang, X. G., & Kainz, M. J. (2018). Feeding strategies for the acquisition of high-quality food sources in stream macroinvertebrates: Collecting, integrating, and mixed feeding. *Limnology and Oceanography, 63*, 1964–1978.

Harrison, K. E. (1990). The role of nutrition in maturation, reproduction and embryonic development of decapod crustaceans: A review. *Journal of Shellfish Research, 9*, 1–28.

Hartwich, M., Martin-Creuzburg, D., Rothhaupt, K. O., & Wacker, A. (2012). Oligotrophication of a large, deep lake alters food quantity and quality constraints at the primary producer-consumer interface. *Oikos, 121*, 1702–1712.

Hazel, J. R. (1995). Thermal adaptation in biological membranes: Is homeoviscous adaptation the explanation? *Annual Review of Physiology, 57*, 19–42.

Hazel, J. R., & Williams, E. E. (1990). The role of alterations in membrane lipid composition in enabling physiological adaptation of organisms to their physical environment. *Progress in Lipid Research, 29*, 167–227.

Koussoroplis, A. M., Pincebourde, S., & Wacker, A. (2017). Understanding and predicting physiological performance of organisms in fluctuating and multifactorial environments. *Ecological Monographs, 87*, 178–197.

Makhutova, O. N., Shulepina, S. P., Sharapova, T. A., Dubovskaya, O. P., Sushchik, N. N., Baturina, M. A., Pryanichnikova, E. G., Kalachova, G. S., & Gladyshev, M. I. (2016). Content of polyunsaturated fatty acids essential for fish nutrition in zoobenthos species. *Freshwater Science, 35*, 1222–1234.

Martin-Creuzburg, D., & von Elert, E. (2009). Good food versus bad food: The role of sterols and polyunsaturated fatty acids in determining growth and reproduction of *Daphnia magna*. *Aquatic Ecology, 43*, 943–950.

Marzetz, V., Koussoroplis, A. M., Martin-Creuzburg, D., Striebel, M., & Wacker, A. (2017). Linking primary producer diversity and food quality effects on herbivores: A biochemical perspective. *Scientific Reports, 7*, 11035.

Müller-Navarra, D. C. (1995). Evidence that a highly unsaturated fatty acid limits *Daphnia* growth in nature. *Archiv für Hydrobiologie, 132*, 297–307.

Müller-Navarra, D. C., Brett, M. T., Liston, A. M., & Goldman, C. R. (2000). A highly unsaturated fatty acid predicts carbon transfer between primary producers and consumers. *Nature, 403*, 74–77.

Rong, Q., Sridhar, K. R., & Bärlocher, F. (1995). Food selection in three leaf-shredding stream invertebrates. *Hydrobiologia, 316*, 173–181.

Schlotz, N., Sørensen, J. G., & Martin-Creuzburg, D. (2012). The potential of dietary polyunsaturated fatty acids to modulate eicosanoid synthesis and reproduction in *Daphnia magna*:

A gene expression approach. *Comparative Biochemistry and Physiology A – Molecular and Integrative Physiology, 162*, 449–454.

Sperfeld, E., & Wacker, A. (2012). Temperature affects the limitation of *Daphnia magna* by eicosapentaenoic acid, and the fatty acid composition of body tissue and eggs. *Freshwater Biology, 57*, 497–508.

Stanley, D. W. (2000). *Eicosanoids in invertebrate signal transduction systems*. Princeton: Princeton University Press.

Torres-Ruiz, M., & Wehr, J. D. (2010). Changes in the nutritional quality of decaying leaf litter in a stream based on fatty acid content. *Hydrobiologia, 651*, 265–278.

Torres-Ruiz, M., Wehr, J. D., & Perrone, A. A. (2007). Trophic relations in a stream food web: Importance of fatty acids for macroinvertebrate consumers. *Journal of the North American Benthological Society, 26*, 509–522.

von Elert, E. (2002). Determination of limiting polyunsaturated fatty acids in *Daphnia galeata* using a new method to enrich food algae with single fatty acids. *Limnology and Oceanography, 47*, 1764–1773.

Vonk, J. A., van Kuijk, B. F., van Beusekom, M., Hunting, E. R., & Kraak, M. H. S. (2016). The significance of linoleic acid in food sources for detritivorous benthic invertebrates. *Scientific Reports, 6*, 35785.

Wacker, A., & Martin-Creuzburg, D. (2007). Allocation of essential lipids in *Daphnia magna* during exposure to poor food quality. *Functional Ecology, 21*, 738–747.

Wacker, A., & von Elert, E. (2001). Polyunsaturated fatty acids: Evidence for non-substitutable biochemical resources in *Daphnia galeata*. *Ecology, 82*, 2507–2520.

Wacker, A., & von Elert, E. (2003a). Effects of larval food conditions on recruitment of *Dreissena polymorpha*. *Ecological Studies, Hazards and Solutions, 6*, 49. Moscow: MAX-Press.

Wacker, A., & von Elert, E. (2003b). Food quality controls reproduction of the zebra mussel (*Dreissena polymorpha*). *Oecologia, 135*, 332–338.

Chapter 18
Total Phenolics

Felix Bärlocher and Manuel A. S. Graça

Keywords AFT-IR · Folin-Ciocalteu reagent · Folin-Denis reagent · GC · HPLC · Inhibitory compounds · LC-MS · Litter quality · NMR · Phenolics extraction · Polyphenolics · Secondary metabolites · Tannic acid · Tannins

1 Introduction

Phenolics are a heterogeneous group of natural substances characterized by an aromatic ring with one or more hydroxyl groups. The number of these compounds identified may now exceed 100,000 (Waterman and Mole 1994; Tissier et al. 2015). Phenolics may occur as monomers with one hydroxyl group (e.g., ferulic acid). Compounds with several phenolic hydroxyl substituents are referred to as polyphenolics (Harborne 2004). Among these, tannins (subdivided into phlorotannins, hydrolyzable and condensed tannins) are of particular interest because of various demonstrated or posited ecological effects (Zucker 1983; Chap. 19). In particular, tannins play a major role in the defense against herbivores and pathogens (Waterman and Mole 1994; Lill and Marquis 2001) or, more generally, in communications between plants and other species (Harborne 2004; Preiss et al. 2015). Other phenolics such as anthocyanins may prevent leaf damage resulting from exposure to excessive light (Gould and Lee 2002).

Larger phenolics are often concentrated in specific tissues or cell structures (e.g., leaves, bark), and overall concentrations of phenolics in green leaves vary widely within a range of 1–25% of dry mass (Hättenschwiler and Vitousek 2000). Since the bulk of phenolics remains present during leaf senescence and after death, these

F. Bärlocher (✉)
Department of Biology, Mount Allison University, Sackville, NB, Canada
e-mail: fbaerloc@mta.ca

M. A. S. Graça
MARE – Marine and Environmental Sciences Centre & Department of Life Sciences, University of Coimbra, Coimbra, Portugal
e-mail: mgraca@ci.uc.pt

F. Bärlocher et al. (eds.), *Methods to Study Litter Decomposition*,
https://doi.org/10.1007/978-3-030-30515-4_18

Table 18.1 Phenolics concentrations in terms of tannic acid or ferulic acid equivalents for selected plant tissues, including senescent leaves (s) and live (l) and yellow-green to brown-dead grass leaves (y)

Species	Common name	Phenolics (% leaf dry mass)	References
Spartina alterniflora (y)	Smooth cordgrass	0.4–1.5	1
		0.2–1.2	2
Qualea sp. (s)		2.7	3
Alnus glutinosa (s)	Alder	2.7	4
		2.7	5
		6.6	6
		6.8–7.6	
Sapium sebiferum (l, s)	Chinese tallow tree	3.0	7
Eucalyptus globulus (s)	Eucalyptus	6.4	4
		9.8	5
Fagus sylvatica (s)	Beech	8.0	7
Carya glabra (s)	Hickory	9.1	9
Platonia (s)		12.5	3
Quercus alba (s)	Oak	16.2	9
Acer saccharum (s)	Sugar maple	15	8

1 = Graça et al. (2000); 2 = Bärlocher and Newell (1994); 3 Coq et al. 2010; 4 = Pereira et al. (1998); 4 = Bärlocher et al. (1995); 6 = Gessner (1991); 7 = Cameron and LaPoint (1978); 8 = Graça and Bärlocher (1998); 9 = Suberkropp et al. (1976)

compounds may also affect microbial decomposers (Harrison 1971) and litter-consuming detritivores and therefore delay decomposition of plant litter (Zucker 1983; Salusso 2000). The amount of phenolics in plant tissues varies with leaf species, age, and decomposition stage. Values for selected plants are summarized in Table 18.1.

A first step in many studies assessing the ecological effects of phenolics is an estimate of the total concentration of phenolic hydroxyl groups. To this end, the material is typically dried and extracted with water or, more commonly, with organic solvents such as acetone or methanol. Pre-drying may affect the efficiency of the extraction process (Chap. 5). In green or senescing material, polyphenol oxidases (PPO) may degrade phenolics. This can be prevented by PPO inhibitors like $K_2S_2O_4$ or by using liquid nitrogen or by boiling the sample for a few seconds to inactivate the enzymes. Solid-phase microextraction, pressurized liquid or fluid extraction, and microwave-assisted extraction are alternatives to conventional solvent extraction (Ainsworth and Gillespie 2007; Ajila et al. 2011).

The most commonly used approach to measure phenolics in the extract was originally designed to quantify the phenolic amino acid tyrosine (Folin and Denis 1912). Folin and Ciocalteu (1927) made the assay more sensitive and less prone to formation of precipitates. Preparation of the Folin-Denis or Folin-Ciocalteu reagent is relatively time-consuming, but these reagents are now commercially available (Waterman and Mole 1994). Here we present the procedure introduced by Folin and Ciocalteu (1927). The Folin-Ciocalteu assay is relatively non-specific, and other reductants in the extract may be inhibitory, additive, or enhance the reaction; for example, additive effects may occur in the presence of aromatic amines, high sugar levels, or ascorbic acid (Ainsworth and Gillespie 2007). More sophisticated approaches to measure phenolics include HPLC, GC, LC-MS, AFT-IR, and NMR (Ajila et al. 2011).

2 Equipment, Chemicals, and Solutions

2.1 Equipment and Material

- Eppendorf pipettes
- Vortex
- Refrigerator
- Dried leaves
- Mill or mortar and pestle
- Analytical balance (±0.1 mg precision)
- Eppendorf tubes
- Centrifuge
- Spectrophotometer

2.2 Chemicals

- Tannic acid standard (other phenolics such as gallic acid could also be used)
- Acetone
- Deionized water
- 2% Na_2CO_3
- 0.1 M NaOH
- Folin-Ciocalteu reagent (e.g., Sigma F-9252; diluted 1:2 with deionized water).

3 Experimental Procedures

3.1 Calibration

1. Prepare a stock solution of 25 mg tannic acid in 100 ml of acetone (30% water, 70% acetone).
2. Transfer 0, 0.2, 0.4, 0.6, 0.8, and 1.0 ml of the stock solution into six Eppendorf tubes and add 1.0, 0.8, 0.6, 0.4, 0.2, and 0 ml of distilled water, respectively. Mix with vortex.

3. Add 5 ml of 2% Na_2CO_3 in 0.1 M NaOH and mix.
4. After 5 min, add 0.5 ml of Folin-Ciocalteu reagent and mix.
5. After 120 min, read absorbance at 760 nm.
6. Plot tannic acid concentration vs. absorbance. The relationship should be linear.

3.2 *Measurement*

1. Grind dried leaves to powder passing through a 0.5 mm mesh size screen.
2. Weigh out approximately 100 mg portions of the ground leaves and transfer to Eppendorf tubes.
3. Extract phenolics in 5 ml of 70% acetone for 1 h at 4 °C.
4. Centrifuge (10,000–20,000 g, 10–20 min).
5. Take 0.5 ml of the supernatant (or another value between 0.1 and 0.8), and make up to 1 ml with distilled water as above.
6. Add Na_2CO_3 and Folin-Ciocalteu reagent as above.
7. After 120 min, read absorbance at 760 nm.
8. Based on the standard curve, determine tannic acid equivalents per mg of leaf powder. Remember that in Step 5, only a fraction (0.5 ml) of the sample was used.

References

Ainsworth, E. A., & Gillespie, K. M. (2007). Estimation of total phenolic content and other oxidation substrates in plant tissues using Folin-Ciocalteu reagent. *Nature Protocols, 2*, 875–877.

Ajila, C. M., Brar, S. K., Verma, M., Tyagi, R. D., Godbout, S., & Valéro, J. R. (2011). Extraction and analysis of polyphenols: Recent trends. *Critical Reviews in Biotechnology, 31*, 227–249.

Bärlocher, F., & Newell, S. Y. (1994). Phenolics and protein affecting palatability of *Spartina* leaves to the gastropod *Littoraria irrorata*. *P.S.Z.N.I.: Marine Ecology, 15*, 65–75.

Bärlocher, F., Canhoto, C., & Graça, M. A. S. (1995). Fungal colonization of alder and eucalypt leaves in two streams in Central Portugal. *Archiv für Hydrobiologie, 133*, 457–470.

Cameron, G. N., & LaPoint, T. W. (1978). Effects of tannins on the decomposition of Chinese tallow leaves by terrestrial and aquatic invertebrates. *Oecologia, 32*, 349–366.

Coq, S., Souquet, J. M., Meudec, E., Cheynier, V., & Hättenschwiler, S. (2010). Interspecific variation in leaf litter tannins drives decomposition in a tropical rain forest of French Guiana. *Ecology, 91*, 2080–2091.

Folin, O., & Ciocalteu, V. (1927). On tyrosine and tryptophane determination in proteins. *Journal of Biological Chemistry, 27*, 239–343.

Folin, O., & Denis, W. (1912). Tyrosine in proteins as determined by a new colorimetric method. *Journal of Biological Chemistry, 12*, 245–251.

Gessner, M. O. (1991). Differences in processing dynamics of fresh and dried leaf litter in a stream ecosystem. *Freshwater Biology, 26*, 387–398.

Gould, K. S., & Lee, D. W. (Eds.). (2002). *Anthocyanins and leaves. The function of anthocyanins in vegetative organs. Advances in Ecological Research* (Vol. 37). London: Academic.

Graça, M. A. S., & Bärlocher, F. (1998). Proteolytic gut enzymes in *Tipula caloptera* – Interaction with phenolics. *Aquatic Insects, 21*, 11–18.

Graça, M. A. S., Newell, S. Y., & Kneib, R. T. (2000). Grazing rates of organic living fungal biomass of decaying *Spartina alterniflora* by three species of salt-marsh invertebrates. *Marine Biology, 136*, 281–289.

Harborne, J. B. (2004). *A guide to modern techniques of plant analysis* (4th ed.). London: Academic.

Harrison, A. F. (1971). The inhibitory effect of oak leaf litter tannins on the growth of fungi, in relation to litter decomposition. *Soil Biology and Biochemistry, 3*, 167–172.

Hättenschwiler, S., & Vitousek, P. M. (2000). The role of polyphenols in terrestrial ecosystem nutrient cycling. *Trends in Ecology & Evolution, 15*, 238–243.

Lill, J. T., & Marquis, R. J. (2001). The effects of leaf quality on herbivore performance and attack from natural enemies. *Oecologia, 126*, 418–428.

Pereira, A. P., Graça, M. A. S., & Molles, M. (1998). Leaf decomposition in relation to litter physico-chemical properties, fungal biomass, arthopod colonization, and geographical origin of plant species. *Pedobiologia, 42*, 316–327.

Preiss, S., Degenhardt, J., & Gershenzon, J. (2015). Plant-animal dialogues. In G.-J. Krauss & D. H. Nies (Eds.), *Ecological biochemistry* (pp. 313–330). Weinheim: Wiley-VCH.

Salusso, M. M. (2000). Biodegradation of subtropical forest woods from north-west Argentina by *Pleurotus laciniatocrenatus*. *New Zealand Journal of Botany, 38*, 721–724.

Suberkropp, K., Godshalk, G. L., & Klug, M. J. (1976). Changes in the chemical composition of leaves during processing in a woodland stream. *Ecology, 57*, 720–727.

Tissier, A. T., Ziegler, J., & Vogt, T. (2015). Specialized plant metabolites: Diversity and biosynthesis. In G.-J. Krauss & D. H. Nies (Eds.), *Ecological biochemistry* (pp. 15–37). Weinheim: Wiley-VCH.

Waterman, P. G., & Mole, S. (1994). *Analysis of phenolic plant metabolites*. Oxford: Blackwell Scientific Publications. Methods in Ecology.

Zucker, W. V. (1983). Tannins: Does structure determine function? An ecological perspective. *American Naturalist, 121*, 335–365.

Chapter 19
Radial Diffusion Assay for Tannins

Manuel A. S. Graça and Felix Bärlocher

Keywords Condensed tannins · Enzyme inhibitors · Feeding deterrents · Food quality · Hydrolysable tannins · Litter quality · Phlorotannins · Proanthocyanidins · Protein precipitation · Secondary metabolites

1 Introduction

Few classes of secondary metabolites have been studied as intensively as tannins (Waterman and Mole 1994; Harborne 2004). There are three chemically distinct types of tannins: phlorotannins (restricted to brown algae), hydrolyzable tannins (some green algae and angiosperms), and condensed tannins (most widely distributed group; Waterman and Mole 1994; Chap. 20). Swain (1979) defines tannins as polymeric compounds (1) having molecular weights between 1000 and 3000; (2) having sufficient phenolic hydroxyl groups to complex with proteins and other macromolecules possessing carbonyl and amino groups; and (3) forming hydrogen bonds with macromolecules that are susceptible to auto-oxidation to form covalent linkages.

Total tannins are often negatively correlated with feeding preferences of vertebrates and invertebrates and with microbial decomposition rates (Rosset et al. 1982; Pennings et al. 2000; Coq et al. 2010). They also reduce nutrient extraction of ingested food and, in the long term, interfere with the reproduction of detritivores

Electronic Supplementary Material The online version of this chapter (https://doi.org/10.1007/978-3-030-30515-4_19) contains supplementary material, which is available to authorized users.

M. A. S. Graça (✉)
MARE – Marine and Environmental Sciences Centre & Department of Life Sciences, University of Coimbra, Coimbra, Portugal
e-mail: mgraca@ci.uc.pt

F. Bärlocher
Department of Biology, Mount Allison University, Sackville, NB, Canada
e-mail: fbaerloc@mta.ca

F. Bärlocher et al. (eds.), *Methods to Study Litter Decomposition*,
https://doi.org/10.1007/978-3-030-30515-4_19

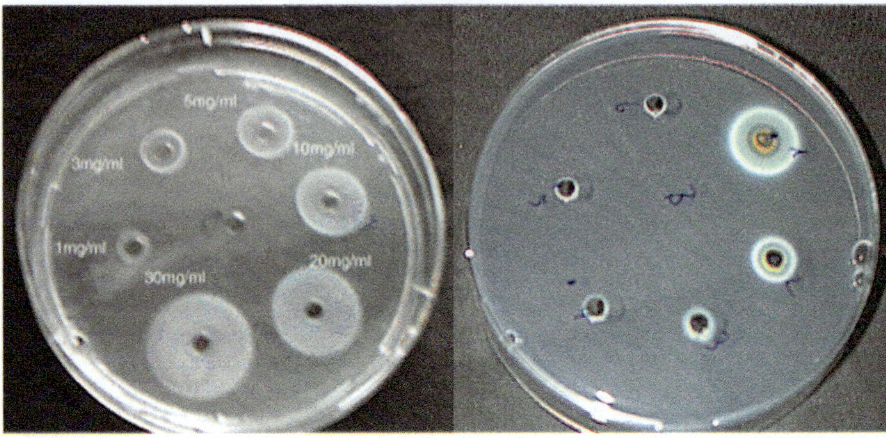

Fig. 19.1 Plaques on an agarose gel resulting from the precipitation of bovine serum albumin (BSA) in the gel by different amounts of commercial tannic acid (left) and different amounts of tannin in a leaf extract (right). (Photos D. Steiner Eawag, Switzerland)

consuming tannins (Harrison 1971; Zimmer and Topp 2000; Simpson and Raubenheimer 2001; Asplund et al. 2013). By retarding microbial degradation and invertebrate feeding, these compounds may influence litter decomposition and hence nutrient recycling rates.

There are two main approaches to measure tannins based on their interaction with proteins (Waterman and Mole 1994). One takes advantage of their ability to inhibit enzymatic reactions. The other assesses the capability of tannins to bind to and precipitate proteins. This capability is exploited in the radial diffusion assay (RDA) introduced by Hagerman (1987) and described here. A standard protein is dissolved in an agar gel. A well is then punched in the gel and a known amount of plant extract added to it. The tannins in the plant extract will bind to and thus precipitate the protein. The phenol-protein complex appears as plaque with an area proportional to the tanning or protein-precipitating activity (Fig. 19.1).

The radial diffusion assay does not quantify the total amount of tannins, but their ability to bind proteins, and therefore addresses only one aspect of the biological significance of these compounds. Some representative values for leaf litter are shown in Table 19.1.

2 Equipment, Chemicals and Solutions

2.1 Equipment and Materials

- Hot plate, with magnetic stirrer and stirring bar
- Eppendorf pipettes

Table 19.1 Tannic acid equivalents in undecomposed senescent leaves (% of leaf dry mass) determined with the radial diffusion assay developed by Hagerman (1987)

Leaf species	Common name	Tannic acid equivalents	References
Alnus glutinosa	Alder	1.5	1
		6.5–7.6	2
		5.2	3
Corylus avellana	Hazel	3.6	3
Eucalyptus globulus	Eucalyptus	3.5	1
Fagus sylvatica	Beech	2.9	3
Fraxinus excelsior	Ash	0.0	3
Platanus hybrida	London plane	2.7	3
Prunus avium	Cherry	3.1	3
Quercus deserticola	Encino Colorado	0.1–0.7	4
Quercus ilex	Evergreen oak	6.7	3
		4.0–12.0	5

1 = Bärlocher et al. (1995); 2 = Gessner (1991); 3 = Gessner and Chauvet (1994); 4 = Cuevas-Reyes et al. (2017); 5 = Solla et al. (2016)

- Polystyrene Petri plates, 8.5 cm diameter
- Cork borer (2–4 mm diameter)
- Refrigerator
- Drying oven (20 °C)
- Mill or mortar and pestle
- Analytical balance (±0.1 mg precision)
- Eppendorf tubes
- Centrifuge (10,000–20,000 *g*)
- Water bath

2.2 Chemicals

- Tannic acid standard
- Bovine serum albumin (BSA)
- Agarose
- Ascorbic acid
- Acetone
- Solution A: 0.2 M acetic acid (11.55 ml glacial acid in 1000 ml deionized water)
- Solution B: 0.2 M sodium acetate (16.4 g sodium acetate in 1000 ml deionized water)
- Solution C: Combine 74 ml of solution A and 176 ml of solution B, and adjust to 500 ml with deionized water.

3 Experimental Procedures

3.1 Preparation of Agarose Plates

1. Dissolve 5 g of agarose and 5.3 mg ascorbic acid (=60 μM) in 500 ml of Solution C on a hotplate. Stir continuously.
2. Cool down to 45 °C in a water bath.
3. Add BSA to a final concentration of 0.1% (500 mg in 500 ml) while stirring.
4. Dispense in exactly 9.5 ml portions into standardized Petri plates while carefully avoiding foam formation and bubbles in the gel.
5. After gelling, punch out 2–4 mm diameter wells.
6. Store plates at 4 °C.

3.2 Calibration

1. Dissolve 250 mg tannic acid in 25 ml of acetone (acetone: water, 70:30, vol:vol).
2. Add 9, 18, 27, 36, 45, or 54 μl of this solution to different wells. This is best done by repeatedly adding portions of 9 μl, allowing 10 min between each application.
3. After incubation for 3–4 days at 20 °C, determine the area of precipitation (Fig. 19.1). Measure the diameter of the plaque twice at right angles, and calculate the surface area from the average diameter. A plot of plaque area vs. amount of tannic acid in well (subtract area of well) should give a linear relationship.

3.3 Measurement

1. Grind up dried leaves to powder passing through a 0.5 mm mesh size screen.
2. Weigh approximately 100 mg portions of ground leaves, and transfer these to Eppendorf tubes.
3. Extract tannins by adding 0.5 ml of 70% acetone to the tubes. Leave for 1 h in the refrigerator for extraction.
4. Centrifuge at 10,000–20,000 *g* for 10–20 min.
5. Transfer aliquots of the supernatant (typically 4 × 9 μl) to the wells in the Petri plates.
6. Incubate for 3–4 days at 20 °C; then measure the area of precipitation plaques as for the standards above.
7. Express the results in tannic acid equivalents by comparing them to the protein-precipitating capacity of the tannic acid standard.

References

Asplund, J., Bokhorst, S., & Wardle, D. A. (2013). Secondary compounds can reduce the soil micro-arthropod effect on lichen decomposition. *Soil Biology and Biochemistry, 66*, 10–16.

Bärlocher, F., Canhoto, C., & Graça, M. A. S. (1995). Fungal colonization of alder and eucalypt leaves in two streams in Central Portugal. *Archiv für Hydrobiologie, 133*, 457–470.

Coq, S., Souquet, J. M., Meudec, E., Cheynier, V., & Hättenschwiler, S. (2010). Interspecific variation in leaf litter tannins drives decomposition in a tropical rain forest of French Guiana. *Ecology, 91*, 2080–2091.

Cuevas-Reyes, P., Pérez-López, G., Maldonado-López, Y., & González-Rodríguez, A. (2017). Effects of herbivory and mistletoe infection by *Psittacanthus calyculatus* on nutritional quality and chemical defense of *Quercus deserticola* along Mexican forest fragments. *Plant Ecology, 218*, 687–697.

Gessner, M. O. (1991). Differences in processing dynamics of fresh and air-dried leaf litter in a stream ecosystem. *Freshwater Biology, 26*, 387–398.

Gessner, M. O., & Chauvet, E. (1994). Importance of stream microfungi in controlling breakdown rates of leaf litter. *Ecology, 78*, 1807–1817.

Hagerman, A. E. (1987). Radial diffusion method for determining tannin in plant extracts. *Journal of Chemical Ecology, 13*, 437–449.

Harborne, J. B. (2004). *A guide to modern techniques of plant analysis* (4th ed.). London: Academic.

Harrison, A. F. (1971). The inhibitory effect of oak leaf litter tannins on the growth of fungi, in relation to litter decomposition. *Soil Biology and Biochemistry, 3*, 167–172.

Pennings, S. C., Carefoot, T. H., Zimmer, M., Danko, J. P., & Ziegler, A. (2000). Feeding preferences of supralittoral isopods and amphipods. *Canadian Journal of Zoology, 78*, 1918–1929.

Rosset, J., Bärlocher, F., & Oertli, J. J. (1982). Decomposition of conifer needles and leaf litter in two Black Forest and two Swiss Jura streams. *Internationale Revue der Gesamten Hydrobiologie, 67*, 695–711.

Simpson, S. J., & Raubenheimer, D. (2001). The geometric analysis of nutrient-allelochemical interactions: A case study using locusts. *Ecology, 82*, 422–439.

Solla, A., Milanović, S., Gallardo, A., Bueno, A., Corcobado, T., Cáceres, Y., Morcuende, D., Quesada, A., Moreno, G., & Pulido, F. (2016). Genetic determination of tannins and herbivore resistance in *Quercus ilex*. *Tree Genetics and Genomes, 12*, 117.

Swain, T. (1979). Tannins and lignins. In G. Rosenthal & D. H. Hanzen (Eds.), *Herbivores: Their interaction with secondary plant metabolites* (pp. 657–682). New York: Academic.

Waterman, P. G., & Mole, S. (1994). *Analysis of phenolic plant metabolites*. Oxford: Blackwell Scientific Publications, Methods in Ecology.

Zimmer, M., & Topp, W. (2000). Species-specific utilization of food sources by sympatric woodlice (Isopoda: Oniscidea). *Journal of Animal Ecology, 69*, 1071–1082.

Chapter 20
Acid Butanol Assay to Determine Bulk Concentrations of Condensed Tannins

Mark O. Gessner and Daniel Steiner

Keywords Food quality · Litter chemistry · Litter quality · Phenolics extraction · Polyphenolics · Proanthocyanidins · Secondary metabolites · Spectrophotometry · Tannin extraction · Tannin standard

1 Introduction

Tannins are a major class of secondary metabolites that are widespread in plants (Waterman and Mole 1994; Kraus et al. 2003a; Schofield et al. 2001). They are polyphenolics with molecular weights typically ranging from 1000 to 3000 (Swain 1979). By definition, tannins are capable of complexing and subsequently precipitating proteins (Chap. 19), and they can also bind to other macromolecules (Zucker 1983; Schofield et al. 2001; Shay et al. 2017). Two main, chemically distinct groups are commonly distinguished in vascular plants: hydrolysable tannins, which are further divided into the gallotannins and ellagitannins, and condensed tannins, which are also called proanthocyanidins and cannot be hydrolysed (Waterman and Mole 1994; Hättenschwiler and Vitousek 2000). Condensed tannins are the most widely distributed tannins in woody plants. They are usually also the most abundant group. Their diversity both within and among species is remarkable, although the complex polymeric structures can all be derived from relatively few building blocks of low-molecular-weight compounds (Schofield et al. 2001). The

M. O. Gessner (✉)
Department of Experimental Limnology, Leibniz Institute of Freshwater Ecology and Inland Fisheries (IGB), Stechlin, Germany

Department of Ecology, Berlin Institute of Technology (TU Berlin), Berlin, Germany
e-mail: gessner@igb-berlin.de

D. Steiner
Department of Aquatic Ecology, Eawag, Centre for Ecology, Evolution, and Biogeochemistry (CEEB), Kastanienbaum, Switzerland
e-mail: steiner@eawag.ch

F. Bärlocher et al. (eds.), *Methods to Study Litter Decomposition*,
https://doi.org/10.1007/978-3-030-30515-4_20

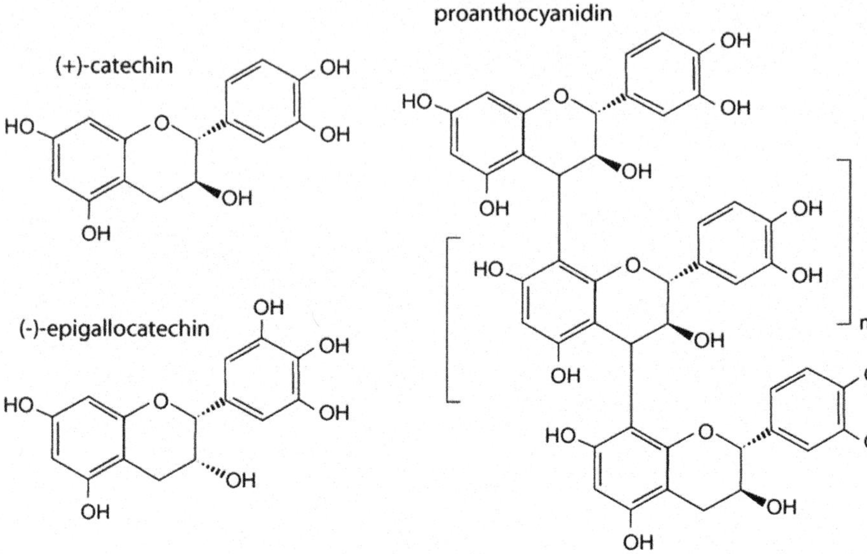

Fig. 20.1 Flavan-3-ols (+)-catechin and (−)-epigallocatechin, examples of monomeric precursors that polymerize to form macromolecular products such as linear proanthocyanidins composed of monomeric flavonoid units connected by C4–C8 linkages

most important monomers are flavan-3-ols such as catechin, epicatechin, gallocatechin and epigallocatechin; they react with one another in various ways, leading to either linear or branched polymers (Fig. 20.1).

Discussions on the ecological functions of tannins have mainly revolved around their capacity to bind to proteins and precipitate them (Zucker 1983). Both vertebrate and invertebrate herbivores can be affected. Herbivores also tend to prefer diets with low tannin concentrations, suggesting that tannins act as feeding deterrents to these consumers, although evidence supporting this tenet is inconclusive (Ayres et al. 1997). A range of additional general ecological functions at both the organismic and ecosystem level have been proposed (Hättenschwiler and Vitousek 2000; Kraus et al. 2003a; Madritch and Lindroth 2015; Chomel et al. 2016). These include the role of tannins as antioxidants, mediators of nutrient availability in soils, and regulating factors of litter decomposition. In addition, as Zucker (1983) pointed out early on, the chemical structure of tannins suggests that there is tremendous scope for specific chemical interactions of tannins both within organisms and in ecosystems. This view of multiple ecological roles for tannins has become widely accepted, although a clear overall picture has not emerged (Hättenschwiler and Vitousek 2000; Kraus et al. 2003a; Chomel et al. 2016).

If tannins remain in leaves following abscission (Table 20.1), similar mechanisms as in plant-herbivore interactions would be expected for trophic interactions between leaf litter and detritivores (e.g. Stout 1989; Ostrofsky 1997; Kraus et al. 2003a), with consequent effects on detritivore performance (Zimmer et al. 2002).

Table 20.1 Range of relative condensed tannin contents of undecomposed leaf litter from woody plant species

Leaf material	Tannin standard	Tannin concentration (mg/g dry mass)	Reference
5 *Acer* species	None	15–130[a]	1
6 *Quercus* species	None	17–110[a]	1
37 woody plant species	None	3–280[a]	1
6 tropical *Eucalyptus* species	None	9–25[b]	2
6 nontropical *Eucalyptus* species	None	8–25[b]	2
6 non-*Eucalyptus* species	None	1–21[b]	2
Populus fremontii, P. angustifolia and hybrids; common garden	Tannins of *P. angustifolia*	2–96	3
Populus fremontii, P. angustifolia and hybrids; field	Tannins of *P. angustifolia*	2–96	3
Populus tremuloides, 5 genotypes; high nitrogen	Purified tannins of *P. tremuloides*	16–89	4
Populus tremuloides, 5 genotypes; low nitrogen	Purified tannins of *P. tremuloides*	33–106	4
3 broad-leaved tree species	Tannins of *Abies balsamea* branch tips	42–172	5
5 conifer species	Tannins of *Abies balsamea* branch tips	0–287	5
Populus tremuloides or *Betula papyrifera*; ambient atmosphere	Purified tannins of senesced aspen and birch leaves	22–26	6
Populus tremuloides or *Betula papyrifera*; elevated CO_2, O_3 or both	Purified tannins of senesced aspen and birch leaves	22–44	6
16 neotropical woody plant species	Tannins of grape seeds	4–97	7
Alnus glutinosa mixed with *Fraxinus angustifolia or Pistacia terebinthus*	Not reported	18–163	8

1 = Ostrofsky (1993); 2 = Campbell and Fuchshuber (1995); 3 = Schweitzer et al. (2004); 4 = Madritch et al. (2006); 5 = Preston et al. (2009); 6= Liu et al. (2009); 7 = Coq et al. (2010); 8 = García-Palacios et al. (2016)

[a]Optical densities per g of extracted dry leaf material

[b]Arbitrary relative numbers

Furthermore, there is evidence that tannins interact with microbial decomposers (Kraus et al. 2003a), indicating significant potential for tannins to affect litter decomposition in both terrestrial (Horner et al. 1988) and aquatic environments (Stout 1989; Ostrofsky 1993; Campbell and Fuchshuber 1995). Tannin concentration thus could be an important indicator of chemical litter quality when addressing a variety of questions relating to litter use and turnover.

The structural diversity of condensed tannins provides challenges for accurate quantitative analyses. Chromatographic characterization of cleavage products is therefore increasingly being used (Waterman and Mole 1994; Hernes and Hedges 2000; Coq et al. 2010) and ^{13}C-NMR, MALDI-TOF and other detection methods have been applied to identify tannins (Schofield et al. 2001; Shay et al. 2017). Nevertheless, two simple methods to determine bulk concentrations of condensed tannins can also provide ecologically meaningful information, the vanillin method and the acid butanol assay (Hagerman and Butler 1989; Waterman and Mole 1994; Schofield et al. 2001; Kraus et al. 2003b). Under strongly acidic conditions, oxidative depolymerization of condensed tannins in alcohols yields anthocyanidins (e.g. cyanidin and delphinidin), which absorb light in the visible range. Condensed tannins can thus be quantified spectrophotometrically following depolymerization. The acid butanol assay recommended by Hagerman & Butler (1989) and Waterman and Mole (1994) for determining total condensed tannins is based on this reaction.

Before tannins can be quantified by wet chemical or chromatographic methods, they need to be extracted from the plant matrix. Various extractants and extraction procedures have been described. Their relative efficiency depends on the analysed material, due to differences in both tannin structure and the sample matrix (Waterman and Mole 1994; Yu and Dahlgren 2000). One of the most common and frequently recommended extraction solvents is 50% methanol (Hagerman 1988; Waterman and Mole 1994). It is used in the procedure described below. However, recent studies have suggested that condensed tannin extraction in acid acetone-butanol-water is preferable (Grabber et al. 2013; Shay et al. 2017). The protocol of the acid butanol assay presented here has been adopted from Porter et al. (1986) and is also described in the detailed review by Waterman and Mole (1994) and the comprehensive compilation of methods for tannin analyses by Hagerman (2011).

2 Equipment, Chemicals and Solutions

2.1 Equipment and Material

- Freeze-dryer
- Mill
- Analytical balance (±0.1 mg precision)
- Glass or polypropylene tubes (10 ml, pressure resistant, with tightly closing screw-caps)
- Multiple-position magnetic stirrer (e.g. Variomag Telesystem HP 15 or Poly 15, or IKAMAG RO 15 Power, all with 15 stirring points)
- Disposable syringes (5 ml)
- Rack holding syringes upright on the magnetic stirrer
- Glass fibre filters (e.g. GF/F, Whatman)
- Cork borer (well sharpened; size matching the inner diameter of syringes)
- Stop cocks with Luer-lock fitting the syringe tips
- Magnetic stirring bars (5 mm length)
- Volumetric flasks (100, 500, 1000 ml)

- Pipettes (e.g. Eppendorf Multipette and/or Varipette; 100–500 μl and 10 ml)
- Glass vials (e.g. 1.6-ml HPLC vials, with Teflon-lined caps), individually weighed to the nearest 0.1 mg
- Test tubes (10 ml)
- Vortex
- Water or dry bath (95 °C)
- Spectrophotometer (set at 550 nm)

2.2 *Chemicals*

- Methanol, reagent grade
- Deionized water (e.g. Nanopure®)
- $FeSO_4 \cdot 7\ H_2O$
- *n*-Butanol, reagent grade
- Concentrated HCl (37%)
- Quebracho tannin (preferably purified; see Hagerman 2011), optional

2.3 *Solutions*

- Solution 1: 50% methanol:H_20 (v/v).
- Solution 2: Dissolve 700 mg $FeSO_4 \cdot 7\ H_2O$ in 50 ml conc. HCl and adjust volume to 1000 ml with *n*-butanol.
- Solution 3: Stock solution of quebracho tannin standard (10–100 mg l^{-1}, depending on purity of tannin): weigh out 10–100 mg of (purified) quebracho tannin to the nearest 0.1 mg and dissolve in 100 ml of Solution 1, then dilute tenfold with Solution 1.
- Standards: Use Solutions 1 (50% methanol) and 3 to prepare quebracho tannin standard solutions in the range 0–2.0 mg ml^{-1} or lower depending on purity of the standard used.

3 Experimental Procedures

3.1 *Tannin Extraction*

1. Dry leaves and grind to powder that passes through a 0.5-mm mesh screen.
2. Cut discs from glass fibre filters with a well-sharpened cork borer and place inside the disposable syringes.
3. Connect syringes to stop cocks with valves closed.
4. Add 50 mg sample material (weighed to the nearest 0.1 mg) to the syringes.

5. Place a small stirring bar in each syringe.
6. Place syringes on custom-built rack on the magnetic stirrer. The Luer ends of the syringes may have to be slightly shortened to minimize the vertical distance between the surface of the magnetic stirrer and the stirring bars in the syringes, so as to ensure continuous movement of the bars during extraction.
7. Add 400 μl of 50% methanol (Solution 1).
8. Connect plungers to the top of syringe barrels.
9. Extract tannins for 30 min with stirring at room temperature.
10. Filter extract directly into tared HPLC vials by slowly pushing plunger into the syringe barrel.
11. Repeat extraction three more times with 350 μl of Solution 1 (50% methanol) each time.
12. Rinse the stop cock with 50 μl methanol (50%) after the first two extraction steps.
13. Cap vials and reweigh them to the nearest 0.1 mg.
14. Calculate the volume of the extract, assuming a density of 0.9266 g ml^{-1} for 50% methanol.

3.2 *Spectrophotometric Analysis*

1. Pipette exact volume of 100–500 μl sample extract in test tube.
2. Add appropriate volume of deionized water to adjust total volume (i.e. sample extract plus water) to 500 μl.
3. Add 7 ml of Solution 2 ($FeSO_4 \cdot 7\,H_2O$) and vortex.
4. Measure absorbance at 550 nm (control to correct for colour of the extract).
5. Place tube in water bath at 95 °C and incubate for exactly 50 min.
6. Let cool to room temperature before measuring absorbance again at 550 nm.
7. Calculate absorbance due to the acid butanol reaction by subtracting the absorbance before heating from that after heating.
8. If (purified) quebracho tannin is available, proceed in the same way with the standard tannin solutions to establish a standard curve.
9. Express results in relative units or, preferably, in equivalents of (purified) quebracho or specifically prepared tannin (see Table 20.1) based on absorbance readings and the standard curve.

4 Final Remarks

The assay is very sensitive to varying amounts of water. Therefore, it is essential to ensure that the volumetric ratio of Solutions 1 and 2 is exactly 1:14 (e.g. 500 μl of Solution 1 plus 7 ml of Solution 2). The water content is then 6.8%, which is close to the water content found by Porter et al. (1986) to yield the highest colour yield.

Acetone has been reported to interfere with the acid butanol assay (Waterman and Mole 1994). Consequently, it was previously recommended not to use the popular acetone-water mixture to extract condensed tannins, unless the extract is completely evaporated and the residue redissolved in a solvent compatible with the assay. However, in sharp contrast to this advice, Grabber et al. (2013) and Shay et al. (2017) found that including 50% acetone in the acid-butanol reagent greatly increases yield, most likely because condensed tannins are completely solubilized under these conditions. This effect is expected to be particularly strong with leaf litter where the fraction of insoluble condensed tannins is increased compared to fresh leaf tissue (Shay et al. 2017). More important than in the procedure described above, it is important to add iron when acetone is included in the reagent mix.

Waterman and Mole (1994) advised against using an unheated reagent-sample mixture because some substances in plant tissue may yield red coloration even without heating. However, in our experience with a wide range of leaf litter from deciduous tree and shrub species, this potential problem was not generally encountered. Conversely, the substitution of HCl by H_2O as recommended by Waterman and Mole (1994) can result in precipitates.

Condensed tannin standards of sufficient purity are not commercially available, limiting quantitative comparisons among studies. To improve this situation, the use of purified quebracho tannin has been recommended. A protocol for purification – along with a wealth of useful information on tannin structural chemistry, other purification methods, biological activities and biosynthesis – can be downloaded at https://www.users.miamioh.edu/hagermae (Hagerman 1998–2011). Others recommend using extracts of the particular plant material that is analysed. Purity of the extracted standard should be checked, although this effort has been rarely made in previous studies (Grabber et al. 2013).

Cyanidin or other commercially available anthocyanidins (e.g. delphinidin, procyanidin or prodelphinidin) can be used as alternative relative standards (Hagerman and Butler 1989). Colour yields vary, however, that of cyaniding, for example, being lower than that of delphinidin, although that difference is small compared to the more than 30-fold lower yield of quebracho tannin (Schofield et al. 2001). Large discrepancies in colour yields evidently complicate the absolute quantification of condensed tannins in diverse samples.

Schofield et al. (2001), Grabber et al. (2013) and Shay et al. (2017) discuss additional weaknesses and pitfalls of the acid butanol assay. One of them is that the standard curve may be slightly curvilinear or discontinuous, the reasons for which are unknown (Waterman and Mole 1994; Grabber et al. 2013). One possibility to circumvent this effect may be to dilute sample extracts and use 5-cm or 10-cm cuvettes instead of standard 1-cm cuvettes for spectrophotometric measurements.

References

Ayres, M. P., Clausen, T. P., MacLean, F., Jr., Redman, A. M., & Reichardt, P. B. (1997). Diversity of structure and antiherbivore activity in condensed tannins. *Ecology, 78*, 1696–1712.

Campbell, I. C., & Fuchshuber, L. (1995). Polyphenols, condensed tannins, and processing rates of tropical and temperate leaves in an Australian stream. *Journal of the North American Benthological Society, 14*, 174–182.

Chomel, M., Guittonny-Larchevêque, M., Fernandez, C., Christiane, G., DesRochers, A., Paré, D., Jackson, B. G., & Baldy, V. (2016). Plant secondary metabolites: A key driver of litter decomposition and soil nutrient cycling. *Journal of Ecology, 104*, 1527–1541.

Coq, S., Souquet, J. M., Meudec, E., Cheynier, V., & Hättenschwiler, S. (2010). Interspecific variation in leaf litter tannins drives decomposition in a tropical rain forest of French Guiana. *Ecology, 91*, 2080–2091.

García-Palacios, P., Shaw, E. A., Wall, D. H., Hättenschwiler, S., & Knops, J. M. H. (2016). Temporal dynamics of biotic and abiotic drivers of litter decomposition. *Ecology Letters, 19*, 554–563.

Grabber, J., Zeller, W. E., & Mueller-Harvey, I. (2013). Acetone enhances the direct analysis of procyanidin- and prodelphinidin-based condensed tannins in *Lotus* species by the butanol-HCl-iron assay. *Journal of Agricultural and Food Chemistry, 61*, 2669–2678.

Hagerman, A. E. (1988). Extraction of tannin from fresh and preserved leaves. *Journal of Chemical Ecology, 14*, 453–461.

Hagerman, A. E. (1998–2011). *The tannin handbook.* Oxford: Miami University. https://www.users.miamioh.edu/hagermae, http://chemistry.muohio.edu/hagerman

Hagerman, A. E., & Butler, L. G. (1989). Choosing appropriate methods and standards for assaying tannin. *Journal of Chemical Ecology, 15*, 1795–1810.

Hättenschwiler, S., & Vitousek, P. M. (2000). The role of polyphenols in terrestrial ecosystem nutrient cycling. *Trends in Ecology and Evolution, 15*, 238–243.

Hernes, P. J., & Hedges, J. I. (2000). Determination of condensed tannin monomers in environmental samples by capillary gas chromatography of acid depolymerization extracts. *Analytical Chemistry, 72*, 5115–5124.

Horner, J. D., Gosz, J. R., & Cates, R. G. (1988). The role of carbon-based plant secondary metabolites in decomposition in terrestrial ecosystems. *The American Naturalist, 132*, 869–883.

Kraus, T. E. C., Dahlgren, R. A., & Zasoski, R. J. (2003a). Tannins in nutrient dynamics of forest ecosystems – A review. *Plant and Soil, 256*, 41–66.

Kraus, T. E. C., Yu, Z., Preston, C. M. R., Dahlgren, A., & Zasoski, R. J. (2003b). Linking chemical reactivity and protein precipitation to structural characteristics of foliar tannins. *Journal of Chemical Ecology, 29*, 703–730.

Liu, L. L., King, J. S., Giardina, C. P., & Booker, F. L. (2009). The influence of chemistry, production and community composition on leaf litter decomposition under elevated atmospheric CO_2 and tropospheric O_3 in a northern hardwood ecosystem. *Ecosystems, 12*, 401–416.

Madritch, M. D., & Lindroth, R. L. (2015). Condensed tannins increase nitrogen recovery by trees following insect defoliation. *New Phytologist, 208*, 410–420.

Madritch, M., Donaldson, J. R., & Lindroth, R. L. (2006). Genetic identity of *Populus tremuloides* litter influences decomposition and nutrient release in a mixed forest stand. *Ecosystems, 9*, 528–537.

Ostrofsky, M. L. (1993). Effect of tannins on leaf processing and conditioning rates in aquatic ecosystems: An empirical approach. *Canadian Journal of Fisheries and Aquatic Sciences, 50*, 1176–1180.

Ostrofsky, M. L. (1997). Relationship between chemical characteristics of autumn-shed leaves and aquatic processing rates. *Journal of the North American Benthological Society, 16*, 750–759.

Porter, L. J., Hrstich, L. N., & Chan, B. G. (1986). The conversion of procyanidin and delphinidin. *Phytochemistry, 25*, 223–230.

Preston, C. M., Nault, J. R., Trofymow, J. A., Smyth, C., & CIDET Working Group (2009). Chemical changes during 6 years of decomposition of 11 litters in some Canadian forest sites. Part 1. Elemental composition, tannins, phenolics, and proximate fractions. *Ecosystems, 12*, 1053–1077.

Schofield, P., Mbugua, D. M., & Pell, A. N. (2001). Analysis of condensed tannins: A review. *Animal Feed Science and Technology, 91*, 21–40.

Schweitzer, J. A., Bailey, J. K., Rehill, B. J., Martinsen, G. D., Hart, S. C., Lindroth, R. L., Keim, P., & Whitham, T. G. (2004). Genetically based trait in a dominant tree affects ecosystem processes. *Ecology Letters, 7*, 127–134.

Shay, P.-E., Trofymow, J. A., & Constabel, C. P. (2017). An improved butanol-HCl assay for quantification of water-soluble, acetone:methanol-soluble, and insoluble proanthocyanidins (condensed tannins). *Plant Methods, 13*, 63.

Stout, R. J. (1989). Effects of condensed tannins on leaf processing in mid-latitude and tropical streams: A theoretical approach. *Canadian Journal of Fisheries and Aquatic Sciences, 46*, 1097–1106.

Swain, T. (1979). Tannins and lignins. In G. Rosenthal & D. H. Janzen (Eds.), *Herbivores: Their interaction with secondary plant metabolites* (pp. 657–682). New York: Academic.

Waterman, P. G., & Mole, S. (1994). *Analysis of phenolic plant metabolites*. Oxford: Blackwell Scientific Publications, Methods in Ecology.

Yu, Z., & Dahlgren, R. A. (2000). Evaluation of methods for measuring polyphenols in conifer foliage. *Jounal of Chemical Ecology, 26*, 2119–2140.

Zimmer, M., Pennings, S. C., Buck, T. L., & Carefoot, T. H. (2002). Species-specific patterns of litter processing by terrestrial isopods (Isopoda: Oniscidea) in high intertidal salt marshes and coastal forests. *Functional Ecology, 16*, 596–607.

Zucker, W. V. (1983). Tannins: Does structure determine function? An ecological perspective. *The American Naturalist, 121*, 335–365.

Chapter 21
Lignin and Cellulose

Mark O. Gessner

Keywords Food quality · Forage fibre analysis · Gravimetry · Litter chemistry · Litter decomposition rate · Litter quality · Plant cell wall · Plant polymers · Van Soest method

1 Introduction

Lignin and cellulose are structural constituents of vascular plants that confer toughness and tensile strength to plant tissues (Chap. 18). Together they typically make up the largest fraction of litter dry mass (Table 21.1). Consequently, plant litter rich in these compounds tends to be highly refractory, with high concentrations particularly of lignin being conducive to slow litter decomposition (e.g. Berg and McClaugherty 2003; Hladyz et al. 2009).

On leaf litter decaying in streams, both biomass accumulation and sporulation activity of fungi decrease as litter lignin concentrations increase, suggesting that the negative effect of lignin is mediated at least partly through an impact on fungal decomposers (Gessner and Chauvet 1994; Maharning and Bärlocher 1996). In addition, lignin and cellulose concentrations may influence litter palatability to leaf-shredding invertebrates and hence litter consumption by these shredders. Freshwater invertebrates typically lack the enzymatic complements to digest cellulose and lignin; therefore, diets rich in these compounds are of poor nutritional quality to shredders, and this may have negative consequences for their survival, growth rate and fecundity (Bärlocher 1985; Suberkropp 1992; Graça 1993; Rong et al. 1995). However, some taxa (e.g. some *Tipula* species) may gain access to at least cellulose by means of a symbiotic cellulose-degrading gut flora (Kukor and Martin 1987; Martin 1987).

M. O. Gessner (✉)
Department of Experimental Limnology, Leibniz Institute of Freshwater Ecology and Inland Fisheries (IGB), Stechlin, Germany

Department of Ecology, Berlin Institute of Technology (TU Berlin), Berlin, Germany
e-mail: gessner@igb-berlin.de

F. Bärlocher et al. (eds.), *Methods to Study Litter Decomposition*,
https://doi.org/10.1007/978-3-030-30515-4_21

Table 21.1 Inter- and intraspecific variation in concentrations of lignin and cellulose in undecomposed leaf litter as determined with the fibre forage method by Goering and Van Soest (1970)

Leaf species	Origin	Lignin (% leaf dry mass)	Cellulose (% leaf dry mass)	References
Fraxinus excelsior	France (autumn 1989)	6.8 ± 0.3	18.6 ± 1.0	1, 2
Prunus avium	France (autumn 1989)	8.4 ± 1.0	16.3 ± 0.3	1, 2
Alnus glutinosa	France (autumn 1989)	8.0 ± 0.7	13.8 ± 1.5	1, 2
Corylus avellana	France (autumn 1989)	13.3 ± 0.9	23.3 ± 1.6	1, 2
Platanus hybrida	France (autumn 1989)	30.9 ± 0.8	24.8 ± 1.0	1, 2
Fagus sylvatica	France (autumn 1989)	25.5 ± 0.8	32.2 ± 2.8	1, 2
Quercus ilex	France (autumn 1989)	18.5 ± 1.0	27.8 ± 3.0	1, 2
Alnus glutinosa	Switzerland (green litter)	8.3 ± 0.6	7.0 ± 0.6	3
Alnus glutinosa	Switzerland (brown litter)	13.0 ± 0.4	7.4 ± 0.4	3
Alnus glutinosa	Sweden (early autumn)	5.5 ± 0.8	4.7 ± 0.4	3
Alnus glutinosa	Sweden (late autumn)	8.1 ± 0.6	6.6 ± 0.8	3
Alnus glutinosa	UK (sun leaves)	5.2 ± 0.2	6.1 ± 0.2	3
Alnus glutinosa	France (autumn 2002)	7.3 ± 0.6	6.0 ± 1.2	3
Alnus glutinosa	France (autumn 2003)	7.0 ± 0.2	5.7 ± 0.6	3
Alnus glutinosa	Portugal (streamside)	13.9 ± 2.0	7.4 ± 1.2	3

Values are means ±1 SD

1 = Gessner and Chauvet (1994); 2 = Gessner (unpublished data); 3 = Lecerf and Chauvet (2008)

A variety of methods have been developed to determine lignin and cellulose in plant tissues (e.g. McLellan et al. 1991; Van Soest et al. 1991; Hatfield et al. 1999), interest in which has recently been rekindled (e.g. Li et al. 2016; Chen et al. 2017). One simple approach, which has been widely used for forage fibre analyses and litter decomposition studies in both terrestrial and aquatic environments, consists of determining the residual weight of samples following successive removal of various tissue constituents. The first step is the extraction of components soluble in an acid detergent. Results by Ryan et al. (1990) suggest that with tree leaves and wood this approach produces similar results as a somewhat more complicated alternative method. Since the approach does not necessarily determine concentrations of cellulose and lignin as defined chemically, the fractions resulting from the forage fibre method are referred to as proximate cellulose and lignin.

The aim of the method presented here is to assess the concentrations of proximate lignin and cellulose in plant litter. Concentrations are determined gravimetrically using the downscaled acid-detergent fibre procedures proposed by Goering & Van Soest (1970) with slight modifications.

2 Equipment, Chemicals and Solutions

2.1 Equipment and Material

- Analytical balance (0.1 mg precision)
- Desiccator
- Dried sample powder ground to pass a 0.5-mm mesh-screen
- Eight screw-cap extraction tubes (approx. 40 ml, pressure-resistant)
- Dry bath or water bath (100 °C) with submersible rack holding at least 8 tubes
- Sixteen Gooch crucibles made of borosilicate glass with fritted glass bases, porosity no. 2
- Filter manifold or individual units adapted for holding 8 crucibles (individual pressure regulation preferable)
- Pump for creating vacuum in filtration systems
- Hot plate or kettle for boiling H_2O
- Eight small trays (e.g. 10 × 15 cm) resistant to 72% sulphuric acid
- Latex gloves
- Eight acid-resistant spatulas or glass rods (about 8 cm long)
- Drying oven (105 °C)
- Muffle furnace (550 °C)

2.2 Chemicals

- Sulphuric acid, 0.5 M (reagent grade)
- Hexadecyltrimethylammonium bromide = Cetyltrimethylammonium bromide (CTAB), 20 g l^{-1}
- Decahydronaphthalene (reagent grade)
- Acetone (reagent grade) in spray bottles
- Sulphuric acid, 72% by weight (reagent grade)

2.3 Solutions

- Solution 1: Acid detergent solution: prepare 0.5 M sulphuric acid from low-molarity stock solution, check molarity by titration, adjust if necessary, then add the detergent CTAB (20 g l^{-1}) and stir.
- Solution 2: Prepare sulphuric acid at 72% by weight as described below. Weigh required amount of water into a volumetric flask and add the calculated amount of H_2SO_4 in *small* portions and *very slowly* with occasional swirling. Because of strong heat production with risk of explosion, constantly cool flask in a water bath (e.g. in a sink). Allow sufficient time for cooling. Do not fill up flask to calibration mark. Finally, let cool to 20 °C and adjust to exact volume.
- Preparation of an acid solution: Given an acid at a concentration of A% and a density, δ, an acid at the concentration of X% is obtained as follows:
  - In mass units (for 100 g of acid solution):

    100 × (X/A) of acid at the concentration A%
    100–100 × (X/A) of H_2O

  - In volumetric units (e.g. in ml):

    100 × (X/A)/δ of acid at the concentration A%
    100–100 × (X/A) of H_2O

- For example, for sulphuric acid at 72% starting with 96% ($\delta = 1.83$ g cm^{-3}):
  - For 100 g of solution:

    100 × (72/96) = 75.0 g of acid at 96%
    100–75.0 = 25.0 g of H_2O

  - Or in volumetric units:

    100 × (72/96)/1.83 = 41.0 ml of acid at 96%
    100–100 × (72/96) = 25.0 ml of H_2O

3 Experimental Procedures

3.1 Sample Preparation

1. Weigh clean and oven-dry crucibles to the nearest 0.1 mg.
2. Weigh air-dry sample ground to pass through a 0.2-mm mesh screen (245–255 mg to the nearest 0.1 mg) and place in extraction tube.
3. Weigh same amount of sample in ignited, tared porcelain or aluminium pans for determining humidity content and ash-free dry mass.
4. Add to tubes 20 ml of acid-detergent solution and 0.4 ml decahydronaphthalene.

3.2 Acid-Detergent Fibre Determination

1. Heat tubes to boiling for 5–10 min in a water bath with occasional swirling.
2. Reduce heat as boiling begins to avoid foaming, and then boil for 60 min at low, even level.
3. Using light suction, filter tube content on a tared Gooch crucible set on a filter manifold.
4. Quantitatively recover particles in the tubes.
5. Break up the filtered mat with a spatula or glass rod and rinse twice with hot water (90–100 °C), including sides of the crucible.
6. Repeat wash with acetone until no more colour is removed, ensuring that all lumps are broken up.
7. Suck the acid-detergent fibre free of acetone and dry overnight at 105 °C.
8. Place the oven-dry crucible in a desiccator for 1 h and then weigh to the nearest 0.1 mg.
9. Calculate acid-detergent fibre (ADF) as follows:

$$\frac{W_0 - W_t}{W_S} \times 100 = ADF \tag{21.1}$$

where
W_0 = weight of the oven-dry crucible including fibre
W_t = tared weight of the oven-dry crucible
W_S = oven-dry sample weight.

10. Correct value for water content of the sample.

3.3 Acid-Detergent Cellulose and Lignin Determination

11. Cover the contents of the crucible with cooled (15 °C; water bath) 72% H_2SO_4 and stir with a spatula or glass rod to a smooth paste, breaking all lumps.

12. Fill crucible about half with acid, stir and keep at 20–23 °C.
13. Leave spatula or glass rod in crucible.
14. Refill with 72% H_2SO_4 and stir at hourly intervals as acid drains away, ensuring that samples are continuously covered, although crucibles do not need to be kept full at all times; three additions of acid suffice.
15. Filter off after 3 h as much acid as possible, starting with a weak vacuum.
16. Wash contents abundantly with hot water until free from acid.
17. Also rinse and then remove stirring rod.
18. Dry crucible overnight at 105 °C.
19. Place crucible in desiccator for 1 h and weigh to the nearest 0.1 mg.
20. Ignite crucible in a muffle furnace at 550 °C for 3 h and then cool to 105 °C.
21. Place in desiccator for 1 h and weigh.
22. Calculate acid-detergent cellulose (ADC) as follows:

$$\frac{L_a}{W_S} \times 100 = ADC \quad (21.2)$$

where
L_a = loss due to 72% H_2SO_4 treatment
W_S = oven-dry sample weight.

23. Caclulate acid-detergent lignin (ADL) as follows:

$$\frac{L_i}{W_S} \times 100 = ADL \quad (21.3)$$

where
L_i = loss upon ignition after 72% H_2SO_4 treatment
W_S = oven-dry sample weight.

24. Correct values for water content of sample.

4 Final Remarks

It is important to wear a laboratory coat, security glasses and latex gloves at all times during handling of acid. Particular caution is needed when preparing the 72% sulphuric acid because of strong heat production with risk of explosion when diluting the concentrated acid.

Glass crucibles may lose some weight when exposed to 550 °C. This can reduce accuracy of the method when not taken into account and sample sizes are small.

References

Bärlocher, F. (1985). The role of fungi in the nutrition of stream invertebrates. *Botanical Journal of the Linnean Society, 91*, 83–94.

Berg, B., & McClaugherty, C. (2003). *Plant litter – Decomposition, humus formation, carbon sequestration*. Berlin: Springer.

Chen, L., Dou, J., Ma, Q., Li, N., Wu, R., Bian, H., Yelle, D. J., Vuorinen, T., Fu, S., Pan, X., & Zhu, J. Y. (2017). Rapid and near-complete dissolution of wood lignin at ≤80°C by a recyclable acid hydrotrope. *Science Advances, 3*, e1701735.

Gessner, M. O., & Chauvet, E. (1994). Importance of stream microfungi in controlling breakdown rates of leaf litter. *Ecology, 75*, 1807–1817.

Goering, H. K., & Van Soest, P. J. (1970). Forage fiber analyses (apparatus, reagents, procedures, and some applications). In *Agriculture handbook 379* (pp. 1–20). Washington DC: U.S. Department of Agriculture.

Graça, M. A. S. (1993). Patterns and processes in detritus-based stream systems. *Limnologica, 23*, 107–114.

Hatfield, R. D., Grabber, J., Ralph, J., & Brei, K. (1999). Using the acetyl bromide assay to determine lignin concentrations in herbaceous plants: Some cautionary notes. *Journal of Agricultural and Food Chemistry, 47*, 628–632.

Hladyz, S., Gessner, M. O., Giller, P. S., Pozo, J., & Woodward, G. (2009). Resource quality and stoichiometric constraints on stream ecosystem functioning. *Freshwater Biology, 54*, 957–970.

Kukor, J. J., & Martin, M. M. (1987). Nutritional ecology of fungus-feeding arthropods. In F. Slansky Jr. & J. G. Rodriguez (Eds.), *Nutritional ecology of insects, mites, spiders, and related invertebrates* (pp. 791–814). New York: Wiley.

Lecerf, A., & Chauvet, E. (2008). Intraspecific variability in leaf traits strongly affects alder leaf decomposition in a stream. *Basic and Applied Ecology, 9*, 598–605.

Li, N., Pan, X., & Alexander, J. (2016). A facile and fast method for quantitating lignin in lignocellulosic biomass using acidic lithium bromide trihydrate (ALBTH). *Green Chemistry, 18*, 5367.

Maharning, A. R., & Bärlocher, F. (1996). Growth and reproduction in aquatic hyphomycetes. *Mycologia, 88*, 80–88.

Martin, M. M. (1987). Acquired enzymes in detritivores. In M. M. Martin (Ed.), *Invertebrate-microbial interactions. Ingested fungal enzymes in arthropod biology* (pp. 49–72). Ithaca: Comstock Publishing Associates.

McLellan, T. M., Aber, J. D., & Martin, M. E. (1991). Determination of nitrogen, lignin and cellulose content of decomposing leaf material by near infrared reflectance spectroscopy. *Canadian Journal of Forest Research, 21*, 1684–1688.

Rong, Q., Sridhar, K. R., & Bärlocher, F. (1995). Food selection in three leaf-shredding stream invertebrates. *Hydrobiologia, 316*, 173–181.

Ryan, M. G., Melillo, J. M., & Ricca, A. (1990). A comparison of methods for determining proximate carbon fractions of forest litter. *Canadian Journal of Forest Reserch, 20*, 166–171.

Suberkropp, K. (1992). Interactions with invertebrates. In F. Bärlocher (Ed.), *The ecology of aquatic hyphomycetes, Ecological Studies* (Vol. 94, pp. 118–133). Berlin: Springer.

Van Soest, P. J., Robertson, J. B., & Lewis, B. A. (1991). Carbohydrate methodology, metabolism, and nutritional implications in dairy cattle. *Journal of Dairy Science, 74*, 3583–3597.

Chapter 22
Physical Litter Properties: Leaf Toughness and Tensile Strength

Manuel A. S. Graça and Martin Zimmer

Keywords Cellulose · Detritivore feeding · Feeding deterrent · Food quality · Leaf mass per area · Lignin · Litter quality · Litter traits · Penetrometer · Specific leaf area · Tear force · Tensiometer

1 Introduction

Studies on plant decomposition have emphasized the role of internal (plant characteristics) and external (environmental) factors in determining decomposition rates of leaves and leaf-litter consumption by detritivores (e.g. Zimmer and Topp 1997, 2000; Gessner et al. 2007; Garcia-Palacios et al. 2016). An important internal factor is leaf chemistry, particularly concentrations of nutrients such as nitrogen and phosphorus (Chap. 11), structural compounds such as lignin and cellulose (Chap. 21) and plant chemical defences such as polyphenolics (Chaps. 18, 19, and 20). Internal factors also include physical leaf attributes such as waxy cuticles, tensile strength and leaf toughness.

Toughness and tensile strength of plant tissues impedes feeding of terrestrial (Cornelissen et al. 1999), marine (Pennings and Paul 1992) and freshwater (Arsuffi and Suberkropp 1984) invertebrates. For example, initial toughness of undecomposed leaf litter was a significant predictor of decomposition rates of a range of leaf species submerged in streams in New Zealand (Quinn et al. 2000) and Hong Kong (Li et al. 2009). Leaf toughness is therefore one of the main factors affecting

M. A. S. Graça (✉)
MARE – Marine and Environmental Sciences Centre & Department of Life Sciences, University of Coimbra, Coimbra, Portugal
e-mail: mgraca@ci.uc.pt

M. Zimmer
Leibniz Centre for Tropical Marine Research (ZMT), University of Bremen, Bremen, Germany
e-mail: martin.zimmer@leibniz-zmt.de

F. Bärlocher et al. (eds.), *Methods to Study Litter Decomposition*,
https://doi.org/10.1007/978-3-030-30515-4_22

decomposition by hampering invertebrate feeding and microbial activity (Pérez-Harguindeguy et al. 2000; Li and Dudgeon 2008).

Leaf toughness correlates with lignin content (Chap. 21) and can be estimated by measuring the force needed to penetrate a leaf sample (Arsuffi and Suberkropp 1984; Boyero et al. 2011). Tensile strength is the force needed to tear apart a leaf sample (Graça et al. 1993; Pérez-Harguindeguy et al. 2000) and is primarily related to cellulose content (Chap. 21). These two leaf properties, tensile strength and toughness, are not necessarily related.

The methodology for measuring leaf toughness was described by Williams (1954), and later modified by Tanton (1962). A leaf disk or leaf is clamped firmly into the base of a penetrometer (Fig. 22.1) consisting of a basal plate and a fixing device, with a central hole, the latter held in space by screws. A transparent fixing device (e.g. made of acrylic) facilitates checking the position of the leaf material to ensure the punch fitting the central hole of the base does not hit a major leaf vein. The punch is best made of a metal rod (e.g. 2 mm diameter) with a rounded tip. The top of the punch is fixed to a tray to carry a beaker that can be filled successively with water to increase mass until a critical value is reached and the punch penetrates the leaf.

Penetrometers can be operated manually or semi-automatically. Manual operation requires manually pouring water into a beaker until the critical mass is reached when the metal rod penetrates the leaf. With a semi-automated instrument, water is pumped from a reservoir into the beaker by an electric pump until the circuit is interrupted when the punch abruptly changes its position (Fig. 22.1). A simplified version of a penetrometer consists of two Teflon (or acrylic) blocks holding the leaf disks and holes to insert the metal rod of the punch (Fig. 22.2).

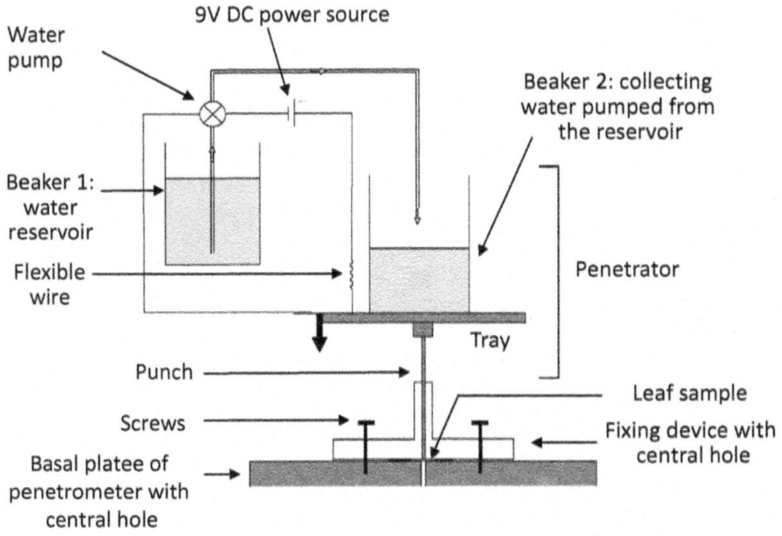

Fig. 22.1 Schematic view of a penetrometer to measure leaf toughness

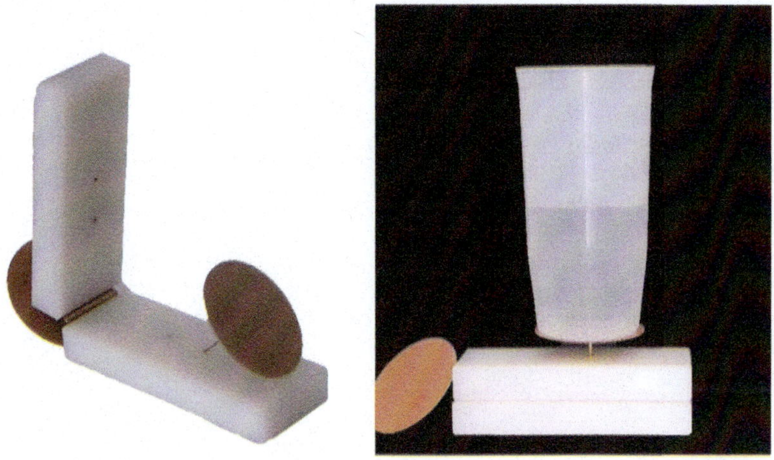

Fig. 22.2 Simple manually operated penetrometer with punches. Leaves or leaf discs are placed in the lower Teflon block to cover the two small holes (left), the penetrometer is then closed, a punch inserted in the corresponding hole, a beaker placed on top and water slowly added until the weight forces the punch through the sandwiched leaf. The two metal rods of the punches are 0.79 and 1.55 mm in diameter (design by N. Connolly, James Cook University, Townsville, Australia; Pearson and Connolly 2000)

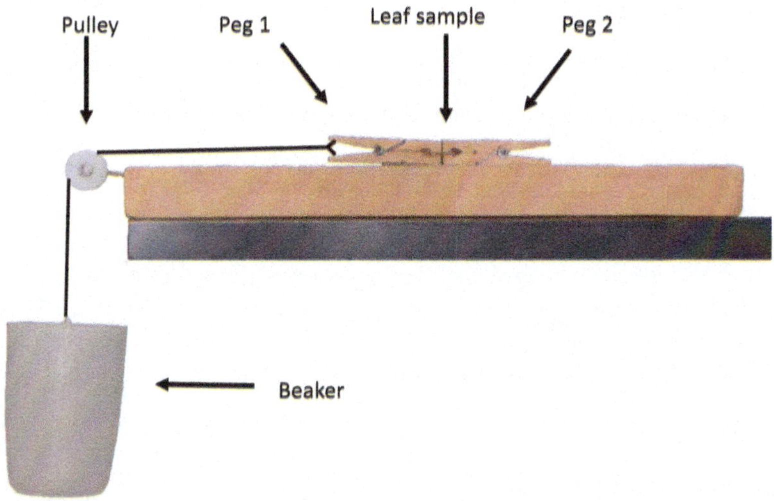

Fig. 22.3 Device to measure tensile strength

Tensiometers essentially consist of two pegs securing a leaf disc. The device represented in Fig. 22.3 has pegs made from wooden cloth-pegs with the anterior end cut out. One of the pegs is fixed to a ring stand by a clamp. The other is connected to a beaker via a string passing through a pulley. To measure tensile strength (or tear resistance), a leaf is secured between the two pegs. Water (or sand) is then

Table 22.1 Leaf toughness of several plants, measured as penetration pressure (kPa), mass required for a punch to penetrate the leaf (g), force to penetrate the leaves (g mm^2) and the maximum force required to tear a strip of leaf per unit width of leaf specimen (N mm^{-1})

Plant material	Measure	Value	References
~400 leaf litter species from 6 continents	Toughness	825–2145 kPa	1
20 species from the Amazon forest (green leaves)	Toughness	~105–135 kPa	2
2 leaf species from Amazonian rain forest	Toughness	130–819 g	3
3 leaf species from the Azores islands	Toughness	98–310 g	4
7 leaf litter species from Hong Kong	Toughness	185–1474 g mm^{-2}	5
4 plant species from tropical Australia, after 2 weeks in a river	Toughness	180–460 g mm^{-2}	6
7 leaf litter species from Hong Kong	Toughness	992–10,026 g mm^{-3}	5
18 British monocots (green leaves)	Tensile strength	0.29–11.48 N mm^{-1}	7
52 species of angiosperms from Argentina	Tensile strength	~1–13 N mm^{-1}	8
20 British Dicotyledons (green leaves)	Tensile strength	0.12–0.78 N mm^{-1}	7

Values are for undecomposed leaves, except in Coughlan et al. (2010)
1 = Boyero et al. (2011); 2 = Fine et al. (2006); 3 = Martins et al. (2015); 43 = Ferreira et al. (2016); 5 = Li et al. (2009); 6 = Coughlan et al. (2010); 7 = Cornelissen and Thompson (1997); 8 = Pérez-Harguindeguy et al. (2000)

gradually added to the beaker via a funnel until the mass of the water or sand exerts a force that tears the leaf apart. The mass needed to reach this point is proportional to leaf tensile strength.

Penetrometers and tensiometers are commercially available, but these instruments are rather expensive. Here we present two devices that can be built in the laboratory with simple, inexpensive materials. Both have been successfully used to measure the toughness and tensile strength of leaves decomposing in streams (Graça et al. 1993; Pearson and Connolly 2000), a salt marsh (Graça et al. 2000) and on forest soils (Zimmer and Topp 1997, 2000), providing complementary information on leaf resistance to physical forces. Some reference values are given in Table 22.1.

2 Equipment and Materials

- Penetrometer
- Tensiometer
- Leaf litter
- Cork borer (e.g., 12 mm diameter)
- Beakers (several sizes, e.g. 50–500 ml)
- Water or sand
- Balance (1 mg precision)

3 Procedures to Measure Leaf Toughness

1. Attach a moistened leaf or leaf disc to the penetrometer base such that no major veins impede leaf penetration by the punch.
2. Gradually add water to the beaker until the punch penetrates the leaf material; beakers of several sizes are useful, since leaf toughness may vary greatly among species.
3. Weigh the water-filled beaker to determine the total mass of the device (punch + tray + beaker + water), which is proportional to leaf toughness.
4. Express leaf toughness as the critical mass (g) needed to penetrate a leaf (Zimmer and Topp 2000), per area of the punch (g/mm^2; Ratnarajah and Barmuta 2009) or as strength (g/mm^3), where leaf thickness is also considered.
5. If desired, transform mass units to penetration pressure (kPa), which corresponds to mass (g) multiplied by gravity (e.g. 9.807 m s^{-2}), and divide by the area (mm^2) of the rod penetrating the leaf (Quinn et al. 2000).

4 Procedures to Measure Tensile Strength

1. Cut discs from a moistened leaf using a cork borer while avoiding main veins.
2. Secure a disc between the two pegs of the penetrometer, ensuring consistency regarding the position of the sample veins (i.e. position the veins parallel to the pegs).
3. Choose a beaker of appropriate size according to the expected volume of sand or water needed, since leaf tensile strength may vary greatly among leaf samples.
4. Gradually add water or sand to the beaker until the leaf disc breaks into two parts.
5. Weigh the beaker with the water (or sand). Tensile strength can be expressed as the critical mass (g) needed to tear apart a lead disk (see 3.4 and 3.5) or in terms of force in Newtons per unit of width of a leaf sample (N/mm).

5 Final Remarks

Specific leaf area (SLA), which is the ratio of leaf area and dry mass (e.g. Cornelissen et al. 1999) and its inverse, leaf mass per area (LMA), are potentially useful proxies of leaf toughness or tensile strength. They can be determined by drying and weighing leaf discs (at least 10) of known surface area. LMA for undecomposed leaf litter of three plant species from the Azores islands ranged from 9 to 18 mm^2 mg^{-1} (Ferreira et al. 2016) and SLA for 155 species from 3 to 30 mm^2 mg^{-1} (Cornelissen et al. 1999; Santiago 2007).

References

Arsuffi, T. L., & Suberkropp, K. (1984). Leaf processing capabilities of aquatic hyphomycetes: Interspecific differences and influence on shredder feeding preferences. *Oikos, 42*, 144–154.

Boyero, L., Pearson, R. G., Dudgeon, D., Graça, M. A. S., Gessner, M. O., Albariño, R. J., Ferreira, V., Yule, C. M., Boulton, A. J., Arunachalam, M., Callisto, M., Chauvet, E., Ramirez, A., Chará, J., Moretti, M. S., Gonçalves, J. F., Helson, J. E., Chará-Serna, A. M., Encalada, A. C., Davies, J. N., Lamothe, S., Cornejo, A., Li, A. O., Buria, L. M., Villanueva, V. D., Zuniga, M. C., & Pringle, C. M. (2011). Global distribution of a key trophic guild contrasts with common latitudinal diversity patterns. *Ecology, 92*, 1839–1848.

Cornelissen, J. H. C., & Thompson, K. (1997). Funtional leaf attributes predict litter decomposition rate in herbaceous plants. *New Phytologist, 135*, 109–114.

Cornelissen, J. H. C., Pérez-Harguindeguy, N., Díaz, S., Grime, J. P., Marzano, B., Cabido, M., Vendramini, F., & Cerabolini, B. (1999). Leaf structure and defence control litter decomposition rate across species and life forms in regional floras on two continents. *New Phytologist, 143*, 191–200.

Coughlan, J. F., Pearson, R. G., & Boyero, L. (2010). Crayfish process leaf litter in tropical streams even when shredding insects are common. *Marine and Freshwater Research, 61*, 541–548.

Ferreira, V., Raposeiro, M. P., Pereira, A., Cruz, A. M., Costa, A. C., Graça, M. A. S., & Gonçalves, V. (2016). Leaf litter decomposition in remote oceanic island streams is driven by microbes and depends on litter quality and environmental conditions. *Freshwater Biology, 61*, 783–799.

Fine, P. V., Miller, Z. M., Mesones, I., Irazuzta, S., Appel, H. M., Stevens, M., Saaksjarvi, I., Schultz, L. C., & Coley, P. D. (2006). The growth-defense trade-off and habitat specialization by plants in Amazonian forests. *Ecology, 87*, S150–S162.

Garcia-Palacios, P., Shaw, E. A., Wall, D. H., & Hättenschwiler, S. (2016). Temporal dynamics of biotic and abiotic drivers of litter decomposition. *Ecology Letters, 19*, 554–563.

Gessner, M. O., Gulis, V., Kuehn, K. A., Chauvet, E., & Suberkropp, K. (2007). Fungal decomposers of plant litter in aquatic ecosystems. In C. P. Kubicek & I. S. Druzhinina (Eds.), *The Mycota, Vol. IV: Environmental and microbial relationships* (2nd ed.). Berlin: Springer-Verlag.

Graça, M. A. S., Maltby, L., & Calow, P. (1993). Importance of fungi in the diet of *Gammarus pulex* and *Asellus aquaticus*. *Oecologia, 96*, 304–309.

Graça, M. A. S., Newell, S. Y., & Kneib, R. T. (2000). Consumption rates of organic matter and fungal mass of the *Spartina alterniflora* decay system by three species of saltmarsh invertebrates. *Marine Biology, 136*, 281–289.

Li, A. O. Y., & Dudgeon, D. (2008). The effects of leaf litter characteristics on feeding and fitness of a tropical stream shredder, *Anisocentropus maculatus* (Trichoptera: Calamoceratidae). *Marine and Freshwater Research, 59*, 897–901.

Li, A. O. Y., Ng, L. C. Y., & Dudgeon, D. (2009). Effects of leaf toughness and nitrogen content on litter breakdown and macroinvertebrates in a tropical stream. *Aquatic Sciences, 71*, 80–93.

Martins, R. T., Melo, A. S., Gonçalves, J. F., & Hamada, N. (2015). Leaf-litter breakdown in urban streams of Central Amazonia: Direct and indirect effects of physical, chemical, and biological factors. *Freshwater Science, 34*, 716–726.

Pearson, R. G., & Connolly, N. M. (2000). Nutrient enhancement, food quality and community dynamics in a tropical rainforest stream. *Freshwater Biology, 43*, 31–42.

Pennings, S. C., & Paul, V. J. (1992). Effect of plant toughness, calcification, and chemistry on hervibory by *Dolabella auricularia*. *Ecology, 73*, 1606–1619.

Pérez-Harguindeguy, N., Díaz, S., Cornelissen, J. H. C., Vendramini, F., Cabido, M., & Castellanos, A. (2000). Chemistry and toughness predict leaf litter decomposition rates over a wide spectrum of functional types and taxa in central Argentina. *Plant and Soil, 218*, 21–30.

Quinn, J. M., Burrel, G. P., & Parkyn, S. M. (2000). Influences of leaf toughness and nitrogen content on in-stream processing and nutrient uptake by litter in a Waikato, New Zealand, pasture stream and streamside channels. *New Zealand Journal of Marine and Freshwater Research, 34*, 253–271.

Ratnarajah, L., & Barmuta, L. A. (2009). The effects of leaf toughness on feeding preference by two Tasmanian shredders. *Hydrobiologia, 636*, 173–178.

Santiago, L. S. (2007). Extending the leaf economics spectrum to decomposition: Evidence from a tropical forest. *Ecology, 88*, 1126–1131.

Tanton, M. T. (1962). The effect of leaf "toughness" on the feeding of larvae of the mustard beetle *Phaedoncochleariae* Fab. *Entomologia Experimentalis et Applicata, 5*, 74–78.

Williams, L. H. (1954). The feeding habits and food preferences of Acrididae and the factors which determine them. *Transactions of the Royal Entomological Society of London, 105*, 423–545.

Zimmer, M., & Topp, W. (1997). Does leaf litter quality influence population parameters of the common woodlouse, *Porcellio scaber* Latr., 1804 (Crustacea: Isopoda)? *Biology and Fertility of Soils, 24*, 435–441.

Zimmer, M., & Topp, W. (2000). Species-specific utilization of food sources by sympatric woodlice (Isopoda: Oniscidea). *Journal of Animal Ecology, 69*, 1071–1082.

Part III
Microbial Decomposers

Chapter 23
Techniques for Handling Ingoldian Fungi

Enrique Descals

Keywords Aquatic hyphomycetes · Conidia · Foam sampling · Fungal colony characteristics · Fungal isolation · Fungal spores · Membrane filtration · Pure cultures · Sporulation · Streams · Teleomorphs

1 Introduction

Numerous taxonomic surveys and ecological studies on Ingoldian fungi (also known as aquatic hyphomycetes) have been published since Ingold's seminal paper of 1942 (Bärlocher 1992). Conidiophores (Shearer and Webster 1985) and, more commonly, detached conidia are used to identify and in some cases to quantify Ingoldian fungi (Gessner et al. 2003; Chap. 26). Identification is facilitated by the characteristic shapes of many conidia (Chap. 25), which are uncommon among fungi in general. However, the identity of a considerable fraction of the conidia encountered in detailed surveys is unknown or ambiguous, often comprising a third of the taxa listed (e.g. Descals 1987, 1998), indicating that Ingoldian fungi remain insufficiently described despite decades of taxonomic and ecological work.

Natural foam is the main source of Ingoldian fungi for biodiversity studies and conidial isolation into pure culture. Lather-like accumulations in front of boulders and other obstacles in small pools in streams can be a rich source, especially in soft waters. In hard waters, natural foam may not readily accumulate and tends to contain few conidia. Likewise, samples from muddy waters tend to contain many bacteria or yeasts but few conidia of Ingoldian fungi. Sources for sampling Ingoldian fungi other than foam are submerged plant litter, particularly decomposing leaves in or near streams, and stream water samples, which may contain up to 30,000 conidia per litre (Webster and Descals 1981). Finally, conidia can accumulate in rain-water throughfall from riparian canopies (and be collected in a funnel), from tree stem

E. Descals (✉)
Esporles, Mallorca, Spain
e-mail: enrique.descals@gmail.com

F. Bärlocher et al. (eds.), *Methods to Study Litter Decomposition*,
https://doi.org/10.1007/978-3-030-30515-4_23

flow, dewdrops beaten off canopies (collected onto an inverted clean umbrella) and waterfall mist onto glass slides (Ando and Tubaki 1984).

Ingoldian fungi typically form conidia under water, but some (e.g. *Calcarispora hiemalis*) also sporulate in Petri dishes without colony submersion when relative humidity is high. Other species sporulate in both air and water, although conidia formed in air often lack branches. Most species, however, require contact with liquid water to form conidia, typically after prolonged incubation (one to several days) in submerged or partly submerged conditions. In spite of the fully aquatic life cycle of the asexual forms of Ingoldian fungi (anamorphs, see Glossary), the species known also to reproduce sexually are amphibious in that their sexual forms (teleomorphs) must release their spores outside water. Meiospores (asco- and basidiospores), which do not share the characteristic shapes of the conidia formed by the Ingoldian anamorphs, are therefore dispersed in air or on the water surface. An exception is *Loramyces juncicola*, a freshwater ascomycete with sigmoid ascospores (Ingold and Chapman 1952). This species does not occur in streams, however. Only leptosphaeriaceous ascospores, probably belonging to *Massarina* species with Ingoldian anamorphs, are regularly recognized in stream drift, but have never been isolated.

Teleomorphs are known for only a minority of species (Webster 1992). However, only a few taxonomists have searched for them, mostly in temperate climates, and the few known cases are spread over a wide range of taxa. This includes *Helotiales*, *Pezizales*, nectriaceous forms, bitunicate fungi and corticiaceous and other basidiomycetes (Shearer et al. 2007). Consequently, many more are likely to be discovered in the future. Ascomycetous teleomorphs, mostly forming pseudothecia and especially apothecia, are commonly found on wood (Willoughby and Archer 1973) collected along riverbanks, especially in the summer, and after moist incubation for several weeks. Teleomorph characters are often associated with Ingoldian species. For example, records of *Miladina lechithina* have been included in distributional studies of *Actinosporella megalospora* by Descals and Rodríguez (2002). Basidiomycetous teleomorphs are inconspicuous, filmy or gelatinous forms known only from pure culture.

Synanamorph conidia are normally tiny asexually produced spores that are occasionally observed in cultures in addition to the normal characteristically shaped conidia of Ingoldian fungi (Chap. 25). Some of these synanamorph conidia are characteristic enough to enable species identification (e.g. *Culicidospora gravida*), but most are small and inconspicuous. They tend to appear after slightly prolonged submergence or semisubmergence and cannot reliably be associated with Ingoldian anamorphs except when they are observed in pure culture. Synanamorph conidia may be functionally spermatial. Inducing synanamorph formation is therefore a critical step for sexual reproduction and hence in the establishment of anamorph-teleomorph relationships.

Apart from synanamorph production and the establishment of anamorph-teleomorph relationships, pure cultures of Ingoldian fungi are needed in descriptive work to record secondary diagnostic characters such as colony pigmentation, morphology, and the presence of sclerotia and microsclerotia. They can also be useful to confirm identifications otherwise based solely on conidia; to produce sufficient material for molecular identification when growth and sporulation are sparse on the natural substrate; and to establish culture collections (Chap. 25) for experimental, commercial or exchange purposes.

This chapter presents common techniques and approaches to handle Ingoldian fungi. These include the analysis of conidia and conidiophores in foam, plant litter and water samples; isolation into pure culture; growth on agar plates and sporulation; and teleomorph induction. More detailed accounts are given in Descals (1997) and Descals and Moralejo (2001). A variety of methods commonly used in ecological studies are summarized in Gessner et al. (2003) and an overview of molecular approaches is given in Bärlocher (2016). Since correct species identification is often fundamental for ecological work, many techniques presented here emphasize preparation of fungal material for this purpose, complementing the taxonomic key on common species from temperate regions in Chap. 25. Standard techniques to isolate fungi present as mycelia, such as particle plating of leaf, wood or root tissues (e.g. Bärlocher and Kendrick 1974; Kirby 1987; Fisher and Petrini 1989), are not included in this chapter.

2 Equipment and Chemicals

2.1 Equipment and Material

- Compound microscope with mechanical stage and 10, 20 and 40× objectives; intermediate magnification lenses (1.25–2×); built-in transformer for field work; phase and differential interference contrast optics for microphotography and as aids for identification; and digital camera and drawing tube are needed for taxonomic (descriptive) purposes
- Inverted microscope for isolation (optional)
- Dissecting microscope (at least up to 100×) with transmitted light; incident light will be needed for teleomorph detection or isolation
- Labelled jars and spoon to collect foam in the field
- Polythene bags (e.g. Ziploc bags) for transport of collected leaves and other substrates colonized by fungi and of teleomorphs
- Chisel, mallet and folding handsaw to sample colonized wood
- Bunsen burner with gas canisters (or alcohol lamp) for isolation
- Fan heater or other source of heat for fast drying of foam slide preparations
- Microtechnique kit: watchmaker's forceps, micro-scalpel consisting of mounted fine sewing needles or 00 insect pins with tip flattened hot with a hammer, surgical scalpel, wide loop for sampling foam for slide preparations, mounted hair, finder slides
- Filtration equipment
- Cellulose-ester membrane filters
- Color transparency or negative films

- Semi-transparent (glassine) envelopes for storage of dried cultures (optional)
- Plankton sedimentation chambers or table centrifuge (optional)
- Tracing satin and India ink pens
- General glassware, including small vials for specimen storage, sterile pipettes (10 and 1 ml), flame-pulled Pasteur pipettes, conical flasks
- Microscope slides (grease-free), coverslips (20 × 20 mm), slide boxes, slide labels
- Sterile disposable syringes for mass transfers of mycelia
- Aquarium aeration system or orbital shaker to induce conidial production
- Near-UV and daylight lamps for anamorph and teleomorph induction
- Autoclave

2.2 Chemicals

- Ethanol (96%)
- Lugol's iodine fixative
- Lactofuchsin: acid fuchsin 0.1 g, lactic acid 100 ml (Kirk et al. 2001)
- Nail varnish (preferably containing nylon) or liquid cover glass such as Merckoglas®
- Dilute bleach (5% NaOCl)
- Sterile distilled water
- Broad-spectrum insecticide and acaricide, preferably containing an ovicide, available in agricultural or gardening supply shops
- Wetting agent (e.g. Tween-80)
- Antibiotics: chloramphenicol (up to 1 g l^{-1}, which may be added before autoclaving the medium), or penicillin (penicillin G sodium salt, 600 mg l^{-1}) plus streptomycin sulphate (1 g l^{-1}) added after autoclaving

2.3 Media

- Stock medium for colony pigment expression: 2% malt extract in 2% agar (2% MEA)
- Sporulation medium: 0.1% malt extract in 2% agar (0.1% MEA)
- Isolation medium: sporulation medium plus antibiotics at the concentrations recommended above, which work well in most situations; however, the best concentration depends on the degree of contamination of the source material. Chloramphenicol controls bacterial growth even in rather dirty foam samples, but may suppress germination in some species of Ingoldian fungi (Gessner et al. 2003). Preparing plates a few days before use allows discarding any contaminated plates. Another advantage is that the agar dries up slightly so that water

drops used in single-sporing techniques (see Sect. 3.4 below) are rapidly adsorbed.

3 Experimental Procedures

3.1 *Foam Samples*

1. Collect stream foam with a spoon and transfer to a jar. Decant excess water. If foam is scanty or breaks up quickly, collect subsamples in separate jars.
2. Preserve samples with a few drops of ethanol if isolation of cultures is not intended. Jars with foam to be used for isolating individual large conidia must be placed on melting ice and processed (see Sects. 3.4 and 3.5) in less than 2 h.
3. With a Pasteur pipette, transfer 3–4 drops of liquefied foam onto a grease-free slide.
4. Air-dry the drop of liquefied foam, remove any sand grains with a needle, add a small drop of fixative, burst the odd gas bubble with a heated needle, cover, seal, label and store in slide box.
5. View preparations under a microscope at 100–500×, preferably with phase contrast optics.
6. Produce line drawings or microphotographs of conidia of doubtful or unknown identity.
7. Save semi-permanent lactofuchsin slide preparations for future identifications.
8. Optionally, obtain pure cultures as described in Sects. 3.4 and 3.5.

3.2 *Plant Litter Samples*

1. Collect decaying plant litter such as leaves in the field and store in polythene bags.
2. Keep samples refrigerated and moist, but not flooded, for transport to the laboratory. Process samples promptly (i.e. within a few days at most) to avoid sporulation by fungal air contaminants.
3. Rinse the plant litter gently to avoid loss of delicate surface mycelia. Use sterile water. Tap water may be used if chlorine toxicity and contamination by *Pythiaceae*, terrestrial hyphomycetes, coelomycetes and other fungi is not a concern.
4. Observe fungal conidia and conidiophores on leaves (Fig. 23.1) directly under a dissecting microscope with high magnification (i.e. 200× or at least 100×). Fungal structures are best seen at the rim of litter pieces.
5. For quantifying sporulation, follow procedures in Chap. 26.

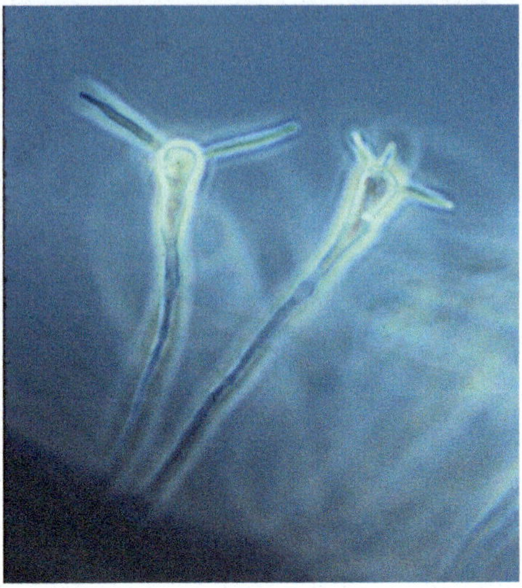

Fig. 23.1 *Clavariopsis aquatica* conidia attached to conidiophores on a maple leaf. (Photo F. Bärlocher)

6. Produce drawings, microphotographs, slide preparations and/or pure cultures as described in sections above and below.

3.3 Water Samples

1. Collect stream water as a source of conidia if their concentration is expected to be high.
2. Filter the water through a membrane filter (typically 200–1000 ml); turn filter translucent by first staining in a water-soluble cytoplasmic stain, such as Waterman's ink diluted tenfold, air-drying and then flooding it in immersion oil (optional); and scan under a compound microscope at 100–500× (see Sect. 3.2, Chap. 26, Gessner et al. 2003).
3. If conidia are scarce, concentrate them in artificial foam. Add a drop of dilute Tween-80 (and a few drops of fixative if the sample is not intended for isolation) to 10 l of stream water, and decant the water back and forth from one bucket into another. Then collect and process foam samples as described in Sect. 3.1.
4. Produce drawings, microphotographs, slide preparations and/or pure cultures as described in sections above and below.

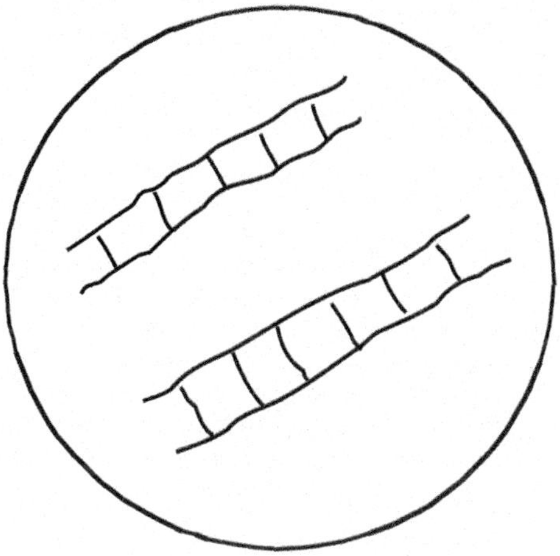

Fig. 23.2 Ladder pattern drawn in the bottom lid of a Petri dish. Individual conidia are placed in the central cells and the germlings transferred to the neighbouring outer cells

3.4 Isolation of Pure Cultures

1. Prepare Petri dishes with an isolation medium.
2. For individual spore lifting techniques, draw a ladder or spoke pattern (Descals 1997) with indelible ink on the bottom lid of the isolation dish (Fig. 23.2).
3. Bore tiny wells with a hot needle in the center of the squares or sectors, where a conidium will be placed. This traps most bacterial or yeast colonies and allows hyphae to grow out into clean agar.
4. Place some leaves collected in the field in Petri dishes with sterilized stream or distilled water in a Petri dish for 1–3 days. Dishes may be aerated or shaken to stimulate sporulation.
5. Lift large conidia (e.g. >400 μm in span) under a dissecting microscope with a mounted hair or Pasteur pipette and place on the agar.
6. Since germination success of floating conidia is low, collect a large loopful from the agar surface containing several conidia and spread it over the medium.
7. Isolate germlings after at least a few hours of incubation.
8. Conidia suspended in water or recently settled can be made to flow into a capillary tube (e.g. a Pasteur pipette pulled over the flame and covered with a perforated teat).
9. After 24–48 h, check isolated conidia for germ tubes.
10. With a flamed microscalpel cut out the section of agar containing the germling and transfer the agar section onto fresh medium.

3.5 Isolation with a Finder Slide

1. Isolate conidia too small to be handled under a dissecting microscope with the aid of a finder slide (Graticules Ltd., Tonbridge, UK; Gessner et al. 2003).

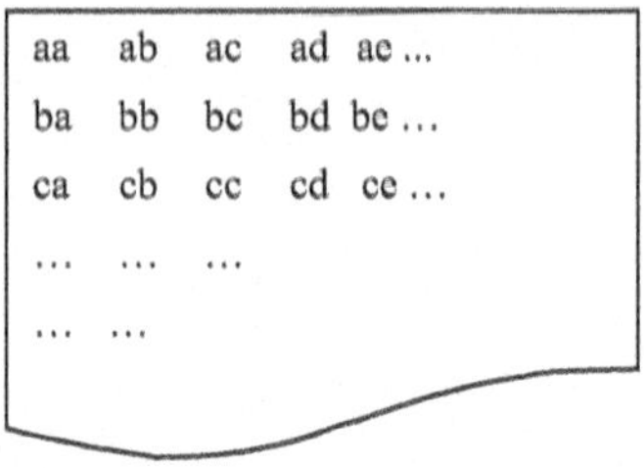

Fig. 23.3 Template printed on a transparency film and placed under a microscope slide to locate conidia

2. A finder can also be improvised by printing a letter-size template with the computer, as indicated in Fig. 23.3. Photograph with a colour transparency film. Cut out negative and attach to microscope slide with nail varnish. Glue coverslip on top of negative with more nail varnish.
3. In the field, flood the isolation medium in a Petri dish with a foam sample. Dilute with sterile distilled water if necessary.
4. Decant the excess suspension back into the sampling jar for fixing.
5. Incubate the plates horizontally in an ice chest for a few hours to initiate germination.
6. Transfer a rectangular portion of the isolation medium onto the finder slide.
7. Scan the surface under the compound microscope at 100×, check at 200× for contaminating spores attached to or lying in the vicinity of the conidium of interest.
8. With a flamed microscalpel cut out the section of agar containing the germling and transfer the agar section onto fresh isolation medium.

3.6 Colony Growth on Agar Plates

1. Let colonies grow under standard conditions to induce pigment production and colony characters needed for comparisons with published descriptions. MEA at 2% is one of the most popular media used for describing fungal cultures.
2. If large amounts of mycelium are rapidly needed, force a piece of agar colony through a sterile syringe (without the needle) onto agar medium. Add some sterile distilled water and spread the suspension evenly over the agar surface with a flamed bent glass rod. This results in rapid colony growth as a single carpet (intra-strain vegetative incompatibility between incipient colonies is occasionally observed and can prevent uniform occupation of the agar surface).
3. Seal cultures growing in Petri dishes with tape to reduce contamination by aerial spores and to limit medium dehydration.
4. To protect cultures from fungal and other contaminations caused by mite intrusion, spray bench surfaces and plates with an acaricide and swab working surfaces with a mineral oil. Sealing dishes with tape is not effective against mites.

3.7 *Induction of Anamorph Sporulation*

1. Place leaf material bearing fungi or fungi growing on agar medium in conical flasks with sterile water to induce anamorph formation. Use colonies grown on 0.1% MEA, since rich media such as 2% MEA may discourage sporulation. Four types of liquid can be used (Descals and Moralejo 2001): (1) sterile distilled water, (2) sterile filtered stream water (chemically variable, but may yield high conidium production), (3) dechlorinated tap water (aerate for at least 24 h) and (4) dilute mineral solutions.
2. Incubate submerged agar or litter pieces for 1–2 days with forced aeration or shaking.
3. View preparations under a microscope at 100–500× and produce line drawings, microphotographs, and semi-permanent lactofuchsin slide preparations for documentation.

3.8 *Teleomorph Induction*

1. Add some water to pure agar cultures or field-collected material.
2. Agitate periodically to facilitate dispersing spermatia over the mycelia.
3. Supplement diffuse daylight with near UV to favour teleomorph induction.
4. Regularly search plate or field material over several months for formation of sexual structures.
5. Identify species using pertinent mycological literature. For temperate species, Ingold (1975) and Descals & Marvanová (ined., available from the authors), and Chap. 25. For tropical species, Nawawi (1985) and Marvanová (1997) are particularly useful. In other cases, the overviews by Webster and Descals (1981) and Descals & Marvanová (ined., available from the authors) are recommended. Consultation of original descriptions is recommended.

4 Final Remarks

Membrane filtration of stream water (Sect. 3.3, Iqbal and Webster 1973; Gessner et al. 2003) has become widely used by ecologists. Potential disadvantages include loss of optical resolution due to incorrect staining, opaque background or mangling of conidia due to excessive vacuum. Techniques for concentrating conidia other than membrane filtration and production of artificial foam (see Sect. 3.3) may include (1) sedimentation by gravity overnight or shorter periods (Chamier and Dixon 1982); (2) evaporation by any combination of vacuum, heat or ventilation; and (3) centrifugation. The last two approaches have not yet been tested.

Aerial conidia of Ingoldian fungi forming on damp-incubated substrates are sticky and cannot be easily lifted individually. They may be first transferred to a drop of sterile water on the isolation medium, spread with a small, sterile loop, allowed to germinate and isolated as in Sects. 3.4 and 3.5.

The following points are worth considering when producing mycelia or spores from field-collected plant litter or pure cultures on agar:

- Turbulence: Although aeration and shaking stimulate sporulation, excess turbulence may cause fragmentation of complex conidia (e.g. *Dendrospora* and *Varicosporium*) and provoke bubble burst which can propel conidia out of the suspension.
- Incubation time: Incubations longer than the 1–2 days recommended in Sect. 3.7 may be needed to observe conidiogenesis and induce synanamorph and teleomorph formation. However, long incubations increase the risk of (a) conidial malformations which interfere with identification and (b) microcycle conidiation (i.e. germination of spores by direct formation of conidia without the intervention of mycelial growth). This may artificially increase conidial counts (e.g. in *Articulospora tetracladia*, *Lemonniera* spp.). Conidia adhering to walls through mucilage production can be a problem even during shorter incubations.
- Temperature and nutrient concentrations: The optimum sporulation temperatures of Ingoldian fungi are usually well above those found in the source streams (e.g. Chauvet and Suberkropp 1998). Such temperatures may therefore be chosen during laboratory incubations (e.g. 10–20 °C), unless sporulation activities under natural conditions are to be estimated. Drastic changes in temperature and nutrient concentrations should be avoided, however, because they may permanently disrupt sporulation.
- Light: In some species, pulses or protracted treatments with near UV light may stimulate growth (e.g. *Mycocentrospora acerina*) or sporulation or both.
- Water pH: The pH of stream water can affect the formation of synanamorphs and hence sexual reproduction. Testing with different pH values including those of the source stream is thus recommended.
- Substrate-to-water volume ratio: This ratio must be kept low in incubations of field material when water is not changed, in order to delay microbiologically induced staling; but also in pure cultures submerged in unchanged water, in order to delay chemically induced staling.

Methods used for anamorph induction other than the technique introduced in Sects. 3.2 and 3.7 include (1) incubation of colonized leaves in continuously renewed distilled water, (2) incubation of pure cultures on 0.1% MEA or water agar, (3) incubation of field material and pure cultures in periodically renewed nutrient solution (e.g. Ciferri 1959; Suberkropp 1991), (4) continuous water drip over pure cultures (Marvanová 1968) and (5) continuous flow chambers (Descals et al. 1976; Descals and Moralejo 2001).

Glossary

Anamorph	A supposedly asexual morph, of which some species may have two or more that can appear simultaneously
Apothecium	Plural: apothecia. A type of teleomorphic fruit body characteristic of a major group of ascomycetes
Ascomycete	A major phylogenetic group of fungi
Ascospore	Meiospore of Ascomycetes
Basidiomycete	A major phylogenetic group of fungi
Basidiospore	Meiospore of Basidiomycetes
Conidium	Plural: conidia, which are asexually produced spores or mitospores
Conidiophore	A part of a thallus bearing one or more conidia
Leptosphaeriaceous	Refers to the Ascomycete family Leptosphaeriaceae
Meiospore	A sexually produced spore whose ontogeny involves at least one meiotic division
Microsclerotium	Plural: microsclerotia. Small sclerotium
Mitospore	Synonym of conidium, which is an asexually produced spore.
Morph or form	A part of a thallus which is usually associated with sexual or asexual reproduction
Mycelium	Plural: mycelia, a mass of filaments (called hyphae), the thallus of most fungi
Pseudothecium	Plural: pseudothecia. A type of teleomorphic fruit body characteristic of a major group of Ascomycetes
Sclerotium	Plural: sclerotia. A discrete, firm mass of hyphae or cells functioning as a resting body in certain fungi. It may give rise to a fruit body or mycelium.
Synanamorph	Any one of two or more anamorphs formed by some fungal species
Teleomorph	The sexual morph of a fungus

References

Ando, K., & Tubaki, K. (1984). Some undescribed hyphomycetes in the rain drops from intact leaf-surface. *Transactions of the Mycological Society of Japan, 25*, 21–37.

Bärlocher, F. (Ed.). (1992). *The ecology of aquatic hyphomycetes* (Ecological studies) (Vol. 94). Berlin: Springer-Verlag.

Bärlocher, F. (2016). Aquatic hyphomycetes in a changing environment. *Fungal Ecology, 19*, 14–27.

Bärlocher, F., & Kendrick, B. (1974). Dynamics of the fungal population on leaves in a stream. *Journal of Ecology, 62*, 761–791.

Chamier, A.-C., & Dixon, P. A. (1982). Pectinases in leaf degradation by aquatic hyphomycetes I – The field-study. The colonization-pattern of aquatic hyphomycetes on leaf packs in a Surrey stream. *Oecologia, 52*, 109–115.

Chauvet, E., & Suberkropp, K. (1998). Temperature and sporulation of aquatic hyphomycetes. *Applied and Environmental Microbiology, 64*, 1522–1525.

Ciferri, R. (1959). Ecological observations on aquatic hyphomycetes. Omagiu lui Traian Savulescu cu prilejui implinirii a 70 de ani, Academia Republicii Populâre Romîne. *Editura Academici Republicii Populâre Romine*, pp. 157–162.

Descals, E. (1987). Muestreo preliminar de hongos ingoldianos de Cataluña. *Revista Ibérica de Micología, 4*, 17–32.

Descals, E. (1997). Ingoldian fungi: Some field and laboratory techniques. *Bolletí de la Societat d'Historia Natural de les Balears, 40*, 169–221.

Descals, E. (1998). Streamborne fungi from Karrantza (Basque Country) and surroundings. *Bolletí de la Societat d'Historia Natural de les Balears, 41*, 191–219.

Descals, E., & Moralejo, E. (2001). Water and asexual reproduction in the Ingoldian fungi. *Botanica Complutensis, 25*, 13–71.

Descals, E., & Rodríguez, J. (2002). *Cuadernos de Trabajo de Flora Micológica Ibérica 18. Bases corológicas*. Madrid: Consejo Superior de Investigaciones Científicas.

Descals, E., Nawawi, A., & Webster, J. (1976). Developmental studies in *Actinospora* and three similar aquatic hyphomycetes. *Transactions of the British Mycological Society, 67*, 207–222.

Fisher, P. J., & Petrini, O. (1989). Two aquatic hyphomycetes as endophytes in *Alnus glutinosa* roots. *Mycological Research, 92*, 367–368.

Gessner, M. O., Bärlocher, F., & Chauvet, E. (2003). Qualitative and quantitative analyses of aquatic hyphomycetes in streams. In C. K. M. Tsui & K. D. Hyde (Eds.), *Freshwater mycology* (pp. 127–157). Hong Kong: Fungal Diversity Press.

Ingold, C. T. (1942). Aquatic hyphomycetes of decaying alder leaves. *Transactions of the British Mycological Society, 25*, 339–417.

Ingold, C. T. (1975). An illustrated guide to aquatic and waterborne hyphomycetes (Fungi Imperfecti) with notes on their biology. *Freshwater Biological Association Scientific Publications, 30*, 1–96.

Ingold, C. T., & Chapman, B. (1952). Aquatic ascomycetes: *Loramyces juncicola* Weston and *L. macrospora* n. sp. *Transactions of the British Mycological Society, 35*, 268–272.

Iqbal, S. H., & Webster, J. (1973). Aquatic hyphomycete spora of the River Exe and its tributaries. *Transactions of the British Mycological Society, 61*, 331–346.

Kirby, J. J. H. (1987). A comparison of serial washing and surface sterilization. *Transactions of the British Mycological Society, 88*, 559–562.

Kirk, P. M., Cannon, P. F., David, J. C., & Stalpers, J. A. (2001). *Dictionary of the fungi*. Wallingford: CAB International.

Marvanová, L. (1968). *Lemonniera centrosphaera* sp. nov. *Transactions of the British Mycological Society, 51*, 613–616.

Marvanová, L. (1997). Freshwater hyphomycetes: A survey with remarks on tropical taxa. In K. K. Janardhanan, C. Rajendran, K. Natarajan, & D. L. Hawksworth (Eds.), *Tropical mycology* (pp. 169–226). Enfield: Science Publishers, Inc..

Nawawi, A. (1985). Aquatic hyphomycetes and other water-borne fungi from Malaysia. *Malayan Nature Journal, 39*, 75–134.

Shearer, C. A., & Webster, J. (1985). Aquatic hyphomycete communities in the River Teign. III. Comparison of sampling techniques. *Transactions of the British Mycological Society, 84*, 509–518.

Shearer, C. A., Descals, E., Kohlmeyer, B., Kohlmeyer, J., Marvanová, L., Padgett, D., Porter, D., Raja, H. A., Schmit, J. P., Thorton, H. A., & Voglmayr, H. (2007). Fungal biodiversity in aquatic habitats. *Biodiversity and Conservation, 16*, 49–67.

Suberkropp, K. (1991). Relationships between growth and sporulation of aquatic hyphomycetes on decomposing leaf litter. *Mycological Research, 95*, 843–850.

Webster, J. (1992). Anamoph-teleomorph relationships. In F. Bärlocher (Ed.), *The ecology of aquatic hyphomycetes* (Ecological studies) (Vol. 94, pp. 99–117). Berlin: Springer.

Webster, J., & Descals, E. (1981). Morphology, distribution, and ecology of conidial fungi in freshwater habitats. In G. T. Cole & B. Kendrick (Eds.), *Biology of conidial fungi* (Vol. 1). New York: Academic.

Willoughby, L. G., & Archer, J. F. (1973). The fungal spora of a freshwater stream and its colonization pattern on wood. *Freshwater Biology, 3*, 219–239.

Chapter 24
Maintenance of Aquatic Hyphomycete Cultures

Ludmila Marvanová

Keywords Culture storage · Deionized water preservation · Fungal culture preservation · Fungal genetic resources · Aquatic fungi · Liquid nitrogen preservation · Pure culture databases · Pure cultures · World culture collections

1 Introduction

Pure cultures of fungi provide important information on morphological and physiological features necessary for reliable identification. Gross colony characters, such as texture, radial growth rate, colour of the front and back side, pigments in the agar medium, and the presence of sclerotia, can be observed by the unaided eye on solid agar media (Fig. 24.1), whereas microscopic features are studied mostly in submerged cultures: conidiophores, conidiogenous cells, the events of conidiogenesis and conidial morphogenesis, morphology of mature detached conidia, microconidial synanamorphs, and the presence and kind of chlamydospores and hyphopodia. All these characters may contribute to the accurate classification of a specimen (Chap. 25).

Immunological and molecular identification methods, including monoclonal antibody-based immunoassays (e.g. Bermingham et al. 2001), fluorescence in situ hybridization (McArthur et al. 2001; Baschien et al. 2008), qPCR (Chap. 36), and genome sequencing (Seena et al. 2018, Chap. 62), likewise require work with pure cultures, as does the biochemical characterization of fungi (e.g. Thornton 1963; Brosed et al. 2017) and various kinds of ecological experiments (Suberkropp 1991; Duarte et al. 2006; Dang et al. 2009; Ferreira and Chauvet 2011). Furthermore, molecular studies aimed at finding teleomorph-anamorph connections and phylogenetic relationships of aquatic hyphomycete taxa or strain variation are also best based on pure cultures (Marvanová et al. 2002; Campbell et al. 2009; Baschien et al.

L. Marvanová (✉)
Faculty of Science, Czech Collection of Microorganisms, Masaryk University, Brno, Czech Republic
e-mail: ludmila@sci.muni.cz

F. Bärlocher et al. (eds.), *Methods to Study Litter Decomposition*,
https://doi.org/10.1007/978-3-030-30515-4_24

Fig. 24.1 Agar culture of *Tetracladium setigerum* CCM F-20987 grown on 2% malt agar for 2 weeks

2013; Seena et al. 2018). Finally, several antibiotic substances produced by aquatic hyphomycetes were discovered in pure cultures (Gulis and Stephanovich 1999; Oh et al. 1999; Kaida et al. 2001).

Pure cultures of taxa unknown to the investigator and of taxa difficult to identify on the basis of detached conidia alone are invaluable to identify species, particularly in taxonomic surveys of unexplored areas. However, other field studies involving species identification benefit from cultures as well. It is recommended to keep voucher specimens and picture documentation of at least those species whose identification is ambiguous. Collaboration with an experienced taxonomist is often unavoidable, not least because even well-studied areas contain many species that are still undescribed or difficult to identify. For example, Marvanová (2001) cited about 20% undescribed taxa in a protected area in the Czech Republic, and Gönczöl et al. (2001) noted some 40% undescribed taxa in a single small stream in Hungary. Although molecular methods are now technically well advanced to help identify aquatic hyphomycetes directly in environmental samples, the number of fungal sequences deposited in data bases such as GenBank is still small, which continues to hamper reliable identification by molecular approaches (Duarte et al. 2015; Hibbett et al. 2016). Establishing such data bases requires pure cultures, preferably derived from type specimens and beginning with types of genera, that are characterized both morphologically and by DNA sequence data (Crous et al. 2014).

If cultures prove important for future work or are referred to in publications, they should be deposited in public culture collections of microorganisms (Smith and Onions 1994), to be generally available. Culture collections have become especially important after the Convention on Biological Diversity was ratified in many states (Hawksworth 1996). They ensure conservation of microbial biodiversity outside the natural habitat (Glowka 1996). This is important for new or extremely rare fungal species, which may not be restricted to the pristine locations usually favoured for nature conservation. In October 2014, the Nagoya Protocol on Access to Genetic Resources and Fair and Equitable Sharing of Benefits Arising from their Utilization came into force. For fungi, it concerns the collection and maintenance of material from territories of states that ratified the Convention of Biological Diversity (Anonymous 2015). Since fungi, like other microorganisms, are considered genetic resources, they are to be protected on a national basis like other natural resources (Anonymous 1996).

1.1 Location of Culture Collections and Information About Strains

General information on culture collections and their holdings can be obtained from the World Data Centre for Microorganisms, a component of the World Federation of Culture Collections (www.wfcc.info/wdcmdb). Cultures of aquatic hyphomycetes are currently preserved mainly in the major public culture collections, such as the Westerdijk Fungal Biodiversity Institute Utrecht (www.westerdijkinstitute.nl/collections), formerly Centraalbureau voor Schimmelcultures Baarn (CBS), CABI Bioscience Genetic Resource Collection (www.cabi.org/services/microbial-services/culture-collection-microorganism-supply), BCCM/MUCL (Agro)Industrial Fungi & Yeasts Collection (www.bccm.belspo.be), the Leibniz Institute DSMZ – German Collection of Microorganisms and Cell Cultures (www.dsmz.de), and the American Type Culture Collection (www.atcc.org).

Maintenance of a culture means keeping it viable, pure, authentic, and genetically unchanged. Long-term maintenance implies preserving cultures for one to tens of years without reinoculation. This can be achieved by reducing or even suspending fungal metabolism. There are many conservation methods to effect this, but none is universally applicable to all microorganisms (Kirsop and Doyle 1991; Smith and Onions 1994; Hubálek 1996; Nakasone et al. 2004; Ryan and Smith 2004; Day and Stacey 2007; Homolka 2013, and www.atcc.org/~/media/PDFs/Technical%20 Bulletins/tb02.ashx).

As a general rule, aquatic hyphomycetes do not produce aerial macroconidia on solid media. Therefore, methods should be employed that enable conservation of nonsporulating mycelia. For all forms of conservation, it is necessary to start with well-grown, vigorous cultures, preferably at the early to mid-stationary phase of growth. They should be cultivated at optimum conditions for growth. However, every method of conservation contains some danger of clone selection, when only a small proportion of the cells survives freezing and thawing (Hubálek 1996). Repeated freezing and thawing is not recommended.

1.2 Conservation at Reduced Metabolism in Deionized Water

Conservation in deionized water is a method that was first used by Castellani (1939) for human pathogenic fungi and successfully adopted for various groups of fungi (e.g. Ellis 1979). Reduced fungal metabolism during storage in deionized water is probably caused by low nutrient concentrations in the medium and by restricted oxygen availability under conditions of submergence. Advantages of the method are the cheap equipment and that cultures require little space for storage (an example is shown in Fig. 24.2), morphology, and sporulation capacity are usually well preserved and that transfer to fresh agar media is generally sufficient to revive cultures. Disadvantages are the risk of genome change and of contamination during storage. The shelf life is generally 2–5 years, although sufficient viability after 20 years of storage was encountered with some isolates of aquatic hyphomycetes in the Czech Collection of Microorganisms (unpubl. data).

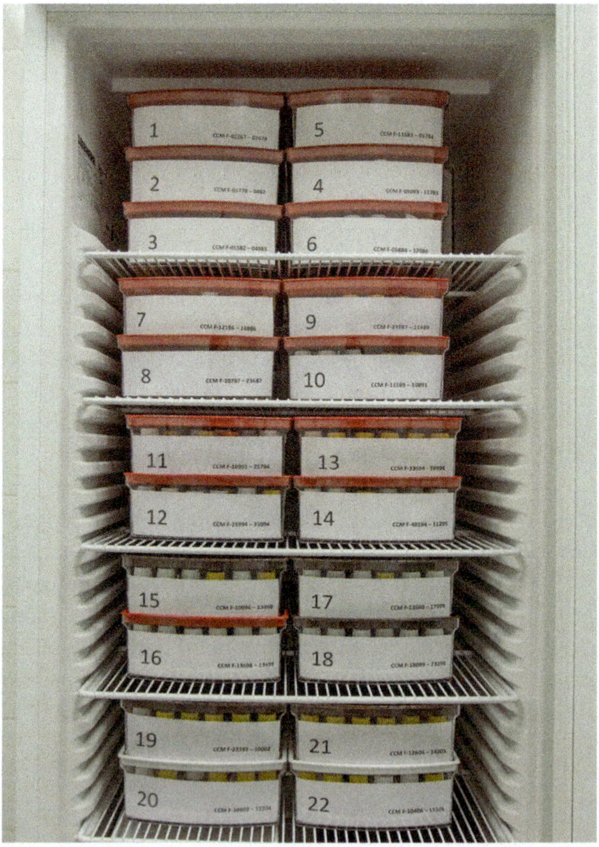

Fig. 24.2 Storage of aquatic hyphomycetes in deionized water at 10 °C

1.3 *Conservation at Reduced Metabolism Under Mineral (Paraffin) Oil*

The first report on extensive use of this method is by Buell and Weston (1947), and further successful applications were reported by Fennell (1960) and Onions (1971). Advantages are the cheap equipment and that the sporulation capacity is moderately well preserved. The shelf life is usually 5–10 years, occasionally even up to 30 years. Disadvantages are that reviving cultures involves repeated transfer because of the presence of oil on the inoculum, in addition to the risk of genome change and contamination during storage.

1.4 *Conservation at Suspended Metabolism in Liquid Nitrogen*

Cryopreservation is achieved by storage in polypropylene straws submerged in liquid nitrogen at ca −196 °C or placed in nitrogen vapour (i.e. the vapour phase above liquid nitrogen) at −130 to −170 °C. This conservation method is based on the classical method of long-term maintenance of fungi in glass vials in liquid nitrogen, which had been developed in the 1970s (Elliott 1976) and successfully adopted by several large culture collections (e.g. Stalpers et al. 1987; Hoffman 1992) for a wide range of both filamentous fungi and yeasts. Cryopreservation is suitable for a wide range of aquatic fungi and other microorganisms participating in litter decomposition. Moreover, minimum genome and phenotype changes have been observed, sporulation capacity of aquatic hyphomycetes is very well preserved, little space is required, the shelf life is long (practically indefinite) without the need for regular transfer, and contamination during storage is practically excluded when fungi are preserved in liquid nitrogen vapours. Disadvantages are relatively high costs of the initial equipment and moderate maintenance costs (requiring reliable liquid nitrogen supply), comparatively laborious procedures, the risk of contamination of liquid nitrogen from a leaking straw in case of submerged storage, and potential health hazards during liquid nitrogen manipulation. In case of failure of the liquid nitrogen supply, the whole fungal collection can be lost. For such eventualities a back-up system is recommended.

1.5 *Conservation at Suspended Metabolism in Deep Freeze*

An alternative and simpler method is preservation at −70 to −80 °C in a deep freezer. Advantages are similar to preservation in liquid nitrogen, but the procedure is shorter, the equipment is cheaper, and the operating costs are lower. Contamination during storage is practically excluded. Disadvantages are that the shelf life is shorter than in liquid nitrogen (tens of years, up to 40). A back-up system is recommended for this conservation method, in case of electricity failure.

2 Equipment, Chemicals, and Cultivation Medium

2.1 Equipment and Material

- Glass bottles with screw caps, approx. 10 ml capacity
- Glass test tubes with inner diameter of ≥12 mm, and cotton plugs, or pressure-resistant plastic screw-cap bottles of at least 30 ml capacity, preferably wide-necked
- Adhesive labels, permanent markers, pens resistant to liquid nitrogen
- Inoculation needles or hooks
- Pasteur pipettes
- Bunsen burners
- Polypropylene or polyvinyl chloride drinking straws, inner diameter 3–4 mm, pressure resistant
- Forceps
- Scalpels
- Polypropylene cryotubes, ca. 2 ml capacity, pressure resistant (e.g. Nunc™)
- Canes for holding a set of cryotubes
- Cylindrical holders for canes
- Transport vessel and storage container for liquid nitrogen
- Small cardboard or plastic boxes (resistant to −70 °C) for cryotubes
- Petri dishes (plastic or glass)
- Biohazard II laminar flow box
- Deep freezer set
- Autoclave

2.2 Chemicals

- Deionized water
- Glycerol, analytical grade
- Mineral oil (medicinal paraffin oil, density of 0.83–0.89 g cm^{-3})
- Liquid nitrogen

2.3 Cultivation Medium

- Malt agar 2% (e.g. Difco, Oxoid)

3 Procedures

3.1 Preservation in Deionized Water

1. Prepare Petri dishes with vigorous, well-grown pure cultures.
2. Sterilize bottles with deionized water (ca. 5 ml in each) with caps loosely screwed on.
3. Cool to room temperature.
4. With the broad end of a sterile Pasteur pipette or drinking straw, aseptically cut disks from a colony (Fig. 24.3). Alternatively, cut small squares with a sterile scalpel.
5. Transfer several disks or squares to a bottle with sterile water and tightly screw on cap.
6. Since the basic unit in a culture collection is a strain (= isolate), label each bottle with strain number and date.
7. Store, preferably at 10 °C in the dark.
8. Revive culture by aseptically lifting a disk or square of the culture from the bottle and placing it mycelium down on a fresh agar medium plate.
9. Incubate at 15–20 °C.

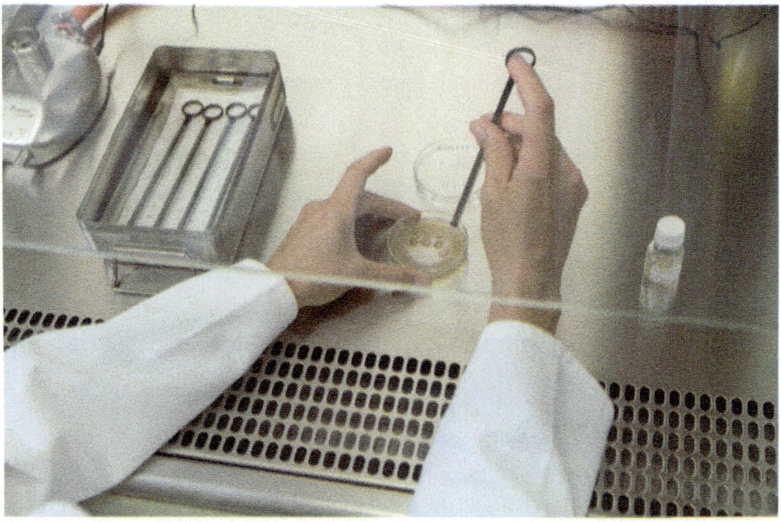

Fig. 24.3 Cutting out disks in a laminar flow box from an agar culture of *Tetracladium apiense* CCM F-23199

3.2 *Preservation Under Paraffin Oil*

1. Prepare glass test tubes with cotton plugs or bottles with screw caps filled up to about one third with melted agar medium.
2. Sterilize at 125 °C for 20 min.
3. Keep in tilted position until the medium solidifies to get the slant surface.
4. To inoculate the agar with a culture, use several pieces of inoculum and distribute them over the entire slant surface when handling slow-growing cultures.
5. Incubate until the agar surface is well covered with mycelium (usually 1–2 weeks at 15–20 °C).
6. Prepare test tubes with sufficient paraffin oil to amply cover the top of a slanted agar surface, one for each culture.
7. Autoclave paraffin oil in test tubes (121 °C for 20 min) and dry them at 160 °C for 2 h. Alternatively, autoclave twice (121 °C for 15 min).
8. Pour the cooled paraffin oil aseptically into tubes or bottles with cultures, always using one tube with paraffin oil per tube or bottle with a culture. The paraffin oil meniscus should be ca. 1 cm above the highest point of the agar slope or fungal growth.
9. Label each tube or bottle with strain number and date.
10. Store at 10–18 °C in the dark.
11. For retrieval, aseptically cut out a piece of the culture with an inoculation needle or hook.
12. Remove paraffin oil by pressing the agar piece against the tube wall and place it on fresh agar medium.
13. Incubate at 15–20 °C.
14. Keep the inoculated Petri dishes tilted at a 45° angle. This allows the remaining paraffin oil to drain. The first subculture is usually slimy, and a second transfer is often necessary.

3.3 *Preservation Submerged in Liquid Nitrogen or in Liquid Nitrogen Vapour*

1. Prepare cultures in Petri dishes with agar medium supplemented with 5% (w/v) of glycerol before autoclaving. Alternatively, flood the culture with 10% sterile glycerol (w/v in distilled water) 1 h before processing.
2. Cut drinking straws into 25 mm pieces and sterilize in autoclave (121 °C, 15 min) or with gamma-radiation.
3. Hold one straw piece with sterile forceps and punch out a disk from the agar culture (Fig. 24.4). Repeat the procedure until the straw is filled.
4. For storage in liquid nitrogen, seal both ends of the straw by squeezing the open ends with forceps and briefly holding them above a flame. This step is not necessary for storage in the vapour phase of liquid nitrogen.

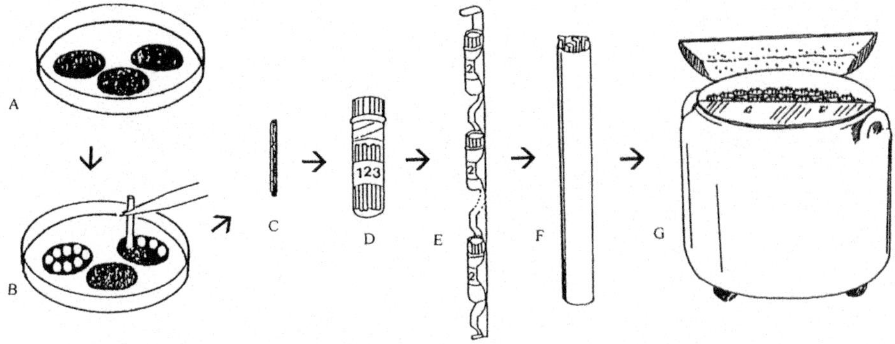

Fig. 24.4 Cryopreservation of fungi in liquid nitrogen vapour. (**A**) pure culture of a filamentous fungus; (**B**) punching the culture with a straw; (**C**) straw filled with pieces of culture; (**D**) cryotube with straws; (**E**) cryotubes clamped to aluminium cane; (**F**) cylindrical holder for canes; (**G**) liquid nitrogen vessel

5. Place four or five straws filled with one strain into one cryotube (Fig. 24.4) and label it with the strain number and date. To differentiate easily between strains, various colours of caps and straws may be used.
6. Place the cryotubes into numbered canes and store canes with cryotubes at −70 °C for 2 h, protected by insulation in polystyrene boxes.
7. Place the canes into numbered cylindrical holders and submerge these into liquid nitrogen or place them into liquid nitrogen vapour (Fig. 24.4). Wear full face, hand, and clothing protection during all work with liquid nitrogen, and avoid risk of anoxia resulting from possible oxygen deficiency caused by vapourization of liquid nitrogen.
8. For easy retrieval, prepare a diagram showing the location of cryotubes with straws within the numbered canes and of canes within the numbered holders.
9. For revival of a strain, remove a cryotube from the cane and put it into a polystyrene block in a box. This will prevent thawing for ca. 15 min.
10. Open the cryotube and place one straw aseptically on a Petri dish with 2% malt agar (Fig. 24.5). If necessary, aseptically cut off the sealed ends.
11. Incubate at 15–20 °C until growth appears from the open ends of the straw.
12. Transfer the growing culture to a fresh agar slant. Alternatively, place the straw directly on agar slant and incubate as above. However, aquatic hyphomycetes usually recover better when first placed on Petri dishes.

3.4 Preservation in Cryotubes at Deep Freeze

1. Follow the procedure for preservation in liquid nitrogen up to and including step 6 above, but skipping step 4.
2. Place the filled cryotubes into numbered small boxes and store them in a deep freezer at −70 to −80 °C.

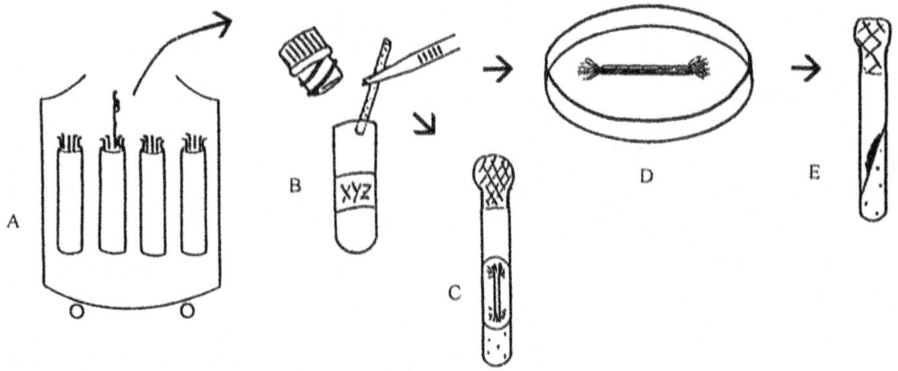

Fig. 24.5 Removal and thawing of frozen fungal cultures. (**A**) liquid nitrogen container; (**B**) removing one straw from the cryotube; (**C**) incubating the straw directly on agar slant; (**D**) incubating the straw on agar medium in a Petri dish; (**E**) subculture from the straw on agar slant

3. Prepare a diagram showing the location of the boxes with cryotubes on shelves in the deep freezer and the position of cryotubes in the boxes.
4. For revival of a strain, remove a cryotube from the box and put it into a polystyrene block. This will prevent thawing for ca. 15 min.
5. Open the cryotube and place one straw aseptically on a Petri dish with 2% malt agar. Incubate at 15–20 °C until growth appears from the open ends of the straw.

References

Anonymous. (1996). *Information document on access to ex-situ microbial genetic resources within the framework of the convention on biological diversity*. World Federation for Culture Collections, Yata, Mishima, Shizuoka, Japan, 25 pp.

Anonymous. (2015). News. *IMA Fungus, 6*, 3–5.

Baschien, C., Manz, W., Neu, T. R., Marvanová, L., & Szewzyk, U. (2008). In situ detection of freshwater fungi in an alpine stream by new taxon-specific fluorescence in situ hybridization probes. *Applied and Environmental Microbiology, 74*, 6427–6436.

Baschien, C., Tsui, C. K.-M., Gulis, V., Szewzyk, U., & Marvanová, L. (2013). The molecular phylogeny of aquatic hyphomycetes with affinity to the Leotiomycetes. *Fungal Biology, 117*, 660–672.

Bermingham, S., Dewey, F. M., Fisher, P. J., & Maltby, L. (2001). Use of a monoclonal antibody based immunoassay for the detection and quantification of *Heliscus lugdunensis* colonizing alder leaves and roots. *Microbial Ecology, 42*, 506–512.

Brosed, M., Jabiol, J., & Gessner, M. O. (2017). Nutrient stoichiometry of aquatic hyphomycetes: Interstrain variation and ergosterol conversion factors. *Fungal Ecology, 29*, 96–102.

Buell, C. B., & Weston, W. H. (1947). Application of the mineral oil conservation method to maintaining collections of fungus cultures. *American Journal of Botany, 34*, 555–561.

Campbell, J., Marvanová, L., & Gulis, V. (2009). Evolutionary relationships between aquatic anamorphs and teleomorphs: *Tricladium* and *Varicosporium*. *Mycological Research, 113*, 1322–1334.

Castellani, A. (1939). Maintenance and cultivation of common pathogenic fungi of man in sterile distilled water. *Journal of Tropical Medicine and Hygiene, 42*, 225–226.

Crous, P. W., Giraldo, A., Hawksword, D. L., Robert, W., Kirk, P. M., Guarro, J., Robertse, B., Schoch, C. L., Damm, U., Trakunyingcharoen, T., & Groenewald, J. Z. (2014). The genera of fungi: Fixing the application of type species of generic names. *IMA Fungus, 5*, 141–160.

Dang, C. K., Schindler, M., Chauvet, E., & Gessner, M. O. (2009). Temperature oscillation coupled with fungal community shifts can modulate warming effects on litter decomposition. *Ecology, 90*, 122–131.

Day, J. G., & Stacey, G. N. (2007). *Cryopreservation and freeze-drying protocols* (2nd ed.). Totowa: Humana Press.

Duarte, S., Pascoal, C., Cássio, F., & Bärlocher, F. (2006). Aquatic hyphomycete diversity and identity affect leaf litter decomposition in microcosms. *Oecologia, 147*, 658–666.

Duarte, S., Bärlocher, F., Trabulo, J., Cássio, F., & Pascoal, C. (2015). Stream-dwelling fungal decomposer communities along a gradient of eutrophication unraveled by 454 pyrosequencing. *Fungal Diversity, 70*, 127–148.

Elliott, T. J. (1976). Alternative ampoule for storing fungal cultures in liquid nitrogen. *Transactions of the British Mycological Society, 67*, 545–546.

Ellis, J. J. (1979). Preserving fungus strains in sterile water. *Mycologia, 71*, 1072–1075.

Fennell, D. I. (1960). Conservation of fungus cultures. *Botanical Review, 26*, 79–141.

Ferreira, V., & Chauvet, E. (2011). Synergistic effects of water temperature and dissolved nutrients on litter decomposition and associated fungi. *Global Change Biology, 17*, 551–564.

Glowka, L. (1996). The convention on biological diversity: Issues of interest to the microbial scientist and microbial culture collections. In R. A. Samson, J. A. Stalpers, D. van der Mei, & A. H. Stouthamer (Eds.), *Culture collections to improve the quality of life* (pp. 36–60). Baarn: Centraalbureau voor Schimmelcultures.

Gönczöl, J., Révay, Á., & Csontos, P. (2001). Effect of sample size on the detection of species and conidial numbers of aquatic hyphomycetes collected by membrane filtration. *Archiv für Hydrobiologie, 150*, 677–691.

Gulis, V. I., & Stephanovich, A. I. (1999). Antibiotic effects of some aquatic hyphomycetes. *Mycological Research, 103*, 111–115.

Hawksworth, D. L. (1996). Microbial collections as a tool in biodiversity and biosystematic research. In R. A. Samson, J. A. Stalpers, D. van der Mei, & A. H. Stouthamer (Eds.), *Culture collections to improve the quality of life* (pp. 26–35). Baarn: Centraalbureau voor Schimmelcultures.

Hibbett, D., Abarenkov, K., Kõljalg, U., Öpik, M., Chai, B., Cole, J., Wang, Q., Crous, P., Robert, V., Helgason, T., Herr, J. R., Kirk, P., Lueschow, S., O'Donnell, K., Nilsson, R. H., Oono, R., Schoch, C., Smyth, C., Walker, D. M., Porras-Alfaro, A., Taylor, J. W., & Geiser, D. M. (2016). Sequence based classification and identification of Fungi. *Mycologia, 108*, 1049–1068.

Hoffman, P. (1992). Technical Information Sheet No. 5. Cryopreservation of fungi. In K. A. Malik (Ed.), *Technical Information for Culture Collections. Curators in Developing Countries* (pp. 17–20). Braunschweig: Leibniz Institut DSMZ = German Collection of Microorganisms and Cell Cultures.

Homolka, L. (2013). Methods of kryopreservation in fungi. In V. K. Gupta, M. G. Tuohy, M. G. Ayyachamy, A. Turner, & A. O'Donovan (Eds.), *Laboratory protocols in fungal biology – Current methods in fungal biology* (pp. 9–16). New York: Springer.

Hubálek, Z. (1996). *Cryopreservation of microorganisms at ultra-low temperatures*. Prague: Academia.

Kaida, K., Fudou, R., Kameyama, T., Tsubaki, Y., Suzuki, Y., Ojika, M., & Sakagami, Y. (2001). New cyclic depsipeptide antibiotics, clavariopsins A and B, produced by an aquatic hyphomycete, *Clavariopsis aquatica*. I. Taxonomy, fermentation, isolation and biological properties. *Journal of Antibiotics, 54*, 17–21.

Kirsop, B. A., & Doyle, A. (1991). *Maintenance of microorganisms and cultured cells. A manual of laboratory methods* (2nd ed.). London: Academic.

Marvanová, L. (2001). Streamborne fungal spora in running waters of the Bohemian Forest. *Silva Gabreta, 7*, 147–164.

Marvanová, L., Landvik, S., Fisher, P. J., Moss, S. T., & Ainsworth, A. M. (2002). A new fungus with arthroconidia from foam. *Nova Hedwigia, 75*, 255–269.

McArthur, F. A., Baerlocher, M. O., MacLean, N. A. B., Hiltz, M. D., & Bärlocher, F. (2001). Asking probing questions: Can fluorescent in situ hybridization identify and localise aquatic hyphomycetes on leaf litter. *International Review of Hydrobiology, 86*, 429–438.

Nakasone, K.-K., Peterson, S. W., & Jong, S.-C. (2004). Preservation and distribution of fungal cultures. In G. M. Mueller, G. F. Bills, & M. S. Foster (Eds.), *Biodiversity of fungi: Inventory and monitoring methods* (pp. 37–47). Amsterdam: Elsevier Academic.

Oh, H., Kwon, T. O., Gloer, J. B., Marvanová, L., & Shearer, C. A. (1999). Tenellic acids A-D: New bioactive diphenyl ether derivatives from the aquatic fungus *Dendrospora tenella*. *Journal of Natural Products, 62*, 580–583.

Onions, A. H. S. (1971). Preservation of fungi. In C. Booth (Ed.), *Methods in microbiology* (pp. 113–151). London: Academic.

Ryan, M. J., & Smith, D. (2004). Fungal genetic resource centres and the genomic challenge. *Mycological Research, 108*, 1351–1362.

Seena, S., Marvanová, L., Letourneau, A., & Bärlocher, F. (2018). *Articulospora* – Phylogeny vs morphology. *Fungal Biology, 122*, 965–976.

Smith, D., & Onions, A. H. S. (1994). *The preservation and maintenance of living Fungi* (2nd ed.). Wallingford: CAB International.

Stalpers, J. A., de Hoog, S., & Vlug, I. J. (1987). Improvement of the straw technique for the preservation of fungi in liquid nitrogen. *Mycologia, 79*, 82–89.

Suberkropp, K. (1991). Relationships between growth and sporulation of aquatic hyphomycetes on decomposing leaf litter. *Mycological Research, 95*, 843–850.

Thornton, D. R. (1963). The physiology and nutrition of some aquatic hyphomycetes. *Journal of General Microbiology, 33*, 23–31.

Chapter 25
An Illustrated Key to the Common Temperate Species of Aquatic Hyphomycetes

Vladislav Gulis, Ludmila Marvanová, and Enrique Descals

Keywords Aquatic fungi · Conidia · Conidiogenesis · Fungal biodiversity · Fungal identification · Fungal spore shapes · Fungal taxonomy · Ingoldian fungi

1 Introduction

Aquatic hyphomycetes play a key role in the decomposition of allochthonous plant litter and food webs in lotic ecosystems (e.g., Bärlocher 1992; Hieber and Gessner 2002; Gessner et al. 2007; Bärlocher and Sridhar 2014). Soon after colonizing a substrate, many species produce vast amounts of conidia that enter the water column and are transported downstream. Aquatic hyphomycetes can invest up to 80% of their production into conidia, which can account for up to 8–12% of leaf litter mass loss (Suberkropp 1991). Most aquatic hyphomycetes form tetraradiate, variously branched or scolecoid (worm-like) conidia adapted for dispersal in flowing waters (Webster and Descals 1981; Descals 2005). Since most conidia are characteristically shaped (Figs. 25.1–10), it is often possible to identify them to species, count them, and thus gain insight into the structure of the fungal community developing on submerged substrates (Bärlocher 2005; Gulis and Bärlocher 2017). This facilitates ecological studies that link fungal biodiversity with functional aspects of ecosystems such as fungal production, microbial respiration, and leaf litter decomposition.

V. Gulis (✉)
Department of Biology, Coastal Carolina University, Conway, SC, USA
e-mail: vgulis@coastal.edu

L. Marvanová
Faculty of Science, Czech Collection of Microorganisms, Masaryk University, Brno, Czech Republic
e-mail: ludmila@sci.muni.cz

E. Descals
Esporles, Mallorca, Spain
e-mail: enrique.descals@gmail.com

F. Bärlocher et al. (eds.), *Methods to Study Litter Decomposition*,
https://doi.org/10.1007/978-3-030-30515-4_25

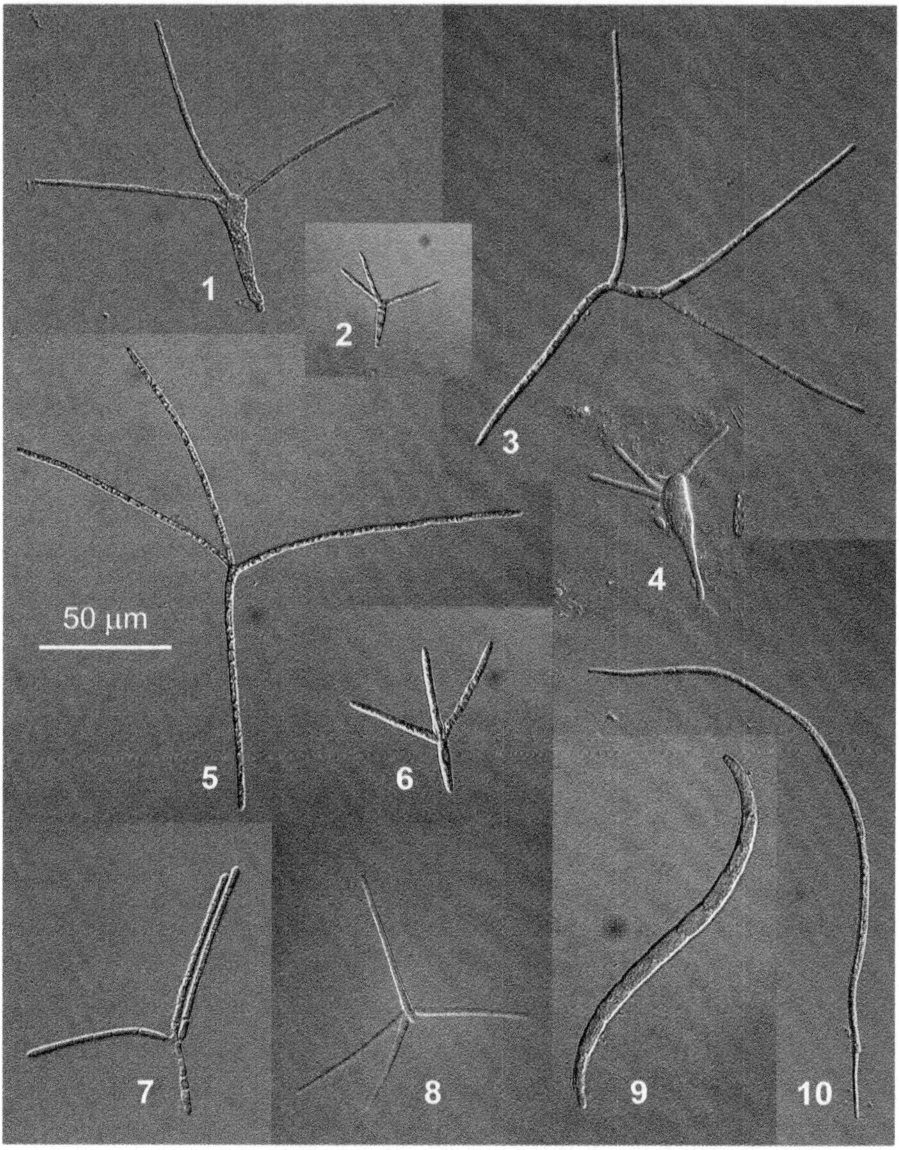

Figs. 25.1–10 Conidia of aquatic hyphomycetes. 1. *Clavariopsis aquatica*. 2. *Clavatospora longibrachiata*. 3. *Hydrocina chaetocladia*. 4. *Culicidospora gravida*. 5. *Tetrachaetum elegans*. 6. *Articulospora tetracladia*. 7. *Fontanospora eccentrica*. 8. *Triscelophorus acuminatus*. 9. *Anguillospora crassa*. 10. *Anguillospora filiformis*

The objective of this chapter is to provide assistance for the fast identification of aquatic hyphomycete conidia in ecological studies carried out in temperate climates. Conidia can be sampled in transport (water, foam), from naturally colonized submerged substrates after inducing sporulation in the laboratory, or from pure cultures (Gessner et al. 2003; Chaps. 23 and 26).

It is important to acknowledge that the presented key includes only 64 of ca. 180 species of aquatic hyphomycetes occurring in temperate climate, and over 300 species have been described worldwide. A key to aquatic hyphomycete genera was provided by L. Marvanová in Seifert et al. (2011, pp. 874–884), while tropical species were treated in Marvanová (1997). Here, along with common species, we also included some less frequent ones, whose conidia may be confused with others. The key is primarily based on morphological characters of detached conidia; however, for species with similar or less differentiated conidia, the mode of conidiogenesis may also be diagnostic and is hence illustrated. Only typical conidia developed under submerged or semi-submerged conditions are considered. Although we include drawings of conidia for all treated species, we encourage the reader to consult with taxonomic experts and additional pertinent literature (e.g., Petersen 1962, 1963a,b; Nilsson 1964; Dudka 1974; Ingold 1975; Webster and Descals 1981), as well as original species descriptions, since some aquatic hyphomycetes (especially scolecosporous species) cannot be identified with certainty based on detached conidia alone. A glossary of some terms that may cause difficulties follows the key.

The development of novel molecular techniques introduced new options for the analysis of fungal communities associated with decaying submerged plant litter. DGGE (Chap. 35, Gulis and Bärlocher 2017) and next-generation sequencing (Chaps. 37, 62, Duarte et al. 2015a) can be used with fungal DNA amplified from environmental samples. In analyses based on spores of aquatic hyphomycetes, pure cultures can be obtained and then DNA used for identification. In both cases, sequences of rDNA loci are commonly analyzed, especially ITS rDNA, which is widely used for barcoding of fungi in general (Schoch et al. 2011); the data base of sequences for aquatic hyphomycetes is constantly growing (e.g., Baschien et al. 2013; Duarte et al. 2013, 2015b).

2 Key to the Common Temperate Species of Aquatic Hyphomycetes (Based on Conidia)

1. Conidia variously branched or appearing tri-, tetra-, or multiradiate..........2
1a. Conidia of simple shape (scolecoid, globose, ellipsoid, fusiform, clavate, etc., some with short outgrowths or basal extensions)..................50
2. Conidia appearing triradiate..3

2a. Conidia of different morphology....4
3. Conidia spanning 8–13 μm, 3-celled, ends obtuse....*Tricellula aquatica* (Fig. 11)
3a. Conidial span 20–46 μm, apices acute.... *Ypsilina graminea* (Fig. 12)
4. Conidia small, spanning up to 25 μm, outline triangular, with a short axis and 3 laterals....*Lateriramulosa uni-inflata* (Fig. 13)
4a. Conidia of different shape....5
5. Conidia with a clamp connection on axis....6
5a. Clamp connections absent....7
6. Conidial elements cylindrical, axis gently curved or sigmoid, with an excentric basal extension, branch insertions unconstricted....*Taeniospora gracilis* var. *enecta* (Fig. 14a)
6a. Conidial elements long-fusoid, axis strongly curved or sigmoid, basal extension absent, branch insertions subconstricted....*Taeniospora gracilis* var. *gracilis* (Fig. 14b)
7. Conidia relatively large with typically numerous primary and secondary (sometimes tertiary) branches, elements cylindrical....8
7a. Conidia of different shape....11
8. Conidia resembling a fir tree with a more or less straight axis and perpendicular branches tending to aggregate near its base....9
8a. Conidial elements gently curved, branches distributed along the axis length, caducous (breaking off readily)....10
9. Conidia ca. 200 μm long, with more than 15 branches, elements 4–5 μm wide....*Dendrospora erecta* (Fig. 15)
9a. Conidia with up to 14 branches, elements 3–4 μm wide....*Dendrospora tenella* (Fig. 16)
10. Branches typically on one side of the axis, branch insertions abruptly constricted....*Varicosporium elodeae* (Fig. 17)
10a. Conidial elements delicate, gently constricted at septa, branches on both sides of the axis, insertions gradually narrowed....*Varicosporium delicatum* (Fig. 18)
11. Conidia with an axis, 1–2 primary and one secondary branch, elements tapering distally....12
11a Conidia of different shape....15
12. Conidial elements straight or nearly so, branching dorsal....13
12a Conidial elements strongly curved or sigmoid, branching ventral....14
13. Axis 26–51 μm long, 3–4(−5) septa.... *Pleuropedium tricladioides* (Fig. 19)
13a. Axis 38–75(−100) μm long, elements multiseptate....*Pleuropedium multiseptatum* (Fig. 20)
14. Conidia with one primary and one secondary branch....*Gyoerffyella gemellipara* (Fig. 21)

14a. Conidia with two primary and one secondary branch...*Gyoerffyella rotula* (Fig. 22)
15. Conidia with recurved axis and 2–3 branches (4–5 ends)..........................16
15a Conidia of different shape...17
16. Conidia with up to 5 ends, stout...*Tripospermum myrti* (Fig. 23)
16a. Conidia with up to 4 ends, more slender in appearance...................................*Tripospermum camelopardus* (Fig. 24)
17. Conidia spanning over 70 μm, with 4 long filiform extensions, conidial body of 2 parts......................*Campylospora chaetocladia* (Fig. 25)
17a Conidia of different shape...18
18. Conidia with a stalk bearing elements of 2 different shapes: filiform (subulate) and digitiform, filiform (subulate) and globose or only digitiform...19
18a Conidia of different morphology...23
19. Conidia with 2 globose and 3 filiform (subulate) elements...*Tetracladium marchalianum* (Fig. 26)
19a. Conidia with filiform (subulate) and digitiform or only digitiform elements...20
20. Filiform (subulate) elements lacking, 2 digitiform elements furcate, conidia with 5–6 apices................................*Tetracladium apiense* (Fig. 27)
20a Both filiform (subulate) and digitiform elements present..........................21
21. Conidia with 2(−3) filiform (subulate) and 2 digitiform elements, basal extension absent..................................*Tetracladium maxilliforme* (Fig. 28)
21a Conidia with 5–6 apices, basal extension typically present......................22
22. Conidia with 2 digitiform and 3 filiform (subulate) elements...*Tetracladium furcatum* (Fig. 29)
22a. Conidia with 3 digitiform and 3 filiform (subulate) elements..*Tetracladium setigerum* (Fig. 30)
23. Conidia with relatively broad body (clavate, fusoid, etc.) and 2–4 thinner branches or if all elements are of similar width then conidia spanning up to 15 μm...24
23a. Conidia with all elements of similar width, spanning more than 25 μm...30
24. Conidial body short-clavate, 10–15 μm long, with 3 conoid branches, conidia appear stellate......................................*Heliscella stellata* (Fig. 31)
24a. Conidial body long-clavate, navicular or obclavate, branches filiform, coronate or one terminal and 2–3 lateral...25
25. Branches coronate...26
25a One branch terminal and 2–3 lateral...28
26. Conidial body 35–50 × 10–12 μm..............*Clavariopsis aquatica* (Fig. 32)
26a Conidial body 3–5 μm wide...27

27. Conidial body 15–25 μm long..........*Clavatospora longibrachiata* (Fig. 33)
27a Conidial body 45–70 μm long.........................*Heliscus tentaculus* (Fig. 34)
28. Conidial body straight, navicular to obclavate, lateral branches 2–3, basal extension absent..*Classicula fluitans* (syn. *Naiadella fluitans*) (Fig. 35)
28a. Conidial body curved distally, long-clavate, lateral branches 2, basal extension percurrent, conidia have "mosquito" or "penguin" appearance..29
29. Conidial body hyaline, branches 40–80(−100) μm long..*Culicidospora aquatica* (Fig. 36)
29a. Conidial body hyaline or subfuscous, branches less than 40 μm long...*Culicidospora gravida* (Fig. 37)
30. Conidia tetraradiate in a broad sense, i.e., appearing as 4 arms radiating from a common point or from a central cell, or basiverticillate, or with terminal branches on a stalk, or with paired or subopposite laterals on a geniculate or curved axis..31
30a. Conidia with elongate axis and 2 alternate, NOT subopposite, branches..45
31. Conidia with a distinct globose central cell and 4 radiating arms..............32
31a Conidia of different shape..33
32. Arms as broad as central cell, long-obclavate, insertions constricted..*Lemonniera pseudofloscula* (Fig. 38)
32a. Arms thinner than central cell, cylindrical with subclavate ends, insertions unconstricted.......................*Lemonniera centrosphaera* (Fig. 39)
33. Conidia with 4 arms radiating from a common point, i.e., truly tetraradiate (indistinct central cell sometimes present).............................34
33a Conidia of different morphology...35
34. Arms cylindrical, 50–100 × 3–4 μm........... *Lemonniera aquatica* (Fig. 40)
34a. Arms conoid or obclavate, 20–45 × 4–9 μm... *Lemonniera terrestris* (Fig. 41)
35. Conidia basiverticillate..36
35a. Conidia with a stalk bearing terminal branches or with paired or subopposite laterals on a geniculate or curved axis...............................38
36. Conidial axis up to 105 μm long, elements cylindrical, branch insertions constricted..................................*Lemonniera filiformis* (Fig. 42)
36a Conidial axis long-obclavate, up to 70 μm long.......................................37
37. Axis 2-celled, lower cell often inflated, branches cylindrical, insertions subconstricted, septa indistinct or lacking...*Triscelophorus monosporus* (Fig. 43)
37a. Axis and branches long-obclavate, multiseptate, branch insertions abruptly constricted............................*Triscelophorus acuminatus* (Fig. 44)

38. Conidia with 3 terminal branches (or 2, one of them forking again), elements constricted at insertions .... *Articulospora tetracladia* (Fig. 45)
38a. Conidia with geniculate or curved axis and 2 branches attached near its middle..39
39. Branches subopposite, axis subconstricted at a septum between branch insertions..40
39a Branches paired, axis not constricted..41
40. Axis typically over 90 μm long, elements cylindrical...*Fontanospora eccentrica* (Fig. 46)
40a. Axis less than 90 μm long, branches long-obclavate..*Fontanospora fusiramosa* (Fig. 47)
41. Conidia spanning over 90 μm..42
41a. Conidia spanning less than 70 μm...43
42. Branches submedian, insertions subconstricted, axis slightly swollen and bent at branch insertions....................... *Geniculospora inflata* (Fig. 48)
42a. Elements of equal length, branches gently curved backward, insertions unconstricted..*Tetrachaetum elegans* (Fig. 49)
43. Conidial elements straight, axis bent at branch insertions, lower part of axis often subclavate, distal part thinner, cylindrical, branch-like, often twice as long....................................*Stenocladiella neglecta* (Fig. 50)
43a Conidial elements gently curved..44
44. Lower element of axis cylindrical to subclavate, distal elements narrow-obclavate, branch insertions strongly constricted..*Alatospora pulchella* (Fig. 51)
44a. Elements cylindrical or long-fusoid or branches (0–2) subulate, insertions unconstricted to constricted..*Alatospora acuminata* (Fig. 52)
45. Branch insertions unconstricted or subconstricted....................................46
45a Branch insertions abruptly constricted...48
46. Axis 150–200 μm long, elements gently curved, branch insertions subconstricted ....*Hydrocina chaetocladia* (syn. *Tricladium chaetocladium*) (Fig. 53)
46a Axis up to 120 μm long...47
47. Axis 50–120 μm long, geniculate, branch insertions unconstricted..*Tricladium angulatum* (Fig. 54)
47a. Axis 40–60 μm long, often curved in lower part, base swollen, branch insertions subconstricted.. *Tricladium curvisporum* (Fig. 55)

48. Axis geniculate or curved, elements cylindrical..*Tricladium patulum* (Fig. 56)
48a. Axis fusoid or long-fusoid, straight or gently curved, branches long-obclavate..49
49. Axis 50–75 × 2.5–3.5 µm, apices acute...*Tricladium attenuatum* (Fig. 57)
49a. Axis 60–140 × 5–7 µm.................................*Tricladium splendens* (Fig. 58)
50. Conidia scolecoid or filiform, i.e., length-to-width ratio > 10....................51
50a Conidia of different shape...58
51. Detachment scar lateral, dorsal (or basal extension integrated), conidia aseptate, lunate or sigmoid............................*Lunulospora curvula* (Fig. 59)
51a. Detachment scar at the base of conidia, basal extension excentric, percurrent or absent, conidia septate, variously curved.............................52
52. Basal extension excentric...53
52a Basal extension percurrent or lacking..55
53. Conidia filiform, 2.5–3.5 µm wide........... *Anguillospora filiformis* (Fig. 60)
53a Conidia 6–15 µm wide...54
54. Conidia hyaline, 110–190 × 6–13 µm, with 4–6 septa, one middle cell typically larger..*Mycofalcella calcarata* (Fig. 61)
54a. Conidia (or central cells) sometimes fuscous, 150–200 × 8–15 µm, with 7–11 septa, cells in the broad part of conidium of similar size..*Mycocentrospora acerina* (Fig. 62)
55. Conidia arcuate or sigmoid, 150–250 × 5–6 µm, 7–13 septate, basal extension growing through a frill (remnants of separating cell), which is usually difficult to observe ....*Amniculicola longissima* (syn. *Anguillospora longissima*) (Fig. 63)
55a Frill at the base of conidia absent..56
56. Conidia sigmoid, long-fusoid, over 5 µm wide...57
56a. Conidia filiform, 90–120 × 1.5–2.5 µm....... *Flagellospora curvula* (Fig. 64)
57. Conidia 150–300 × 5–7 (−9) µm, 10–23 septate, base truncate or with a subulate extension................................... *Anguillospora furtiva* (Fig. 65)
57a. Conidia 120–180 × 8–14 µm, base truncate or with a blunt extension...*Anguillospora crassa* (Fig. 66)
58. Conidia isodiametric, clavate or fusiform to rhomboid, with short outgrowths..59
58a. Conidia ellipsoid to reniform, without outgrowths, aseptate, (13–16–20(−24) × 8–10 µm....................*Dimorphospora foliicola* (Fig. 67)
59. Conidia globose, cubic to almost stellate in appearance, with 4–6 more or less equidistant, sometimes indistinct outgrowths................................60
59a Conidia clavate or fusiform to long-rhomboid..61

60. Conidia (10–)11–17(−21) μm in diam.......... *Goniopila monticola* (Fig. 68)
60a. Conidia (8–)9–13(−14) μm in diam. (conidia can also be limoniform to fusiform with 0–2 septa)..................... *Margaritispora aquatica* (Fig. 69)
61. Conidia fusiform to long-rhomboid, 3-celled, central cell inflated, with short equatorial outgrowths.....................*Tumularia aquatica* (Fig. 70)
61a Conidia clavate..62
62. Conidia curved, subclavate, 25–50 × 3–4 μm, outgrowths (1–)2 ....*Aquanectria submersa* (syn. *Heliscus submersus*) (Fig. 71)
62a Conidia straight, outgrowths 0–4...63
63. Distal cell swollen, outgrowths scattered over its surface, conidial body 25–40 × 10–12 μm................*Tumularia tuberculata* (Fig. 72)
63a Distal cell not inflated, outgrowths coronate...64
64. Conidia broadly clavate to obcampanulate, 13–25 × 7–13 μm, conidial base with a denticle *Heliscina campanulata* (Fig. 73)
64a. Conidia subclavate or clove-shaped, 20–45 × 4–6 μm, with 3(−4) conoid outgrowths or apex oblique..........*Neonectria lugdunensis* (syn. *Heliscus lugdunensis*) (Fig. 74)

3 Glossary

Arcuate	Curved like a bow or arch
Basiverticillate	Having similar elements arranged in a whorl at the lowermost portion of the parent element
Clavate	Gradually broadening toward the distal part, club-shaped
Constricted	Strongly and often abruptly narrowed
Coronate	Having elements arranged in a crown-like fashion
Denticle	Small tooth-like projection
Digitiform	Finger-shaped
Excentric	Located off the center, referring here to basal extensions on the side of conidial scar
Filiform	Resembling a thread or filament
Furcate	Divided into two elements, forked
Fuscous	Having a brownish grey color
Fusiform, fusoid	Tapering toward each end, spindle-shaped
Geniculate	Bent abruptly at an angle like a knee joint
Navicular	Resembling a boat
Obclavate	Gradually broadening toward the proximal part, cf. clavate

Obcampanulate	Shaped like an inverted bell
Percurrent	Refers to basal extensions growing through a scar, cf. excentric
Recurved	Curved backward
Reniform	Kidney-shaped
Scar (conidial)	Part of a septum involved in secession; it forms the conidial base, but is sometimes replaced by a percurrent basal extension
Scolecoid	Worm-like
Sigmoid	Curved like the Greek letter sigma when standing at the end of a word (ς) or the Latin letter S
Stellate	Star-shaped, consisting of short elements radiating from a common center
Sub	Prefix signifying inferior position or degree: under, below, almost, not completely; e.g., *submedian*, situated below the middle; *subopposite*, arranged in pairs but not exactly on the same level
Subulate	Tapering gradually to a point, awl-shaped
Tetraradiate	Having four radiating elements
Triradiate	Having three radiating elements
Truncate	Terminating abruptly as if having the end cutoff

4 Final Remarks

Most of the illustrations of conidia used in the key are published or unpublished drawings by the authors. Twenty-seven were taken from Bärlocher and Marvanová (2010). Illustrations of the genera *Dendrospora*, *Gyoerffyella*, and *Tetracladium* are from Descals and Webster (1980), Marvanová (1975), and Roldán et al. (1989), respectively. *Anguillospora furtiva* is from the original description by Descals et al. (1998).

Details of conidiogenesis should be observed to identify species with simply shaped conidia such as *Dimorphospora foliicola*, *Goniopila monticola*, *Margaritispora aquatica*, and many species with filiform conidia. Ideally, single-spore isolates are obtained for detailed morphological examinations and to sequence ITS rDNA or other loci to confirm the species identity.

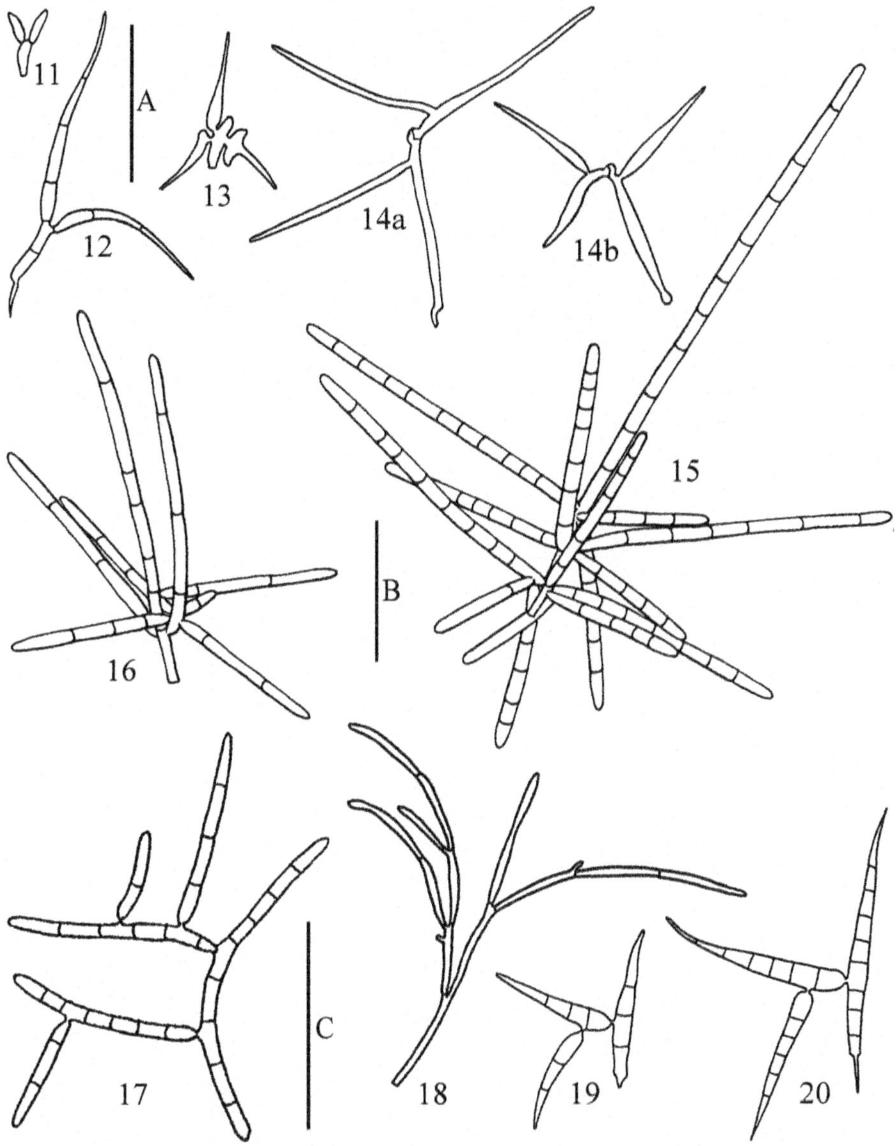

Figs. 25.11–20 Conidia of aquatic hyphomycetes. 11. *Tricellula aquatica*. 12. *Ypsilina graminea*. 13. *Lateriramulosa uni-inflata*. 14a. *Taeniospora gracilis* var. *enecta*. 14b. *Taeniospora gracilis* var. *gracilis*. 15. *Dendrospora erecta*. 16. *Dendrospora tenella*. 17. *Varicosporium elodeae*. 18. *Varicosporium delicatum*. 19. *Pleuropedium tricladioides*. 20. *Pleuropedium multiseptatum*. Scale bar A (Figs. 11–14) = 25 μm, B (Figs. 15–16) = 50 μm, C (Figs. 17–20) = 50 μm

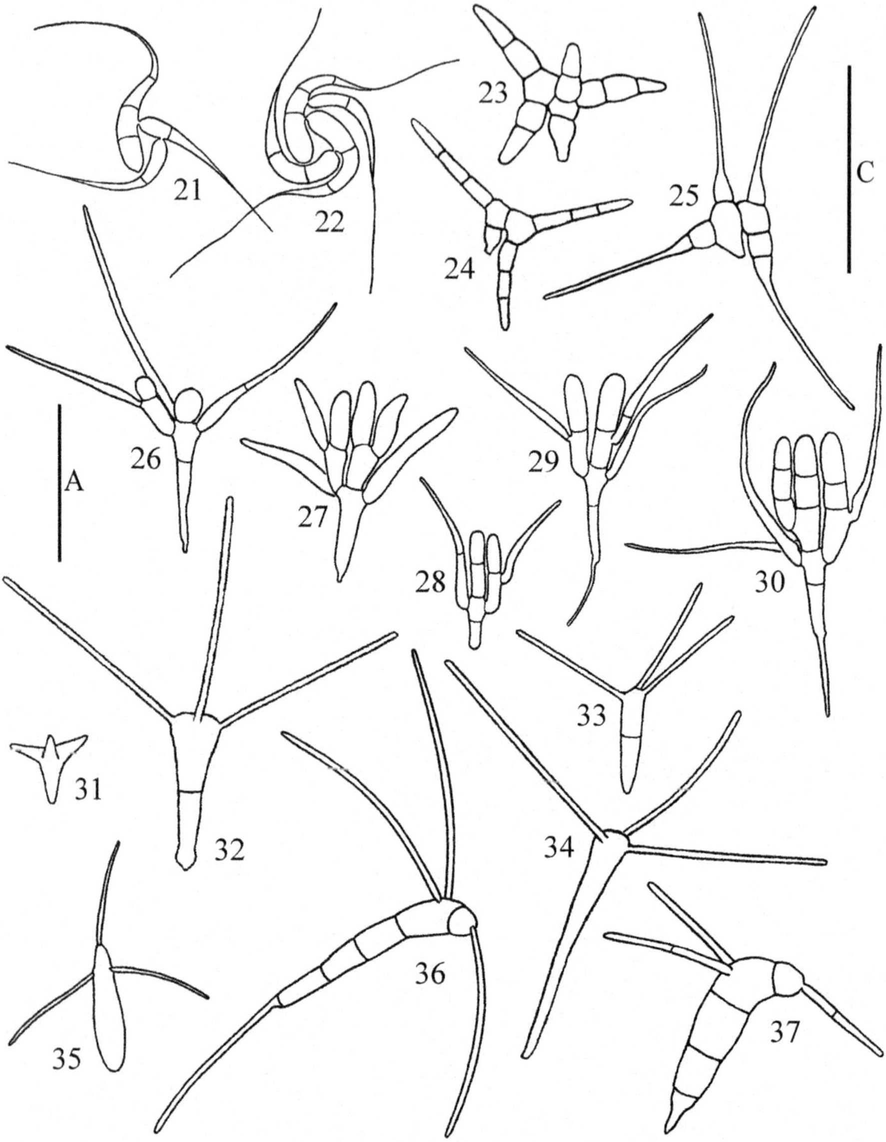

Figs. 25.21–37 Conidia of aquatic hyphomycetes. 21. *Gyoerffyella gemellipara*. 22. *Gyoerffyella rotula*. 23. *Tripospermum myrti*. 24. *Tripospermum camelopardus*. 25. *Campylospora chaetocladia*. 26. *Tetracladium marchalianum*. 27. *Tetracladium apiense*. 28. *Tetracladium maxilliforme*. 29. *Tetracladium furcatum*. 30. *Tetracladium setigerum*. 31. *Heliscella stellata*. 32. *Clavariopsis aquatica*. 33. *Clavatospora longibrachiata*. 34. *Heliscus tentaculus*. 35. *Classicula fluitans*. 36. *Culicidospora aquatica*. 37. *Culicidospora gravida*. Scale bar A (Figs. 21, 22, 26–31, 33) = 25 μm, C (Figs. 23–25, 32, 34–37) = 50 μm

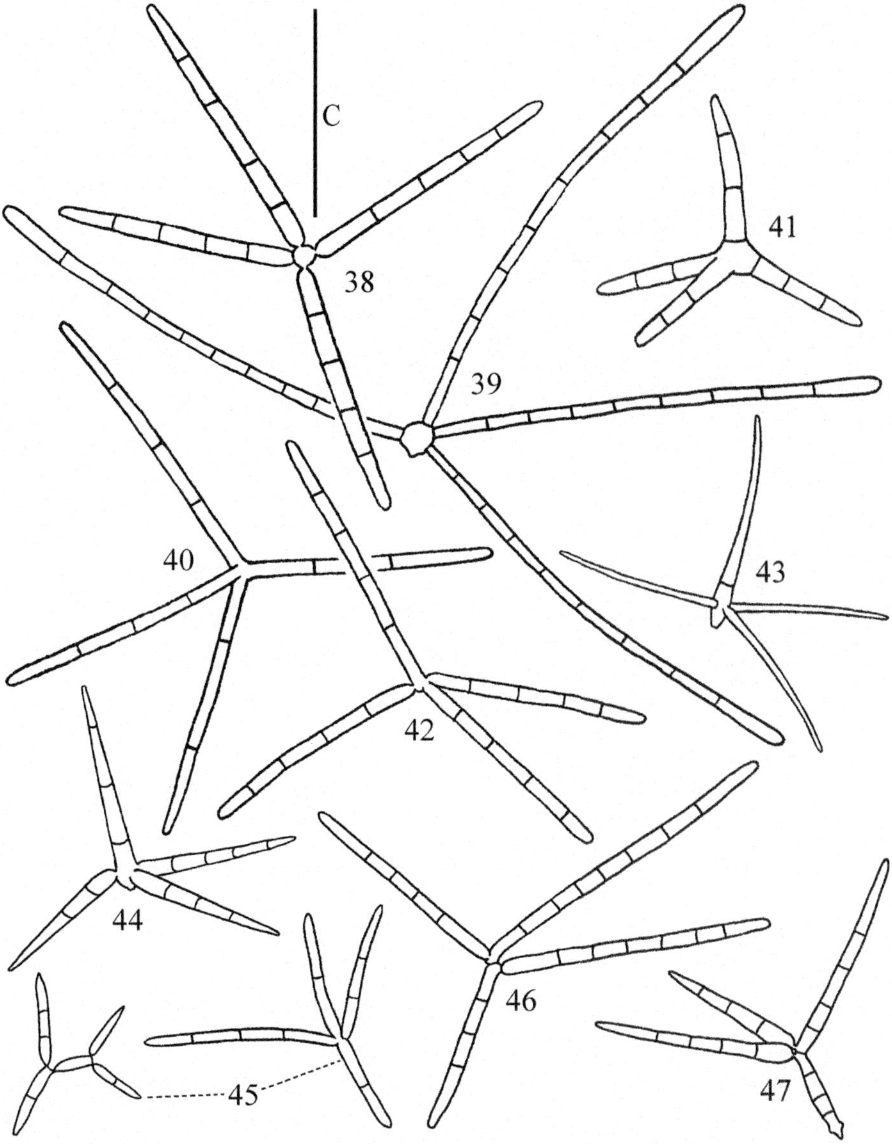

Figs. 25.38–47 Conidia of aquatic hyphomycetes. 38. *Lemonniera pseudofloscula*. 39. *Lemonniera centrosphaera*. 40. *Lemonniera aquatica*. 41. *Lemonniera terrestris*. 42. *Lemonniera filiformis*. 43. *Triscelophorus monosporus*. 44. *Triscelophorus acuminatus*. 45. *Articulospora tetracladia*. 46. *Fontanospora eccentrica*. 47. *Fontanospora fusiramosa*. Scale bar C (Figs. 38–47) = 50 μm

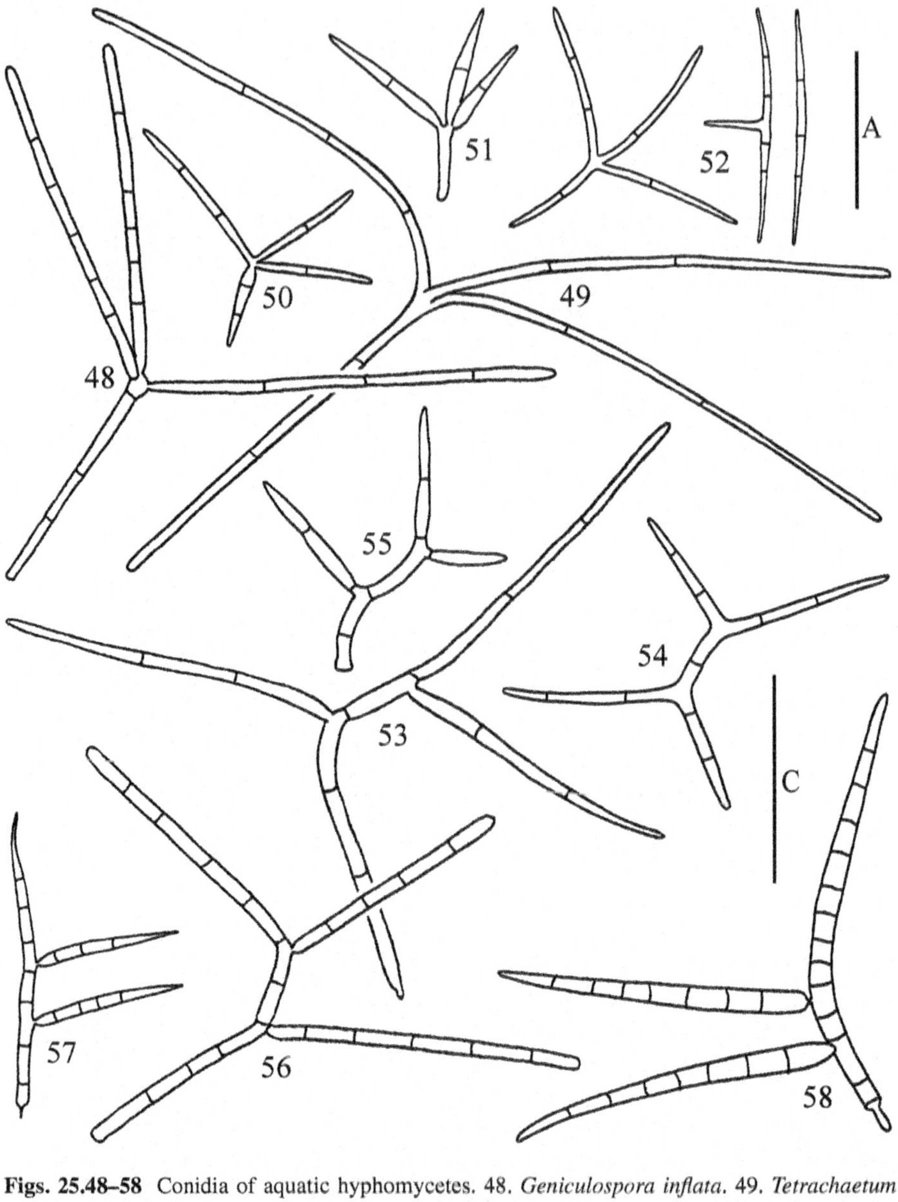

Figs. 25.48–58 Conidia of aquatic hyphomycetes. 48. *Geniculospora inflata*. 49. *Tetrachaetum elegans*. 50. *Stenocladiella neglecta*. 51. *Alatospora pulchella*. 52. *Alatospora acuminata*. 53. *Hydrocina chaetocladia*. 54. *Tricladium angulatum*. 55. *Tricladium curvisporum*. 56. *Tricladium patulum*. 57. *Tricladium attenuatum*. 58. *Tricladium splendens*. Scale bar A (Figs. 50–52, 55) = 25 μm, C (Figs. 48, 49, 53, 54, 56–58) = 50 μm

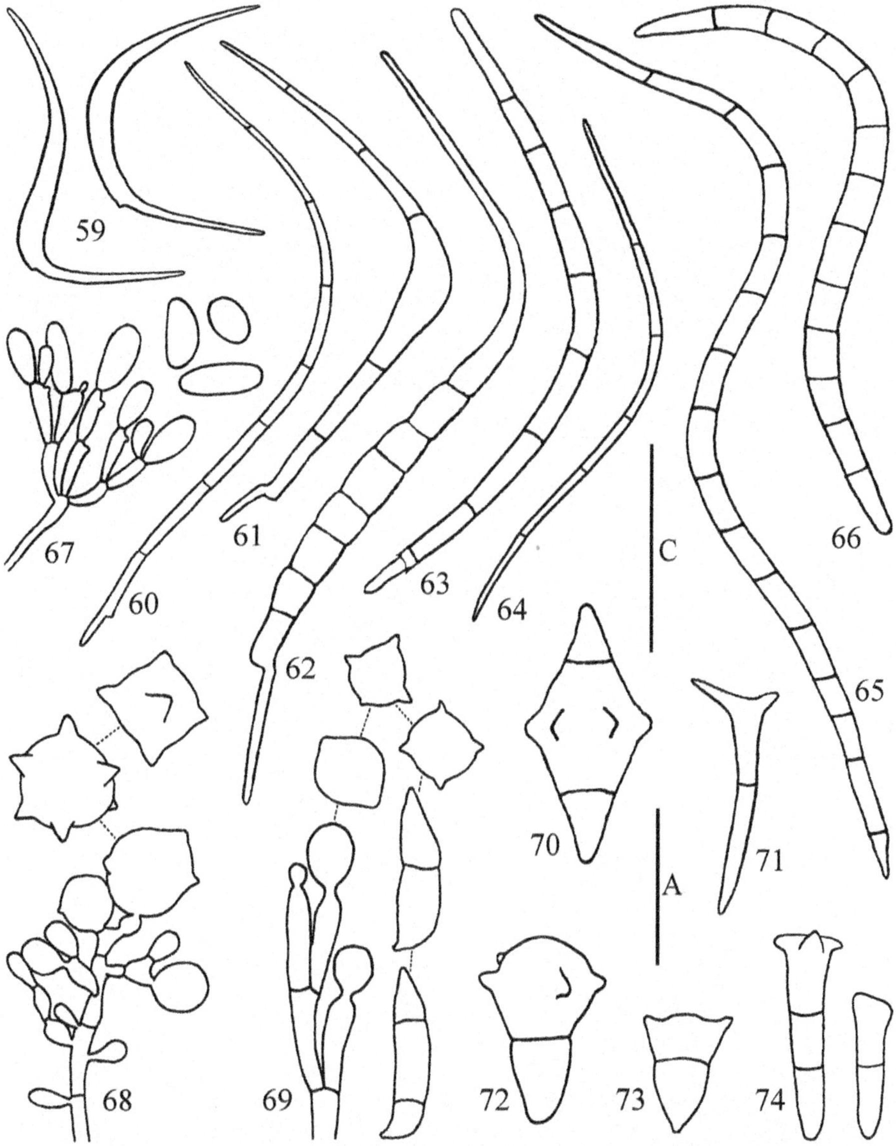

Figs. 25.59–74 Conidia of aquatic hyphomycetes. Some details of conidiogenesis are shown in Figs. 67–69. 59. *Lunulospora curvula*. 60. *Anguillospora filiformis*. 61. *Mycofalcella calcarata*. 62. *Mycocentrospora acerina*. 63. *Amniculicola longissima*. 64. *Flagellospora curvula*. 65. *Anguillospora furtiva*. 66. *Anguillospora crassa*. 67. *Dimorphospora foliicola*. 68. *Goniopila monticola*. 69. *Margaritispora aquatica*. 70. *Tumularia aquatica*. 71. *Aquanectria submersa*. 72. *Tumularia tuberculata*. 73. *Heliscina campanulata*. 74. *Neonectria lugdunensis*. Scale bar A (Figs. 68–74) = 25 μm, C (Figs. 59–67) = 50 μm

References

Bärlocher, F. (1992). *The ecology of aquatic hyphomycetes*. Berlin: Springer.
Bärlocher, F. (2005). Freshwater fungal communities. In J. Dighton, J. F. White, & P. Oudemans (Eds.), *The fungal community: Its organization and role in the ecosystem* (3rd ed., pp. 39–60). Boca Raton: CRC Press.
Bärlocher, F., & Marvanová, L. (2010). Aquatic hyphomycetes (Deuteromycotina) of the Atlantic maritime ecozone. In D. F. McAlpine & I. M. Smith (Eds.), *Assessment of species diversity in the atlantic maritime ecozone* (pp. 71–106). Ottawa: NRC Research Press.
Bärlocher, F., & Sridhar, K. R. (2014). Association of animals and fungi in leaf decomposition. In E. B. G. Jones, K. D. Hyde, & K.-L. Pang (Eds.), *Freshwater fungi and fungal-like organisms* (pp. 412–441). Berlin: De Gruyter.
Baschien, C., Tsui, C. K. M., Gulis, V., Szewzyk, U., & Marvanová, L. (2013). The molecular phylogeny of aquatic hyphomycetes with affinity to the Leotiomycetes. *Fungal Biology, 117*, 660–672.
Descals, E. (2005). Diagnostic characters of propagules of Ingoldian fungi. *Mycological Research, 109*, 545–555.
Descals, E., Marvanová, L., & Webster, J. (1998). New taxa and combinations of aquatic hyphomycetes. *Canadian Journal of Botany, 76*, 1647–1659.
Descals, E., & Webster, J. (1980). Taxonomic studies on aquatic hyphomycetes II. The *Dendrospora* aggregate. *Transactions of the British Mycological Society, 74*, 135–158.
Duarte, S., Bärlocher, F., Trabulo, J., Cássio, F., & Pascoal, C. (2015a). Stream-dwelling fungal decomposer communities along a gradient of eutrophication unraveled by 454 pyrosequencing. *Fungal Diversity, 70*, 127–148.
Duarte, S., Batista, D., Bärlocher, F., Cássio, F., & Pascoal, C. (2015b). Some new DNA barcodes of aquatic hyphomycetes species. *Mycoscience, 56*, 102–108.
Duarte, S., Seena, S., Bärlocher, F., Pascoal, C., & Cássio, F. (2013). A decade's perspective on the impact of DNA sequencing on aquatic hyphomycete research. *Fungal Biology Reviews, 27*, 19–24.
Dudka, I. O. (1974). *Aquatic hyphomycetes of the Ukraine*. Kiev: Naukova Dumka. [In Ukrainian].
Gessner, M. O., Bärlocher, F., & Chauvet, E. (2003). Qualitative and quantitative analyses of aquatic hyphomycetes in streams. In C. K. M. Tsui & K. D. Hyde (Eds.), *Freshwater mycology* (pp. 127–157). Hong Kong: Fungal Diversity Press.
Gessner, M. O., Gulis, V., Kuehn, K. A., Chauvet, E., & Suberkropp, K. (2007). Fungal decomposers of plant litter in aquatic ecosystems. In C. P. Kubicek & I. S. Druzhinina (Eds.), *The mycota, Vol. IV, environmental and microbial relationship* (pp. 301–324). Berlin: Springer-Verlag.
Gulis, V., & Bärlocher, F. (2017). Fungi: Biomass, production, and community structure. In F. R. Hauer & G. A. Lamberti (Eds.), *Methods in stream ecology* (Vol. 1, pp. 177–192). San Diego: Academic Press.
Hieber, M., & Gessner, M. O. (2002). Contribution of stream detritivores, fungi, and bacteria to leaf breakdown based on biomass estimates. *Ecology, 83*, 1026–1038.
Ingold, C. T. (1975). *An illustrated guide to aquatic and water-borne hyphomycetes (Fungi Imperfecti) with notes on their biology*. Freshwater Biological Association Scientific Publication No. 30. Ambleside, UK.
Marvanová, L. (1975). Concerning *Gyoerffyella* Kol. *Transactions of the British Mycological Society, 65*, 555–565.
Marvanová, L. (1997). Freshwater hyphomycetes: A survey with remarks on tropical taxa. In K. K. Janardhanan, C. Rajendran, K. Natarajan, & D. L. Hawksworth (Eds.), *Tropical mycology* (pp. 169–226). Enfield: Science Publishers.
Nilsson, S. (1964). Freshwater hyphomycetes. Taxonomy, morphology and ecology. *Symbolae Botanicae Upsaliensis, 18*, 1–130.
Petersen, R. H. (1962). Aquatic hyphomycetes from North America. I. Aleuriosporae (part 1) and key to the genera. *Mycologia, 54*, 117–151.

Petersen, R. H. (1963a). Aquatic hyphomycetes from North America. II. Aleuriosporae (part 2), and Blastosporae. *Mycologia, 55*, 18–29.

Petersen, R. H. (1963b). Aquatic hyphomycetes from North America III. Phialosporae and miscellaneous species. *Mycologia, 55*, 570–581.

Roldán, A., Descals, E., & Honrubia, M. (1989). Pure culture studies of *Tetracladium. Mycological Research, 93*, 452–465.

Seifert, K. A., Morgan-Jones, J., Gams, W., & Kendrick, B. (2011). *The genera of Hyphomycetes.* Utrecht: CBS-KNAW Fungal Biodiversity Centre.

Schoch, C. L., Seifert, K. A., Huhndorf, S., Robert, V., Spouge, J. L., Levesque, C. A., Chen, W., & Fungal Barcoding Consortium. (2011). Nuclear ribosomal internal transcribed spacer (ITS) region as a universal DNA barcode marker for Fungi. *Proceedings of the National Academy of Sciences, USA, 109*, 6241–6246.

Suberkropp, K. (1991). Relationships between growth and sporulation of aquatic hyphomycetes on decomposing leaf litter. *Mycological Research, 95*, 843–850.

Webster, J., & Descals, E. (1981). Morphology, distribution, and ecology of conidial fungi in freshwater habitats. In G. T. Cole & B. Kendrick (Eds.), *Biology of conidial Fungi* (Vol. 1, pp. 295–355). New York: Academic Press.

Chapter 26
Sporulation by Aquatic Hyphomycetes

Felix Bärlocher

Keywords Conidia · Fungal propagules · Fungal reproductive potential · Induction of sporulation · Ingoldian fungi · Membrane filtration · Mitospores · Spore production · Sporulation activity

1 Introduction

Fungi are instrumental in leaf decomposition in streams, and their biomass accumulating on leaves improves substrate palatability and nutritional value to shredders (Bärlocher 1985; Suberkropp 1992; Graça 2001; Gessner et al. 2007; Canhoto and Graça 2008; Krauss et al. 2011; Bärlocher and Sridhar 2014). The preferred method to measure fungal biomass and production is based on the indicator molecule ergosterol, which occurs at a relatively constant concentration in living mycelia (Chap. 27). A very substantial proportion of fungal production, often in excess of 50% (Findlay and Arsuffi 1989; Chauvet and Suberkropp 1998; Sridhar and Bärlocher 2000), is invested in propagules that are released from leaves. Asexually produced spores (mitospores, conidia) dominate. On leaves freshly recovered from a stream, only a few conidia can be observed. However, if such leaves are incubated for 1–2 days under conditions that stimulate sporulation (low to intermediate nutrient levels, high turbulence), newly formed conidia will be released. They can be trapped on a membrane filter, stained, and counted and identified under a microscope (Fig. 26.1). There is often a significant correlation between maximum fungal biomass on the leaf and maximum spore production over the course of decomposition (Gessner and Chauvet 1994; Maharning and Bärlocher 1996; Bärlocher et al. 2012). However, at any given point during decomposition, high sporulation rate by a species does not imply the presence of high mycelial biomass belonging to the same

F. Bärlocher (✉)
Department of Biology, Mount Allison University, Sackville, NB, Canada
e-mail: fbaerloc@mta.ca

F. Bärlocher et al. (eds.), *Methods to Study Litter Decomposition*,
https://doi.org/10.1007/978-3-030-30515-4_26

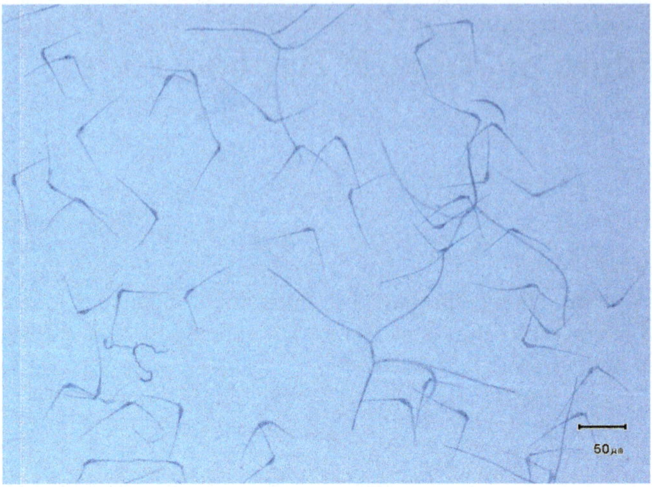

Fig. 26.1 Stained filter with conidia (*Lunulospora* spp., *Tricladium chaetocladium*) trapped from aerated *Eucalyptus viminalis* leaves (Bärlocher et al. 2012)

Table 26.1 Maximum spore production rates (no. day^{-1} mg^{-1} of leaf dry mass) reported from leaves decomposing in streams, from selected studies

Sporulation rate	Length	Leaf species	Condition	References
75	93	*Liriodendron tulipifera*	Softwater	Suberkropp (2001)
425	21	*Liriodendron tulipifera*	Hardwater	Suberkropp (2001)
1500	28	*Alnus glutinosa*	Softwater, 10 °C	Hieber and Gessner (2002)
4000	14	*Alnus glutinosa*	15 °C	Bärlocher et al. (1995)
7000	28	*Eucalyptus globulus*	15 °C	Bärlocher et al. (1995)

Length = number of days of incubation in stream before maximum was reached

species on the leaf (Bermingham et al. 1997). Sporulation under laboratory conditions can be as high as 7000 spores produced day^{-1} mg^{-1} of leaf dry mass (for reviews, see Gessner 1997; Gessner et al. 2003; selected values are shown in Table 26.1).

The aim of the described procedure is an estimate of the reproductive potential of the mycelia present in leaves recovered from a stream, following procedures based on Bärlocher (1982). This and similar procedures are also described in Gessner et al. (2003). The data can be used to estimate the amount of conidial biomass released from leaves or to describe the diversity and composition of the fungal community.

2 Equipment, Chemicals, and Solutions

2.1 Equipment and Material

- Autumn-shed leaves, air dried
- Litter bags (10 × 10 cm, 10, 1, or 0.5 mm mesh size)
- Erlenmeyer flasks, 250 ml, with 100 ml of deionized, sterile water
- Membrane filters, 5 or 8 μm pore size, and filtering apparatus
- Supply of pressurized air (e.g., aquarium pumps), tubing and Pasteur pipettes, or shaker
- Drying oven (50 °C)
- Balance (±1 mg precision)
- Microscope (with 16, 40 and 100 × objectives)

2.2 Chemicals and Solutions

- Lactic acid
- Deionized water
- Phenol and glycerol for long-term storage
- 0.1% Trypan Blue or Cotton Blue in 60% lactic acid (Trypan Blue in lactophenol is preferable for long-term storage: 10 ml phenol, 10 ml lactic acid, 20 ml glycerol, 10 ml H_2O)

3 Experimental Procedures

1. Prepare litter bags to be placed in a stream as indicated in Chaps. 5 and 6.
2. Anchor leaf bags to stream bed by means of bricks, steel pegs, or other devices. Be careful not to place too many bags close to each other, because this may greatly change flow patterns and thereby affect fungal colonization of leaves.
3. Recover bags at appropriate intervals (fungal colonization proceeds faster on leaf species that decompose more rapidly; see Chap. 6).
4. Rinse leaves to remove silt, sand, and invertebrates.
5. Place some leaf material (ca. 9 cm^2) in an Erlenmeyer flask with sterile, deionized water or filtered stream water.
6. Induce turbulence by placing the flask on a shaker (100–150 rpm) or by aerating it (connect Pasteur pipettes with tubing to source of pressurized air and adjust airflow to approx. 1 ml s^{-1}).
7. After 24–48 h, remove leaf material and determine its dry mass (40–50 °C, 2 days or until constant weight is reached).
8. Filter supernatant through membrane filter.

9. Add a few drops of Trypan Blue solution to filter; incubate for 30–60 min at 40–50 °C.
10. Scan the surface of the filter under the light microscope. Count and identify all conidia, or, if they are very numerous, all conidia in 2–30 randomly chosen microscope fields or on a defined fraction of the filter (cf. Gönczöl et al. 2001). If you count in some fields, you will need to know the area of the field and the retention zone of the filter to extrapolate your counts to the entire filter. If the filter dries before analysis, the conidial shapes can usually be restored by adding small amounts of lactic acid and briefly heating the slide.
11. Express the number of spores produced during laboratory incubations per leaf dry mass or ash-free dry mass.
12. To estimate biomass of the spores, determine total volume of spores, and assume a density of 500 fg μm^{-3} (Findlay and Arsuffi 1989). Volumes of selected species are listed in Bärlocher and Schweizer (1983). Chauvet and Suberkropp (1998) provide average spore masses of additional species. Or, assume an average conidial biomass of 200 pg (conservative estimate; Gessner 1997).
13. Analyze fungal community structure as described in Chap. 61 or by other means (e.g., multivariate analyses).

References

Bärlocher, F. (1982). Conidium production from leaves and needles in four streams. *Canadian Journal of Botany, 60*, 1487–1494.

Bärlocher, F. (1985). The role of fungi in the nutrition of stream invertebrates. *Botanical Journal of the Linnean Society, 91*, 83–94.

Bärlocher, F., & Schweizer, M. (1983). Effects of leaf size and decay rate on colonization by aquatic hyphomycetes. *Oikos, 41*, 205–210.

Bärlocher, F., & Sridhar, K. R. (2014). Association of animals and fungi in leaf decomposition. In E. B. Gareth Jones, K. D. Hyde, & K.-L. Pang (Eds.), *Freshwater fungi and fungal-like organisms* (pp. 412–441). Berlin/Boston: De Gruyter.

Bärlocher, F., Canhoto, C., & Graça, M. A. S. (1995). Fungal colonization of alder and eucalypt leaves in two streams in Central Portugal. *Archiv für Hydrobiologie, 133*, 457–470.

Bärlocher, F., Stewart, M., & Ryder, D. S. (2012). Processing of *Eucalyptus viminalis* leaves in Australian streams – Importance of aquatic hyphomycetes and zoosporic fungi. *Fundamental and Applied Limnology, 179*, 305–319.

Bermingham, S., Maltby, L., & Dewey, F. M. (1997). Use of immunoassays for the study of natural assemblages of aquatic hyphomycetes. *Microbial Ecology, 33*, 223–229.

Canhoto, C., & Graça, M. A. S. (2008). Interactions between fungi and stream invertebrates: Back to the future. In K. R. Sridhar, F. Bärlocher, & K. D. Hyde (Eds.), *Novel techniques and ideas in mycology* (Fungal diversity research series) (Vol. 20, pp. 305–325).

Chauvet, E., & Suberkropp, K. (1998). Temperature and sporulation of aquatic hyphomycetes. *Applied and Environmental Microbiology, 64*, 1522–1525.

Findlay, S. E. G., & Arsuffi, T. L. (1989). Microbial growth and detritus transformations during decomposition of leaf litter in a stream. *Freshwater Biology, 21*, 261–269.

Gessner, M. O. (1997). Litter breakdown in rivers and streams. *Limnetica, 13*, 33–44.

Gessner, M. O., & Chauvet, E. (1994). Importance of stream microfungi in controlling breakdown rates of leaf litter. *Ecology, 75*, 1807–1817.
Gessner, M. O., Bärlocher, F., & Chauvet, E. (2003). Qualitative and quantitative analyses of aquatic hyphomycetes in streams. In C. K. M. Tsui & K. D. Hyde (Eds.), *Freshwater mycology* (pp. 120–157). Hong Kong: Fungal Diversity Press.
Gessner, M. O., Gulis, V., Kuehn, K. A., Chauvet, E., & Suberkropp, K. (2007). Fungal decomposers of plant litter in aquatic ecosystems. In C. P. Kubicek & I. S. Druzhinina (Eds.), *The mycota: Environmental and microbial relationships* (Vol. IV, 2nd ed., pp. 301–321). Berlin: Springer.
Gönczöl, J., Révay, A., & Csontas, P. (2001). Effect of sample size on the detection of species and conidial numbers of aquatic hyphomycetes collected by membrane filtration. *Archiv für Hydrobiologie, 150*, 677–691.
Graça, M. A. S. (2001). The role of invertebrates on leaf litter decomposition in streams – A review. *International Review of Hydrobiology, 86*, 383–393.
Hieber, M., & Gessner, M. O. (2002). Contribution of stream detritivores, fungi and bacteria to leaf breakdown based on biomass estimates. *Ecology, 83*, 1026–1038.
Krauss, G.-J., Solé, M., Krauss, G., Schlosser, D., Wesenberg, D., & Bärlocher, F. (2011). Fungi in freshwaters: Ecology, physiology and biochemical potential. *FEMS Microbiology Reviews, 35*, 620–651.
Maharning, A. R., & Bärlocher, F. (1996). Growth and reproduction in aquatic hyphomycetes. *Mycologia, 88*, 80–88.
Sridhar, K. R., & Bärlocher, F. (2000). Initial colonization, nutrient supply and fungal activity on leaves decaying in streams. *Applied and Environmental Microbiology, 66*, 1114–1119.
Suberkropp, K. (1992). Interactions with invertebrates. In F. Bärlocher (Ed.), *The ecology of aquatic hyphomycetes* (Ecological studies) (Vol. 94, pp. 118–133). Berlin: Springer.
Suberkropp, K. (2001). Fungal growth, production, and sporulation during leaf decomposition in two streams. *Applied and Environmental Microbiology, 67*, 5063–5068.

Chapter 27
Ergosterol as a Measure of Fungal Biomass

Mark O. Gessner

Keywords Aquatic hyphomycetes · Biomass conversion factor · Extraction efficiency · Fungal biomarker · Fungal growth · HPLC · Leaf litter · Lipid extraction · Litter decomposers · Solid-phase extraction

1 Introduction

Fungi are important decomposers associated with plant litter in streams and other aquatic and terrestrial environments. Fungal biomass accumulating in decomposing litter can be substantial (Table 27.1), and this has been used as one line of evidence that fungi are instrumental in litter decomposition (Gessner and Chauvet 1994; Maharning and Bärlocher 1996). Total annual biomass production can also be substantial, both in streams (Suberkropp et al. 2010) and other aquatic environments (Buesing and Gessner 2006). Furthermore, by colonizing and degrading litter, fungi enhance its palatability to litter-consuming detritivores by providing an attractive food source to these consumers (Suberkropp 1992; Graça 2001). Thus, fungi play multiple important roles in streams and other ecosystems relying on plant litter inputs. The method presented here is a means to assess their quantitative importance in decomposing plant litter.

Determining fungal biomass in litter and other solid substrates has long proven difficult (Newell 1992; Gessner and Newell 2002), partly because fungal hyphae spread within their substrate rather than growing at surfaces. As a result, the fungal mycelium is not easily separated from the leaf tissue by either optical or mechanical methods. One way to circumvent this problem is to quantify a cell constituent that

M. O. Gessner (✉)
Department of Experimental Limnology, Leibniz Institute of Freshwater Ecology and Inland Fisheries (IGB), Stechlin, Germany

Department of Ecology, Berlin Institute of Technology (TU Berlin), Berlin, Germany
e-mail: gessner@igb-berlin.de

F. Bärlocher et al. (eds.), *Methods to Study Litter Decomposition*,
https://doi.org/10.1007/978-3-030-30515-4_27

Table 27.1 Maximum ergosterol concentrations and corresponding estimates of fungal biomass (mean ± SD) in leaf litter decomposing in softwater mountain streams of the Pyrenees (France) and Black Forest (Germany). A conversion factor of 5.5 mg ergosterol g^{-1} fungal biomass was used for the study in the Pyrenees, whereas species-specific conversion factors were applied in the Black Forest stream (see Gessner and Chauvet 1993)

Leaf species	Location	Peak time (week)	Ergosterol (μg g^{-1} litter dry mass)	Fungal dry mass (% litter dry mass)	Reference
Fraxinus excelsior	Pyrenees	2	896 ± 105	15.5 ± 3.6	Gessner and Chauvet (1994)
Prunus avium	Pyrenees	2	619 ± 97	11.7 ± 2.9	Gessner and Chauvet (1994)
Corylus avellana	Pyrenees	4	586 ± 26	10.8 ± 1.6	Gessner and Chauvet (1994)
Alnus glutinosa	Pyrenees	4	575 ± 55	9.0 ± 1.0	Gessner and Chauvet (1994)
Platanus hybrida	Pyrenees	12	443 ± 32	7.7 ± 0.7	Gessner and Chauvet (1994)
Fagus sylvatica	Pyrenees	8	416 ± 25	6.3 ± 0.2	Gessner and Chauvet (1994)
Quercus ilex	Pyrenees	24	391 ± 14	6.1 ± 1.1	Gessner and Chauvet (1994)
Alnus glutinosa	Black Forest	8	411 ± 44	7.6 ± 0.9	Hieber and Gessner (2002)
Salix fragilis	Black Forest	8	401 ± 35	7.3 ± 0.6	Hieber and Gessner (2002)

is specific to the target fungi and occurs in rather constant amounts in fungal mycelium. Chitin, a major cell wall component of many fungi, and ergosterol, a major membrane component, are the two main constituents that have been used to this end, but ergosterol is the preferred choice when metabolically active biomass is to be determined (Newell 1992; Charcosset and Chauvet 2001; Gessner and Newell 2002). With advances in quantitative PCR, biomass can be increasingly well quantified for individual fungal species (Baudy et al. 2019; Chap. 36). However, applying this approach to diverse fungal communities remains a challenge at present.

The purpose of the method presented in this chapter is to determine the biomass of fungal communities in decomposing plant litter, which can exceed 10% of litter dry mass (Table 27.1). Pure fungal mycelium may also be analysed so as to establish conversion factors relating ergosterol to fungal biomass (e.g., Gessner and Chauvet 1993) and nutrient contents (Brosed et al. 2017; Gulis et al. 2017). Determination of ergosterol is achieved by high-performance liquid chromatography (HPLC) after lipid extraction in alkaline methanol and purification of the extract by means of solid-phase extraction (SPE). The method is discussed in detail by Gessner and Newell (2002). The procedures adopted here have been slightly modified from Gessner and Schmitt (1996).

2 Equipment, Chemicals, and Solutions

2.1 Equipment and Material

- Screw-cap glass extraction tubes (40–50 ml, thick-walled, pressure-resistant)
- Water or dry bath
- Magnetic stirrer and stirring bars
- Vacuum manifold for solid-phase extraction (SPE; available from most suppliers of chromatographic equipment and supplies)
- Pump to create vacuum in manifold
- Solid-phase extraction cartridges (Waters Sep-Pak®, Vac RC, tC18, 500 mg sorbent)
- HPLC for isocratic operation (1 pump, injector, UV detector set to 282 nm, recording unit)
- HPLC column (e.g., LiChrospher RP18, 25 cm × 4.6 mm; Merck, Darmstadt, Germany)
- Gas-tight syringe (1 ml)
- Glassware, pipettes, and plastic syringes

2.2 Chemicals

- Methanol (HPLC or analytical grade)
- Propanol-2 = isopropanol (HPLC or analytical grade)
- Ergosterol standard (high purity, e.g., ≥98%; Cayman Chemical, Ann Arbor, MI, USA; Acros Organics distributed by Thermo Fisher Scientific, Geel, Belgium)
- KOH (pellets, analytical grade)
- HCl (analytical grade)
- Boiling chips

2.3 Solutions

- Solution 1 – Methanol
- Solution 2 – Storage and extraction solvent: KOH in methanol, 8 g l^{-1} (e.g., 4 g in 0.5 l are sufficient for >30 samples)
- Solution 3 – 0.65 M HCl (ca. 100 ml for 30 samples)
- Solution 4 – Conditioning solution: methanol (one volumetric part) + KOH in methanol (five parts) + 0.65 M HCl (one part); e.g., 30 ml Solution 1 + 150 ml Solution 2 + 30 ml Solution 3, which is sufficient for >30 samples; check before use whether pH is between 2 and 3

- Solution 5 – Washing solution: 0.4 M KOH in methanol: H_2O (6:4; vol/vol); e.g., 1.8 g KOH + 32 ml H_20 + 48 ml methanol; sufficient for >30 samples
- Solution 6 – Isopropanol
- Solution 7 – Ergosterol standard in isopropanol: weigh ca. 10 mg ergosterol to nearest 0.1 mg in volumetric flask (50 ml); dissolve in isopropanol, adjust volume, and transfer to tightly closing 50-ml glass bottle; store in refrigerator (4 °C), where the solution is stable for several months
- Solution 8 – Ergosterol standard in KOH/methanol (ca. 200 mg l^{-1}): dissolve ergosterol under stirring and gentle heating (50 °C) in volumetric flask, let cool, remove magnetic stirrer, adjust volume, and store at 4 °C

3 Experimental Procedures

3.1 Sample Preparation

1. Collect decomposing leaves from streams and transport to laboratory in ice chest.
2. Clean litter of adhering debris and macroinvertebrates.
3. Cut set of leaf discs from leaves with cork borer and blot lightly on filter paper.
4. Place samples in extraction tubes, preserve in 10 ml KOH/methanol (Solution 2), and store in refrigerator overnight; alternatively, freeze-dry samples, and analyse immediately when dry and weighed (next step is then unnecessary).
5. Prepare replicate sets of samples, dry overnight at 105 °C, and weigh to the nearest 0.1 mg to determine sample dry mass; optionally ash and reweigh to determine ash-free dry mass (AFDM).

3.2 Lipid Extraction and Saponification

1. Add a boiling chip to samples in KOH/methanol.
2. Close tube tightly and heat to 80 °C for 30 min.
3. Let extract cool down to room temperature (ca. 20 min).
4. To estimate recovery rates, include in each extraction series one leaf sample known to contain no ergosterol but spiked with 250 μl of ergosterol stock solution in KOH/methanol (Solution 8).

3.3 Conditioning of SPE Cartridges

1. Connect stopcocks and cartridges to manifold and close stopcocks.
2. Open pressure regulation valve of vacuum manifold.
3. Add 7.5 ml of methanol to each cartridge, start pump, and open stopcocks.

4. Increase vacuum by closing the valve, if necessary to initiate flow.
5. Close stopcocks when about 5 mm of methanol remain above the sorbent bed.
6. Add 7.5 ml of conditioning solvent (Solution 4).
7. Open stopcocks to suck solvent through cartridge, but leave about 5 mm above the sorbent bed.
8. Close stopcocks and stop pump; never let cartridge fall dry during conditioning, restart from beginning if this happens accidentally.

3.4 Loading of Lipid Extract onto SPE Cartridge

1. Quantitatively transfer lipid extract to cartridge.
2. Rinse tube 4× with 0.5 ml methanol (e.g., small syringe or Pasteur pipette).
3. Adjust volume to 12 ml with methanol if necessary.
4. Add 2 ml 0.65 M HCl (Solution 3) and make sure solutions are well mixed.
5. Open pressure regulation valve of vacuum manifold.
6. Start pump and open stopcocks to apply vacuum.
7. Set flow rate to $\leq$0.5 ml min^{-1} for fastest cartridge by adjusting pressure regulation valve.
8. Continuously control flow rate and adjust if necessary.

3.5 Washing and Drying of Sorbent in SPE Cartridge

1. After complete loading of sample onto cartridge, wash sorbent bed with 2.5 ml washing solution (Solution 5).
2. Dry sorbent bed for 60 min under stream of nitrogen or air (valve completely open to achieve maximum vacuum) while ensuring temperature remains at or below 18 °C; verify that sorbent is completely dry as indicated by the cartridge attaining ambient temperature.

3.6 Elution of Ergosterol

1. Weigh HPLC vial with its cap to the nearest 0.1 mg.
2. Place weighed HPLC vial in vacuum manifold.
3. Apply gentle vacuum and elute ergosterol with 4 × 400 μl of isopropanol.
4. Set flow rate to about 1 ml min^{-1} during elution.
5. Close stopcocks, cautiously open pressure regulation valve to reduce vacuum slowly, stop pump, and remove vials form manifold.
6. Close vials tightly with corresponding cap and reweigh to the nearest 0.1 mg.
7. Calculate fluid volume in vial (i.e., multiply sample weight by 1.27, given a density of isopropanol at 25 °C of 0.786 g cm^{-3}).

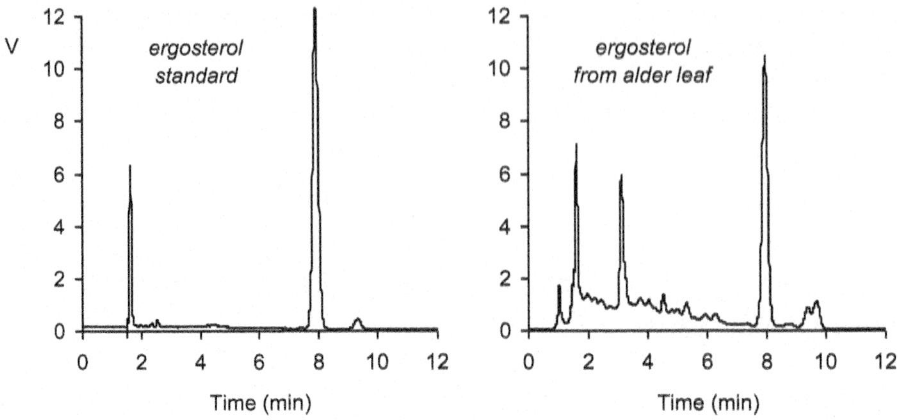

Fig. 27.1 Chromatograms of an ergosterol standard and a lipid extract prepared from leaf litter that had been intensely colonized by fungi in a stream

3.7 HPLC Analysis

1. Set chromatograph to the following conditions:
   - Mobile phase: 100% methanol
   - Flow rate: 1.4 ml min^{-1}
   - Column temperature: 33 °C
   - Detection wavelength: 282 nm
   - Injection volume: 10 µl
2. Slightly adjust conditions (flow rate, column temperature) to ensure a retention time of ergosterol of about 8 min.
3. Prepare standard curve (concentrations of 0, 2.5, 5, 10, 20, and 40 µg ml^{-1}) from ergosterol stock solution in isopropanol.
4. Run standards on HPLC, then inject samples in duplicate.
5. Check identity of putative ergosterol peaks (1) by co-injection of the ergosterol standard with sample extract (Fig. 27.1) (2) by UV spectrometry (Fig. 27.2).
6. Measure area and/or height of ergosterol peaks.
7. Calculate ergosterol concentration in extract and leaves, based on concentration in final extract, total sample volume, and sample dry mass or ash-free dry mass.

4 Final Remarks

The method to quantify ergosterol presented here for leaf litter can be adapted to a wide variety of environmental samples, mostly with little modification. These types of samples include decomposing wood and other types of plant litter, soils (e.g.,

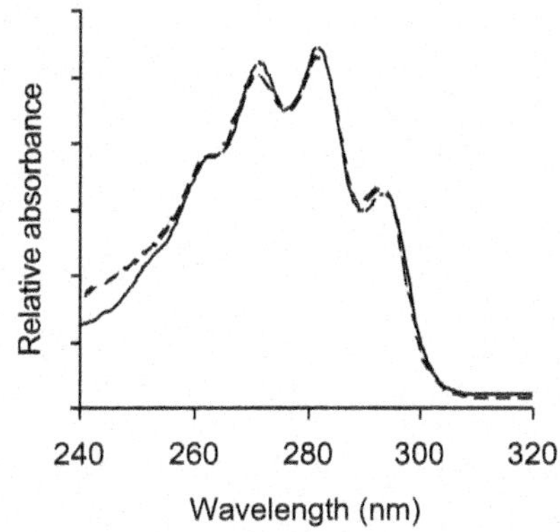

Fig. 27.2 UV absorbance spectrum of ergosterol in methanol showing the characteristic absorbance maxima of the provitamins D at about 262, 271, 282, and 294 nm. The solid line indicates a commercially available ergosterol standard; the broken line corresponds to the eluted HPLC fraction of a lipid extract prepared from leaf litter that had been colonized by fungi in a stream (from Gessner and Newell 2002)

Beni et al. 2014), and gut contents or faeces of litter-consuming invertebrates. The main challenges with some of these samples are small sample sizes and low ergosterol contents.

Pure ergosterol is remarkably stable when protected from light. Nevertheless, mechanical, chemical, or enzymatic losses can easily occur when ergosterol, a molecule that has a reactive conjugated double bond, is extracted from complex environmental samples. Therefore, it is advisable to routinely include an external ergosterol standard in each run. Recovery rates are typically above 90%. Lower values indicate avoidable losses. A precautionary measure to ensure high recovery is to carry out the drying step after sample loading and washing of the SPE cartridge at an ambient temperature ≤ 18 °C, especially when the cartridges are dried under a stream of air rather than nitrogen. Poor recovery is also likely when the characteristics of the SPE cartridge do not match the conditions described here (solvent polarity, pH, etc.), since details in column characteristics differ markedly among suppliers. Consequently, it is necessary either to adhere strictly to the materials and conditions described in this chapter or to adjust materials and procedures such that high recovery rates are achieved.

Microwave-assisted extraction (MAE) can notably accelerate the extraction of ergosterol (Young 1995), and the approach has also proved useful in studies of fungi associated with decomposing plant litter (Reis et al. 2018). A saponification step to analyse ergosterol esters and other bound ergosterol fractions was not considered in those MAE procedures. However, it could be readily included (Zhang et al. 2008) to make results directly comparable to those of the conventional extraction method presented in this chapter.

Conversion factors to relate ergosterol concentrations to fungal biomass in terms of dry mass or carbon have been established for aquatic hyphomycetes (e.g., Gessner and Chauvet 1993; Charcosset and Chauvet 2001) and various other fungi (e.g., Klamer and Bååth 2004; Wallander et al. 2013). A commonly used factor derives from an average ergosterol content in mycelium (5.5 μg/g fungal dry mass) of a range of aquatic hyphomycete strains grown in liquid culture (Gessner and Chauvet 1993; Brosed et al. 2017). Equivalent factors have been established to relate ergosterol also to amounts of nitrogen and phosphorus contained in fungal biomass (e.g., Brosed et al. 2017; Gulis et al. 2017).

A recurring criticism of the ergosterol method is significant variation in ergosterol contents among growth conditions and fungal strains (>10×). This variability must be acknowledged when interpreting ergosterol data. However, as Klamer and Bååth (2004) point out, fungal communities are commonly diverse and include fungi in various growth stages, suggesting that the application of average conversion factors results in much more accurate estimates of fungal biomass than the level of variation in ergosterol contents among individual strains and growth stages might imply.

References

Baudy, P., Zubrod, J. P., Röder, N., Baschien, C., Feckler, A., Schulz, R., & Bundschuh, M. (2019). A glance into the black box: Novel species-specific quantitative real-time PCR assays to disentangle aquatic hyphomycete community composition. *Fungal Ecology, 42*, 100858.

Beni, A., Soki, E., Lajtha, K., & Fekete, I. (2014). An optimized HPLC method for soil fungal biomass determination and its application to a detritus manipulation study. *Journal of Microbiological Methods, 103*, 124–130.

Brosed, M., Jabiol, J., & Gessner, M. O. (2017). Nutrient stoichiometry of aquatic hyphomycetes: Interstrain variation and ergosterol conversion factors. *Fungal Ecology, 29*, 96–102.

Buesing, N., & Gessner, M. O. (2006). Benthic bacterial and fungal productivity and carbon turnover in a freshwater marsh. *Applied and Environmental Microbiology, 72*, 596–605.

Charcosset, J.-Y., & Chauvet, E. (2001). Effect of culture conditions on ergosterol concentration in mycelium of aquatic hyphomycetes. *Applied and Environmental Microbiology, 67*, 2051–2055.

Gessner, M. O., & Chauvet, E. (1993). Ergosterol-to-biomass conversion factors for aquatic hyphomycetes. *Applied and Environmental Microbiology, 85*, 377–384.

Gessner, M. O., & Chauvet, E. (1994). Importance of stream microfungi in controlling breakdown rates of leaf litter. *Ecology, 75*, 1807–1817.

Gessner, M. O., & Newell, S. Y. (2002). Biomass, growth rate and production of filamentous fungi in plant litter. In C. J. Hurst, G. Knudsen, M. McInerney, L. D. Stetzenbach, & M. Walter (Eds.), *Manual of environmental microbiology* (2nd ed., pp. 295–308). Washington: American Society for Microbiology.

Gessner, M. O., & Schmitt, A. L. (1996). Use of solid-phase extraction to determine ergosterol concentrations in plant tissue colonized by fungi. *Applied and Environmental Microbiology, 62*, 415–419.

Graça, M. A. S. (2001). The role of invertebrates on leaf litter decomposition in streams – A review. *International Review of Hydrobiology, 86*, 383–393.

Gulis, V., Kuehn, K. A., Schoettle, L. N., Leach, D., Benstead, J. P., & Rosemond, A. D. (2017). Changes in nutrient stoichiometry, elemental homeostasis and growth rate of aquatic litter-associated fungi in response to inorganic nutrient supply. *The ISME Journal, 11*, 2729–2739.

Hieber, M., & Gessner, M. O. (2002). Contribution of stream detritivores, fungi and bacteria to leaf breakdown based on biomass estimates. *Ecology, 83*, 1026–1038.

Klamer, M., & Bååth, E. (2004). Estimation of conversion factors for fungal biomass determination in compost using ergosterol and PLFA 18:2ω6,9. *Soil Biology and Biochemistry, 36*, 57–65.

Maharning, A. R., & Bärlocher, F. (1996). Growth and reproduction in aquatic hyphomycetes. *Mycologia, 88*, 80–88.

Newell, S. Y. (1992). Estimating fungal biomass and productivity in decomposing litter. In G. C. Carroll & D. T. Wicklow (Eds.), *The fungal community. Its organization and role in the ecosystem* (2nd ed., pp. 521–561). New York: Marcel Dekker.

Reis, F., Nascimento, E., Castro, H., Canhoto, C., Gonçalves, A. L., Simões, S., García-Palacios, P., Milla, R., Sousa, J. P., & Martins da Silva, P. (2018). Land management impacts on the feeding preferences of the woodlouse *Porcellio dilatatus* (Isopoda: Oniscidea) via changes in plant litter quality. *Applied Soil Ecology, 132*, 45–52.

Suberkropp, K. (1992). Interactions with invertebrates. In F. Bärlocher (Ed.), *The ecology of aquatic hyphomycetes*, (Ecological studies, Vol. 94, pp. 118–133). Berlin: Springer.

Suberkropp, K., Gulis, V., Rosemond, A. D., & Benstead, J. P. (2010). Ecosystem and physiological scales of microbial responses to nutrients in a detritus-based stream: Results of a 5-year continuous enrichment. *Limnology and Oceanography, 55*, 149–160.

Wallander, H., Ekblad, A., Godbold, D. L., Johnson, D., Bahr, A., Baldrian, P., Björk, R. G., Kieliszewska-Rokicka, B., Kjøller, R., Kraigher, H., Plassard, C., & Rudawska, M. (2013). Evaluation of methods to estimate production, biomass and turnover of ectomycorrhizal mycelium in forests soils – A review. *Soil Biology and Biochemistry, 57*, 1034–1047.

Young, J. C. (1995). Microwave-assisted extraction of the fungal metabolite ergosterol and total fatty-acids. *Journal of Agricultural and Food Chemistry, 43*, 2904–2910.

Zhang, H., Wolf-Hall, C., & Hall, C. (2008). Modified microwave-assisted extraction of ergosterol for measuring fungal biomass in grain cultures. *Journal of Agricultural and Food Chemistry, 56*, 11077–11080.

Chapter 28
Fungal Growth Rates and Production

K. Suberkropp, M. O. Gessner, and K. A. Kuehn

Keywords ^{14}C · Carbon budget · Ergosterol · Fungal growth · Fungal litter decomposers · HPLC · Leaf litter · Radiolabelled acetate · Radiotracer · Secondary production

1 Introduction

Fungi are a key component of litter decomposer communities in aquatic ecosystems (Gessner et al. 2007; Kuehn 2016) and important secondary producers (Suberkropp 1997; Newell 2001a; Suberkropp et al. 2010). Estimates of fungal growth and biomass production associated with decaying leaves can hence serve to assess the quantitative importance of fungi in natural habitats (Suberkropp 2001; Gulis and Bärlocher 2017). In addition, such information is useful to construct carbon budgets for the litter decomposition process (Kuehn et al. 2011; Su et al. 2015), determine organic matter turnover at the ecosystem level and quantify the food base available to detritivorous consumers, which prefer leaves colonized by fungi.

The principal method available to determine growth rates and biomass production of fungi in leaf litter was originally developed for salt marsh fungi (Newell and Fallon 1991) and has been modified for, and applied to, fungi associated with decomposing litter in streams, freshwater marshes and soils (Suberkropp and

K. Suberkropp
Department of Biological Sciences, University of Alabama, Tuscaloosa, AL, USA
e-mail: kellersuber@gmail.com

M. O. Gessner (✉)
Department of Experimental Limnology, Leibniz Institute of Freshwater Ecology and Inland Fisheries (IGB), Stechlin, Germany

Department of Ecology, Berlin Institute of Technology (TU Berlin), Berlin, Germany
e-mail: gessner@igb-berlin.de

K. A. Kuehn
School of Biological, Environmental and Earth Sciences, The University of Southern Mississippi, Hattiesburg, MS, USA
e-mail: kevin.kuehn@usm.edu

F. Bärlocher et al. (eds.), *Methods to Study Litter Decomposition*,
https://doi.org/10.1007/978-3-030-30515-4_28

Table 28.1 Maximum fungal growth rates (μ) and biomass production rates of fungi associated with decomposing leaf litter

Leaf species	Environment	Instantaneous growth rate (% d^{-1})	Production rate (mg DM g^{-1} AFDM d^{-1})[a,b]	Reference
Liriodendron tulipifera	Softwater stream	19.7 ± 17.3	6.0 ± 4.8	Suberkropp (1995)
Liriodendron tulipifera	Hardwater stream	11.4 ± 4.7	15.9 ± 1.9	Suberkropp (1995)
Mixed natural leaf litter	Softwater stream	8.0 ± 3.8	2.2 ± 1.9	Gulis et al. (2008)[c]
Mixed natural leaf litter	Nutrient-enriched softwater stream	11.0 ± 6.0	6.3 ± 3.1	Gulis et al. (2008)[c]
Mixed natural wood litter	Softwater stream	1.1 ± 0.6	0.2 ± 0.1	Gulis et al. (2008)[c]
Mixed natural wood litter	Nutrient-enriched softwater stream	1.4 ± 1.0	0.3 ± 0.2	Gulis et al. (2008)[c]
Mixed natural leaf litter	Softwater stream	10.4 ± 4.2	2.8 ± 1.3	Suberkropp et al. (2010)
Mixed natural leaf litter	Nutrient-enriched softwater stream	11.0 ± 4.9	6.7 ± 3.1	Suberkropp et al. (2010)
Carex walteriana	Freshwater marsh, standing dead leaves	4.5 ± 2.2	1.1 ± 0.1	Newell et al. (1995)[d]
Typha angustifolia	Freshwater marsh, standing dead leaves	5.0 ± 1.5	3.8 ± 1.4	Kuehn et al. (2011)
Typha domingensis	Freshwater marsh, standing dead leaves	1.5 ± 0.5	0.5 ± 0.2	Su et al. (2015)
Phragmites australis	Freshwater marsh, submerged leaves	1.9 ± 0.1	0.07 ± 0.02	Kominková et al. (2000)
Juncus effusus	Freshwater marsh, submerged leaves	16.9 ± 5.8	6.6 ± 1.6	Kuehn et al. (2000)
Spartina alterniflora	Salt marsh, standing dead leaves	3.2 ± 1.9	2.6 ± 1.0	Newell (2001a)[d,e]
Spartina alterniflora	Salt marsh, standing dead leaves	8.2 ± 4.8	4.4 ± 1.4	Newell (2001b)[d,f]

DM dry mass, *AFDM* ash-free dry mass. Measures of spread are SD

[a]Values reported per gram of litter carbon or dry mass were converted by assuming 90% organic matter (AFDM) and 45% carbon in litter dry mass

[b]If needed, values were converted by assuming 43% carbon in fungal dry mass (Findlay et al. 2002)

[c]Values reported have been corrected for an error in the original publication (V. Gulis, pers. comm)

[d]Estimates of growth rates (μ) were determined by first calculating P/B and then converting these ratios to growth rate according to the relation $\mu = \ln(1 + P/B)$ (e.g. Gessner and Chauvet 1997). Corresponding errors were estimated by using the delta method (Salkind 2007). Values reported here per day were converted from hourly estimates by assuming fungi were active for 12 h per day

[e]Estimates derived from values reported in Tables 4 and 6 for autumn-spring cohorts in 1998 and 1999 (Newell 2001a)

[f]Estimates derived from values reported in Tables 1 for spring (season) and 2 for spring cohorts in 1998 and 1999 (Newell 2001b)

Weyers 1996, Gessner and Chauvet 1997, Rousk and Bååth 2011; Table 28.1). An alternative method to estimate fungal production in ecosystems consists of tagging leaves in situ (Chap. 7) or placing leaves in the natural environment and retrieving them periodically (Chap. 6) to follow increases in fungal biomass (Hieber and Gessner 2002; Baldy et al. 2007). However, this method only works if losses of fungal biomass are negligible or can be accounted for and thus tends to yield rather conservative estimates.

The method introduced by Newell and Fallon (1991) involves determining the rate of incorporation of radiolabelled acetate into ergosterol, a membrane sterol specific to fungi (Gessner and Newell 2002; Gulis and Bärlocher 2017). Fungal growth rates are directly proportional to acetate incorporation rates and can be calculated from incorporation rates using either empirical or theoretical conversion factors (Gessner and Newell 2002). Growth rates can then be multiplied by biomass (determined from ergosterol concentrations) to obtain fungal production. Gessner and Chauvet (1997) proposed a mathematically more accurate calculation method under the assumption of exponential growth, which should be applied when growth rates are greater than about 10% per day. Since fungal biomass is generally determined as a concentration in decomposing plant litter, production is initially calculated as the amount of fungus produced per gram of plant litter dry mass (DM), ash-free dry mass (AFDM) or carbon per unit time (Table 28.1). If fungal production per stream, marsh or soil surface area is to be determined, then the amount of leaf litter (Suberkropp 1997; Suberkropp et al. 2010) or various size classes of wood (Gulis et al. 2008) per m^2 in the ecosystem must also be determined at the same time (Table 28.2; Chap. 4).

The specific procedure presented here has been adopted from Suberkropp and Weyers (1996). Pieces of decomposing leaves are incubated with aeration in a solution containing radiolabelled acetate. Depending on the level of activity, the incubation is ended after 2–5 h, and the samples are placed in methanol. Ergosterol is then extracted using proper precautions for handling radioactive samples and waste. Extracted samples are injected into a high-pressure liquid chromatograph (HPLC) and the concentration of ergosterol in the sample determined. The radioactivity in the ergosterol eluting from the HPLC is then determined with a scintillation counter.

2 Equipment, Chemicals and Solutions

2.1 *Equipment and Materials*

- Equipment and material for ergosterol extraction (see Chap. 27) plus fraction collector for HPLC to collect ergosterol fraction
- Incubation tubes (12 mm diameter) fitted with two-holed rubber stoppers containing glass tubing for aeration. One tube for the killed control is fitted with a screw cap. Autoclave tubes before use
- Filter apparatus with membrane filter (0.45 μm pore size) to filter-sterilize stream water
- Flow metres for each incubation tube

Table 28.2 Annual fungal production (P) per area of stream bottom, mean fungal biomass (B), P/B ratio and turnover times associated with naturally colonized leaf litter in streams

Stream	pH	Mean leaf litter (g AFDM m^{-2})	Annual fungal production (g DM m^{-2} $year^{-1}$)	Fungal biomass (g DM m^{-2})	Annual P/B ($year^{-1}$)	Turnover time (d)	Reference
Payne Creek	6.3	27 ± 25	16 ± 6	1.4 ± 0.6	11.5	32	Carter and Suberkropp (2004)
Hendrick Mill	8.2	20 ± 16	27 ± 9	1.4 ± 0.4	20.8	18	Methvin and Suberkropp (2003)
Basin Creek	6.3	42 ± 29	32 ± 11	2.4 ± 1.0	13.3	27	Methvin and Suberkropp (2003)
Walker Branch	8.0	64 ± 33	34 ± 8	4.2 ± 0.8	8.2	44	Suberkropp (1997)
Lindsey Spring Branch	8.2	51 ± 65	46 ± 25	3.5 ± 2.1	13.0	28	Carter and Suberkropp (2004)
Coweeta C-53[a]	6.6	305 ± 46	193 ± 54	12.3 ± 3.3	15.6	23	Suberkropp et al. (2010)
Coweeta C-54[a]	6.6	442 ± 70	243 ± 84	14.8 ± 4.7	16.4	22	Suberkropp et al. (2010)

DM dry mass, *AFDM* ash-free dry mass. Measures of spread are 95% confidence intervals
[a]Values of the year before stream nutrient enrichment

- Battery-operated air pump for field incubations; fishing supply stores often carry air pumps that can be operated with alkaline batteries (e.g. size D) that provide sufficient air flow for these studies
- Set of adjustable automatic pipettes (5, 1 and 0.2 ml)

2.2 Chemicals

- Chemicals for ergosterol extraction (see Chap. 27)
- Scintillation fluid (e.g. Ecolume, MP Biomedicals, Santa Ana, CA, USA)
- [1-^{14}C]Acetate, sodium salt

2.3 Solutions

- Solutions for ergosterol extraction (see Chap. 27)
- Solution of [1-^{14}C]acetate plus non-radioactive sodium acetate. Adjust final concentration of acetate for incubations to 5 mM. The final specific activity should be about 50 MBq mmol^{-1}. Stock solution of acetate is made so that 50 µl contains about 1 MBq of [^{14}C]acetate and the total acetate concentration is 0.4 M.

3 Experimental Procedures

3.1 Sample Preparation and Incubation with Radiolabel

1. For each replicate tube, collect 3–5 decomposing leaves, and cut two disks (12 mm diameter) from each leaf.
2. Place one disk from each of the 3–5 leaves in a container with stream water while avoiding changes in temperature and other ambient conditions.
3. Combine the second set of leaf disks from each of the 3–5 leaves, dry, weigh and combust the disks to determine ash-free dry mass.
4. Filter-sterilize stream water and pipette 3.95 ml into each of the incubation tubes and the control tube, and then place the tubes in a rack in the stream.
5. Alternatively, if handling radioactivity in the field is not possible for safety, legal or other reasons, place tubes and leaf disks in an ice chest maintained at stream temperature, and transport them back to the laboratory where they are placed in a water bath, chamber or room adjusted to stream temperature.
6. Connect aeration tubes to a battery-operated air pump, and aerate each tube with 30–40 ml air min^{-1}; alternatively, gently shake tubes during the incubation.
7. To the control tube, add formalin to reach a final concentration of 2%, and use this tube to determine the background radioactivity in the samples; do not

aerate this tube during incubation to avoid volatilization of formaldehyde and potential contamination of live samples.

8. Add the leaf disks for each replicate to the incubation and control tubes, and allow 10–20 min equilibration time.
9. Add 50 μl of [^{14}C]acetate solution to each tube at timed (e.g. 30 s) intervals. Incubate for an exact length of time (e.g. 180 min).
10. At timed intervals, remove tubes from stream (or water bath) in the same order as above, and place in an ice bath to slow further uptake of acetate; alternatively, add formalin to stop acetate uptake.
11. Filter fluid and leaf disks immediately through glass fibre filters (e.g. Whatman GF/D, 25 mm diameter). Rinse well and place filter and leaf disks in 5 ml methanol.
12. Store at −20 °C until ergosterol is extracted.

3.2 Ergosterol Extraction and Determination

1. Extract ergosterol from samples following the protocol in Chap. 27, except for adjustments given below.
2. Keep the final volume of the sample extract as small as practicable (e.g. about 500 μl).
3. Collect the ergosterol peak eluting from the HPLC in a scintillation vial, either manually or with an automatic fraction collector, taking care with manual collection to avoid radioactive contamination, even though radioactivity in final extracts is typically very low (<500 Bq).
4. Add 10 ml of scintillation fluid to the combined ergosterol fractions.
5. Determine radioactivity with scintillation counter and correct for quenching.
6. Calculate ergosterol concentrations from peak area and standard curve as in Chap. 27.
7. Calculate the rate of acetate incorporation into ergosterol (mmole mg^{-1} d^{-1}) as the corrected radioactivity (Bq; 1 Bq = 1/60 dpm; dpm = disintegrations per minute) in the sample (radioactivity of sample − radioactivity of formalin-treated control) divided by the product of the specific activity of the acetate (Bq mmol^{-1}), the fraction of the sample volume injected, time of the incubation (d) and biomass of the fungus in the sample (mg):

$$\frac{R_{sample} - R_{control}}{SA \cdot F_{sample} \cdot t \cdot B_{fungus}} \left[\frac{Bq - Bq}{Bq / mmol \cdot d \cdot mg} \right] \tag{28.1}$$

where R_{sample} is the radioactivity (Bq) in the sample, $R_{control}$ is the radioactivity (Bq) in the control, SA is the specific activity (Bq mmol^{-1}), F_{sample} is the fraction of the sample injected into the HPLC, t is the incubation time in d, and B_{fungus} is the biomass of the fungus in the sample (mg).

8. Calculate fungal growth rate (mg mg^{-1} d^{-1}) by multiplying the rate of acetate incorporation (as calculated above) by 19,300 mg $mmol^{-1}$, which is an empirically determined conversion factor (Suberkropp and Weyers 1996), or apply another suitable conversion factor (see below and Gessner and Newell 2002).

4 Final Remarks

As with all procedures using radioactivity, proper precautions must be used for purchasing, handling and disposing of radioactive materials.

Some leaf samples may have very low incorporation rates, requiring high specific activity of the radiolabelled acetate to measure incorporation reliably. As a result, assays can become expensive. To increase sensitivity of the assay and reduce costs and radioactive waste, it is advisable to make several adjustments to the standard protocol for ergosterol determination (Chap. 27). These include reducing the volume of the final ergosterol extract as much as feasible, injecting (and collecting) as large a fraction of the extract as possible (e.g. 100 µl) and combining multiple (2–4) injections to determine radioactivity.

If formalin is used to stop acetate incorporation, samples must be filtered immediately and washed abundantly, because prolonged exposure to formalin can reduce ergosterol concentrations.

Gessner and Chauvet (1997) calculated a theoretical conversion factor of 6.6 mg mycelial biomass μmol^{-1} of incorporated acetate to convert rates of acetate incorporation into growth rates of the aquatic hyphomycete *Articulospora tetracladia*, and Newell (2000) suggested empirically determined conversion factors of 7.0–17.8 mg mycelial biomass μmol^{-1} of incorporated acetate for salt marsh fungi (Gessner and Newell 2002).

For samples from calm lentic environments, where turbulence is not needed to simulate conditions in the environment and may even curb fungal activity, aeration or shaking of incubation tubes should be avoided.

References

Baldy, V., Gobert, V., Guérold, F., Chauvet, E., Lambrigot, D., & Charcosset, J.-Y. (2007). Leaf litter breakdown budgets in streams of various trophic status: Effects of dissolved inorganic nutrients on microorganisms and invertebrates. *Freshwater Biology, 52*, 1322–1335.

Carter, M. D., & Suberkropp, K. (2004). Respiration and annual fungal production associated with decomposing leaf litter in two streams. *Freshwater Biology, 49*, 1112–1122.

Findlay, S. E. G., Dye, S., & Kuehn, K. A. (2002). Microbial growth and nitrogen retention in litter of *Phragmites australis* compared to *Typha angustifolia*. *Wetlands, 22*, 616–625.

Gessner, M. O., & Chauvet, E. (1997). Growth and production of aquatic hyphomycetes in decomposing leaf litter. *Limnology and Oceanography, 42*, 496–505.

Gessner, M. O., & Newell, S. Y. (2002). Biomass, growth rate, and production of filamentous fungi in plant litter. In C. J. Hurst, R. L. Crawford, G. Knudsen, M. McInerney, & L. D. Stetzenbach (Eds.), *Manual of environmental microbiology* (2nd ed., pp. 390–408). Washington, DC: ASM Press.

Gessner, M. O., Gulis, V., Kuehn, K. A., Chauvet, E., & Suberkropp, K. (2007). Fungal decomposers of plant litter in aquatic ecosystems. In C. Kubicek & I. Druzhinina (Eds.), *The Mycota IV: Environmental and microbial relationships* (2nd ed., pp. 301–324). Berlin: Springer.

Gulis, V., & Bärlocher, F. (2017). Fungi: Biomass, production, and community structure. In F. Hauer & G. Lamberti (Eds.), *Methods in stream ecology: Volume 1: Ecosystem structure* (3rd ed., pp. 177–192). Amsterdam: Elsevier.

Gulis, V., Suberkropp, K., & Rosemond, A. D. (2008). Comparison of fungal activities on wood and leaf litter in unaltered and nutrient-enriched headwater streams. *Applied and Environmental Microbiology, 74*, 1094–1101.

Hieber, M., & Gessner, M. O. (2002). Contribution of stream detrivores, fungi, and bacteria to leaf breakdown based on biomass estimates. *Ecology, 83*, 1026–1038.

Kominková, D., Kuehn, K. A., Büsing, N., Steiner, D., & Gessner, M. O. (2000). Microbial biomass, growth and respiration associated with submerged litter of *Phragmites australis* decomposing in a littoral reed stand of a large lake. *Aquatic Microbial Ecology, 22*, 271–282.

Kuehn, K. A. (2016). Lentic and lotic habitats as templets for fungal communities: Traits, adaptations, and their significance to litter decomposition within freshwater ecosystems. *Fungal Ecology, 19*, 135–154.

Kuehn, K. A., Lemke, M. J., Suberkropp, K., & Wetzel, R. G. (2000). Microbial biomass and production associated with decaying leaf litter of the emergent macrophyte *Juncus effusus*. *Limnology and Oceanography, 45*, 862–870.

Kuehn, K. A., Ohsowski, B. M., Francoeur, S. N., & Neely, R. K. (2011). Contributions of fungi to carbon flow and nutrient cycling from standing dead *Typha angustifolia* leaf litter in a temperate freshwater marsh. *Limnology and Oceanography, 56*, 529–539.

Methvin, B. R., & Suberkropp, K. (2003). Annual production of leaf-decaying fungi in 2 streams. *Journal of the North American Benthological Society, 22*, 554–564.

Newell, S. Y. (2000). Methods for determining biomass and productivity of mycelial marine fungi. In K. D. Hyde & S. B. Pointing (Eds.), *Marine mycology – A practical approach* (pp. 68–90). Hong Kong: Fungal Diversity Press.

Newell, S. Y. (2001a). Multiyear patterns of fungal biomass dynamics and productivity within naturally decaying smooth cordgrass shoots. *Limnology and Oceanography, 46*, 573–583.

Newell, S. Y. (2001b). Fungal biomass and productivity in standing-decaying leaves of black needlerush (*Juncus roemerianus*). *Marine and Freshwater Research, 52*, 249–255.

Newell, S. Y., & Fallon, R. D. (1991). Toward a method for measuring fungal instantaneous growth rates in field samples. *Ecology, 72*, 1547–1559.

Newell, S. Y., Moran, M. A., Wicks, R., & Hodson, R. E. (1995). Productivities of microbial decomposers during early stages of decomposition of leaves of a freshwater sedge. *Freshwater Biology, 34*, 135–148.

Rousk, J., & Bååth, E. (2011). Growth of saprotrophic fungi and bacteria in soil. *FEMS Microbiology Ecology, 78*, 17–30.

Salkind, N. J. (2007). *Encyclopedia of measurements and statistics*. Thousand Oaks: SAGE Publications.

Su, R., Kuehn, K. A., & Phipps, S. W. (2015). Fungal contributions to carbon flow and nutrient cycling during decomposition of standing *Typha domingensis* leaves in a subtropical freshwater marsh. *Freshwater Biology, 60*, 2100–2112.

Suberkropp, K. (1995). The influence of nutrients on fungal growth, productivity, and sporulation during leaf breakdown in streams. *Canadian Journal of Botany, 73*(Suppl. 1), S1361–S1369.

Suberkropp, K. (1997). Annual production of leaf-decaying fungi in a woodland stream. *Freshwater Biology, 38*, 169–178.

Suberkropp, K. (2001). Fungal growth, production, and sporulation during leaf decomposition in two streams. *Applied and Environmental Microbiology, 67*, 5063–5068.

Suberkropp, K., & Weyers, H. (1996). Application of fungal and bacterial production methodologies to decomposing leaves in streams. *Applied and Environmental Microbiology, 62*, 1610–1615.

Suberkropp, K., Gulis, V., Rosemond, A. D., & Benstead, J. P. (2010). Ecosystem and physiological scales of microbial responses to nutrients in a detritus-based stream: Results of a 5-year continuous enrichment. *Limnology and Oceanography, 55*, 149–160.

Chapter 29
Bacterial Abundance and Biomass Determination in Plant Litter by Epifluorescence Microscopy

Nanna Buesing and Mark O. Gessner

Keywords Bacterial cell detachment · Bacterial counts · Biomass conversion factor · Biovolume determination · FISH · Flow cytometry · Image analysis · Leaf litter · Microscopy · SYBRGreen

1 Introduction

Accurate estimates of bacterial abundance and biomass are a critical prerequisite for assessing the roles of bacteria in biogeochemical cycles and food webs. In addition, they are important for understanding bacterial population dynamics in natural systems, including litter decomposition systems in streams, wetlands and other environments. The most widely used approach to obtain such estimates is to pass a bacterial suspension through a membrane filter, stain the trapped cells with a fluorescent dye and count them under an epifluorescence microscope (Kepner and Pratt 1994). When bacteria are associated with particles such as sediments and decomposing litter, it is best to first detach cells quantitatively from their substrate before counting them (Fry 1988), which is generally efficient with ultrasonic probes (e.g. Velji and Albright 1986; Epstein and Rossel 1995; Buesing and Gessner 2002).

Specific and intense staining of bacteria is required to facilitate clear differentiation between bacterial cells and other particles. This is especially critical when

N. Buesing
Department of Aquatic Ecology, Eawag, Center for Ecology, Evolution, and Biogeochemistry (CEEB), Kastanienbaum, Switzerland

M. O. Gessner (✉)
Department of Experimental Limnology, Leibniz Institute of Freshwater Ecology and Inland Fisheries (IGB), Stechlin, Germany

Department of Ecology, Berlin Institute of Technology (TU Berlin), Berlin, Germany
e-mail: gessner@igb-berlin.de

F. Bärlocher et al. (eds.), *Methods to Study Litter Decomposition*,
https://doi.org/10.1007/978-3-030-30515-4_29

samples are rich in organic matter. Acridine orange (AO) and 4′,6-diamidino-2-phenylindole (DAPI) have traditionally been used for that purpose (Kepner and Pratt 1994). DAPI staining is still applied today, but a variety of generally superior RNA- and DNA-binding dyes have become commercially available, including PicoGreen, SYTO Green and particularly SYBR Green dyes. The specificity and staining intensity of these dyes are much greater than those of DAPI and AO, thus significantly facilitating recognition and quantification of bacterial cells in environmental samples. However, viruses and free nucleic acids, especially when absorbed to particles, also produce a fluorescent signal that could be mistakenly interpreted as bacterial cells. If not differentiated by shape, these counts need to be excluded from the total tally by setting an appropriate size threshold that takes the halo around fluorescent particles into account (e.g. 0.2 or 0.3 μm; Filippini et al. 2006).

The detached and stained bacterial cells are viewed with an epifluorescence microscope, counted, sized and assigned to size classes (e.g. Baldy et al. 1995). However, better estimates of biomass (Table 29.1) are obtained by taking digital photographs that are analysed with an image analysis system such as the freeware ImageJ (Schneider et al. 2012). In addition to increasing sample size and reducing

Table 29.1 Range of bacterial biomass associated with decomposing leaf litter in freshwater environments. *DM* dry mass, *AFDM* ash-free dry mass

Leaf species	Environment	Bacterial biomass (mg DM g^{-1}AFDM)[a]
Liriodendron tulipifera	Softwater stream	0.04–0.89
Liriodendron tulipifera	Hardwater stream	0.04–0.48
Alnus glutinosa	Hardwater stream	0.09–0.48
Populus gr. *nigra*	Hardwater stream	0.03–0.58
Mixed natural leaf litter	Softwater stream	0.42–1.84
Mixed natural leaf litter	Nutrient-enriched softwater stream	0.35–2.02
Mixed natural macrophyte litter	Freshwater marsh, submerged litter	0.28–1.05
Carex walteriana	Freshwater marsh, standing-dead leaves	0.28–0.44
Carex walteriana	Freshwater marsh, leaves on dried-up sediment	0.60–1.91
Phragmites australis	Freshwater marsh, submerged leaves	0.51–6.28
Juncus effusus	Freshwater marsh, submerged leaves	0.13–0.50
Phragmites australis	Freshwater marsh, submerged leaves	1.20–3.00
Typha angustifolia	Freshwater marsh, submerged leaves	1.44–2.11
Schoenoplectus acutus	Freshwater marsh, submerged leaves	1.11–1.33

[a]Compilation adapted from Kuehn (2016) by assuming 45% carbon in bacterial dry mass and 50% carbon in litter ash-free dry mass

observer bias, image analysis offers the advantage that cell dimensions and shapes to calculate biovolumes and biomass are measured for each individual bacterial cell, thus circumventing the need to delineate a limited number of size classes. In addition, image analysis facilitates high sample throughput, if the procedures are automated or semiautomated.

The detachment method described here has been adopted from Buesing and Gessner (2002). The staining procedures follow protocols developed for counting viruses in water samples by Noble and Fuhrman (1998), Weinbauer et al. (1998) and Lebaron et al. (1998) and applied to decomposing litter by Buesing and Gessner (2006).

2 Equipment, Chemicals and Solutions

2.1 Equipment and Material

- Epifluorescence microscope equipped with a high-pressure mercury lamp (HPO 50 W or, preferably, HPO 100 W)
- Optical filter set for detection of stained cells (e.g. Chroma light filter set no. 41001: excitation filter 480 nm, beam splitter 505 nm, emission filter 530 nm)
- Charged-coupled device (CCD) microscope camera with frame grabber
- Image analysis software (e.g. freeware ImageJ; https://imagej.nih.gov/ij/download.html)
- Aluminium oxide membrane filters (Whatman Anodisc, 0.2 µm pore size, 25 mm diameter)
- Membrane filters of cellulose nitrate or mixed cellulose esters (e.g. Millipore HAWP, 0.45 µm pore size, 25 mm diameter)
- Forceps for handling filters
- Clean boxes to collect leaf material
- Sterile 20-ml glass vials for leaf disks or litter pieces
- Sterile Eppendorf tubes
- Micropipettes and sterile tips for dispensing volumes of 10–1000 µl
- Ultrasonic probe (e.g. Branson Sonifier 250)
- Vortex
- Filter manifold with straight filtration funnels
- Vacuum pump
- Slides and cover slips
- Petri dishes (60 mm diameter)
- Small cardboard box

2.2 Chemicals

- SYBR Green II (Molecular Probes, Eugene, Oregon, USA)
- Ethanol, technical grade

- Formaldehyde (37%), analytical grade
- Non-fluorescing immersion oil (e.g. Zeiss, Immersol 518 N)
- Sodium pyrophosphate ($Na_4P_2O_7$), analytical grade
- Glycerol, analytical grade
- NaCl, analytical grade
- NaH_2PO_4, analytical grade
- *p*-Phenylenediamine ($C_6H_8N_2$)

2.3 Solutions

- Particle-free water (e.g. 0.2 μm filtered autoclaved nanopure water)
- Staining solution 1 (original SYBR Green II solution diluted 1:10 in nanopure water)
- Staining solution 2 (2.5% working solution from staining solution 1, freshly prepared)
- Fixation solution (2% formalin, 0.1% sodium pyrophosphate)
- Antifade mounting solution: 50% glycerol, 50% PBS (120 mM NaCl, 10 mM NaH_2PO_4, pH 7.5), 0.1% p-phenylenediamine

3 Experimental Procedures

3.1 Sample Preparation

1. Place small litter pieces such as leaf disks (~100–500 mg wet mass; corresponding to about 20–100 mg dry mass) into glass scintillation vials, and add 10 ml of fixation solution.
2. Keep at 4 °C while minimising storage time because cell losses can be quick in some types of samples.
3. Take a second set of identical subsamples for determining relationships between sample wet mass and dry mass or between surface area and dry mass.
4. Place sample on ice, and sonicate for 1 min with an ultrasonic probe (output: 80 W and 76 μm amplitude).
5. Clean ultrasonic probe carefully with ethanol before treating the next sample.
6. Mount cellulose filter onto the filtration manifold, and rinse with a small volume of nanopure water.
7. Place the Anodisc filter flat on top of the moist cellulose filter.
8. Connect filtration funnel, and add 1 ml nanopure water.
9. Vortex sample, wait for 10 s, take a 10–400 μl aliquot from 2 mm below the surface, and add the aliquot to the nanopure water in the filtration funnel.
10. Add another 1 ml of nanopure water to ensure good mixing of the sample suspension prior to filtration.

11. Filter sample through the Anodisc filter by applying a vacuum of max. 20 kPa (200 mbar).
12. Dry filter carefully by placing it on a cleansing tissue.

3.2 Staining

1. Pipette 100 μl of staining solution 2 in a clean Petri dish, and place Anodisc filter on top.
2. Cover Petri dish with the cardboard box to keep the sample in the dark during staining.
3. Stain for 15 min.
4. Dry filter again by placing it on a cleansing tissue.
5. Mount filter on a clean slide, add 30 μl of antifading solution and place a cover slip on top.
6. Press gently on the cover slip with forceps until the antifading solution is evenly distributed.

3.3 Counting and Cell Sizing

1. Determine cell numbers in a minimum of 10–20 randomly selected microscopic fields (>400 cells; Kirchman 1993), although the use of automated or semiautomated image analysis enables counting a much larger number.
2. Take a digital image of each microscopic field, and store the images.
3. Begin the image analysis by enhancing contrast to make the bacterial cells more distinguishable in a process referred to as contrast stretching, which extends the range of grey values of the captured photographs (e.g. to 256 grey levels in a 16-bit image).
4. Apply a digital filter algorithm to sharpen the cell edges.
5. Manually set the optimal threshold defining the objects (i.e. bacterial cells).
6. Edit the resulting binary picture by erode and dilate functions, which will separate touching cells and fill small holes.
7. Complete the editing interactively in overlay mode with the originally captured image.
8. Once the edited binary image is complete, determine the area and perimeter of each cell to derive cell lengths (l) and widths (w); this indirect procedure to determine l and w greatly improves precision, because imaging software generally overestimates real cell dimensions severely, especially when bacteria are curved (Massana et al. 1997).

3.4 Calculations

1. Calculate volumes (V) of individual cells under the assumption that cells are cylinders with hemispherical ends (Fry 1988), which works for both rods and cocci.
2. Calculate the total biovolume (BV) of bacterial cells per g of litter dry mass according to:

$$\frac{BV}{DM_l} = \frac{(\Sigma bv_i) \cdot V_s \cdot A_f}{S_f \cdot A_c \cdot DM_l} \tag{29.1}$$

where bv_i is the biovolume of individual bacterial cell, V_s the sample volume, A_f the filtration area, S_f the volume of the sample passed over the filter, A_c the total filter area scanned and DM_l the dry mass of the sample.

3. Calculate total bacterial dry mass or bacterial carbon from bacterial BV based on empirically established relationships such as:

$$dm_i = 435 \cdot bv_i^{0.86} \tag{29.2}$$

obtained for freshwater pelagic bacteria (Loferer-Krößbacher et al. 1998), where dm_i is the dry mass and bv_i the biovolume of an individual bacterial cell.

4 Final Remarks

Control counts of nanopure water without sample must be run daily and for every new batch of filters and stain to check for possible contamination. Controls without samples should also be prepared every time samples are preserved for later counts.

Care must be taken to ensure that all filtered bacteria appear in a single optical layer. Since Anodisc filters have a distinctive ring around the actual filtration area, there is a risk of producing multiple layers when adding the antifading solution. This is best avoided by adding the antifading solution after the filter has completely dried.

If bacterial cells are counted manually, then classes need to be defined according to their shape and size. Bacteria in each optical field are counted in each of these established size classes. Again, >400 cells should be counted in each size class (Kirchman 1993). This number is based on the assumption that counts follow a Poisson distribution, where the standard deviation is equal to the square root of the mean. Consequently, with 400 cell counts, the standard deviation is 20/400 = 5% of the mean, suggesting that the true value lies within an interval of about ±10% around the estimated mean.

A number of factors to convert bacterial biovolume to biomass have been published (e.g. Fagerbakke et al. 1996; Theil-Nielsen and Søndergaard 1998; Loferer-Krößbacher et al. 1998; Vrede et al. 2002). Most of these have been derived from either *E. coli* in culture, which favours large cells, or from bacteria sampled in marine pelagic environments, where cells tend to be small. In the size range of cells expected for bacteria associated with organic matter, most published conversion factors result in slightly lower estimates of bacterial biomass than the conversion factors derived from the relationship found by Loferer-Krößbacher et al. (1998). A specific conversion factor for bacteria associated with litter is not available. Some general aspects of choosing factors for converting bacterial biovolume to biomass are discussed by Norland (1993).

Flow cytometry offers tremendous potential to increase sample throughput compared to manual counts of bacterial cells by epifluorescence microscopy (Wang et al. 2010; Deng et al. 2019). However, the approach has rarely been applied to bacteria associated with decomposing leaf litter (Frossard et al. 2012; Halvorson et al. 2019) and requires effective separation from nonbacterial particles while avoiding bacterial cell lysis. Density gradient centrifugation using Histodenz as density gradient medium (also marketed as Nycodenz, Accudenz and under other brand names) has proved effective to separate bacteria from sediments once the cells have been detached (Amalfitano and Fazi 2008; Frossard et al. 2016; Deng et al. 2019), although the effectiveness with samples rich in organic matter is less clear (Frossard et al. 2016). Therefore, because flow cytometry can result in markedly reduced counts compared to epifluorescence microscopy (Frossard et al. 2016; Halvorson et al. 2019), it is advisable to determine recovery rates by spiking samples with fluorescent microbeads (e.g. Halvorson et al. 2019). One general disadvantage of flow cytometry is that determining cell size to estimate bacterial biomass in addition to abundance is much less precise than with epifluorescence microscopy.

A promising option to enhance microscopic counts of bacteria is to discriminate between life and dead cells (Berney et al. 2007) and particularly to distinguish bacterial taxa by fluorescence in situ hybridization (FISH; Bouvier and Del Giorgio 2003; Amann and Fuchs 2008). However, apart from one study (Fazi et al. 2005), the potential of FISH, which can also be combined with flow cytometry (Amann and Fuchs 2008), has not been explored with bacteria attached to decomposing plant litter.

References

Amalfitano, S., & Fazi, S. (2008). Recovery and quantification of bacterial cells associated with streambed sediments. *Journal of Microbiological Methods, 75*, 237–243.

Amann, R., & Fuchs, B. M. (2008). Single-cell identification in microbial communities by improved fluorescence *in situ* hybridization techniques. *Nature Reviews Microbiology, 6*, 339–348.

Baldy, V., Gessner, M. O., & Chauvet, E. (1995). Bacteria, fungi and the breakdown of leaf litter in a large river. *Oikos, 74*, 93–102.

Berney, M., Hammes, F., Bosshard, F., Weilenmann, H. U., & Egli, T. (2007). Assessment and interpretation of bacterial viability by using the LIVE/DEAD BacLight Kit in combination with flow cytometry. *Applied and Environmental Microbiology, 73*, 3283–3290.

Bouvier, T., & Del Giorgio, P. A. (2003). Factors influencing the detection of bacterial cells using fluorescence *in situ* hybridization (FISH): A quantitative review of published reports. *FEMS Microbiology Ecology, 44*, 3–15.

Buesing, N., & Gessner, M. O. (2002). Comparison of detachment procedures for direct counts of bacteria associated with sediment particles, plant litter and epiphytic biofilms. *Aquatic Microbial Ecology, 27*, 29–36.

Buesing, N., & Gessner, M. O. (2006). Benthic bacterial and fungal productivity and carbon turnover in a freshwater marsh. *Applied and Environmental Microbiology, 72*, 596–605.

Deng, L., Fiskal, A., Han, X., Dubois, N., Bernasconi, S. M., & Lever, M. A. (2019). Improving the accuracy of flow cytometric quantification of microbial populations in sediments: Importance of cell staining procedures. *Frontiers in Microbiology, 10*, 720.

Epstein, S. S., & Rossel, J. (1995). Enumeration of sandy sediment bacteria: Search for optimal protocol. *Marine Ecology Progress Series, 117*, 289–298.

Fagerbakke, K. M., Heldal, M., & Norland, S. (1996). Content of carbon, nitrogen, oxygen, sulfur and phosphorus in native aquatic and cultured bacteria. *Aquatic Microbial Ecology, 10*, 15–27.

Fazi, S., Amalfitano, S., Pernthaler, J., & Puddu, A. (2005). Bacterial communities associated with benthic organic matter in headwater stream microhabitats. *Environmental Microbiology, 7*, 1633–1640.

Filippini, M., Buesing, N., Bettarel, Y., Sime-Ngando, T., & Gessner, M. O. (2006). Infection paradox: High abundance but low impact of freshwater benthic viruses. *Applied and Environmental Microbiology, 72*, 4893–4898.

Frossard, A., Gerull, L., Mutz, M., & Gessner, M. O. (2012). Fungal importance extends beyond litter decomposition in experimental early-successional streams. *Environmental Microbiology, 14*, 2971–2983.

Frossard, A., Hammes, F., & Gessner, M. O. (2016). Flow cytometric assessment of bacterial abundance in soils, sediments and sludge. *Frontiers in Microbiology, 7*, 903.

Fry, J. C. (1988). Determination of biomass. In B. Austin (Ed.), *Methods in aquatic bacteriology* (pp. 27–72). New York: Wiley.

Halvorson, H. M., Barry, J. R., Lodato, M. B., Findlay, R. H., Francoeur, S. N., & Kuehn, K. A. (2019). Periphytic algae decouple fungal activity from leaf litter decomposition via negative priming. *Functional Ecology, 33*, 188–201.

Kepner, R. L. Jr., & Pratt, J. R. (1994). Use of fluorochromes for direct enumeration of total bacteria in environmental samples: Past and present. *Microbiological Reviews, 58*, 603–615.

Kirchman, D. L. (1993). Statistical analysis of direct counts of microbial abundance. In P. F. Kemp, B. F. Sherr, E. B. Sherr, & J. J. Cole (Eds.), *Handbook of methods in aquatic microbial ecology* (pp. 117–119). Boca Raton: Lewis Publishers.

Kuehn, K. A. (2016). Lentic and lotic habitats as templets for fungal communities: Traits, adaptations, and their significance to litter decomposition within freshwater ecosystems. *Fungal Ecology, 19*, 135–154.

Lebaron, P., Parthuisot, N., & Catala, P. (1998). Comparison of blue nucleic acid dyes for flow cytometric enumeration of bacteria in aquatic systems. *Applied and Environmental Microbiology, 64*, 1725–1730.

Loferer-Krößbacher, M., Klima, J., & Psenner, R. (1998). Determination of bacterial cell dry mass by transmission electron microscopy and densiometric image analysis. *Applied and Environmental Microbiology, 64*, 688–694.

Massana, R., Gasol, J. M., Bjørnsen, P. K., Blackburn, N., Hagström, Å., Hietanen, S., Hygum, B. H., Kuparinen, J., & Pedrós-Alió, C. (1997). Measurement of bacterial size via image analysis of epifluorescence preparations: Description of an inexpensive system and solutions to some of the most common problems. *Scientia Marina, 61*, 397–407.

Noble, R. T., & Fuhrman, J. A. (1998). Use of SYBR Green I for rapid epifluorescence counts of marine viruses and bacteria. *Aquatic Microbial Ecology, 14*, 113–118.

Norland, S. (1993). The relationship between biomass and volume of bacteria. In P. F. Kemp, B. F. Sherr, E. B. Sherr, & J. J. Cole (Eds.), *Handbook of methods in aquatic microbial ecology* (pp. 303–307). Boca Raton: Lewis Publishers.

Schneider, C. A., Rasband, W. S., & Eliceiri, K. W. (2012). NIH image to ImageJ: 25 years of image analysis. *Nature Methods, 9*, 671–675.

Theil-Nielsen, J., & Søndergaard, M. (1998). Bacterial carbon biomass calculated from biovolumes. *Archiv für Hydrobiologie, 2*, 195–207.

Velji, M. I., & Albright, L. J. (1986). Microscopic enumeration of attached marine bacteria of seawater, marine sediment, fecal matter, and kelp blade samples following pyrophosphate and ultrasound treatments. *Canadian Journal of Microbiology, 32*, 121–126.

Vrede, K., Heldal, M., Norland, S., & Bratbak, G. (2002). Elemental composition (C, N, P) and cell volume of exponentially growing and nutrient-limited bacterioplankton. *Applied and Environmental Microbiology, 68*, 2965–2971.

Wang, Y., Hammes, F., De Roy, K., Verstraete, W., & Boon, N. (2010). Past, present and future applications of flow cytometry in aquatic microbiology. *Trends in Biotechnology, 28*, 416–424.

Weinbauer, M. G., Beckmann, C., & Höfle, M. G. (1998). Utility of green fluorescent nucleic acid dyes and aluminum oxide membrane filters for rapid epifluorescence enumeration of soil and sediment bacteria. *Applied and Environmental Microbiology, 64*, 5000–5003.

Chapter 30
Growth and Production of Litter-Associated Bacteria

Nanna Buesing, Mark O. Gessner, and Kevin A. Kuehn

Keywords Bacterial growth rate · Conversion factor · Isotope dilution · Leaf litter · Leucine method · Leucine saturation · Protein extraction · Quenching · Radiotracer · Secondary production

1 Introduction

Bacterial secondary production (BSP) can constitute a significant fraction of total secondary production associated with decomposing litter (Buesing and Gessner 2006; Kuehn et al. 2014), suggesting that BSP on plant litter can be a crucial component of carbon flow in ecosystems where litter is an important source of organic matter. Quantifying bacterial production and growth rate is therefore important for addressing many ecological questions. This includes assessing the role and dynamics of bacteria in decomposing litter. Similar to fungal production (Chap. 28), bacterial production is calculated as the amount of bacterial biomass produced per gram of ash-free dry mass or carbon of plant litter per unit time (Table 30.1). Furthermore, bacterial production per stream, marsh or soil surface area can be determined if the

N. Buesing
Department of Aquatic Ecology, Eawag, Center for Ecology, Evolution, and Biogeochemistry (CEEB), Kastanienbaum, Switzerland

M. O. Gessner (✉)
Department of Experimental Limnology, Leibniz Institute of Freshwater Ecology and Inland Fisheries (IGB), Stechlin, Germany

Department of Ecology, Berlin Institute of Technology (TU Berlin), Berlin, Germany
e-mail: gessner@igb-berlin.de

K. A. Kuehn
School of Biological, Environmental and Earth Sciences, The University of Southern Mississippi, Hattiesburg, MS, USA

F. Bärlocher et al. (eds.), *Methods to Study Litter Decomposition*,
https://doi.org/10.1007/978-3-030-30515-4_30

Table 30.1 Maximum bacterial growth rates (μ) and biomass production rates of bacteria associated with decomposing leaf litter

Leaf species	Environment	Instantaneous growth rate (d^{-1})[c]	Production rate (mg DM g^{-1} AFDM d^{-1})[a, b]	Reference
Liriodendron tulipifera	Softwater stream	0.05 ± 0.07	0.01 ± 0.002	Weyers and Suberkropp (1996)[d]
Liriodendron tulipifera	Hardwater stream	0.54 ± 0.95	0.07 ± 0.04	Weyers and Suberkropp (1996)[d]
Alnus glutinosa	Hardwater stream	1.4 ± 3.3	1.4 ± 1.0	Baldy and Gessner (1997)[d]
Populus gr. *nigra*	Hardwater stream	1.1 ± 0.4	0.46 ± 0.09	Baldy et al. (2002)[d]
Mixed natural leaf litter	Softwater stream	0.24 ± 0.05	0.07 ± 0.01	Suberkropp et al. (2010)
Mixed natural leaf litter	Nutrient-enriched softwater stream	0.30 ± 0.14	0.08 ± 0.04	Suberkropp et al. (2010)
Liriodendron tulipifera	Softwater stream mesocosm	1.3 ± 0.2	3.5 ± 0.5	Halvorson et al. (2019)[e]
Quercus nigra	Softwater stream mesocosm	1.3 ± 0.1	1.4 ± 0.3	Halvorson et al. (2019)[e]
Juncus effusus	Freshwater marsh, submerged leaves	0.14 ± 0.01	0.06 ± 0.01	Kuehn et al. (2000)
Phragmites australis	Freshwater marsh, submerged litter	8.6 ± 3.2	19.5 ± 5.7	Buesing and Gessner (2006)[d]
Typha angustifolia	Freshwater marsh, submerged leaves	4.8 ± 2.9	4.1 ± 1.9	Gillies et al. (2006)[f]
Schoenoplectus acutus	Freshwater marsh, submerged leaves	7.6 ± 7.6	3.4 ± 0.3	Gillies et al. (2006)[f]
Phragmites australis	Freshwater marsh, submerged leaves	1.7 ± 0.2	4.9 ± 0.9	Gillies et al. (2006)[f]
Typha angustifolia	Freshwater marsh, submerged leaves	4.0 ± 0.6	5.1 ± 2.2	Kuehn et al. (2014)
Schoenoplectus acutus	Freshwater marsh, submerged leaves	7.5 ± 2.4	8.9 ± 2.0	Kuehn et al. (2014)

DM dry mass, *AFDM* ash-free dry mass; measures of spread are SD

[a]Values reported per gram of litter carbon or dry mass were converted by assuming 90% organic matter (AFDM) and 45% carbon in litter dry mass

[b]Values reported in mg bacterial biomass were converted by assuming 45% carbon in bacterial dry mass

[c]Estimates of bacterial growth rates (μ) were determined by first calculating P/B and then converting these ratios to growth rate according to the relation $\mu = \ln(1 + P/B)$ (Gessner and Chauvet 1997); corresponding errors were estimated by using the delta method (Salkind 2007)

[d]Values and associated errors derived from published figures using the data extraction programme, DataThief v. 3 (www.datathief.org)

[e]Estimates of bacterial biomass for calculating growth rates (μ) were determined by flow cytometry, with counts corrected for an empirically determined 69-fold underestimation of cell counts by flow cytometry relative to epifluorescence microscopy (see also underestimation reported by Frossard et al. 2016), and using an average conversion factor of 95.5 fg C per cell based on measured cell area and perimeter converted to biovolume (Bjørnsen 1986) and biomass (Loferer-Krößbacher et al. 1998)

[f]Bacterial production and estimates of bacterial growth rates (μ) assuming a total leucine concentration (i.e. saturation) of 2.5 μM (see Table 5 in Gillies et al. 2006)

amount of plant litter per square metre is quantified (Buesing and Gessner 2006; Suberkropp et al. 2010). In conjunction with estimates of bacterial biomass, BSP allows calculating bacterial growth rates, which is another key parameter to describe bacterial population dynamics.

Various methods have been used to estimate bacterial production. The two most common approaches are the thymidine (Fuhrman and Azam 1980) and leucine (Kirchman et al. 1985) methods. Both are based on measuring the incorporation of radiolabelled precursor molecules into macromolecules of the bacterial cell over a known period. [Methyl-^{3}H]thymidine is used to determine rates of DNA synthesis. [^{3}H]leucine or [^{14}C]leucine and sometimes other radiolabelled amino acids are used to estimate rates of protein synthesis. Originally developed for pelagic environments, these methods have later been applied also to litter in freshwater (Findlay et al. 1984; Buesing and Gessner 2003) and terrestrial environments (Rousk and Bååth 2011). The underlying assumption of both approaches is that the synthesis rate of macromolecules is directly proportional to growth. Riemann and Bell (1990), Robarts and Zohary (1993), Chin-Leo (2002) and others discuss general theoretical and practical aspects of these methods.

The present chapter presents a procedure that is applicable to bacteria associated with particulate organic matter, such as plant litter derived from macrophyte tissues or leaves from riparian trees (Buesing and Gessner 2003). The method is based on the incorporation of radiolabelled leucine into protein. It has several advantages over the thymidine method. In particular, protein represents a very large and rather constant fraction of bacterial carbon (Simon and Azam 1989). As a result, conversion factors for calculating bacterial biomass production from leucine incorporation rates vary only twofold or less (Simon and Azam 1989; Moran and Hodson 1992; Buesing and Marxsen 2005) compared to a tenfold variation in conversion factors for thymidine incorporation rates (Riemann and Bell 1990). The main potential shortcoming is that specificity of the leucine method relies solely on the capacity of bacteria to take up organic molecules efficiently at very low concentrations. Therefore, if leucine concentrations needed to achieve rate-saturating conditions are in the micromolar range (see below), eukaryotic organisms such as fungi could notably contribute to the synthesis of radiolabelled protein (Gillies et al. 2006).

In practice, radioactive leucine is added to litter samples, and leucine is incorporated into bacterial protein during a short incubation period. The incorporation is stopped by adding trichloroacetic acid (TCA). Samples are sonicated, and the liquid is removed from the litter sample and passed through a membrane filter. Both the filter and plant litter are then successively washed with TCA, a nonradioactive leucine solution, ethanol and water. Finally, the filter and litter are combined, and the protein is dissolved in hot alkaline solution and radioassayed. The specific protocol presented here is taken from Buesing and Gessner (2003).

2 Equipment, Chemicals and Solutions

2.1 Equipment and Material

- Polycarbonate filter (e.g. Millipore GTTP, 0.2 μm pore size, 25 mm diameter)
- Cellulose membrane filters (e.g. Millipore HAWP, 0.45 μm pore size, 25 mm diameter)
- Cork borer, cutter or scissors for subsampling plant litter
- Sterile 20-ml glass vials
- Sterile glass bottle for storing lake or stream water
- Analytical balance (preferably 0.01 mg precision)
- Aluminium or porcelain dishes for determining sample fresh mass, dry mass and ash mass
- Drying oven (50 °C)
- Micropipettes and sterile tips for dispensing volumes between 10 and 1000 μl
- Screw-cap microcentrifuge tubes (2 ml)
- Ultrasonic probe (e.g. Branson Sonifier 250, output 80 W, amplitude 76 μm)
- Filter manifold (e.g. Millipore 1225 sampling manifold)
- Dry block heater (90 °C)
- Benchtop centrifuge (14,000 *g*)
- Plastic scintillation vials (20 ml)
- Scintillation counter

2.2 Chemicals

- [^{3}H]leucine (sterile; specific activity 4.4–7 TBq mmol^{-1}) or [^{14}C]leucine (specific activity >11 GBq mmol^{-1})
- L-leucine, analytical grade
- Trichloroacetic acid (TCA)
- Ethanol, analytical grade
- NaOH, analytical grade
- Ethylenediaminetetraacetic acid (EDTA), analytical grade
- Sodium dodecyl sulfate (SDS), analytical grade
- Scintillation cocktail for counting samples containing TCA and NaOH (e.g. Hionic-Fluor™, Packard Bioscience, Meriden, CT, USA)

2.3 Solutions

- 1.5 mM leucine (4.5 μM radioactive leucine plus nonradioactive leucine, specific activity 6–8 · 10^9 Bq mmol^{-1}) for a 3-ml incubation volume per sample, or lower concentration if sufficient to achieve saturation (see below)

- 40 mM L-leucine (nonradioactive)
- 50% TCA
- 5% TCA
- 80% ethanol
- Deionized water (e.g. nanopure water)
- Alkaline extraction solution: 0.5 M NaOH, 25 mM EDTA, 0.1% SDS

3 Experimental Procedures

3.1 Sample Preparation

1. Collect litter samples in the field, and keep them at *in situ* temperature during transport, processing and incubation with radiolabel; avoid also other changes in environmental conditions and process samples as quickly as possible, preferably in the field.
2. Take representative subsample from plant material; e.g., cut discs from leaf litter with a cork borer, or cut bulk litter samples into small pieces with a cutter or scissors.
3. Take three sets of subsamples, one for measuring leucine incorporation rates into bacterial production, one for a control in which microbes have been killed with TCA before the addition of leucine and one for establishing area-dry mass relationships or fresh mass-dry mass relationships.
4. Place five leaf discs or ~75 mg litter wet mass (corresponding to 15 mg dry mass) into a 20-ml glass scintillation vial.
5. Add 2.9 ml of filtered (0.2-μm pore-size membrane) stream or lake water.
6. Place subsamples for determining dry mass in aluminium or porcelain dishes, and dry at 50 °C to constant weight.

3.2 Incubation

1. Add TCA to controls (final concentration of 5%) before incubating samples.
2. Add 0.1 ml of 1.5 mM leucine (mixture of radioactive and nonradioactive leucine at a final concentration of 50 μM) to each sample at timed intervals (e.g. 30 s).
3. Incubate samples for 30 min at *in situ* temperature.
4. Stop leucine incorporation at timed intervals as above by the addition of TCA to a final concentration of 5%.

3.3 *Purification and Extraction*

1. Place samples on ice.
2. Sonicate samples for 1 min.
3. Place samples back on ice for at least 15 min.
4. Transfer the 3-ml sample to the filtration manifold, and filter onto a 0.2-μm polycarbonate filter backed by a cellulose filter.
5. Wash plant litter in the glass vial with 1 ml of 5% TCA.
6. Transfer washing volume to the filtration manifold, and filter.
7. Repeat TCA washing step.
8. Wash sample with 1 ml of 40 mM nonradioactive leucine solution, 1 ml of ethanol and 1 ml of nanopure water, each time transferring the washing volume onto the filter, and apply vacuum.
9. Transfer plant material and polycarbonate filter to a 2-ml screw-cap microcentrifuge tube.
10. Add 1 ml of the alkaline extraction solution.
11. Heat samples for 60 min at 90 °C in a dry block heater to dissolve proteins.
12. Allow samples to cool down to ambient temperature.
13. Pipette a 100–500 μl aliquot into a scintillation vial.
14. Add 5 ml of scintillation cocktail, and determine radioactivity in the scintillation counter.

3.4 *Calculation*

1. Calculate leucine incorporated into bacterial protein (leu_{inc}) in mol per g litter dry mass per day as follows:

$$leu_{inc} = \frac{\left(dpm_{sample} - dpm_{control}\right) \cdot V}{SA \cdot t \cdot v \cdot DM} \tag{30.1}$$

where *dpm* = disintegrations per minute converted from measured *cpm* = counts per minute (see remarks below; 1 Bq = 60 dpm), *V* = total volume of the extract (ml), *SA* = specific activity of the final leucine solution (dpm mol^{-1}), *t* = incubation time (hour; 30 min = 0.5 hours), *v* = aliquot of counted sample (ml) and *DM* = litter dry mass (g).

2. Calculate bacterial secondary production (*BSP*) in g C per g litter dry mass per day is as follows:

$$BSP = leu_{inc} \cdot \frac{b}{a} \cdot c \cdot ID \tag{30.2}$$

where a = molar fraction of leucine in protein (0.073) (Simon and Azam 1989), b = the molecular mass of leucine (131.2 g mol^{-1}), c = mass fraction of cellular carbon in protein (0.86; Simon and Azam 1989) and ID = the isotope dilution (e.g. 1.23 for plant litter; Buesing and Gessner 2003).

4 Final Remarks

It is crucial that preliminary tests be run to establish that the basic assumptions underlying the method are met. First, leucine must be incorporated into protein at a constant rate. This is tested by checking that radioactivity in protein increases linearly with time when samples are incubated with an appropriate concentration of total leucine for different time periods (e.g. 5–90 min).

Secondly, incubations are carried out at saturating total leucine concentrations. This is tested by incubating samples with different total leucine concentrations while keeping the molar ratio of radioactive and nonradioactive leucine constant. Saturation curves are obtained by plotting incorporation rates vs. the total leucine concentration. In litter and sediment samples, saturation may occur at concentrations as high as 50 µM (Marxsen 1996; Fischer and Pusch 1999; Buesing and Gessner 2003), although considerably lower concentrations have also been observed (Suberkropp and Weyers 1996; Gillies et al. 2006). Data must be interpreted with great care if a high total leucine concentration is used, particularly at micromolar leucine concentrations where uptake by metabolically active eukaryotes such as fungi can be significant (Horák 1986; Gillies et al. 2006).

Saturation cannot always be reached at acceptably low concentrations of leucine. One possible, though laborious, remedy in that situation is to incubate samples at several leucine concentrations below the saturating concentration and construct a saturation curve by regression analysis to infer the maximum leucine incorporation rate. The calculated asymptote is equal to the leucine incorporation rate at saturating leucine concentrations (e.g. according to Monod or Michaelis-Menten kinetics) and informs about the degree to which the measured leucine incorporation at any lower concentration underestimates the maximum incorporation rate so that a conversion factor can be applied.

Isotope dilution is the dilution of the added radiotracer by the extracellular and/or intracellular pools of that substance. It may be determined either by regressing the reciprocal of incorporated radioactivity against leucine concentration (Bird 1999) or by nonlinear regression analysis of leucine saturation curves (van Looij and Riemann 1993; Buesing and Gessner 2003; Gillies et al. 2006).

Quenching by NaOH can be very important, requiring use of a scintillation cocktail that minimizes quenching. We recommend establishing a quench curve for each type of sample by extracting nonradioactive organic matter using the same procedure as described above. Constant amounts of radioactivity are then added to an increasing volume of a sample extract. Total volumes are kept constant by adding appropriate volumes of NaOH to the sample extract before measuring radioactivity in the scintillation counter. Subsequently, the 'transformed spectral index of the external standard' (tSIE, Packard scintillation counter) is plotted against the counting efficiency, and the resulting quench curve is used to convert cpm to dpm.

Tritiated leucine has been used in most applications to date, probably because it is much cheaper than [^{14}C]leucine. However, the use of [^{14}C]leucine is preferable from a theoretical point of view, because the decay energy of ^{14}C is much higher (beta maximum energy of 156 keV) than that of ^{3}H (18.6 keV), resulting in a higher counting efficiency (~90% compared to ~60%). When only few bacterial cells are active, leucine with a high specific activity may have to be used. In this case, the use of [^{3}H]leucine may be advantageous, because it can be purchased at a specific activity (4.4–7 TBq mmol^{-1}) nearly 1000 times higher than [^{14}C]leucine with only ~11 GBq mmol^{-1}. Reducing the incubation volume from 3 to 1 ml can help save costs particularly when large amounts of radiolabelled leucine are required.

Centrifugation may be performed in lieu of filtration during the purification and extraction (see Sect. 3.3 above). Samples (i.e. plant litter and solution) are transferred to 15-ml polypropylene centrifuge tubes where they are successively washed (i.e. 5% TCA, ethanol, nanopure water, etc.) and centrifuged for 10 min (9000–13,000 *g*) instead of being filtered. After each washing and centrifugation step, the supernatant is carefully removed. After the final washing step with nanopure water, the alkaline extraction solution can be added directly to centrifuge tubes and then heated to dissolve proteins. It is recommended that a carrier protein (e.g. ~1–2 mg bovine serum albumin) be added to each sample prior to any washing and centrifugation, which facilitates precipitation and pelleting of the radiolabelled proteins.

References

Baldy, V., & Gessner, M. O. (1997). Towards a budget of leaf litter decomposition in a first-order woodland stream. *Comptes Rendus de l'Académie des Sciences - Série III - Sciences de la Vie, 320*, 747–758.

Baldy, V., Chauvet, E., Charcosset, J. Y., & Gessner, M. O. (2002). Microbial dynamics associated with leaves decomposing in the mainstem and floodplain pond of a large river. *Aquatic Microbial Ecology, 28*, 25–36.

Bird, D. F. (1999). A critical examination of substoichiometric isotope dilution analysis using thymidine and leucine. *Scientia Marina, 63*, 61–70.

Bjørnsen, P. K. (1986). Automatic determination of bacterioplankton biomass by image analysis. *Applied and Environmental Microbiology, 51*, 1199–1204.

Buesing, N., & Gessner, M. O. (2003). Incorporation of radiolabeled leucine into protein to estimate bacterial production in plant litter, sediment, epiphytic biofilms, and water samples. *Microbial Ecology, 45*, 291–301.

Buesing, N., & Gessner, M. O. (2006). Benthic bacterial and fungal productivity and carbon turnover in a freshwater marsh. *Applied and Environmental Microbiology, 72*, 596–605.

Buesing, N., & Marxsen, J. (2005). Theoretical and empirical conversion factors for determining bacterial production in freshwater sediments via leucine incorporation. *Limnology and Oceanography: Methods, 3*, 101–107.

Chin-Leo, G. (2002). Bacterial secondary productivity. In C. J. Hurst, R. L. Crawford, G. R. Knudsen, M. J. McInerney, & L. D. Stetzenbach (Eds.), *Manual of environmental microbiology* (2nd ed., pp. 354–363). Washington, DC: ASM Press.

Findlay, S. E. G., Meyer, J. L., & Edwards, R. T. (1984). Measuring bacterial production via rate of incorporation of [^{3}H]thymidine into DNA. *Journal of Microbiological Methods, 2*, 57–72.

Fischer, H., & Pusch, M. (1999). Use of the [^{14}C]leucine incorporation technique to measure bacterial production in river sediments and the epiphyton. *Applied and Environmental Microbiology, 65*, 4411–4418.

Frossard, A., Hammes, F., & Gessner, M. O. (2016). Flow cytometric assessment of bacterial abundance in soils, sediments and sludge. *Frontiers in Microbiology, 7*, 903.

Fuhrman, J. A., & Azam, F. (1980). Bacterioplankton secondary production estimates for coastal waters of British Columbia, Antarctica, and California. *Applied and Environmental Microbiology, 39*, 1085–1095.

Gessner, M. O., & Chauvet, E. (1997). Growth and production of aquatic hyphomycetes in decomposing leaf litter. *Limnology and Oceanography, 42*, 496–505.

Gillies, J. E., Kuehn, K. A., Francoeur, S. N., & Neely, R. K. (2006). Application of the [^{3}H]leucine incorporation technique for quantification of bacterial secondary production associated with decaying wetland plant litter. *Applied and Environmental Microbiology, 72*, 5948–5956.

Halvorson, H. M., Barry, J. R., Lodato, M. B., Findlay, R. H., Francoeur, S. N., & Kuehn, K. A. (2019). Periphytic algae decouple fungal activity from leaf litter decomposition via negative priming. *Functional Ecology, 33*, 188–201.

Horák, J. (1986). Amino acid transport in eukaryotic microorganisms. *Biochimica et Biophysica Acta, 864*, 223–256.

Kirchman, D. L., K'Nees, E., & Hodson, R. (1985). Leucine incorporation and its potential as a measure of protein synthesis by bacteria in natural aquatic systems. *Applied and Environmental Microbiology, 49*, 599–607.

Kuehn, K. A., Francoeur, S. N., Findlay, R. H., & Neely, R. K. (2014). Priming in the microbial landscape: Periphytic algal stimulation of litter-associated microbial decomposers. *Ecology, 95*, 749–762.

Kuehn, K. A., Lemke, M. J., Suberkropp, K., & Wetzel, R. G. (2000). Microbial biomass and production associated with decaying leaf litter of the emergent macrophyte *Juncus effusus*. *Limnology and Oceanography, 45*, 862–870.

Loferer-Krößbacher, M., Klima, J., & Psenner, R. (1998). Determination of bacterial cell dry mass by transmission electron microscopy and densitometric image analysis. *Applied and Environmental Microbiology, 64*, 688–694.

Marxsen, J. (1996). Measurement of bacterial production in stream-bed sediments via leucine incorporation. *FEMS Microbiology Ecology, 21*, 313–325.

Moran, M. A., & Hodson, R. E. (1992). Contributions of three subsystems of a freshwater marsh to total bacterial secondary productivity. *Microbial Ecology, 24*, 161–170.

Riemann, B., & Bell, R. T. (1990). Advances in estimating bacterial biomass and growth in aquatic systems. *Archiv für Hydrobiologie, 118*, 385–402.

Robarts, R. D., & Zohary, T. (1993). Fact or fiction – Bacterial growth rates and production as determined by [methyl-^{3}H]thymidine? *Advances in Microbial Ecology, 13*, 371–425.

Rousk, J., & Bååth, E. (2011). Growth of saprotrophic fungi and bacteria in soil. *FEMS Microbiology Ecology, 78*, 17–30.

Salkind, N. J. (2007). *Encyclopedia of measurements and statistics*. Thousand Oaks: SAGE Publications.

Simon, M., & Azam, F. (1989). Protein content and protein synthesis rates of planktonic marine bacteria. *Marine Ecology Progress Series, 51*, 201–213.

Suberkropp, K., Gulis, V., Rosemond, A. D., & Benstead, J. P. (2010). Ecosystem and physiological scales of microbial responses to nutrients in a detritus-based stream: Results of a 5-year continuous enrichment. *Limnology and Oceanography, 55*, 149–160.

Suberkropp, K., & Weyers, H. (1996). Application of fungal and bacterial production methodologies to decomposing leaves in streams. *Applied and Environmental Microbiology, 62*, 1610–1615.

van Looij, Q., & Riemann, B. (1993). Measurements of bacterial production in coastal marine environments using leucine: Application of a kinetic approach to correct for isotope dilution. *Marine Ecology Progress Series, 102*, 97–104.

Weyers, H., & Suberkropp, K. (1996). Fungal and bacterial production during the breakdown of yellow poplar leaves in 2 streams. *Journal of the North American Benthological Society, 15*, 408–420.

Chapter 31
Isolation of Cellulose-Degrading Bacteria

Jürgen Marxsen

Keywords Bacterial cell detachment · Bacterial culture storage · Bacterial cultures · Bacterial isolation · *Bacteroidetes* · Dilution plate method · Dilution series · Leaf litter · Pure culture · Streak plate method

1 Introduction

Despite the tremendous developments in molecular biological techniques over the past decades, the analysis of many microbial processes can still greatly benefit from studies with microbial cultures. Such processes include symbiotic activities during the decomposition of cellulose and other polymeric plant constituents. Thus, to assess the specific roles of microbial species in plant litter decomposition, it is important that these organisms are available as pure cultures.

The term pure culture means that all its constituent cells are descendants of the same individual. In principle, a pure culture is therefore genetically pure, although mutations can lead to genetic changes during storage, especially in growing cultures. Axenic cultures, in contrast, contain cells of a single species, free of any other living organisms, but not necessarily consisting of genetically identical individuals (Pelczar and Chan 1977).

A pure culture can be obtained by using a micromanipulator in combination with a microscope, but in most cases, indirect methods are applied (Rodina 1972; Schneider and Rheinheimer 1988; Overmann 2003). Samples are commonly inoculated on selective media that allow the target organism to multiply while inhibiting most other organisms or preventing their growth. However, a pure culture will result from this approach only if the microbial colony of interest has grown from a single cell. This may not always be the case. Thus, it is necessary to examine

J. Marxsen (✉)
iFZ Research Centre for Biosystems, Land Use and Nutrition, Department of Animal Ecology and Systematics, Justus Liebig University, Giessen, Germany
e-mail: jmar@zo.jlug.de

F. Bärlocher et al. (eds.), *Methods to Study Litter Decomposition*,
https://doi.org/10.1007/978-3-030-30515-4_31

the culture further by detailed microscopic, cultural, biochemical or molecular biological methods to ascertain its genetic purity (Trüper et al. 2001).

The basic technique to obtain a pure bacterial culture is illustrated in this chapter for cellulose-degrading bacteria, since cellulose is the most common plant polymer in nature (Klemm et al. 2005). Cellulose is rather resistant to biological attack and only degraded by a small number of bacteria. Numerous descriptions for cultivating and isolating cellulose-degrading bacteria have been published (Reichenbach and Dworkin 1981; Reichenbach 1999). The method presented here has been adopted from Reichenbach and Dworkin (1981) and Schneider and Rheinheimer (1988).

2 Equipment, Chemicals, Solutions and Media

2.1 Equipment

- Safety cabinet or laminar flow hood
- Autoclave
- Incubator (preferably with cooling system)
- Drying oven for sterilization of glassware (160–180 °C)
- Sterile Petri dishes (90–100 mm diameter)
- Glass spreaders (Drigalski spatula)
- Inoculation loops
- Test tubes
- Vortex mixer
- Erlenmeyer flasks
- Pipettes (1 and 10 ml)

2.2 Chemicals (at Least Reagent-Grade Quality)

- Deionized water
- Agar
- Mineral salts (see Sect. 2.3.)
- Ethylenediaminetetraacetic acid (EDTA), iron(III) sodium salt (trihydrate)
- Cycloheximide
- KOH
- HCl
- Powdered cellulose (e.g. native fibrous cellulose MN 300 from Macherey & Nagel, Düren, Germany, as suggested by Reichenbach and Dworkin 1981)
- Filtered water from sampling site

2.3 Solutions and Media

- Trace element solution (Reichenbach and Dworkin 1981): 100 mg $MnCl_2 \cdot 4H_2O$, 20 mg $CoCl_2$, 10 mg $CuSO_4$, 10 mg $Na_2MoO_4 \cdot 2H_2O$, 20 mg $ZnCl_2$, 5 mg LiCl, 5 mg $SnCl_2 \cdot 2H_2O$, 10 mg H_3BO_3, 20 mg KBr, 20 mg KI, 8000 mg EDTA iron(III) sodium salt (trihydrate), distilled water to 1 l. Sterilize by filtration. Many other trace element solutions, which can be used instead, have been described in the literature.
- Stanier mineral agar (Reichenbach and Dworkin 1981): 1.0 g $(NH_4)_2SO_4$, 1.0 g K_2HPO_4 (to be separately autoclaved), 0.2 g $MgSO_4 \cdot 7H_20$, 0.1 g $CaCl_2 \cdot 2H_20$, 0.02 g $FeCl_3$, 1 ml trace element solution, 1% agar, distilled water to 1 l. After sterilization, adjust pH to 7.0–7.5 with 1 M NaOH or HCl, if necessary. To suppress fungal growth, 25 mg l^{-1} cycloheximide (filter-sterilized, final concentration) can be added to the medium after autoclaving (Brockman 1967).
- Cellulose overlay agar: 0.4% powdered cellulose in Stanier mineral agar, poured as thin layer on top of Stanier mineral agar without cellulose.

3 Experimental Procedures

3.1 Preparation of Media and Incubation of Agar Plates

1. Pour the mineral medium without cellulose into sterile Petri dishes. After it has gelled, pour a thin layer of cellulose overlay agar on top.
2. Separate bacterial cells from the plant litter. This can be done by cutting the plant material into small pieces and agitating them in sterile water from the sampling site containing 0.1% sodium pyrophosphate or detergents (e.g. Tween). Additionally, samples can be treated with an ultrasonic probe, a tissue homogenizer (e.g. Ultra-Turrax or Polytron) or a laboratory blender (Buesing and Gessner 2002, Chap. 29). Special care is needed during the detachment step to apply stringent aseptic procedures to ensure that native bacteria in the plant litter, rather than contaminants, are recovered.
3. Depending on the cell concentration, it may be necessary to dilute the bacterial suspension resulting from the detachment procedure. To this end, prepare at least five test tubes each with 9 ml of autoclaved water from the sampling site. Use a sterile pipette to transfer 1 ml of the original cell suspension to the first dilution tube. Vortex and transfer 1 ml of the diluted suspension to a new tube containing 9 ml of sterile water. Repeat this procedure several times to obtain a dilution series.
4. Use a sterile loop to place a drop of the inoculum suspension in the centre of an agar plate. Alternatively, transfer the inoculum (e.g. 0.5 ml) to the medium with a sterile pipette. In both cases, spread the drop over the entire agar surface of the dish with a sterile Drigalski spatula. Use suspensions from different dilution steps for different plates.

5. Incubate the Petri dishes at temperatures between 12 and 30 °C (or lower), depending on the origin of the sample. Incubation at *in situ* temperature is generally preferable, as it favours the development of typical strains active in the natural environment. A drawback of incubation at low temperature is that colonies develop much more slowly than colonies incubated at room or higher temperature.
6. Check the plates at regular intervals. The first colonies may appear within 24 h or after several weeks or even months if the incubation temperature is low.
7. Continue to check plates regularly to prevent colonies from mixing. Many gliding organisms, mainly from the phylum *Bacteroidetes*, typically occur when attempting to isolate cellulose-degrading bacteria. As they may spread over the agar surface, they are often difficult to recognize because they are mostly yellowish to whitish and form a thin, translucent cover that lacks sharp borders. When colony mixing is suspected to occur, immediately transfer the colonies of interest to fresh medium.

3.2 Isolation

1. Use a sterile loop to transfer bacterial cells from a clearly separated colony to a new plate, and spread the inoculum over an area A of the plate as shown in Fig. 31.1. After flaming the loop or by using a new sterile disposable loop, make a streak through a small section of A to spread some of the inoculum to area B. Flame the loop again, or use a new sterile disposable loop, and streak through a section of B to further spread the inoculum over C. When using a metal loop, let it cool down after flaming before using it again to avoid killing cells by heat.
2. Repeat this step several times by streaking material from newly grown, clearly separated, colonies on a fresh plate, until all colonies on the final plate look identical.
3. A purity test of the culture is essential. The classical approach is to regard isolates as pure, if (i) the same uniform colony type develops in subsequent subcultures and if (ii) the cells are morphologically uniform (Schneider and Rheinheimer 1988).

Fig. 31.1 Example scheme for streaking inocula on agar surfaces. *A*, *B* and *C* = first and subsequent steps

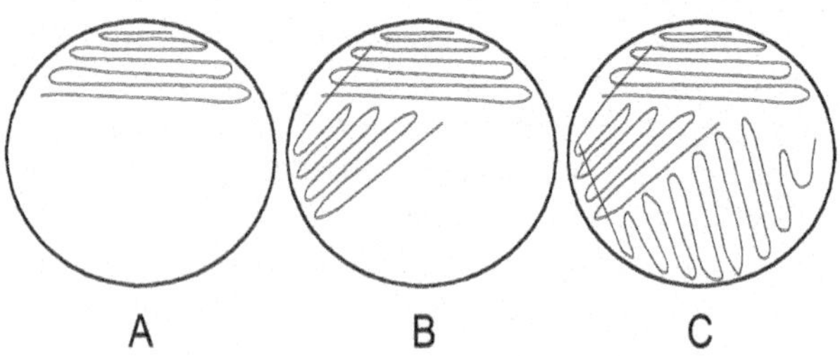

Advice by an experienced microbiologist is recommended. In addition, a range of molecular biological techniques can be used (e.g. Trüper et al. 2001) to enhance confidence in the identity and purity of cultures.

4. After a pure bacterial culture has been obtained, it is normally necessary to maintain it in a viable state for an extended period (weeks to years). Short-term preservation during ongoing work is possible by periodic transfer to fresh medium. Lyophilization and subsequent storage under vacuum (Schneider and Rheinheimer 1988) or storage in liquid nitrogen (Hespell and Bryant 1981; Pfennig and Trüper 1989) are useful techniques for long-term preservation (years to decades).

4 Final Remarks

Cellulose agar overlays are used to prevent the cellulose from settling on the bottom of the dish where it is out of reach for the developing colonies. Cell–fibre contact is essential for cellulose attack by bacteria (Reichenbach and Dworkin 1981).

It is crucial to maintain stringent aseptic conditions during all steps of the isolation procedure. Otherwise there is a high risk of cultivating contaminants instead of the desired native bacteria.

Instead of detaching cells from plant litter, it is possible to place small (e.g. 2–3 mm diameter) plant pieces directly on cellulose agar plates. A mixture of bacterial strains develops around such pieces. From this mixture, a small amount can be transferred to a new agar plate where the cells are streaked out as described in Sect. 3.2.

Reichenbach and Dworkin (1981) described several other approaches to isolate cellulose-degrading bacteria from water samples, which also might be adapted to litter samples.

By using other growth media, many different types of microorganisms can be isolated. Such media and alternative procedures are readily found in the microbiological literature (e.g. Rodina 1972; Starr et al. 1981; Austin 1988; Dworkin et al. 1999–2003). It is important to realize, however, that most bacteria resist isolation and culturing by current techniques (Head et al. 1998) despite the development of several sophisticated isolation techniques (Overmann 2003).

References

Austin, B. (Ed.). (1988). *Methods in aquatic bacteriology*. Chichester: Wiley.

Brockman, E. R. (1967). Fruiting myxobacteria from the South Carolina coast. *Journal of Bacteriology, 94*, 1253–1254.

Buesing, N., & Gessner, M. O. (2002). Comparison of detachment procedures for direct counts of bacteria associated with sediment particles, plant litter and epiphytic biofilms. *Aquatic Microbial Ecology, 27*, 29–36.

Dworkin, M. (Ed.). (1999–2003). *The prokaryotes: An evolving electronic resource for the microbiological community* (3rd ed.). New York: Springer.

Head, I. M., Saunders, J. R., & Pickup, R. W. (1998). Microbial evolution, diversity, and ecology: A decade of ribosomal RNA analysis of uncultivated microorganisms. *Microbial Ecology, 35*, 1–21.

Hespell, R. B., & Bryant, M. P. (1981). The genera *Butyrivibrio*, *Succinivibrio*, *Lachnospira* and *Selenomonas*. In M. P. Starr, H. Stolp, H. G. Trüper, A. Balows, & H. G. Schlegel (Eds.), *The prokaryotes: A handbook on habitats, isolation, and identification of bacteria* (Vol. 1, pp. 1479–1494). Berlin: Springer.

Klemm, D., Heublein, B., Fink, H.-P., & Bohn, A. (2005). Cellulose: Fascinating biopolymer and sustainable raw material. *Angewandte Chemie International Edition, 44*, 3358–3393.

Overmann, J. (2003). Principles of enrichment, isolation, cultivation, and preservation of prokaryotes. In M. Dworkin (Ed.), *The prokaryotes: An evolving electronic resource for the microbiological community* (3rd ed.). New York: Springer.

Pelczar, M. J., & Chan, E. C. S. (1977). *Laboratory exercises in microbiology* (4th ed.). New York: McGraw-Hill.

Pfennig, N., & Trüper, H. G. (1989). Anoxygenic phototrophic bacteria. In J. T. Staley, M. P. Bryant, N. Pfennig, & J. G. Holt (Eds.), *Bergey's manual of systematic bacteriology* (Vol. 3, pp. 1635–1709). Baltimore: Williams and Wilkins.

Reichenbach, H. (1999). The order cytophagales. In M. Dworkin (Ed.), *The prokaryotes: An evolving electronic resource for the microbiological community* (3rd ed.). New York: Springer.

Reichenbach, H., & Dworkin, M. (1981). The order cytophagales (with addenda on the genera *Herpetosiphon*, *Saprospira*, and *Flexithrix*). In M. P. Starr, H. Stolp, H. G. Trüper, A. Balows, & H. G. Schlegel (Eds.), *The prokaryotes: A handbook on habitats, isolation, and identification of bacteria* (Vol. 1, pp. 356–379). Berlin: Springer.

Rodina, A. G. (1972). *Methods in aquatic microbiology*. Translated, edited and revised by R. R. Colwell & M. S. Zambruski. Baltimore/London: University Park Press/Butterworth.

Schneider, J., & Rheinheimer, G. (1988). Isolation methods. In B. Austin (Ed.), *Methods in aquatic bacteriology* (pp. 73–94). Chichester: Wiley.

Starr, M. P., Stolp, H., Trüper, H. G., Balows, A., & Schlegel, H. G. (Eds.). (1981). *The prokaryotes: A handbook on habitats, isolation, and identification of bacteria* (Vol. 1–2). Berlin: Springer.

Trüper, H. G., Stackebrandt, E., & Schleifer, K.-H. (2001). Prokaryote characterization and identification. In M. Dworkin (Ed.), *The prokaryotes: An evolving electronic resource for the microbiological community* (3rd ed.). New York: Springer.

Chapter 32
ATP as a Measure of Microbial Biomass

Manuela Abelho

Keywords ATP extraction · Bioluminescence · Biomass conversion factors · Cold acid extraction · Firefly assay · Fungal biomass · Leaf litter · Luciferin-luciferase reaction · Luminometry · Microbial biomass

1 Introduction

The ubiquitous distribution of ATP in living cells, the rapid loss from dead cells, the ease of extraction and measurement, and relatively low costs have favoured the use of ATP as an indicator of active microbial biomass. As a result, the method has been applied to a variety of biological and environmental systems, such as surface waters (Hammes et al. 2010), soils and sediments (Wen et al. 2005; Frossard et al. 2016), and different types of organic matter, including decomposing leaf litter in streams (Abelho et al. 2005, 2010) and forests (Siegenthaler et al. 2010).

Numerous methods are available to extract ATP from microbial cells. Important general requisites for the efficient extraction are rapid cell death and lysis, complete ATP release, complete and irreversible inactivation of ATP-degrading enzymes, long-term stability of the extracted ATP (Karl 1980), and lack of inhibition of the firefly reaction (Gregg 1991). Extractions may be carried out by homogenization with boiling buffers or cold acids, but cold acids (H_2SO_4) have provided more consistent results (Amyot et al. 1992) and higher recovery rates in microorganisms associated with nonliving organic matter (e.g., Holm-Hansen and Karl 1978; Karl 1980), such as decomposing leaves in streams.

Methods for the quantification of ATP include high-performance liquid chromatography (HPLC) with phosphate buffer as the mobile phase and ion-exchange chromatography, both with UV detection (Ally and Park 1992; Maguire et al. 1992). However, the firefly luciferin-luciferase bioluminescence method – based on the

M. Abelho (✉)
Instituto Politécnico de Coimbra, Escola Superior Agrária, Centre for Functional Ecology, Bencanta, Coimbra, Portugal

Centre for Functional Ecology, Department of Life Sciences, University of Coimbra, Coimbra, Portugal
e-mail: abelho@esac.pt

F. Bärlocher et al. (eds.), *Methods to Study Litter Decomposition*,
https://doi.org/10.1007/978-3-030-30515-4_32

bioluminescent reaction catalyzed by the firefly luciferase – is the most rapid, sensitive (detection limit <0.1 fmol or 50 fg ATP; Gregg 1991), and reproducible assay:

$$\text{D-luciferin} + \text{ATP} + \text{O}_2 \underset{\text{luciferase}+\text{Mg}^{2+}}{\rightarrow} \text{Oxyluciferin} + \text{PP}_\text{i} + \text{AMP} + \text{CO}_2 \quad (32.1)$$

The rapid loss of energy of the oxyluciferin molecule when it turns from an excited to a stable state emits light in the wavelength range of 470–700 nm, with the maximum intensity occurring at 562 nm. Optimal conditions for the reaction include a pH of 7.75 and a temperature of 25–28 °C (DeLuca 1976). Lower pH increases the wavelength of peak emission, while higher temperatures slow the reaction. Under optimal conditions and at low ATP concentrations, light intensity is directly proportional to the ATP concentration in the sample.

Although light intensity can be measured at peak intensity (Lundin and Thore 1975), integration over an extended period has three major advantages: (1) increased sensitivity, (2) ease and reliability of mixing, and (3) independence from measuring the response exactly at peak height (Holm-Hansen and Karl 1978). An integration period of 30 s has shown a linear relationship between ATP concentrations and light emission over at least three orders of magnitude and has been successfully used to quantify ATP associated with decomposing leaves in streams (Suberkropp et al. 1983). The measured light intensity can be accurately related to ATP concentrations by means of internal standards (Holm-Hansen and Karl 1978), i.e., known amounts of ATP added to the sample at various steps during the analytical procedure. Internal standardization also compensates for potential interferences with bioluminescence by several components other than ATP, which typically occur in complex samples (Lundin and Thore 1975; Holm-Hansen and Karl 1978).

The aim of the procedure presented here is to describe the quantitative analysis of ATP in decomposing leaves as a measure of microbial leaf colonization. ATP is quantified by bioluminescence after extraction in cold sulfuric acid and the buffer HEPES (e.g., Holm-Hansen and Karl 1978; Karl 1980; Suberkropp et al. 1983). Some ATP values of decomposing leaves in streams are shown in Table 32.1. Values vary with leaf species, stream characteristics, and location.

ATP concentrations in cells vary with species, metabolic activity, and growth stage (Karl 1980), but estimates of microbial biomass derived from ATP measurements are often a valuable indicator of real microbial colonization. Thus, ATP concentrations can be converted into microbial biomass by means of appropriate conversion factors (Table 32.2).

ATP has been found to correlate with ergosterol, a specific indicator of fungal biomass (Chap. 27) during early stages of leaf decomposition in streams (Suberkropp et al. 1993; Abelho 2009). However, this relationship only held when leaf tissue did not account for a significant fraction of the total ATP pool (Suberkropp et al. 1993). This suggests that ATP can be confidently used as an indicator of fungal biomass in situations where neither bacteria, benthic algae, or other microorganisms attached to plant litter nor the plant material itself, contribute significantly to the ATP pool in

Table 32.1 Maximum concentrations of ATP associated with decomposing leaf litter in streams

Leaf species	Stream type	Geographic coordinates	ATP (nmol g^{-1} AFDM)[a]	References
Carya glabra	Hardwater	42°N, 85°E	347	Suberkropp and Klug (1976)
Quercus alba	Hardwater	42°N, 85°E	130	Suberkropp and Klug (1976)
Alnus glutinosa	Softwater	42°N, 01°E	271	Suberkropp et al. (1993)
Liriodendron tulipifera	Hardwater	33°N, 87°W	99–394	Suberkropp et al. (1993), Suberkropp and Chauvet (1995)
Liriodendron tulipifera	Softwater	33°N, 87°W	49–74	Suberkropp and Chauvet (1995)
Picea abies	Softwater	47°N, 08°E	150	Rosset et al. (1982)
Quercus petraea	Softwater	47°N, 08°E	250	Rosset et al. (1982)
Larix decidua	Softwater	47°N, 08°E	300	Rosset et al. (1982)
Larix decidua	Hardwater	47°N, 07°E	300	Rosset et al. (1982)
Picea abies	Hardwater	47°N, 07°E	300	Rosset et al. (1982)
Quercus petraea	Hardwater	47°N, 07°E	500	Rosset et al. (1982)
Hura crepitans	Hardwater	10°N, 66°W	750	Abelho et al. (2005)
Castanea sativa	Softwater	40°N, 08°W	200	Abelho (2009)
Baccharis platypoda	Softwater	19°S, 43°W	1236	Alvim et al. (2015)
Coccoloba cereifera	Softwater	19°S, 43°W	3315	Alvim et al. (2015)

AFDM ash-free dry mass
[a]Values in μg were converted to moles based on the molecular mass of ATP (507.18 g mol^{-1})

Table 32.2 ATP-to-microbial biomass conversion factors

Conditions	ATP content (nmol mg^{-1})	Estimate	Multiplication factor	References
Two species of fungi in leaves decomposing in stream microcosms	3.45	Fungal DM	0.29	Suberkropp et al. (1993)
Fungal community in leaves decomposing in a stream[a]	0.59	Fungal DM	1.70	Abelho (2009)
Bacterial community in leaves decomposing in stream[a]	1.30	Bacterial DM	0.77	Abelho (2009)

DM dry mass
[a]Bacteria and fungi distinguished by applying antibacterial and antifungal compounds

decomposing plant litter. Estimates of fungal biomass based on ATP are lower than those calculated from ergosterol concentrations (Suberkropp et al. 1993; Abelho 2009), but the magnitude of the difference depends on the conversion factor (Table 32.2).

2 Equipment, Chemicals, and Solutions

2.1 Equipment and Material

ATP Extraction

- Latex gloves
- Forceps
- Analytical balance (±0.1 mg precision)
- Pipettes (0.1 and 5 ml)
- Homogenizer (e.g., Polytron® PT 6100, Kinematica AG, Lucerne, Switzerland)
- Sterile centrifuge tubes (50 ml)
- Refrigerated centrifuge (10,000 *g*)
- Sterile filters (0.2 μm) and filter holders
- Sterile syringes (20 ml)
- Stirring plate and sterile stirring bars
- pH meter
- Sterile glassware (20 ml beakers, 20 ml volumetric cylinders)
- Drying oven (50 °C)

ATP Quantification

- Luminometer (e.g., Turner Designs TD20/20, San Jose, CA, USA; Fig. 32.1), consisting of a reaction chamber mounted vertically, such that the light emitted from the reaction vessel inside the chamber reaches the photosensitive surface of a photomultiplier tube

Fig. 32.1 Example of a luminometer used to quantify ATP concentrations by measuring bioluminescence from the firefly reaction. Photo by M.A.S. Graça

- Corresponding reaction vessels: polypropylene test tubes (1.6 ml; 8 × 50 mm) with hydrophobic inner surfaces (e.g., Turner Designs)
- Pipettes (200, 100, 20, and 10 μl)
- Eppendorf tubes
- Vortex

2.2 Chemicals for ATP Extraction and Quantification

- Adenosine-5′-triphosphate (ATP disodium salt: $C_{10}H_{14}N_5Na_2O_{13}P_3$; MW 551.1 anhydrous; e.g., Creative Enzymes, Sigma-Aldrich)
- HEPES sodium salt buffer (N-[2-hydroxyethyl]-piperazine-N′-[2-ethanesulfonic acid] sodium salt: $C_8H_{17}N_2O_4SNa$; MW 260.3; e.g., Biomol, Sigma-Aldrich)
- Oxalic acid dihydrate (ethanedioic acid: $C_2H_2O_4{\cdot}2H_2O$; MW 126.07; e.g., Biosolve Chemical, Sigma-Aldrich)
- Sulfuric acid, 95–97% (H_2SO_4)
- Ammonium hydroxide (NH_4OH)
- Hydrochloric acid, 37% (HCl)
- Ultrapure water (e.g., Seral Pur PRO 90 CN)
- Firefly luciferase, lyophilized powder (e.g., SRE0045, Sigma-Aldrich) and stored at −20 °C

2.3 Solutions

- Solution 1: 100 μM ATP prepared in ultrapure water. Store in Eppendorf tubes. The aqueous solution is stable for 2 months at −20 °C or for 1 week at 0 °C.
- Solution 2: 0.05 M HEPES prepared in ultrapure water and adjusted to pH 7.5 with HCl. Store at 4 °C for up to 2 months.
- Solution 3: 0.6 M sulfuric acid containing 8 g l^{-1} oxalic acid. Store at 4 °C for up to 2 months.
- Solution 4: 1 μM ATP prepared in ultrapure water. Store in Eppendorf tubes for up to 2 months at −20 °C or for 1 week at 0 °C.
- Solution 5: luciferase solution. Dissolve the contents of one sealed vial in 5 ml of HEPES buffer, or follow the manufacturer's instructions. Protect from light. To maximize solubility, it is important to reconstitute the enzyme at a high salt concentration and at pH 7–8. After reconstitution, the enzyme solution can be kept at 4–8 °C for up to 2 days. Alternatively, working aliquots frozen at −20 °C can be kept for up to 1 month. Repeated freezing and thawing are not recommended. Freshly prepared enzyme emits some light without the addition of ATP; however, the endogenous light usually falls below detectable levels upon storage overnight in the refrigerator (Lundin and Thore 1975).

3 Experimental Procedures

3.1 Sample Preparation

1. Collect leaves decomposing in the field, and transport to the laboratory in an ice chest.
2. Clean the leaves from adhering debris and invertebrates with sterile water.
3. Cut duplicate sets of leaf discs with a cork borer from each side of the leaf (suggested diameter 12–14 mm).
4. Oven-dry (50 °C, 48 h) one set of leaf discs, and weigh to the nearest 0.1 mg to determine dry mass. Ash (500 °C, 5 h), weigh to the nearest 0.1 mg, and subtract from dry mass to determine ash-free dry mass (AFDM).
5. Use the second set of leaf discs to determine ATP concentration of the sample.

3.2 ATP Extraction

1. To avoid contamination, it is advisable to wear gloves and use sterile materials throughout the analytical procedure.
2. Homogenize the leaf discs for 15 s in 5 ml of cold Solution 3 and 5 ml of Solution 2 (Polytron homogenizer set to position 30).
3. Centrifuge for 20 min at 4 °C and 10,000 *g*.
4. Filter with sterile 0.2 μm pore-size filters in filter holders.
5. Adjust pH of the supernatant to 7.0–7.5 with ammonium hydroxide, using a magnetic stirrer and pH meter.
6. Note the final volume, and store extract in 20 ml scintillation vials.
7. Freeze at −20 °C until ATP is measured.
8. To determine recovery rates of the extracted ATP, carry through the whole procedure a parallel set of sample discs known to contain no living organisms and spiked with a known amount of ATP (e.g., 50 μl of Solution 1, which corresponds to 5 nmol).

3.3 ATP Quantification

1. Set up the luminometer in a low-light environment according to the specific instructions of the instrument.
2. Choose an integration period of 30 s, and if available, choose a double measurement mode.
3. Add 50 μL of luciferase, 130 μl HEPES buffer, and 20 μl of the sample in a reaction vessel.

4. Vortex lightly, or mix by hand, since vigorous agitation may denature the luciferase.
5. Insert the reaction vessel into the chamber, and record the light intensity.
6. For internal calibration, remove the reaction vessel as fast as possible, and add 2 μl of the standard 1 μM ATP solution (Solution 4).
7. Record light intensity.
8. Repeat all steps once, and calculate the average ATP concentration of the extract.

3.4 Calculations

1. To determine the amount of ATP present in the assay sample, divide the light emission before addition of the internal standard (*A*) by the difference in light emission after (*B*) and before the addition of the internal standard (*A*), and multiply by the amount of ATP of the internal standard (*C*):

$$\text{ATP in assay sample (nmol)} = \frac{A}{(B-A)} \times C \text{ (nmol)} \tag{32.2}$$

2. To determine the total amount of ATP in the sample, multiply the volume of the sample (*D*) by the amount of ATP measured in 20 μl of the sample extract (*E*), and divide by the volume of the analyzed sample (*F*):

$$\text{ATP in sample (nmol)} = \frac{D \text{ (mL)} \times E \text{ (nmol)}}{F\ (\mu\text{L}) \times 1000} \tag{32.3}$$

3. To determine the extraction efficiency, divide the measured values of the uncolonized samples spiked with ATP (*G*) by the amount of ATP added before the extraction (*H*):

$$\text{Extraction efficiency} = \frac{G \text{ (nmol)}}{H \text{ (nmol)}} \tag{32.4}$$

4. To account for losses of ATP during sample analysis, divide the ATP values measured on each sample (*I*) by the average extraction efficiency (*J*):

$$\text{Final amount of } ATP\text{(nmol)} = \frac{I\text{(nmol)}}{J} \tag{32.5}$$

5. To convert moles of ATP to microbial dry biomass, select a conversion factor in Table 32.2. Be aware that additional calibration with pure fungal cultures will be essential to establish robust conversion factors.

4 Final Remarks

ATP analysis has been streamlined in recent years by the advent of reagent kits supplied by commercial companies such as Promega Corp. (Madison, WI, USA), which markets the BacTiter-Glo reagent. This method is also based on the firefly reaction and has been successfully applied to organic matter collected in both technical (Hammes et al. 2010) and environmental (Frossard et al. 2016) systems. The approach also holds potential for analyses of ATP in decomposing leaf litter, but has not yet been critically tested for application in this context.

References

Abelho, M. (2009). ATP and ergosterol as indicators of fungal biomass during leaf decomposition in streams: A comparative study. *International Review of Hydrobiology, 94*, 3–15.

Abelho, M., Cressa, C., & Graça, M. A. S. (2005). Microbial biomass, respiration, and decomposition of *Hura crepitans* l. (Euphorbiaceae) leaves in a tropical stream. *Biotropica, 37*, 397–402.

Abelho, M., Moretti, M., França, J., & Callisto, M. (2010). Nutrient addition does not enhance leaf decomposition in a southeastern Brazilian stream (Espinhaço mountain range). *Brazilian Journal of Biology, 70*(suppl), 747–754.

Ally, A., & Park, G. (1992). Rapid-determination of creatine, phosphocreatine, purine-bases and nucleotides (ATP, ADP, AMP, GTP, GDP) in heart biopsies by gradient ion-pair reversed-phase liquid-chromatography. *Journal of Chromatography – Biomedical Applications, 575*, 19–27.

Alvim, E. A. C. C., Medeiros, A. O., Rezende, R. S., & Gonçalves Júnior, J. F. (2015). Leaf breakdown in a natural open tropical stream. *Journal of Limnology, 74*, 248–260.

Amyot, M., Pinelalloul, B., Bastien, C., Methot, G., Blaise, C., Vancoillie, R., & Thellen, C. (1992). Firefly assay of ATP from freshwater zooplankton – Comparison of extraction methods. *Environmental Toxicology and Water Quality, 7*, 295–311.

DeLuca, M. (1976). Firefly luciferase. *Advances in Enzymology, 44*, 37–68.

Frossard, A., Hammes, F., & Gessner, M. O. (2016). Flow cytometric assessment of bacterial abundance in soils, sediments and sludge. *Frontiers in Microbiology, 7*, 903.

Gregg, C. T. (1991). Bioluminescence in clinical microbiology. In W. H. Nelson (Ed.), *Physical methods for microorganisms detection* (pp. 1–28). Boca Raton: CRC Press.

Hammes, F., Goldschmidt, F., Vital, M., Wanga, Y., & Egli, T. (2010). Measurement and interpretation of microbial adenosine tri-phosphate (ATP) in aquatic environments. *Water Research, 44*, 3915–3923.

Holm-Hansen, O., & Karl, D. M. (1978). Biomass and adenylate energy charge determination in microbial cell extracts and environmental samples. *Methods in Enzymology, 57*, 73–85.

Karl, D. M. (1980). Cellular nucleotide measurements and applications in microbial ecology. *Microbiological Reviews, 44*, 739–796.

Lundin, A., & Thore, A. (1975). Analytical information obtainable by evaluation of the time course of firefly bioluminescence in the assay of ATP. *Analytical Biochemistry, 66*, 47–63.

Maguire, M. H., Szabo, I., Slegel, P., & King, C. R. (1992). Determination of concentrations of adenosine and other purines in human term placenta by reversed-phase high-performance liquid chromatography with photodiode-array detection – Evidence for pathways of purine metabolism in the placenta. *Journal of Chromatography – Biomedical Application, 575*, 243–253.

Rosset, J., Bärlocher, F., & Oertli, J. J. (1982). Decomposition of conifer needles and deciduous leaves in two Black Forest and two Swiss Jura streams. *Internationale Revue der gesamten Hydrobiologie, 67*, 695–711.

Siegenthaler, A., Buttler, A., Bragazza, L., van der Heijden, E., Grosvernier, P., Gobat, J.-M., & Mitchell, E. A. D. (2010). Litter- and ecosystem-driven decomposition under elevated CO_2 and enhanced N deposition in a *Sphagnum* peatland. *Soil Biology and Biochemistry, 42*, 968e977.

Suberkropp, K., & Chauvet, E. (1995). Regulation of leaf breakdown by fungi in streams: Influences of water chemistry. *Ecology, 76*, 1433–1445.

Suberkropp, K., & Klug, M. J. (1976). Fungi and bacteria associated with leaves during processing in a woodland stream. *Ecology, 57*, 707–719.

Suberkropp, K., Arsuffi, T. L., & Anderson, J. P. (1983). Comparison of degradative ability, enzymatic activity, and palatability of aquatic hyphomycetes grown on leaf litter. *Applied and Environmental Microbiology, 46*, 237–244.

Suberkropp, K., Gessner, M. O., & Chauvet, E. (1993). Comparison of ATP and ergosterol as indicators of fungal biomass associated with decomposing leaves in streams. *Applied and Environmental Microbiology, 59*, 3367–3372.

Wen, G., Voroney, R. P., Curtin, D., Schoenau, J. J., Qian, P. Y., & Inanaga, S. (2005). Modification and application of a soil ATP determination method. *Soil Biology and Biochemistry, 37*, 1999–2006.

Chapter 33
Respiration of Litter-Associated Microbes and Invertebrates

Manuel A. S. Graça and Manuela Abelho

Keywords Aquatic fungi · Bacteria · Invertebrate litter consumers · Leaf litter · Litter decomposition · Microbial activity · Oxygen consumption · Shredders

1 Introduction

Microbial community respiration is a measure of biological activity, reflecting the microbial use of organic matter and, therefore, the functional significance of microbes in decomposition. In invertebrates, too, respiration is an indicator of energy allocated to metabolism (see Chap. 53). In terrestrial systems, respiration rates are usually calculated from CO_2 fluxes measured with infrared gas analysers (e.g. Eriksen and Jensen 2001; Kuehn and Suberkropp 1998; Conant et al. 2008). In aquatic ecosystems, in contrast, respiration rates both of microorganisms associated with decomposing leaves and of invertebrates are generally determined by measuring oxygen consumption (e.g. Ramirez et al. 2003; Griffiths and Tiegs 2016). Respiration rates associated with decomposing leaves are typically <1 μg O_2 h^{-1} mg^{-1} AFDM, while slightly higher values have been measured for invertebrates feeding on leaf litter (Table 33.1).

Oxygen consumption can be measured in closed or open systems. The simplest closed system consists of a water-filled flask without headspace and well-sealed to prevent gas exchange with the environment (e.g. Iversen 1979; Griffiths and Tiegs

M. A. S. Graça (✉)
MARE – Marine and Environmental Sciences Centre & Department of Life Sciences, University of Coimbra, Coimbra, Portugal
e-mail: mgraca@ci.uc.pt

M. Abelho
Instituto Politécnico de Coimbra, Escola Superior Agrária, Centre for Functional Ecology, Bencanta, Coimbra, Portugal

Centre for Functional Ecology, Department of Life Sciences, University of Coimbra, Coimbra, Portugal
e-mail: abelho@esac.pt

F. Bärlocher et al. (eds.), *Methods to Study Litter Decomposition*,
https://doi.org/10.1007/978-3-030-30515-4_33

Table 33.1 Respiration rates of microorganisms associated with decomposing plant litter in streams

Sample[b]	Method	Temperature (°C)	Respiration (µg O_2 h^{-1} mg^{-1} AFDM)	References
Acer rubrum, Rhododendron maximum	Respiration chambers	Ambient	0.03–0.60	1
Fagus sylvatica, Corylus avellana, Fraxinus excelsior	Respiration chambers	10	0.04–0.14[a]	2
Alnus glutinosa	Respiration chambers	10	0.20–0.31[a]	3
Salix sp.	Closed chambers	10	0.3	4
Acer rubrum, Rhododendron maximum	Closed chambers in stream	4–15	≤0.5	5
Acer rubrum, Liriodendron tulipifera and *Quercus alba*	Closed chambers in stream	Ambient	0.1–0.5	6
Cornus stolonifera, Populus tremuloides, Betula occidentalis	In situ recirculating chambers	Ambient	0.75–1.5	7
Ficus insipida	Closed recirculating chambers	24–27	1.90–2.49	8
Sericostoma personatum	Closed flasks	n.r.	0.05–0.65	9
Sericostoma vittatum	Flow-through system	15	0.51–4.99	10
Lepidostoma unicolor	Gilson respirometer	5–15	0.93–2.07	11
Asellus aquaticus	Gilson respirometer	n.r.	0.44–2.22[a]	12

AFDM ash-free dry mass, *n.r* not reported

[a]Converted from dry mass (DM) values by assuming that AFDM is 90% of DM

[b]Plant names refer to decomposing leaves

1 = Kominoski et al. (2015); 2 = Sanpera-Calbet et al. (2009); 3 = Arroita (2015); 4 = Niyogi et al. (2002); 5 = Gulis and Suberkropp (2003); 6 = Griffiths and Tiegs (2016); 7 = Royer and Minshall (2001); 8 = Ramirez et al. (2003); 9 = Grafius and Anderson (1980); 10 = Feio and Graça (2000); 11 = Iversen (1979); 12 = Adock (1982)

2016). Water circulation can be achieved by magnetic stirrers, isolated from the leaf material by a wire mesh (Quinn et al. 2000) or by a pump creating a unidirectional flow in a recirculating chamber (Royer and Minshall 2001). Oxygen concentrations can be measured continuously in respiration chambers with an oxygen electrode (e.g. Komínková et al. 2000) or at intervals. Another system frequently used in both soil and aquatic environments is the Gilson differential respirometer, which measures respiration as the change in pressure in respiration chambers due to the production or consumption of gases (e.g. Tank and Webster 1998; Simon and Benfield 2001).

As an alternative to closed systems, respiration may be measured in a flow-through system, with a leaf sample placed inside a chamber through which oxygenated water flows. Respiration is calculated as the difference in oxygen concentration in the incoming and outflowing water multiplied by the flow rate. The flow-through method described here was adapted from Wrona and Davies (1984). In addition, a procedure is presented for closed chambers using a six-channel dissolved oxygen measuring system (Strathkelvin 929 System, North Lanarkshire, UK).

2 Equipment, Materials, Chemicals and Solutions

2.1 Equipment

- Analytical balance (0.01 mg precision).
- Drying oven (50 °C).
- Constant temperature room (e.g. 18 °C).
- For the flow-through method: oxygen meter, oxygen electrode and measuring cell (approx. 0.1 ml; Fig. 33.1 and 33.2).
- Peristaltic pump connected to multiple silicon tubes (e.g. 0.38-mm bore).
- Respiration chambers consisting of 5-ml glass syringes with the plungers replaced by rubber or silicon stoppers and the stopper pierced by a blunted hypodermic syringe connected to a Luer Lock.
- For the closed-chamber method: closed-chamber system consisting of respiration chambers placed in a water bath and individually closed with multiple oxygen meters connected to a computer via an interface (e.g. Strathkelvin 929 System with six chambers, North Lanarkshire, United Kingdom; Fig. 33.3).

2.2 Materials

- Stream-incubated leaves or invertebrates.
- Cork borer (e.g. 1 cm diameter).
- Micro-syringe (0.5 ml).

Fig. 33.1 Oxygen meter with electrode and measuring cell (left) and an opened measuring cell (right). Arrows point to the holes where water is injected for measurements

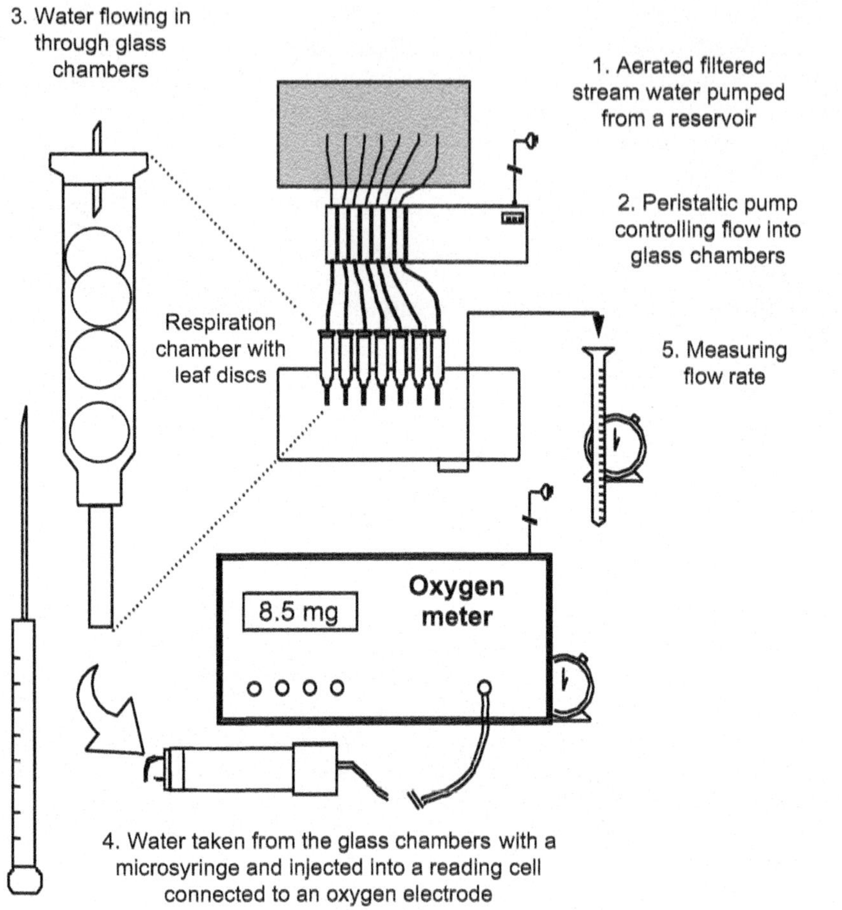

Fig. 33.2 Respiration system. (Modified from Graça 1990)

- Water tank (capacity 5 l).
- Measuring cylinder (5 ml).
- Aeration system (e.g. aquarium pump or pressurized air).

2.3 *Chemicals and Solutions*

- Filtered stream water.
- Oxygen-free calibration solution for the oxygen meter: sodium borate (3.8 g l^{-1}) saturated with crystalline sodium sulphite.

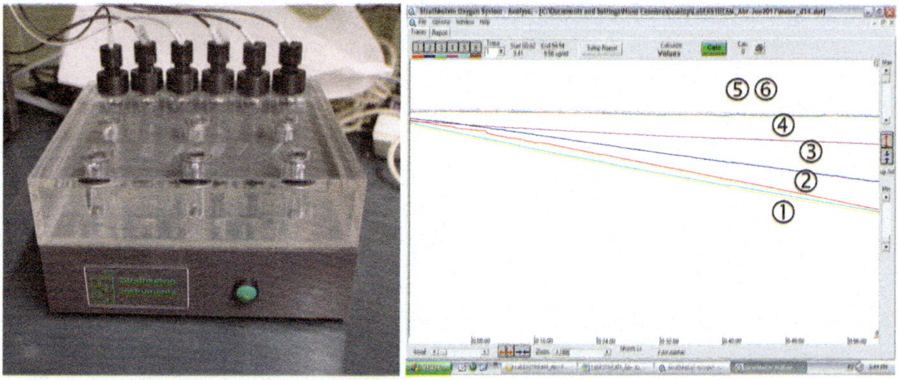

Fig. 33.3 Respiration chambers in a water bath (left) and a screenshot of the Strathkelvin 929 software showing changes in oxygen over time. Chambers 1–4 had five leaf discs each and chambers 5 and 6 are controls with no leaf discs

3 Experimental Procedures: Flow-through Method

1. Calibrate the oxygen meter with oxygen saturated (100% O_2; bubble air into a water sample) and oxygen-free (0% O_2; sodium borate solution).
2. With a cork borer, punch out discs from leaf samples collected from a stream, and insert 5–8 leaf discs in each respiration chamber.
3. Fill the respiration chamber with water by pumping filtered aerated stream water from the tank into the respiration chambers via tubes connected to the Luer Lock of the hypodermic needle (Figs. 33.1 and 33.2).
4. After complete renewal of the syringe volume, take water sample from the respiration chambers with a 0.5-ml micro-syringe, and transfer to a measuring cell where an oxygen electrode is inserted.
5. Read the oxygen concentration after a 30-s stabilization period.
6. Measure initial oxygen concentrations in the water entering the chambers between readings by taking samples from a control respiration chamber or from a valve at the inflow to chambers.
7. Check the oxygen meter, and if necessary, recalibrate between readings to adjust it to any drift.
8. Calculate the oxygen consumption as the difference in oxygen concentrations in the outflow of a chamber containing a biological sample, $[O_2]_b$, and the outflow of an empty control chamber, $[O_2]_c$.
9. Measure the rate of water flow through the respiration chambers (ν) by collecting the outflowing water for 10–20 min in a 5-ml measuring cylinder.
10. Since high flow rates decrease the sensitivity of the measurement, and low levels of oxygen can affect respiration, best adjust the peristaltic pump to deliver a rate at which oxygen concentration of the water leaving the chamber is approximately 70% of the concentration of the inflowing water at saturation level.

11. After at least three separate readings per chamber, transfer the leaf discs to aluminium pans and dry in the oven to constant weight (typically for 48 h).
12. Determine the dry mass (*DM*) of leaf discs to the nearest 0.01 mg.
13. Express respiration rates (*R*) as mg O_2 consumed per mg leaf dry mass per time in hours (t):

$$R = \frac{[O_2]_b - [O_2]_c \times v}{DM \cdot t} \quad (33.1)$$

Alternatively, express respiration rates per leaf area (*A*) as mg O_2 consumed per cm^2 per hour:

$$R = \frac{[O_2]_b - [O_2]_c \times v}{A \cdot t} \quad (33.2)$$

4 Experimental Procedures: Closed-Chamber Method

1. Carefully regulate the temperature of the water bath of the respiration measurement system, because temperature greatly influences oxygen solubility and respiration rates.
2. Calibrate the oxygen meters against oxygen-saturated medium (e.g. filtered stream water, nutrient solution, etc.) and oxygen-depleted solution as above.
3. Repeat the calibration every time incubation conditions are changed (e.g. to a different temperature or a different medium).
4. Insert a magnetic stirring bar to ensure a homogeneous oxygen distribution in the chambers during incubations. If the respiration of invertebrates is measured, separate the invertebrates from the stirring bar with a fine mesh.
5. Add O_2-saturated water to the chambers (e.g. 3 ml).
6. Add the samples (leaf discs, invertebrates) to five chambers, and close with the oxygen meter without leaving air bubbles inside. One chamber without sample will serve as a control for non-biological oxygen uptake.
7. Continuously (e.g. every second) record oxygen concentrations (mg O_2 l^{-1}) for up to 1 h or until the oxygen concentration in at least one chamber reaches 5 mg l^{-1}, at which point some organisms could experience stress.
8. Determine oxygen consumption (mg O_2 g^{-1} DM h^{-1}) for a time interval where the decline in oxygen concentration is linear, discarding the first 10–20 min of the incubation when the system stabilizes.
9. Calculate oxygen consumption as the difference between the initial (O_i) and final (O_f) oxygen concentration in a chamber with the biological sample, corrected for changes in the control chamber (O_i^c and O_f^c, as the initial and final oxygen con-

centrations in the control chamber), corrected for the chamber volume in litres (*V*), the incubation interval in minutes (*t*) and the mass of biological sample in grams (*M*):

$$R = \frac{\left((O_i - O_f) - (O_i^c - O_f^c)\right) \times V}{\left(\frac{t}{60}\right) \times M} \qquad (33.3)$$

5 Final Remarks

In the flow-through system, leaf discs must be acclimated to the chamber conditions for ~1.5 h before readings. Moreover, readings should be taken only after replacement of all water in the respiration chambers.

Six respiration chambers placed in a water bath and individually connected to oxygen meters, one of them used as a control chamber, enable processing of five samples in 1–2 h.

Since oxygen solubility in water and respiration are temperature dependent, and activities of consumers are also affected by photoperiod (e.g. Adock 1982; Fuss and Smock 1996; Feio and Graça 2000), measurements must be made at a constant temperature, and the photoperiod must be controlled.

Multiplying concentrations given in ml l^{-1} by 1.33 yields values in mg l^{-1}. Multiplying concentrations given in ml l^{-1} by 4.16 10^{-4} yields mol O_2 at standard conditions of 20 °C and 1 MPa.

References

Adock, J. A. (1982). Energetics of a population of *Asellus aquaticus* (Crustacea, Isopoda): Respiration and energy budgets. *Freshwater Biology, 12*, 257–269.

Arroita Azkarate, M. (2015). *Effects of water abstraction on stream ecosystem functioning*. Ph.D. thesis. Universidad del País Vasco.

Conant, R. T., Drijber, R. A., Haddix, M. L., Parton, W. J., Paul, E. A., Plante, A. F., Six, J., & Steinweg, J. M. (2008). Sensitivity of organic matter decomposition to warming varies with its quality. *Global Change Biology, 14*, 868–877.

Eriksen, J., & Jensen, L. S. (2001). Soil respiration, nitrogen mineralization and uptake in barley following cultivation of grazed grassland. *Biology and Fertility of Soils, 33*, 139–145.

Feio, M. J., & Graça, M. A. S. (2000). Food consumption by the larvae of *Sericostoma vittatum* (Trichoptera), an endemic species of the Iberian Peninsula. *Hydrobiologia, 439*, 7–11.

Fuss, C. L., & Smock, L. A. (1996). Spatial and temporal variation of microbial respiration rates in a blackwater stream. *Freshwater Biology, 36*, 339–349.

Graça, M. A. S. (1990). *Observations on the feeding biology of two stream-dwelling detritivores: Gammarus pulex (L.) and Asellus aquaticus (L.)*. Ph.D. Thesis, University of Sheffield, U.K.

Grafius, E., & Anderson, N. H. (1980). Population dynamics and the role of two species of *Lepidostoma* (Trichoptera: Lepidostomatidae) in an Oregon coniferous forest stream. *Ecology, 61*, 808–816.

Griffiths, N. A., & Tiegs, S. D. (2016). Organic-matter decomposition along a temperature gradient in a forested headwater stream. *Freshwater Science, 35*, 518–533.

Gulis, V., & Suberkropp, K. (2003). Leaf litter decomposition and microbial activity in nutrient-enriched and unaltered reaches of a headwater stream. *Freshwater Biology, 48*, 123–134.

Iversen, T. M. (1979). Laboratory energetics of larvae of *Sericostoma personatum* (Trichoptera). *Holarctic Ecology, 2*, 1–5.

Komínková, D., Kuehn, K. A., Buesing, N., Steiner, D., & Gessner, M. O. (2000). Microbial biomass, growth, and respiration associated with submerged litter of *Phragmites australis* decomposing in a littoral reed stand of a large lake. *Aquatic Microbial Ecology, 22*, 271–282.

Kominoski, J. S., Rosemond, A. D., Benstead, J. P., Gulis, V., Maerz, J. C., & Manning, D. W. P. (2015). Low-to-moderate nitrogen and phosphorus concentrations accelerate microbially driven litter breakdown rates. *Ecological Applications, 25*, 856–865.

Kuehn, K. A., & Suberkropp, K. (1998). Diel fluctuations in rates of CO_2 evolution from standing dead leaf litter of the emergent macrophyte *Juncus effusus*. *Aquatic Microbial Ecology, 14*, 171–182.

Niyogi, D. K., McKnight, D. M., & Lewis, W. M., Jr. (2002). Fungal communities and biomass in mountain streams affected by mine drainage. *Archiv für Hydrobiologie, 155*, 255–271.

Quinn, J. M., Burrell, G. P., & Parkyn, S. M. (2000). Influences of leaf toughness and nitrogen content on in-stream processing and nutrient uptake by litter in a Waikato, New Zealand, pasture stream and streamside channels. *New Zealand Journal of Marine Freshwater Research, 34*, 353–271.

Ramirez, A., Pringle, C. M., & Molina, L. (2003). Effects of stream phosphorus levels on microbial respiration. *Freshwater Biology., 48*, 88–97.

Royer, T. V., & Minshall, G. W. (2001). Effects of nutrient enrichment and leaf quality on breakdown of leaves in a hardwater stream. *Freshwater Biology, 46*, 603–610.

Sanpera-Calbet, I., Lecerf, A., & Chauvet, E. (2009). Leaf diversity influences in-stream litter decomposition through effects on shredders. *Freshwater Biology, 54*, 1671–1168.

Simon, K. S., & Benfield, E. F. (2001). Leaf and wood breakdown in cave streams. *Journal of the North American Benthological Society, 20*, 550–563.

Tank, J. L., & Webster, J. R. (1998). Interaction of substrate and nutrient availability on wood biofilm processes in streams. *Ecology, 79*, 2168–2179.

Wrona, F., & Davies, R. (1984). An improved flow-through respirometer for aquatic macroinvertebrate bioenergetic research. *Canadian Journal of Fisheries and Aquatic Sciences, 41*, 380–385.

Part IV
Molecular Microbial Community Analyses

Chapter 34
Terminal Restriction Fragment Length Polymorphism (T-RFLP) to Estimate Fungal Diversity

Liliya G. Nikolcheva and Felix Bärlocher

Keywords Fungal molecular identification · Fungal-specific primers · DNA extraction · DNA amplification · Endonuclease digestion · DNA fragment analysis · Restriction enzymes · Fungal community analysis

1 Introduction

The major ecological functions of aquatic hyphomycetes centre around the decomposition of leaves and other plant detritus in streams and rivers (Bärlocher 1992; Gessner et al. 2003). To fully characterize the fungal contribution to leaf decomposition and invertebrate nutrition, it is essential to subdivide the community into its constituent species.

The most commonly used approach to describe community structure of aquatic hyphomycetes is based on inducing sporulation in mycelia present in the substrate (Gessner et al. 2003, Chapter 26). Released conidia are captured on a membrane filter, stained, counted and identified. This method will miss mycelia that do not sporulate, because they are either too small (recently established), too old (e.g. because they are dormant) or too slow to form conidia during standard incubation periods (commonly 48 h but up to 96 h for root endophytes; Sridhar and Bärlocher 1992; Gessner et al. 2003). As a result, the number of spores released by aquatic hyphomycetes is not necessarily correlated with mycelial biomass of the species present in a substrate (Bermingham et al. 1997; Bärlocher et al. 2012). Furthermore, reliable species identification may be hampered in some cases because spore morphologies of some aquatic hyphomycetes broadly overlap (Chap. 25). Finally,

L. G. Nikolcheva
Mackenzie Orthodontics, Saint John, NB, Canada

F. Bärlocher (✉)
Department of Biology, Mount Allison University, Sackville, NB, Canada
e-mail: fbaerloc@mta.ca

F. Bärlocher et al. (eds.), *Methods to Study Litter Decomposition*,
https://doi.org/10.1007/978-3-030-30515-4_34

the conventional approach to assessing aquatic hyphomycete community structure by microscopically identifying released spores may miss the presence of fungal or fungal-like taxa other than aquatic hyphomycetes (e.g. other Ascomycota and Basidiomycota, Chytridiomycota, Zygomycota and Oomycota) which do not normally sporulate under water (Bärlocher et al. 2012).

Molecular approaches, based on characterizing the sequence and diversity of nucleic acids (most notably DNA), potentially circumvent these problems. The advantage of molecular techniques in general is their extreme sensitivity (very low microbial biomass can be detected) and their applicability to all stages of the microbial life cycle. Consequently, they can overcome many of the limitation of the microscopic analyses of spores.

When applied to microbial community diversity, molecular techniques most often rely on the amplification of DNA with taxon-specific primers (Borneman and Hatrin 2000) and subsequent characterization of the diversity of amplified DNA (Head et al. 1998). A high-throughput technique originally developed in bacterial ecology is terminal restriction fragment length polymorphism (T-RFLP). In this method, the extracted DNA is amplified with one or both of two primers fluorescently labelled at the 5′ end. The products of the polymerase chain reaction (PCR) are then digested with a restriction enzyme, and the labelled terminal fragments are separated and detected on a DNA sequencer. The number of DNA fragments of different sizes gives an estimate of the minimum number of strains present in the analyzed community (Liu et al. 1997). Thus, the number of fragments detected is indicative of the number of phylogenetically different strains (phylotypes) present in a sample. The pattern of fragments obtained from T-RFLP is compared to the fragment lengths generated from digestion of DNA from pure cultures of aquatic hyphomycetes (Nikolcheva et al. 2003). Although this technique cannot be used to identify fungal species or strains, it provides information on the number of fungal phylotypes, which is related to fungal species richness. This fast, high-throughput technique has been used to characterize salt-marsh (Buchan et al. 2002), stream (Zeglin 2015) and soil fungi (Klamer et al. 2002).

The method for T-RFLP presented here is modified from Nikolcheva et al. (2003) and describes the extraction, amplification and purification of fungal DNA, followed by endonuclease digestion and separation of the fragments. The technique has been adapted to the specific materials and equipment available at our laboratory but can easily be adapted to other laboratory set-ups.

2 Equipment, Chemicals and Solutions

2.1 Equipment and Material

- Pure fungal cultures
- Autumn-shed leaves, air-dried
- Litter bags (10 × 10 cm; 10, 1 or 0.5 mm mesh size)

- Filtering apparatus and 8 μm membrane filters (e.g. Millipore Corporation, Bedford, MA, USA)
- Freeze-dryer
- Mortar and pestle
- MoBio UltraClean Soil DNA extraction kit (MoBio, Solana Beach, CA, USA) or Nucleon PhytoPure Plant DNA extraction kit (Amersham Biosciences, Piscataway, NJ, USA)
- Automatic pipettes: ranges of 0.5–10, 2–20 and 20–200 μl
- Pipette tips: 0.1–20 μl (white) and 2–200 μl (yellow)
- Ready-To-Go PCR beads (Amersham Biosciences, Piscataway, New Jersey, USA). Any other available PCR mix should work as well
- Chambers for running horizontal agarose gels
- PCR thermocycler
- Parafilm
- DNA-automated sequencer with fragment analysis software (e.g. Long-Read Tower by Visible Genetics, Suwanee, GA, USA)
- Water bath at 37 °C
- Ice bath

2.2 Chemicals

- Agarose (molecular biology grade)
- Malt broth
- Liquid nitrogen
- Tris(hydroxymethyl)aminomethane (TRIS, analytical grade)
- Ethylenediaminetetraacetic acid (EDTA, analytical grade)
- Sodium acetate
- Boric acid
- Acryl amide (6% for sequencing gels), which must be compatible with the type of automated sequencer used
- Bromophenol blue
- Glycerol
- Autoclaved deionized water (e.g. Milli-Q quality)
- Primers for the PCR reaction, specifically the fungal-specific primer F1300 (5′ GATAACGAACGAGACCTTAAC 3′; Nikolcheva et al. 2003) labelled at the 5′ end with Cy5.5 and the primer D (5′ CYGCAGGTTCACCTAC 3′; Elwood et al. 1985) labelled at the 5′ end with Cy5. Primer NS8 (5′ TCCGCAGGTTCACCTACGGA 3′; White et al. 1990) can be used instead of primer D
- Fluorescent DNA stain, preferably GelStar (BioWhittaker Molecular Applications, Rockland, ME, USA), but other stains such as Gel Red and SYBR Safe can also be used

- DNA ladder with 100 or 250 base pairs (bp; Amersham Biosciences, Little Chalfont, UK)
- Dimethyl sulfoxide (DMSO, analytical grade)
- Formamide loading dye, which is included with the sequencing kit; it keeps the DNA single stranded before separation of the fragments by gel electrophoresis
- GFX PCR and gel band purification kit (Amersham Biosciences)
- Restriction enzymes CfoI, RsaI, HinfI and DdeI (Roche Diagnostics, Indianapolis, IN, USA)
- Buffers for restriction enzymes: SuRE/Cut buffer L and buffer H (Roche Diagnostics); these buffers are supplied with the restriction enzymes
- ALF Express Sizer 50–500 bases (Amersham Biosciences). This DNA standard works only if the automated sequencer can detect Cy5 and Cy5.5 fluorescent dyes. If the sequencer has a different detection system, DNA fragments of standard sizes labelled with other appropriate dyes must be used.

2.3 Solutions

- 5 × TBE (450 mM Tris, 450 mM boric acid, 13 mM EDTA)
- 1 × TBE (100 ml 5 × TBE and 400 ml deionized water)
- 0.5 × TBE (100 ml 5 × TBE and 900 ml deionized water)
- Agarose gel loading buffer (5 ml glycerol, 5 ml water, 5 mg bromophenol blue).
- 100 × GelStar DNA stain (2 µl concentrated GelStar, 198 µl DMSO).

3 Experimental Procedures

3.1 Sample Preparation

1. Grow pure fungal cultures in malt broth for 3 weeks. Filter through an 8 µm filter, and freeze-dry overnight. These samples can be stored indefinitely at –80 °C.
2. Expose leaves in litter bags in the field and collected as described in Chap. 6.
3. Rinse leaves to remove any adhering particles, freeze-dry and store at −80 °C.

3.2 DNA Extraction

1. Use 30 mg of the dry pure fungal culture or up to 50 mg of dry plant tissue. Gently grind the samples with a mortar, and pestle in liquid nitrogen for no more than 1 min.

2. Use the MoBio UltraClean Soil DNA extraction kit, and follow the manufacturer's instruction for DNA isolation. The Nucleon PhytoPure kit may also be used. Elute or resuspend the extracted DNA in 50 μl autoclaved deionized water.
3. The extracted DNA can be stored for up to 2 months at 4 °C or indefinitely at –20 °C.

3.3 DNA Amplification

1. To a Ready-To-Go PCR bead, add 19 μl of deionized autoclaved water, 2 μl of 10 μM forward primer (F1300; Nikolcheva et al. 2003), 2 μl of 10 μM reverse primer (D; Elwood et al. 1985) and 2 μl of extracted DNA.
2. Place all amplification reactions in a thermocycler programmed for 2 min of 95 °C initial denaturation followed by 35 cycles of denaturation at 95 °C for 30s, primer annealing at 55 °C for 30 s and primer extension at 72 °C for 1 min. End the program with 72 °C for 5 min of final extension. Store the PCR products at 4 °C.

3.4 Agarose Gel Electrophoresis

1. To check the concentration and quality of the PCR products, run an agarose gel.
2. Mix 30 ml 0.5 × TBE buffer with 300 mg agarose. Microwave until the agarose has melted, and then cast the gel. Let it solidify for 20 min.
3. Mix 1 μl loading dye, 0.7 μl 100 × GelStar and 2 μl of PCR product (or 0.5 μg DNA ladder) on a piece of Parafilm.
4. Load the entire reaction in a well of the gel. Run the gel in 0.5 × TBE buffer at 160 V for 15 min.
5. View on a UV transilluminator (wear protective glasses!).

3.5 DNA Purification

1. Purify the PCR products from solution (there should be 23 μl left) with GFX DNA and the gel band purification kit. Use the manufacturer's protocol for purification of PCR products from solution.
2. If the calculated DNA concentration from the agarose gel is between 0.1 and 0.3 μg ml^{-1}, then elute the DNA with 50 μl deionized water. If the concentration is lower, elute in 25 μl.

3.6 Restriction Digest

1. Each digestion reaction contains 4.5 μl purified PCR product, 0.5 μl 10 × enzyme buffer (buffer L is suitable for CfoI and RsaI, buffer H is suitable for DdeI and HinfI) and 5 U restriction enzyme.
2. If the restriction enzyme cleavage of DNA fragments is performed with a single enzyme, use 5 U of that enzyme. If the restriction is performed with two enzymes acting simultaneously, use 2.5 U of each. The enzymes CfoI and RsaI are compatible with each other, because both require SuRE/Cut buffer L; DdeI and HinfI are also compatible with one another, as they require SuRE/Cut buffer H.
3. Incubate the digestion reactions in a water bath or in the thermocycler at 37 °C for 2 h. The digestion products should be used immediately or stored at −20 °C.

3.7 Separation of Terminal Restriction Fragments

1. Prepare digestion products for loading on sequencing gel or capillary.
2. Combine 2 μl digestion product, 3 μl deionized water and 3 μl formamide loading dye.
3. For each set of eight reaction mixtures, use 2 μl of the ALF Express 50–500 bp sizer and 3 μl loading dye to run an external control.
4. Incubate all mixtures in a preheated thermocycler at 90 °C for 3 min to denature the DNA, and then place them immediately on ice to keep the DNA single stranded.
5. Load 2 μl of each sample on a sequencing gel within 1 h after denaturation.
6. Run the gel in 1 × TBE buffer at 1500 V and 52 °C for 45 min. The conditions for electrophoresis are specific to the automated sequencer used.

4 Final Remarks

For calibration, the DNA from pure fungal cultures can be analyzed before the environmental samples. Each fungal species should theoretically yield one fragment, and ideally, the fragment length is species specific. However, some primers will amplify conserved genes, and some enzymes will cut at a conserved region in DNA; when this happens, fragments of different species or genera may be identical. The results of a T-RFLP analysis (i.e. number of fragments differing in length) therefore provide an estimate of how much variability can be detected with a given primer/enzyme combination, and true diversity will generally be underestimated. In our experience, the enzyme DdeI gives the highest variability among pure fungal cultures. This restriction enzyme was therefore used to analyze environmental samples (Fig. 34.1). The fragment lengths obtained from pure cultures can be

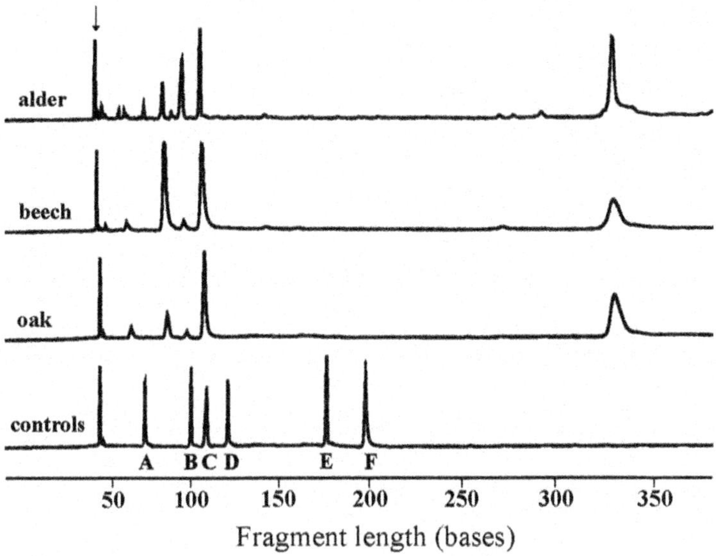

Fig. 34.1 T-RFLP chromatograms from fungal communities associated with alder, beech and oak leaves after 7 days of exposure in a stream. The 3′ end of the 18S rRNA gene was amplified with a fungal-specific primer pair and digested with DdeI. The peak designated by the arrow is unextended primer. The lane labelled 'controls' contains fragments from the digestion of *A Anguillospora furtiva*, *B Colispora elongata*, *C Articulospora tetracladia*, *D Anguillospora rubescens*, *E Tetracladium marchalianum* and *F Heliscus lugdunensis*

compared to observed lengths in environmental samples; if the peaks on the chromatograms coincide, the environmental samples may contain the particular fungal species. However, the fragments in the environmental samples cannot be identified with certainty.

Rather than using extraction kits, DNA can be extracted from samples with a standard SDS- or CTAB-based phenol-chloroform procedure. The DNA yield from these types of extractions can be very high, but in our experience, the DNA quality is often poor, making amplification by PCR impossible.

In our experience, the 18S rRNA restriction enzymes CfoI and DdeI detect the highest community variability (Nikolcheva et al. 2003), and a combination of restriction enzymes may result in an even higher interspecific variability (unpubl. data).

For more detailed and accurate descriptions of fungal diversity, next-generation sequencing (NGS) is now commonly being used (Chaps. 37, 62; Duarte et al. 2015), often in combination with barcodes (Duarte et al. 2013). It requires considerably greater investment in time and equipment. Regardless, traditional, microscopy-based approaches (Chap. 26) continue to provide otherwise inaccessible information on the reproductive status of fungi on leaves decaying in streams (Duarte et al. 2010).

References

Bärlocher, F. (1992). *The ecology of aquatic hyphomycetes*. Berlin: Springer.

Bärlocher, F., Stewart, M., & Ryder, D. S. (2012). Processing of *Eucalyptus viminalis* leaves in Australian streams – Importance of aquatic hyphomycetes and zoosporic fungi. *Fundamental and Applied Limnology, 179*, 305–319.

Bermingham, S., Maltby, L., & Dewey, F. M. (1997). Use of immunoassays for the study of natural assemblages of aquatic hyphomycetes. *Microbial Ecology, 33*, 223–229.

Borneman, J., & Hatrin, R. J. (2000). PCR primers that amplify fungal rRNA genes from environmental samples. *Applied and Environmental Microbiology, 66*, 4356–4360.

Buchan, A., Newell, S. Y., Moreta, J. I. L., & Moran, M. A. (2002). Analysis of internal transcribed spacer (ITS) regions of rRNA genes in fungal communities in a southeastern U.S. salt marsh. *Microbial Ecology, 43*, 329–340.

Duarte, S., Pascoal, C., Alves, A., Correia, A., & Cássio, F. (2010). Assessing the dynamic of microbial communities during leaf decomposition in a low-order stream by microscopic and molecular techniques. *Microbial Research, 165*, 351–362.

Duarte, S., Seena, S., Bärlocher, F., Pascoal, C., & Cássio, F. (2013). A decade's perspective on the impact of DNA sequencing on aquatic hyphomycete research. *Fungal Biology Reviews, 27*, 19–24.

Duarte, S., Bärlocher, F., Trabulo, J., Cássio, F., & Pascoal, C. (2015). Stream-dwelling fungal decomposer communities along a gradient of eutrophication unraveled by 454 pyrosequencing. *Fungal Diversity, 70*, 127–148.

Elwood, H. J., Olsen, G. J., & Sogin, M. L. (1985). The small-subunit ribosomal RNA gene sequences from the hypotrichous ciliates *Oxytricha nova* and *Stylonychia pustulata*. *Molecular and Biological Evolution, 2*, 399–410.

Gessner, M. O., Bärlocher, F., & Chauvet, E. (2003). Qualitative and quantitative analyses of aquatic hyphomycetes in streams. In C. K. M. Tsui & K. D. Hyde (Eds.), *Freshwater mycology* (pp. 127–157). Hong Kong: Fungal Diversity Press.

Head, I. M., Sanders, J. R., & Pickup, R. W. (1998). Microbial evolution, diversity, and ecology: A decade of ribosomal RNA analysis of uncultivated organisms. *Microbial Ecology, 35*, 1–21.

Klamer, M., Roberts, M. S., Levine, L. H., Drake, B. G., & Graland, J. L. (2002). Influence of elevated CO_2 on the fungal community in a coastal scrub oak forest soil investigated with terminal-restriction fragment length polymorphism analysis. *Applied and Environmental Microbiology, 68*, 4370–4376.

Liu, W.-T., Marsh, T. L., Cheng, H., & Forney, L. J. (1997). Characterization of microbial diversity by determining terminal restriction fragment length polymorphisms of genes encoding 16S rRNA. *Applied and Environmental Microbiology, 63*, 4516–4522.

Nikolcheva, L. G., Cockshutt, A. M., & Bärlocher, F. (2003). Diversity of freshwater fungi on decaying leaves: Comparing traditional and molecular approaches. *Applied and Environmental Microbiology, 69*, 2548–2554.

Sridhar, K. R., & Bärlocher, F. (1992). Endophytic aquatic hyphomycetes of roots of spruce, birch and maple. *Mycological Research, 96*, 305–308.

White, T. J., Bruns, T., Lee, S., & Taylor, J. W. (1990). Amplification and direct sequencing of fungal ribosomal RNA genes for phylogenetics. In M. A. Innis, D. H. Gelfand, J. J. Sninsky, & T. J. White (Eds.), *PCR protocols: A guide to methods and applications* (pp. 315–322). New York: Academic.

Zeglin, L. H. (2015). Stream microbial diversity in response to environmental changes: Review and synthesis of existing research. *Frontiers in Microbiology, 6*, 454.

Chapter 35
Denaturing Gradient Gel Electrophoresis (DGGE) to Estimate Fungal Diversity

Liliya G. Nikolcheva and Felix Bärlocher

Keywords DNA amplification · DNA denaturation · DNA extraction · DNA sequencing · Electrophoresis · Fungal community analysis · Fungal molecular identification · Fungal-specific primers · Polyacrylamide gel

1 Introduction

In addition to T-RFLP (Chap. 34), a range of other molecular methods have been developed to determine the species composition of microbial communities. For example, taxon-specific primers are used to amplify a gene of interest (Borneman and Hatrin 2000), which is then cloned into a bacterial vector. Each bacterial clone is grown on a plate to produce an individual colony, and the cloned gene from each colony is sequenced (Head et al. 1998). The sequence is then compared to sequences published in a genomic database such as GenBank (http://www.ncbi.nlm.nih.gov/). This technique accurately determines the identity and phylogeny of microorganisms in diverse communities. However, it is both time-consuming and expensive. Moreover, since clones are randomly sampled before sequencing, it is likely that some clones are never sampled, and the diversity of the community is thus underestimated. The cloning and sequencing approach has been used for analyzing fungal communities on plant roots in soil (Vandenkoornhuyse et al. 2002), on decomposing leaves of the salt-marsh grass *Spartina alterniflora* (Buchan et al. 2002) and of the common reed *Phragmites australis* (Buesing et al. 2009), on deciduous leaves decaying in streams (Seena et al. 2008), and on feces of stream invertebrates (Sridhar et al. 2011).

L. G. Nikolcheva
Mackenzie Orthodontics, Saint John, NB, Canada

F. Bärlocher (✉)
Department of Biology, Mount Allison University, Sackville, NB, Canada
e-mail: fbaerloc@mta.ca

F. Bärlocher et al. (eds.), *Methods to Study Litter Decomposition*,
https://doi.org/10.1007/978-3-030-30515-4_35

Denaturing gradient gel electrophoresis (DGGE) combines the advantages of cloning and sequencing and T-RFLP. A gene of interest present in all members of a community, such as the aquatic hyphomycete community on a leaf, is targeted with appropriate primers and amplified through PCR. If the genes of the various species differ significantly in length, they can be separated by conventional electrophoresis. However, species- or strain-specific differences may be more subtle, because they can involve a small change in base composition (AT versus GC), while maintaining the same overall length. Double-stranded DNA sequences of identical lengths but of different base composition cannot normally be separated by electrophoresis; however, they differ in their ease of denaturation (i.e., separation of double-stranded DNA into single strands). DGGE exploits this difference to separate amplified DNA on a high-resolution polyacrylamide gel along a denaturing gradient (Fischer and Lerman 1983; Muyzer et al. 1993). Sequences differing in base composition denature at different locations on this gel. When denaturation is initiated, a "bubble" forms at the site where the two strands separate. This dramatically slows migration of the DNA in the gel. The number of bands on the gel is therefore indicative of the gene diversity in the original sample, such as a DNA extract from a decomposing leaf colonized by microorganisms.

One advantage of the DGGE technique over T-RFLP is that DNA from each separate band on the gel can be isolated and sequenced to identify all genes in the sample (Buesing et al. 2009). The sequences can be compared to those published in GenBank, which, depending on the gene, allows identification at the level of phylum, genus, or even species.

The method presented here is modified from Nikolcheva et al. (2003). The procedures have been adapted to the specific materials and equipment available at our laboratory but can easily be modified for other laboratory setups (Duarte et al. 2010, 2012). With the primers described here, the method does not discriminate among members of different fungal phyla (e.g., Ascomycota, Basidiomycota). To achieve this, taxon-specific primers have to be used (Nikolcheva and Bärlocher 2004).

2 Equipment, Chemicals, and Solutions

2.1 Equipment and Materials

- Pure fungal cultures
- Autumn-shed leaves, air-dried
- Litter bags (10 × 10 cm; 10, 1, or 0.5 mm mesh size)
- Filtering apparatus and membrane filters with 8 μm pore size (e.g., Millipore Corporation, Bedford, MA, USA)
- Freeze dryer
- Mortar and pestle

- MoBio UltraClean Soil DNA extraction kit (MoBio, Solana Beach, CA, USA) or Nucleon PhytoPure Plant DNA extraction kit (Amersham Biosciences, Piscataway, NJ, USA)
- Automatic pipettes: ranges of 0.5–10, 2–20, and 20–200 μl
- Pipette tips: 0.1–20 μl (white) and 2–200 μl (yellow)
- DCode Mutation Detection System (BioRad, Hercules, CA, USA) with model 475 Gradient Former (BioRad) or any other DGGE system
- Ready-to-Go PCR beads (Amersham Biosciences); any other PCR mix should work as well
- Chambers for running horizontal agarose gels
- Direct current (DC) power supply
- PCR thermocycler
- Parafilm
- Ice bath

2.2 Chemicals

- Agarose (molecular biology grade)
- Malt broth
- Liquid nitrogen
- Tris(hydroxymethyl)aminomethane (TRIS, analytical grade)
- Ethylenediaminetetraacetic acid (EDTA, analytical grade)
- Sodium acetate
- Boric acid
- Acryl amide (40%), 37.5:1 acrylamide/bisacrylamide
- Formamide (100%, analytical grade)
- Urea (analytical grade)
- N,N,N′,N′-Tetramethylethylenediamine (TEMED, analytical grade)
- Ammonium persulfate (analytical grade)
- Bromophenol blue
- Glycerol
- Autoclaved deionized water (e.g., Milli-Q quality)
- Primers for the PCR reaction: NS1 (5′-GTAGTCATATGCTTGTCTC-3′, White et al. 1990) and GCfung sequence (May et al. 2001)
- Fluorescent DNA stain, preferably GelStar (BioWhittaker Molecular Applications, Rockland, ME, USA), but other stains such as Gel Red and SYBR Safe can also be used
- DNA ladder with 100 or 250 base pairs (bp; Amersham Biosciences, Little Chalfont, UK)
- Dimethyl sulfoxide (DMSO)

2.3 Solutions

- 5× TBE (450 mM Tris, 450 mM boric acid, 13 mM EDTA)
- 0.5× TBE (100 ml 5× TBE and 900 ml water)
- 50× TAE (2.0 M Tris, 1.0 M acetic acid, 50 mM EDTA, pH 8.3)
- 1× TAE (20 ml 50× TAE and 980 ml water)
- 8% acryl amide and 0% denaturant (20 ml 40% acryl amide, 2 ml 50× TAE, 78 ml water)
- 8% acryl amide and 100% denaturant (20 ml 40% acryl amide, 2 ml 50× TAE, 40 ml formamide, 42 g urea, 8 ml water)
- Agarose gel loading buffer (5 ml glycerol, 5 ml water, 5 mg bromophenol blue)
- 100× GelStar DNA stain (2 μl concentrated GelStar, 198 μl DMSO)

3 Experimental Procedures

3.1 Sample Preparation

1. Grow pure fungal cultures in malt broth for 3 weeks. Filter through a membrane filter (8 μm pore size), and freeze-dry overnight. These samples can be stored indefinitely at −80 °C.
2. Expose leaves in litter bags in the field, and retrieve, as described in Chaps. 5 and 6.
3. Rinse leaves to remove any adhering particles, freeze-dry, and store at −80 °C.

3.2 DNA Extraction

1. Use 30 mg of the dry pure fungal culture or up to 50 mg of dry plant tissue. Gently grind the samples with a mortar, and pestle in liquid nitrogen for up to 1 min.
2. Use the MoBio UltraClean Soil DNA extraction kit, and follow the manufacturer's instruction for DNA isolation. The Nucleon PhytoPure DNA extraction kit can also be used. Elute or resuspend the DNA in 50 μl autoclaved deionized water. In our experience, SDS- or CTAB-based phenol-chloroform DNA extraction can give very high yields, but the quality of DNA is often poor, making amplification by PCR difficult.
3. The extracted DNA can be stored for up to 2 months at 4 °C or indefinitely at –20 °C.

3.3 DNA Amplification

1. To a Ready-to-Go PCR bead, add 19 μl of autoclaved deionized water, 2 μl of 10 μM forward primer (NS1; White et al. 1990), 2 μl of 10 μM reverse primer (GCfung; May et al. 2001), and 2 μl of extracted DNA.
2. Place all amplification reactions in a thermocycler programmed for 2 min of 95 °C initial denaturation followed by 35 cycles of denaturation at 95 °C for 30s, primer annealing at 55 °C for 30 s, and primer extension at 72 °C for 1 min. End the program at 72 °C for 5 min for final extension. Store the PCR products at 4 °C.

3.4 Agarose Gel Electrophoresis

1. To check the concentration and quality of the PCR products, run an agarose gel.
2. Mix 30 ml 0.5× TBE buffer with 300 mg agarose. Microwave until the agarose has melted, and then cast the gel. Let it solidify for 20 min.
3. Mix 1 μl loading dye, 0.7 μl 100x GelStar, and 2 μl of PCR product (or 0.5 μg DNA ladder) on a piece of Parafilm.
4. Load the entire reaction in a well of the gel. Run the gel in 0.5× TBE buffer at 160 V for 15 min.
5. View on a UV transilluminator (wear protective glasses!).

3.5 Denaturing Gradient Gel Electrophoresis (DGGE)

1. Pour 7 l of 1× TAE buffer in the gel-running chamber, and start pre-warming to 56 °C. Depending on the room temperature, this will take 45–60 min.
2. Clean all glass plates with ethanol, and assemble them with 1 mm spacers according to the manufacturer's instructions for parallel DGGE.
3. In a 50 ml plastic tube, prepare 16 ml of a low-denaturant solution containing 20% denaturing gradient (100% denaturing gradient corresponds to 7 M urea and 40% formamide). To make the low-denaturant solution, mix 12.8 ml of 0% denaturant solution with 3.2 ml of 100% denaturant solution. In a different tube, mix 7.2 ml of 0% denaturant solution and 8.8 ml of 100% denaturant solution (this makes 55% denaturant). Keep both solutions on ice.
4. Optionally, add 60 μl of gel dye to the high-denaturant solution and mix. To both high- and low-denaturant solutions, add 14.4 μl TEMED and 144 μl 10% APS and mix. The solutions will take about 10 min to polymerize if kept on ice; once they are at room temperature, they polymerize more quickly. All steps involving the casting of the gel should be performed as quickly as possible.

5. Fill two 30 ml syringes with tubing with the solutions, place on the Gradient Former, and assemble the Tygon tubing as described by the manufacturer. At the end of the tubing, add an 18-gauge needle.
6. Cast the gel. The blue high-denaturant solution should be on the bottom and should form a uniform transition into the transparent low-denaturant solution on the top. Add a 16-well comb, and let the gel polymerize for 45 min at room temperature.
7. If a second gel is made, wash the plastic tubes, syringes, and tubing before repeating all steps.
8. Once the gels have polymerized, assemble the gel-running gasket according to the manufacturer's instructions. Place the gasket in the gel-running chamber, and pre-warm the gel for 10 min.
9. Prepare the samples for loading. Mix 12 μl PCR product with 12 μl gel loading dye. Load lanes 2–15 with a yellow 2–200 μl pipette tip. Use a new tip for each lane. The outer lanes on the gel do not run at a consistent rate, thus forming crescent-shaped rather than straight DNA bands.
10. Turn on the pump and the heater; attach the gel-running apparatus to a DC power supply. There can be slight fluctuations in the current during the run, probably due to temperature fluctuations. In our experience, an older, less sensitive power supply works more consistently than an electronic power supply.
11. Run the gel at 50 V and 56 °C for 16 h. The voltage can be increased up to 80 V, and the running time should then be decreased accordingly to 9 h.
12. The gel dye and the loading dye migrate out of the gel and into the buffer during the run. The gel should be transparent. Switch off the power supply, the pump, and the heater. Take the gel gasket out of the chamber, and disassemble the gels. Prepare staining solution for each gel: 30 ml of 1× TAE with 3 μl 10,000× GelStar. Stain each gel in the staining solution for 10 min.
13. View the gels on a UV transilluminator (wear protective glasses!).
14. Analyze the band intensity in each lane using image analysis software (e.g. National Institutes of Health, Bethesda, Maryland, USA).
15. Initially, analyze DNA from pure fungal cultures to determine the variability in the sequences and to generate a set of standards that can be used to calibrate environmental samples (Fig. 35.1).
16. Use a mixture of PCR products from pure fungal cultures whose separation is optimized as a standard on a gel with environmental samples (Fig. 35.2). Some bands from an environmental sample may migrate at the same rate as bands from pure fungal cultures, which suggests that a particular fungal species may be present in the environmental sample. To confirm identity, the DNA from the bands should be extracted and sequenced.

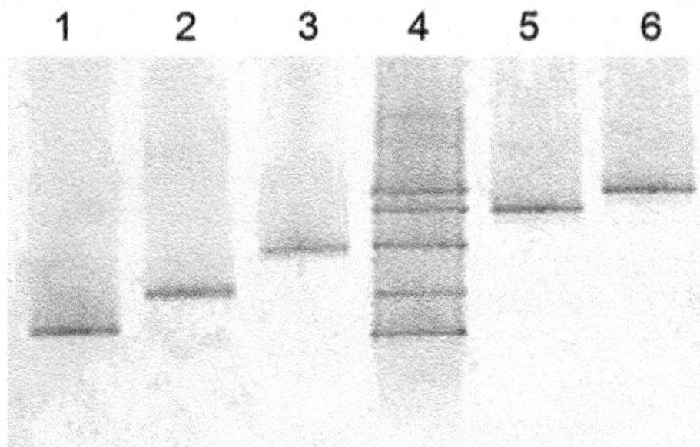

Fig. 35.1 Amplified DNA from pure cultures of aquatic hyphomycetes separated by DGGE. Lanes (1) *Tumularia aquatica*; (2) *Articulospora tetracladia*; (3) *Tetracladium marchalianum*; (4) Mix of all five cultures; (5) *Heliscus lugdunensis*; (6) *Anguillospora furtiva*

Fig. 35.2 Amplified fungal DNA from environmental samples separated by DGGE. Leaves of alder, beech, and oak were submerged in a stream for 7 days; the DNA was extracted, amplified, and separated. Lanes (1) fungal DNA from alder leaves, (2) beech leaves, (3) oak leaves, and (4) amplification products from pure cultures of aquatic hyphomycetes used as standards (same species as in Fig. 23.1)

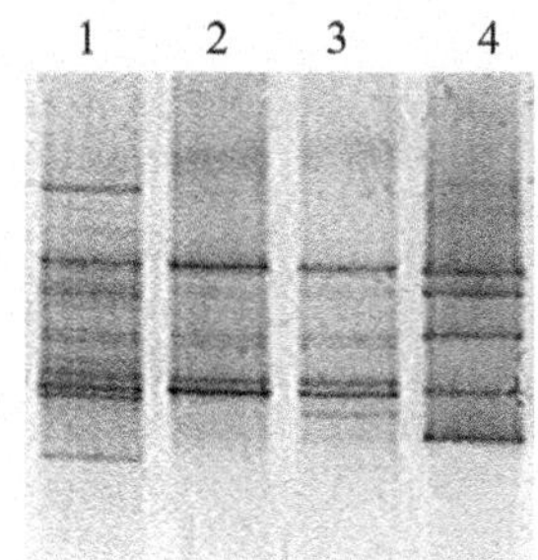

4 Final Remarks

Duarte et al. (2012) provide a thorough discussion of the advantages and disadvantages of DGGE for analyzing aquatic fungal communities. Like the related temperature gradient gel electrophoresis (TGGE) and T-RFLP (Chap. 34), DGGE allows a quick assessment of molecular fungal diversity (Clivot et al. 2014). For more comprehensive and in-depth analyses, a cloning-sequencing approach can be used. However, the most detailed assessment is afforded by next-generation sequencing (NGS) (e.g., Duarte et al. 2015; Wurzbacher et al. 2015).

DGGE can potentially be used to estimate abundances of specific taxa, provided there was no bias in the extraction or amplification procedures (Nikolcheva et al. 2003; Duarte et al. 2012), though quantitative real-time PCR is more sensitive and accurate (Feckler et al. 2017, Chapter 36).

References

Borneman, J., & Hatrin, R. J. (2000). PCR primers that amplify fungal rRNA genes from environmental samples. *Applied and Environmental Microbiology, 66*, 4356–4360.

Buchan, A., Newell, S. Y., Moreta, J. I. L., & Moran, M. A. (2002). Analysis of internal transcribed spacer (ITS) regions of rRNA genes in fungal communities in a southeastern U.S. salt marsh. *Microbial Ecology, 43*, 329–340.

Buesing, N., Filippini, M., Bürgmann, H., & Gessner, M. O. (2009). Microbial communities in contrasting freshwater microhabitats. *FEMS Microbial Ecology, 69*, 84–97.

Clivot, H., Cornut, J., Chauvet, E., Elger, A., Poupin, P., Guérold, F., & Planout, C. (2014). Leaf-associated fungal diversity in acidified streams: Insights from combining traditional and molecular approaches. *Environmental Microbiology, 16*, 2145–2156.

Duarte, S., Pascoal, C., Alves, A., Correia, A., & Cássio, F. (2010). Assessing the dynamic of microbial communities during leaf decomposition in a low-order stream by microscopic and molecular techniques. *Microbial Research, 165*, 351–362.

Duarte, D., Cássio, F., & Pascoal, C. (2012). Denaturing Gradient Gel Electrophoresis (DGGE) in Microbial ecology – Insights from freshwaters. In S. Magdeldin (Ed.), *Gel electrophoresis – Principles and basics*. ISBN: 978-953-51-0458-2.

Duarte, S., Bärlocher, F., Trabulo, J., Cássio, F., & Pascoal, C. (2015). Stream-dwelling fungal decomposer communities along a gradient of eutrophication unraveled by 454 pyrosequencing. *Fungal Diversity, 70*, 127–148.

Feckler, A., Schrimpf, A., Bundschuh, M., Bärlocher, F., Baudy, P., Cornut, J., & Schulz, R. (2017). Quantitative real-time PCR as a promising tool for the detection and quantification of leaf-associated fungal species – A proof-of-concept using *Alatospora pulchella*. *PLoS One, 12*, e0174634.

Fischer, S. G., & Lerman, L. S. (1983). DNA fragments differing by single base-pair substitutions are separated in denaturing gradient gels: Correspondence with melting theory. *Proceedings of the National Academy of Sciences, USA, 80*, 1579–1583.

Head, I. M., Sanders, J. R., & Pickup, R. W. (1998). Microbial evolution, diversity, and ecology: A decade of ribosomal RNA analysis of uncultivated organisms. *Microbial Ecology, 35*, 1–21.

May, L. A., Smiley, B., & Schmidt, M. G. (2001). Comparative denaturing gradient gel electrophoresis of fungal communities associated with whole plant corn silage. *Canadian Journal of Microbiology, 47*, 829–841.

Muyzer, G., de Waal, E. C., & Uitterlinden, A. G. (1993). Profiling of complex microbial populations by denaturing gradient gel electrophoresis analysis of polymerase chain reaction – Amplified genes coding for 16S rRNA. *Applied and Environmental Microbiology, 59*, 695–700.

Nikolcheva, L. G., & Bärlocher, F. (2004). Taxon-specific fungal primers reveal unexpectedly high diversity during leaf decomposition in a stream. *Mycological Progress, 3*, 41–50.

Nikolcheva, L. G., Cockshutt, A. M., & Bärlocher, F. (2003). Diversity of freshwater fungi on decaying leaves: Comparing traditional and molecular approaches. *Applied and Environmental Microbiology, 69*, 2548–2554.

Seena, S., Wynberg, N., & Bärlocher, F. (2008). Fungal diversity during leaf decomposition in a stream assessed through clone libraries. *Fungal Diversity, 30*, 1–14.

Sridhar, K. R., Beaton, M., & Bärlocher, F. (2011). Fungal propagules and DNA in feces of two detritus-feeding amphipods. *Microbial Ecology, 61*, 31–40.

Vandenkoornhuyse, P., Baldauf, S. L., Leyval, C., Straczek, J., & Young, J. P. W. (2002). Extensive fungal diversity in plant roots. *Science, 295*, 2051.

White, T. J., Bruns, T., Lee, S., & Taylor, J. W. (1990). Amplification and direct sequencing of fungal ribosomal RNA genes for phylogenetics. In M. A. Innis, D. H. Gelfand, J. J. Sninsky, & T. J. White (Eds.), *PCR protocols: A guide to methods and applications* (pp. 315–322). New York: Academic.

Wurzbacher, C., Grimmett, I. J., & Bärlocher, F. (2015). Metabarcoding-based fungal diversity on coarse and fine particulate organic matter in a first-order stream in Nova Scotia, Canada. *F1000Research, 4*, 1378.

Chapter 36
Quantitative Real-Time PCR (qPCR) to Estimate Molecular Fungal Abundance

Christiane Baschien and Steffen C. Carl

Keywords Aquatic hyphomycetes · DNA extraction · DNA quantification · Fungal community analysis · PCR inhibition · Primer design · Probe design · LinReg software · Taqman® · Taxon-specific probes

1 Introduction

Fungi are key players in the decomposition of leaves in freshwater ecosystems (Bärlocher 1992; Gessner et al. 2007; Tsui et al. 2016). To assess the contribution of fungal species to ecosystem processes, it is essential to determine their abundance. Quantitative real-time PCR (qPCR) is a useful method for this purpose, because it can detect individual species of aquatic fungi in micro- and mesocosms (Fernandes et al. 2011) as well as in environmental samples taken in the field (Lefèvre et al. 2010; Gehesquière et al. 2013). Therefore, qPCR can be used, for example, to monitor fungal taxa of interest in studies where particular species might serve as indicators of environmental conditions (e.g., Zubrod et al. 2015; Rossi et al. 2017).

Quantitative polymerase chain reactions (PCR) are an extension of the traditional PCR method, which makes it possible to measure the exponential amplification of PCR products via the emission of fluorescence signals in real time (Kubista et al. 2006). Quantification is achieved via a fluorescent dye that binds to

C. Baschien (✉)
Leibniz-Institute DSMZ – German Collection of Microorganisms and Cell Cultures, Braunschweig, Germany
e-mail: chb14@dsmz.de

S. C. Carl
Leibniz-Institute DSMZ – German Collection of Microorganisms and Cell Cultures, Braunschweig, Germany

Umweltbundesamt, Berlin, Germany
e-mail: steffen.carl@uba.de

F. Bärlocher et al. (eds.), *Methods to Study Litter Decomposition*,
https://doi.org/10.1007/978-3-030-30515-4_36

double-stranded DNA (e.g., SYBR Green) or by hybridization of a fluorescently labelled probe to sequence-specific DNA.

Probe-based assays are also referred to as 5′–3′ exonuclease assays since probes are double labelled with a fluorescent reporter at the 5′ end of the probe sequence (e.g., 6-carboxyfluorescein, also known as FAM) and with a quencher at the 3′ end (e.g., "black hole quencher" BHQ-1), which suppresses fluorescence emission by its proximity to the reporter. The TaqMan® or hydrolysis probe (Holland et al. 1991) is complementary to a taxon-specific region of the target sequence and hybridizes between the two primers, which ideally are also species specific. Due to the 5′-3′ exonuclease activity of the DNA polymerase during a PCR (Fig. 36.1), the reporter dye is released, and the probe sequence is cleaved so that the fluorescence signal increases in proportion to the amplification of the target gene.

For each sample, the cycle of quantification (C_q) can be determined where the amplification enters the exponential phase and the fluorescent signal clearly exceeds the background noise. The quantification of unknown DNA samples is then performed on the basis of standards with a defined copy number of the target sequence or with a known DNA concentration of the target organism. A linear calibration curve is derived from the standards (see Sect. 3.4), which enables calculating the DNA concentration based on the measured C_q values of the unknown samples.

1.1 Primer and Probe Design

Taxon-specific primers and probes must be designed for each target fungus. This requires knowledge of the target sequences (e.g., ITS regions of rDNA or other loci). The primers and probes can be designed using the NCBI Primer-Blast (https://blast.ncbi.nlm.nih.gov/Blast.cgi) according to the following criteria:

- Specificity, i.e., no match with sequences of other fungal species
- Melting temperature (T_m), which should be close to the annealing temperature used in the qPCR reactions (50–60 ± 3 °C)
- Product length, which should not significantly exceed 250 base pairs

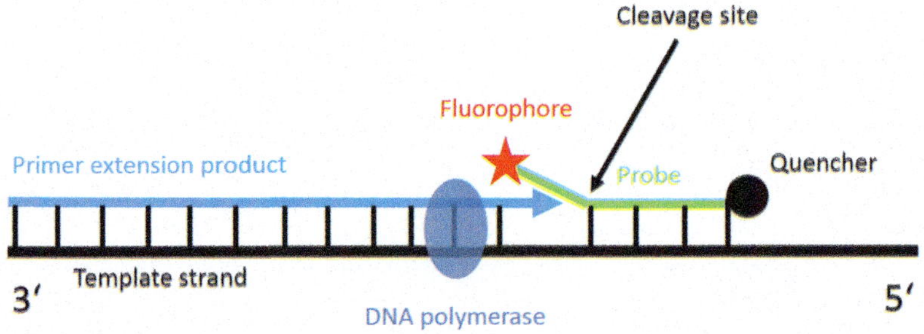

Fig. 36.1 Schematic diagram illustrating the principle of TaqMan® qPCR after Holland et al. (1991)

Further general recommendations for primer design include particularly the following points:

- The proportion of guanine and cytosine (GC content) should not exceed 65% of the total nucleotide content as higher proportions could inhibit the PCR, because the binding energy of GC pairs (three hydrogen bonds) is higher than for adenine and thymine pairs (two hydrogen bonds).
- Primers and probes should be checked for their secondary structure to avoid PCR inhibition caused, for example, by hairpin or loop formation. This check can be done by programs such as Oligo Calc (http://biotools.nubic.northwestern.edu/OligoCalc.html).
- *In silico* specificity of a new assay should be tested by aligning target sequences, primers, and probes.

For more information about primer design for qPCR applications and validation of primers, see Bustin and Huggett (2017). The design tools of the companies offering qPCR instruments often provide useful information.

1.2 Evaluation of New qPCR Assays

According to the Minimum Information for Publication of Quantitative Real-Time PCR Experiments (MIQE) guidelines (Bustin et al. 2009, 2010), newly designed assays should be evaluated based on five criteria, in addition to those listed in Sect. 1.1. First, the specificity of primers and probes must be empirically verified in every run of the qPCR assay by including positive and negative controls. The positive control (PC) is the target species, and the two negative controls are (1) DNA of nontarget fungi (non-target controls) and (2) samples without template DNA (no-template controls).

Secondly, linearity needs to be tested by establishing calibration curves of each run. Standards for calibration can be DNA of target fungi in known concentrations, oligonucleotides of the complete PCR amplicon, plasmid DNA constructs in known copy numbers, etc. Ideally, the linear part of the calibration curve covers more than the interval for the target amplicons to be quantified.

Thirdly, repeatability and reproducibility of qPCR assays are assessed by comparing the C_q values of replicates within a given run (intra-assay variance or repeatability) and between different runs (inter-assay variance or reproducibility), and precision can be tested by comparing qPCR results for pure-culture DNA (i.e., standards and positive controls) to DNA measurements with Qubit™ or PicoGreen®. Precision decreases with DNA concentration and copy number. Replicates inform about experimental variation and enable estimating the statistical significance of differences between samples.

Fourthly, the limit of detection (LOD) must be determined. It is defined as the lowest concentration where 95% of the target samples can be detected (Bustin et al. 2009). This is necessary because a zero value for the logarithmic calibration curve is not defined and, thus, meaningless when interpreting qPCR results. The detection limit is determined for each qPCR assay by diluting standards and samples to 10^{-8} ng/µl DNA or less. This is repeated at least twice in different runs. Usually, this will result in an on-off amplification signal at a certain concentration. The LOD is the next higher concentration level.

Finally, the overall efficiency of a qPCR run is determined from the log-linear slope of the calibration curve. Efficiency can also be determined more precisely by calculating the slope of every single amplification curve with LinRegPCR (Ruijter et al. 2009, 2013, 2014; Tuomi et al. 2010; software freely available at *linregpcr.nl*).

This software determines the fluorescence threshold (as the qPCR cycler software does) and also a window of linearity (WoL) over all samples of a qPCR run. The best-fitting linear regression is calculated within this WoL for every single curve, and the efficiency is derived from the slope of the curve (Fig. 36.2).

The qPCR efficiency is affected not only by the binding chemistry of primers and probes but also by impurities in DNA extracts. In addition, the quality of qPCR results depends on whether the amplification is impaired by remnants of proteins, sugars, foreign DNA, or humic substances. In environmental samples such as leaves, impurities are usually abundant and can lead to low PCR efficiencies (e.g., Kontanis and Reed 2006). This PCR inhibition can be visualized in the course of the amplification, which ideally produces sigmoid curves (Fig. 36.3).

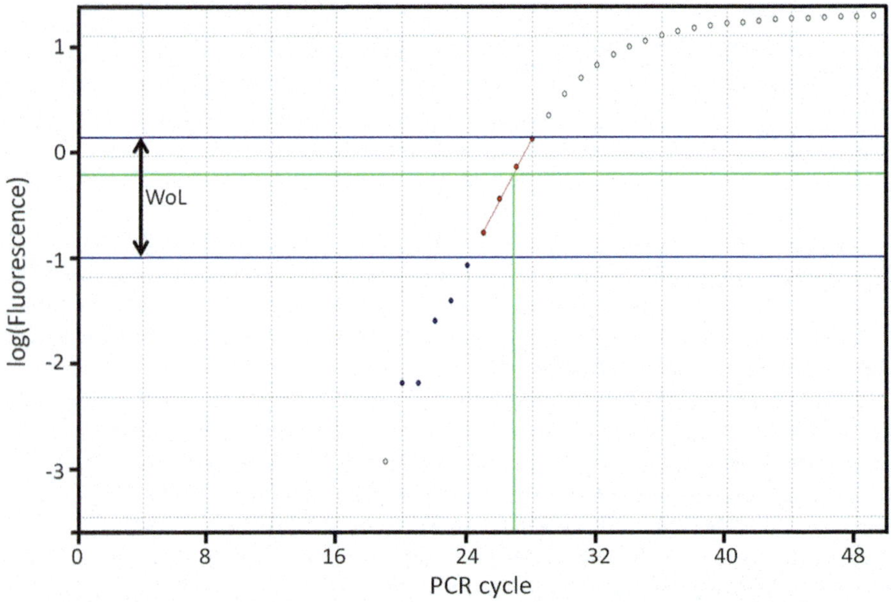

Fig. 36.2 Edited screenshot produced with the software LinRegPCR showing how PCR efficiency is determined by linear regression within a window of linearity (WoL). Green lines indicate the fluorescence threshold (intersection with y-axis) and the cycle of quantification C_q (intersection with x-axis)

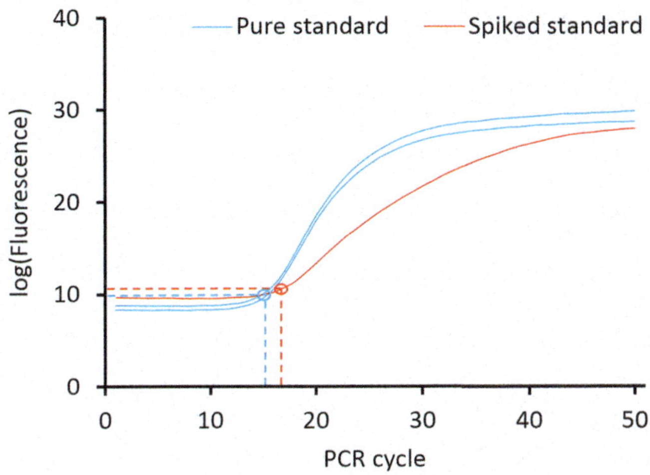

Fig. 36.3 Inhibition of a qPCR standard curve caused by spiking with nontarget DNA (here leaf DNA in a qPCR assay for *Cladosporium ramotenellum*). Curve of the contaminated standard (red) is significantly flatter than the sigmoid curves of the pure standard (blue), which leads to a difference in the calculation of the C_q value (circles with dashed lines). Fluorescence was measured at 465–510 nm

The degree of inhibition can be mathematically determined by measuring PCR efficiency as described above. A slope of 2 corresponds to 100% efficiency, because it indicates perfect doubling of the template DNA during amplification. Lower values indicate inhibition. In practice, PCR efficiency and inhibition are tested by comparing pure-culture DNA samples and samples spiked with impurities. Ideally, standards and samples have similar PCR efficiencies; otherwise, significant inaccuracies may occur (Töwe et al. 2010).

2 Equipment, Chemicals, and Solutions

2.1 Equipment and Material

- Pure fungal cultures
- Laminar flow hood
- Autumn-shed leaves, air-dried
- Litter bags (10, 1, 0.5, or 0.25 mm mesh size)
- Autoclave
- Freeze-dryer
- Homogenizer FastPrep®-24 MP Biomedicals (Santa Ana, CA, USA)
- Automatic pipettes (0.1–1000 µl)
- Pipette tips (0.1–1000 µl)

- Multichannel pipettes (0.5–100 μl)
- PCR strips
- Ninety-six-well plates for microplate reader and qPCR cycler
- Sealing foil for qPCR multi-well plates
- Plate centrifuge
- Microplate reader
- qPCR cycler with multichannel fluorescence equipment
- Vortex
- Microcentrifuge
- Thermomixer

2.2 *Chemicals*

- FastDNA™ SPIN Kit for Soil (MP Biomedicals, Santa Ana, CA, USA)
- Quant-iT™ PicoGreen® dsDNA Assay Kit (Thermo Fisher Scientific, Waltham, MA, USA)
- Sterile DNAse and pyrogen-free water
- Primers and probe for the qPCR reaction (probe labelled at the 5′ end with FAM and at the 3′ end with BHQ-1), which can be purchased from Eurofins Genomics (Ebersberg, Germany) or other suppliers
- qPCR Master Mix

3 Experimental Procedures

3.1 *Sample Preparation*

1. Grow pure fungal cultures on malt extract agar for 3 weeks.
2. Expose leaves in litter bags in the field, and collect as described in Chap. 6.
3. Rinse leaves to remove any adhering particles, freeze-dry, and store at −80 °C.

3.2 *DNA Extraction*

1. Transfer up to 500 mg fresh mass of pure fungal mycelium or colonized plant tissue to lysis tubes of the extraction kit, and freeze-dry overnight. These samples can be stored indefinitely at −80 °C.
2. Isolate DNA with FastDNA™ SPIN Kit for Soil following the manufacturer's instructions, including the recommended additional step of 5 min incubation at

55 °C and 550 rpm during the elution of DNA in 50 µl of DNAse and pyrogen-free water.

3. To increase DNA yield further, the extraction protocol can be extended by (i) cooling the lysis tubes with sample material in liquid nitrogen (~1 min) prior to homogenization with lysis buffer in the FastPrep®-24 Instrument (1 min, 6.5 m/s) and (ii) incubating for 2 h at 65 °C and 1400 rpm followed by another homogenization step for 1 min at 6.5 m/s during the lysis step.
4. Check quality of the DNA extracts by gel electrophoresis as in Chap. 34; NanoDrop® measurements allow estimating the contamination of extracts by proteins, sugars, salt, and phenols (Armbrecht 2013).
5. If the required number of analyses is high, distribute aliquots of samples and standards in PCR strips to allow the use of multichannel pipettes when preparing qPCR reactions.
6. Store the extracted DNA for up to 2 months at 4 °C or indefinitely at −80 °C.

3.3 DNA Amplification

1. Ensure that the final DNA concentration in qPCR reactions does not exceed 10 ng/µl (i.e., the final concentration in wells after adding all reagents) by checking the range of DNA concentrations (e.g., by measuring fluorescence of PicoGreen® or Qubit™ DNA) and diluting samples if necessary. Check recommendations of the qPCR Master Mix/Taq polymerase in use, and adapt the following instructions, if necessary, in particular the concentration of PCR reagents in step 1 and the temperature profile in step 5.
2. Prepare the qPCR Master Mix by adding for each 25 µl reaction 4 µl of Master Mix, 0.5 µl of each 10 pM primer, and 0.25 µl of 10 pM probe to 10.75 µl of water suitable for PCR. Minimize light exposure of photosensitive reagents, especially of the probe, and work under laminar flow to avoid contamination.
3. Pipette 18 µl of the Master Mix to each well of a microplate suitable for the qPCR cycler, and add 2 µl of extracted DNA.
4. Seal the microplate with a suitable foil to avoid evaporation, protect from exposure to light, and centrifuge for a few seconds.
5. Place the microplate in a qPCR cycler programmed for 15 min at 95 °C (ramp 4.4 °C/s) for initial denaturation, followed by 50 cycles of denaturation at 95 °C for 15 s and primer-probe annealing at 60 °C for 1 min (ramp 2.2 °C/s). End the program with a cooling step at 40 °C for 30 s (ramp 2.2 °C/s). Amplification is automatically measured in real time for every PCR reaction via a fluorescence signal between 465 and 510 nm.

3.4 Quantification

1. Follow the instructions of the qPCR cycler to determine the cycle of quantification (C_q), which is normally automated. Document which method is used for the C_q determination (e.g., "second derivative maximum").
2. To prepare a calibration curve (Fig. 36.4), choose five or more dilutions of a pure-culture DNA extract within the expected range of the samples. For each dilution, run at least two replicates showing clearly sigmoid and very similar curves (Fig. 36.5). Document which standards are used, and calculate the coefficient of determination (R^2) of the calibration curve.
3. The DNA of unknown samples is quantified against this calibration curve. Normally, this is done automatically by the cycler program, but it can also be calculated separately (e.g., in a spreadsheet) to enhance control over the analysis.

3.5 qPCR Establishment and Validation

- Perform two or more establishment runs, where standards are measured at least in triplicate. The serial dilution of pure-culture DNA should be chosen to yield regular distances between the amplification curves of the different dilutions (Fig. 36.5). For calibration, choose only standards with reliable curves (i.e., two or more nearly identical sigmoid curves).
- Check linearity, variance, and precision according to Sect. 1.2.
- Determine the LOD by identifying the dilution where at least 95% of all qPCR reactions are quantifiable. Always quantify at least one dilution beneath the LOD, to check the sensitivity of every run.

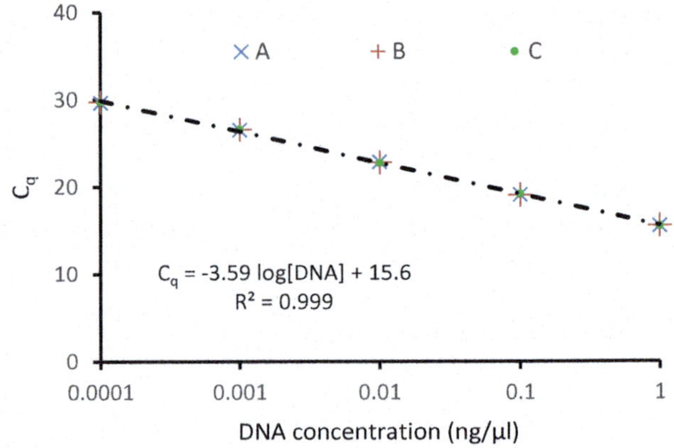

Fig. 36.4 qPCR calibration curve and equation over five dilutions of pure-culture DNA obtained from the fungus *Cladosporium ramotenellum*; A, B, and C refer to technical replicates; C_q: cycle of quantification

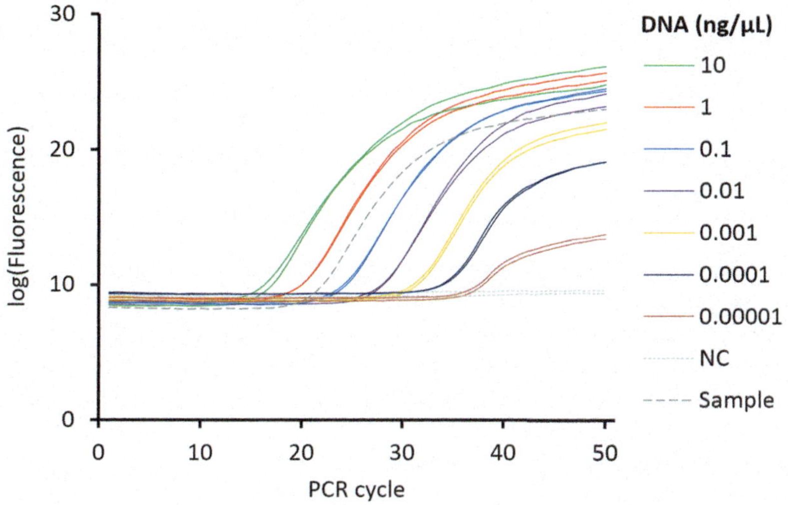

Fig. 36.5 Standard amplification curves of a qPCR assay (10^1–10^{-5} ng/µl) on DNA extracted from *Neonectria lugdunensis*. The curve labelled 'Sample' corresponds to a DNA extract from an alder leaf colonized by *N. lugdunensis* and two other freshwater fungi grown under laboratory conditions. Negative controls (NC) include non-target DNA of other fungi as well as no-template controls

- For the same reason, also document the overall PCR efficiency, which is normally given by the cycler program but can also be manually calculated as the slope of the standard curve:

$$E = 10^{-\frac{1}{slope}} - 1 \tag{36.1}$$

- Include controls and standards in every run at least in duplicate.
- Discard results of a complete run, if any negative control (NC) shows a significant amplification curve.
- As a positive control, include a leaf sample inoculated with the target organism at a DNA concentration close to the LOD of the assay. If this is not possible, pure-culture DNA can be used.
- Include as negative controls sterilized leaves that are treated exactly like the samples.
- To assess potential PCR inhibition, spike pure-culture DNA standards with all possible constituents of the environmental sample except for the target DNA. Check inhibition, and assess PCR efficiency as described in Sect. 1.2.
- Choose an experimental design that ensures enough sample DNA from colonized leaves is available for the establishment runs. This enables initial checking of how well the qPCR assay works, which concentrations to expect, and whether a dilution or concentration is necessary prior to the actual analysis.
- In sample runs, every DNA extract is best measured in triplicate or at least in duplicate, especially standards and controls. Also repeat runs at least twice to

ensure consistent results, but be aware of inter-run variability, and consider variability when interpreting results. Standards, controls, and samples need to be analyzed within the same run.

4 Final Remarks

A qPCR assay is only as good as its specificity. Specificity depends on the size and quality of data bases, which are constantly updated. Therefore, assay specificity should be checked at least every 1–2 months.

Be aware that copy numbers of the amplified target sequences do not directly correlate with the number of cells in a sample, especially when the target sequence belongs to the widely used ITS region of the rRNA operon (Schoch et al. 2012), which has multiple copies in a single fungal genome.

DNA concentrations used for calibration should be measured by means of quantification kits with DNA-binding reagents (e.g., PicoGreen® or Qubit™) following the manufacturer's instructions.

Finally, appropriate documentation is important to ensure publication of clear and complete results. To achieve this, the MIQE standards (Bustin et al. 2009, 2010) should be followed. For field studies, complementary experiments under controlled conditions (e.g., laboratory experiments) may help to validate the field results.

References

Armbrecht, M. (2013). Detection of contamination in DNA and protein samples by photometric measurements. *Eppendorf* Application Note No. 279: 1–6.

Bärlocher, F. (1992). *The ecology of aquatic hyphomycetes*. Berlin: Springer.

Bustin, S., & Huggett, J. (2017). qPCR primer design revisited. *Biomolecular Detection and Quantification, 14*, 19–28.

Bustin, S. A., Benes, V., Garson, J. A., Hellemans, J., Huggett, J., Kubista, M., Mueller, R., Nolan, T., Pfaffl, M., Shipley, G. L., Vandesompele, J., & Wittwer, C. T. (2009). The MIQE guidelines: Minimum information for publication of quantitative real-time PCR experiments. *Clinical Chemistry, 55*, 611–622.

Bustin, S. A., Beaulieu, J. F., Huggett, J., Jaggi, R., Kibenge, F. S., Olsvik, P. A., Penning, L. C., & Toegel, S. (2010). MIQE precis: Practical implementation of minimum standard guidelines for fluorescence-based quantitative real-time PCR experiments. *BMC Molecular Biology, 11*, 74.

Fernandes, I., Pascoal, C., & Cássio, F. (2011). Intraspecific traits change biodiversity effects on ecosystem functioning under metal stress. *Oecologia, 166*, 1019–1028.

Gehesquière, B., D'Haeyer, S., Pham, K. T. K., Van Kuik, A. J., Maes, M., Höfte, M., & Heungens, K. (2013). qPCR assays for the detection of *Cylindrocladium buxicola* in plant, water, and air samples. *Plant Disease, 97*, 1082–1090.

Gessner, M. O., Gulis, V., Kuehn, K. A., Chauvet, E., & Suberkropp, K. (2007). Fungal decomposers of plant litter in aquatic ecosystems. In C. P. Kubicek & I. S. Druzhinina (Eds.), *The Mycota, Vol. 4: Environmental and microbial relationships* (2nd ed., pp. 301–324). Berlin: Springer.

Holland, P. M., Abramson, R. D., Watson, R., & Gelfand, D. H. (1991). Detection of specific polymerase chain reaction product by utilizing the 5′–3′ exonuclease activity of *Thermus aquaticus* DNA polymerase. *Proceedings of the National Academy of Sciences of the USA, 88*, 7276–7280.

Kontanis, E. J., & Reed, F. A. (2006). Evaluation of real-time PCR amplification efficiencies to detect PCR inhibitors. *Journal of Forensic Sciences, 51*, 795–804.

Kubista, M., Andrade, J. M., Bengtsson, M., Forootan, A., Jonák, J., Lind, K., Radek, S., Sjöback, R., Sjögreen, B., Strömbom, L., Ståhlberg, A., & Zoric, N. (2006). The real-time polymerase chain reaction. *Molecular Aspects of Medicine, 27*, 95–125.

Lefèvre, E., Jobard, M., Venisse, J. S., Bec, A., Kagami, M., Amblard, C., & Sime-Ngando, T. (2010). Development of a real-time PCR assay for quantitative assessment of uncultured freshwater zoosporic fungi. *Journal of Microbiological Methods, 81*, 69–76.

Rossi, F., Artigas, J., & Mallet, C. (2017). Structural and functional responses of leaf-associated fungal communities to chemical pollution in streams. *Freshwater Biology, 62*, 1207–1219.

Ruijter, J. M., Ramakers, C., Hoogaars, W. M. H., Karlen, Y., Bakker, O., Van den Hoff, M. J. B., & Moorman, A. F. M. (2009). Amplification efficiency: Linking baseline and bias in the analysis of quantitative PCR data. *Nucleic Acids Research, 37*, e45.

Ruijter, J. M., Pfaffl, M. W., Zhao, S., Spiess, A. N., Boggy, G., Blom, J., Rutledge, R. G., Sisti, D., Lievens, A., De Preter, K., Derveaux, S., Hellemans, J., & Vandesompele, J. (2013). Evaluation of qPCR curve analysis methods for reliable biomarker discovery: Bias, resolution, precision, and implications. *Methods, 59*, 32–46.

Ruijter, J. M., Lorenz, P., Tuomi, J. M., Hecker, M., & van den Hoff, M. J. B. (2014). Fluorescent-increase kinetics of different fluorescent reporters used for qPCR depend on monitoring chemistry, targeted sequence, type of DNA input and PCR efficiency. *Microchimica Acta, 181*, 1689–1696.

Schoch, C. L., Seifert, K. A., Huhndorf, S., Robert, V., Spouge, J. L., Levesque, C. A., Chen, W., & Fungal Barcording Consortium (2012). Nuclear ribosomal internal transcribed spacer (ITS) region as a universal DNA barcode marker for Fungi. *Proceedings of the National Academy of Sciences of the USA, 109*, 6241–6246.

Töwe, S., Kleineidam, K., & Schloter, M. (2010). Differences in amplification efficiency of standard curves in quantitative real-time PCR assays and consequences for gene quantification in environmental samples. *Journal of Microbiological Methods, 82*, 338–341.

Tsui, C. K., Baschien, C., & Goh, T. K. (2016). Biology and ecology of freshwater fungi. In D. W. Li (Ed.), *Biology of Microfungi* (pp. 285–313). Cham: Springer.

Tuomi, J. M., Voorbraak, F., Jones, D. L., & Ruijter, J. M. (2010). Bias in the C_q value observed with hydrolysis probe based quantitative PCR can be corrected with the estimated PCR efficiency value. *Methods, 50*, 313–322.

Zubrod, J. P., Englert, D., Feckler, A., Koksharova, N., Konschak, M., Bundschuh, R., Schnetzer, N., Englert, K., Schulz, R., & Bundschuh, M. (2015). Does the current fungicide risk assessment provide sufficient protection for key drivers in aquatic ecosystem functioning? *Environmental Science & Technology, 49*, 1173–1181.

Chapter 37
Metabarcoding of Litter-Associated Fungi and Bacteria

Sofia Duarte, Christian Wurzbacher, and Sahadevan Seena

Keywords DNA barcoding · High-throughput sequencing · ITS · Metagenomics · Microbial community analysis · Microbial decomposers · Microbiome · Next-generation sequencing · Nucleic acid extraction · PCR

1 Introduction

Litter-associated microorganisms play key roles in litter decomposition in forested streams by decomposing and transferring energy from dead organic matter to higher trophic levels (Suberkropp 1998). Although both fungi and bacteria participate in this process, a greater role has been attributed to fungi, particularly to aquatic hyphomycetes. Much of the knowledge on the diversity of aquatic hyphomycetes has been based in the identification of asexual spores (Chap. 26). For leaf-associated bacteria, traditional cultivation-based diversity studies have been scarce (e.g. Suberkropp and Klug 1976). Regardless, assessments based on reproductive potential or cultivability fail to adequately present the true diversity of fungi or bacteria on decomposing litter.

S. Duarte (✉)
Department of Biology, Centre of Molecular and Environmental Biology (CBMA), University of Minho, Braga, Portugal
e-mail: sduarte@bio.uminho.pt

C. Wurzbacher
Chair of Urban Water Systems Engineering, Technical University of Munich, Garching, Germany
e-mail: c.wurzbacher@tum.de

S. Seena
Department of Life Sciences, MARE – Marine and Environmental Sciences Centre, University of Coimbra, Coimbra, Portugal
e-mail: seena.sahadevan@uc.pt

F. Bärlocher et al. (eds.), *Methods to Study Litter Decomposition*,
https://doi.org/10.1007/978-3-030-30515-4_37

To circumvent the shortcomings of traditional microbiological methods and conventional microscopy, a range of DNA-based techniques have been optimized in the last 15 years to assess the composition of microbial communities on decomposing litter: terminal restriction fragment length polymorphism (T-RFLP; Chap. 34), denaturing gradient gel electrophoresis (DGGE; Chap. 35) and clone libraries (Seena et al. 2008). However, all these molecular techniques suffer from low resolution and incomplete assessment of the present microbial diversity. For instance, the construction of clone libraries accurately determines the identity and phylogeny of litter-associated fungi (Seena et al. 2008), but diversity will generally be underestimated due to the low number of economically analysable clones. Similarly, subsequencing DGGE bands can allow identification of the dominant taxa yet will suffer from band impurities and, due to the inherent limitations of the DGGE, may miss low-abundance taxa.

On the other hand, the combination of DNA barcoding, for which selected DNA marker regions (barcodes) are used as proxies of species identification (Hebert et al. 2003), with high-throughput or "next-generation" sequencing (HTS or NGS), enables researchers to improve the accuracy of current biodiversity assessments. This recently developed method is called metabarcoding (synonyms: metagenomics, microbiota profiling, microbiome analysis) and provides a detailed profile of microbial communities on an unprecedented scale (Roesch et al. 2007) by massively parallel sequencing short reads (100–800 bp) to read lengths of tens of kilobases (kb). With this approach, genetic information can be obtained directly from leaf-litter samples by using standardized DNA barcode markers targeting fungi and bacteria (e.g. Duarte et al. 2015; Heino et al. 2014; Wurzbacher et al. 2016a).

Briefly, analysing the diversity of litter-associated fungi or bacteria by metabarcoding includes the following:

(i) DNA extraction from litter with mixed microbial populations, often by using a commercial kit (easier but not mandatory; e.g. PowerSoil DNA Isolation Kit (Duarte et al. 2015; Wurzbacher et al. 2016a) and RNA/DNA Purification Kit from Norgen (Duarte et al. 2017))
(ii) Amplification of a barcode sequence (a gene fragment specific for fungal or bacterial taxa, present in all members of the community and amplifiable with universal primers in a PCR reaction). Preferred markers include the internal transcribed spacer region (ITS) or 18S ribosomal RNA (rRNA) gene region for fungi and the 16S rRNA gene region for bacteria, respectively
(iii) Parallel sequencing of the amplicons by an HTS platform (e.g. 454 pyrosequencing, Duarte et al. 2015, Wurzbacher et al. 2016a; Illumina MiSeq, Duarte et al. 2017)
(iv) Processing of the generated sequences (~10^4 to 10^8 reads) by an appropriate bioinformatics pipeline (Chap. 62)
(v) Comparison of the processed sequences with reference sequences on genetic databases (e.g. GenBank, UNITE, SILVA) (Fig. 37.1)

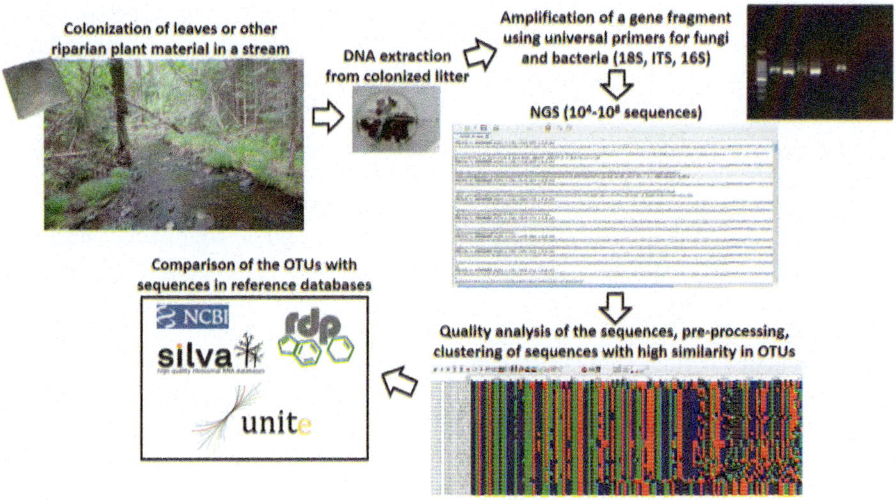

Fig. 37.1 Steps for metabarcoding analysis of the microbial diversity associated with decomposing litter in freshwater ecosystems

In the last decade, there has been an incredible reduction in HTS costs with an exponential increase of outputs, leading to a democratization and broad applicability of sequencing as a tool for biodiversity assessments. Most laboratories analysing litter-associated microbial communities opt to contract the service to external companies that are continuously updating their equipment and using the most up-to-date HTS platforms. Thus, the methodology presented here focuses on sample preparation that will influence the output of HTS (sample collection, DNA extraction, markers used in PCR amplification), rather than the HTS technology, whose sample preparation and library construction may change considerably between platforms (see van Dijk et al. (2014) for details of NGS platforms).

2 Equipment, Chemicals and Solutions

2.1 Equipment and Materials

- Autumn-shed leaves or other plant materials, air-dried
- Litter bags (10 × 10 cm, 10, 1 or 0.5 mm mesh size)
- Freeze dryer
- Deep freezer (−20 °C or − 80 °C)

- Lysis materials or equipment: mortar and pestle or Eppendorf tubes and micropestles or MoBio vortex adapter tube holder (MoBio Laboratories, Solana Beach, CA, USA) or FastPrep FP120 instrument (Qbiogene, Heidelberg, Germany)
- NanoDrop ND-1000 (VWR, International, PA, USA) or Qubit (Thermo Fisher Scientific, MA, USA)
- Automatic pipettes: ranges of 0.1–10, 10–100, 20–200 and 100–1000 μl
- Filter pipette tips: 0.1–10, 10–100, 20–200 and 100–1000 μl
- PCR workstation (preferentially with UV light and a continuous sterilized air flow)
- Microwave oven
- PCR thermocycler
- Equipment for running agarose gels (casting tray, electrophoresis chamber and current power supply)
- UV transilluminator
- NGS sequencer (external company)

2.2 Chemicals and Solutions

- RNAlater (Sigma-Aldrich, MO, USA)
- RNaseZap™ RNase Decontamination Solution (Thermo Fisher Scientific, MA, USA)
- DNAgard® Tissue (Biomatrica, CA, USA)
- DNA and RNA extraction kits: DNeasy PowerSoil Kit (Qiagen, Venlo, Netherlands) or RNA/DNA purification kit from Norgen (Norgen Biotek Corp., Ontario, Canada) or other protocols that have proven to be efficient in extracting DNA from decomposing litter
- PCR reagents: Taq DNA polymerase (5 U/μl), 10× Taq DNA polymerase buffer, dNTPs (10 mM), $MgCl_2$ (50 mM), forward and reverse primers (working solutions at a concentration of 10 μM; see Table 37.1 for details of suitable primers) and nuclease-free water. Nowadays, most suppliers provide premixed formulations containing all necessary components at optimum concentrations, except for primers (e.g. GoTaq Green Master Mix, Promega Corporation, WI, USA). Premixed formulations save time and reduce contaminations due to a reduced number of pipetting steps for PCR set-up. A mix containing a high-fidelity Taq DNA polymerase may be used to prevent GC bias and higher PCR substitution errors (e.g. Accuzyme™ Mix, Bioline, London, UK)
- Fluorescent DNA stain (e.g. GelStar, Lonza, ME, USA; GelRed, Biotium, Inc., CA, USA)
- RT-PCR kit (e.g. OneStep RT-PCR kit, Qiagen, Venlo, Netherlands)
- Agarose (molecular biology grade)
- 50× TAE (2.0 M Tris, 1.0 M acetic acid, 50 mM EDTA, pH 8.3)
- 1× TAE (20 ml 50× TAE and 980 ml water)
- Agarose gel-loading dye 6×
- DNA ladder from 50 to 1500 bp

Table 37.1 Target organisms, primers and NGS platforms used in metabarcoding studies on litter-associated fungi and bacteria

Target organism	Target region	Primers used	NGS platform	References
Bacteria	16S rRNA gene	8F[a]/D'[b]	GS FLX 454	Heino et al. (2014) and Mykrä et al. (2017)
Fungi	ITS	ITS1F[c]/ITS4[d]	GS FLX 454	Tolkkinen et al. (2013, 2015a, b), Heino et al. (2014), and Mykrä et al. (2016, 2017)
	ITS1	ITS1F[c]/58A2R[e]	Illumina MiSeq	Duarte et al. (2017)
	ITS2	ITS3[d]/ITS4[d]	GS FLX 454 titanium	Duarte et al. (2015) and Fernandes et al. (2015)
	18S rRNA gene	nu-SSU-0817[f]/ nu-SSU-1536[f]	GS junior 454	Wurzbacher et al. (2016a)
Eukaryotes	26S rRNA gene (D1/D2 variable region)	NLF184cw[g]/ Euk573rev[g]	GS FLX 454 titanium	Wurzbacher et al. (2016b)

[a]Edwards, U. et al. (1989). Nucleic Acids Research, 17, 7843–7853
[b]Ritari, J. et al. (2012). BMC Microbiology, 12, 121
[c]Gardes, M. and Bruns, T.D. (1993). Molecular Ecology, 2, 113–118
[d]White, T.J. et al. (1990). PCR protocols: a guide to methods and applications (pp. 315–322). Academic Press. New York
[e]Martin, K.J. and Rygiewicz, P.T. (2005). BMC Microbiology, 5, 28
[f]Borneman, J. and Hartin, R.J. (2000). Applied and Environmental Microbiology, 66, 4356–4360
[g]van der Auwera et al. (1994). FEBS Letters, 338, 133–136

3 Experimental Procedures

3.1 Sample Preparation for NGS (Sample Collection, DNA and/or RNA Extraction, PCR Amplification)

1. Expose leaves or other plant materials in litter bags in the field, and collect as in Chap. 6.
2. Rinse litter to remove any adhering particles, freeze-dry and store at −20 °C (up to 6 months) or −80 °C (indefinitely) or store in DNAgard® Tissue for up to 6 months at room temperature.
3. If RNA is to be extracted, store leaf disks or other plant materials in Eppendorf tubes containing RNAlater immediately after collection.
4. To extract DNA, use up to 50 mg of freeze-dried litter tissue for DNA or the tissue stored in RNAlater (e.g. 4 to 8 12 mm-diameter leaf disks), if RNA is also to be extracted.
5. Use one of the commercial kits suggested to extract the DNA and/or RNA, following the manufacturer's instructions or any other protocol that works with your samples.

6. The extracted DNA can be stored for up to 2 months at 4 °C, 6 months at −20 °C or indefinitely at −80 °C. RNA should be immediately reverse transcribed into cDNA.
7. Check DNA and RNA concentrations, as well as A260/280 and A260/230, using NanoDrop. Use Qubit for more precise and accurate measurements.
8. Synthesize the RNA into cDNA using a RT-PCR kit, following the manufacturer's instructions. cDNA can then be stored using the same conditions used for DNA.
9. Optional: HTS providers accept DNA templates. Most of providers offer commonly used primers for an in-house PCR at very low rates (e.g. bacterial 16S, fungal ITS). Primer customization is possible at additional costs.
10. Amplify the DNA and/or cDNA targeting the same region and using the same markers that will be used in HTS. Prepare all PCR reactions on ice and in replicates (triplicates are recommended). Avoid too low and too high template DNA concentration due to the higher probability of contaminations (Salter et al. 2014) and chimera formation (D'Amore et al. 2016), respectively.
11. Prepare a 2% agarose gel to precheck the suitability of the samples to be run in the NGS sequencer (2% agarose gel, 1 g agarose in 50 ml 1x × TAE buffer; microwave, add DNA stain, allow to gel ~30 min, for a gel tray size of 7 × 10 cm, W × L). Load the amplicons in the wells of the solidified gel. Run the gel in 1× TAE buffer at 90 V for 30 min, and view on a UV transilluminator.
12. Pool successfully amplified technical PCR replicates into one well/tube.
13. Ship the samples that have passed the amplification control on dry ice to the company where the HTS service was contracted.
14. Optional: It is also possible to prepare the library in your own lab. It requires an initial investment in barcoded secondary primers but may help to save money if you have multiple projects. There are several publications on this, and the basic principle is outlined by Illumina: https://support.illumina.com/content/dam/illumina-support/documents/documentation/chemistry_documentation/16s/16s-metagenomic-library-prep-guide-15044223-b.pdf. The barcoding kit should be then replaced by your own customized primers.

3.2 *Best Practices for Preparing Samples for Metabarcoding*

1. Cleaning: clean the pre-PCR (DNA and RNA extraction and quantification) and PCR areas on a daily basis. Use 0.5% (w/v) sodium hypochlorite (10% bleach) to discard pipette tips and PCR products that are not needed anymore. If RNA is going to be extracted, clean all areas, including the pipettes, with an RNase decontamination solution, such as RNaseZap.
2. Laboratory space: physically separate the areas where pre-PCR procedures are performed from the space where the PCRs are performed. To avoid cross contaminations, sterilized PCR workstations should be used, and all pipetting steps

(both pre-PCR and PCR) should be conducted using filter tips. Change gloves frequently, in particular during RNA extractions.

3. Controls: use negative and positive controls during the pre-PCR and PCR procedures. Negative controls will detect the putative presence of contaminations and should be integrated in the DNA extraction step following all the procedure including HTS, while positive controls evaluate the efficiency of your metabarcoding approach in detecting the species in your samples. For instance, mock communities can be used as positive controls (see below).
4. Mock communities: evaluate the sensitivity of your metabarcoding approach by using a mock community with mixtures of known species, isolated and maintained in your laboratory, which already have DNA sequences in genetic databases.
5. Environmental samples: in environmental samples, the identity of the sequences is not known a priori and may contain high GC contents. This can interfere with PCR amplification due to the low efficiency of template dissociation. The use of an adapted polymerase or a denaturant such as acetamide or DMSO in the PCR can alleviate such conditions.
6. PCR bias: mixed templates in PCR often cause bias in the final products, which may affect the final results. This bias can be mitigated by (i) using replicate PCR reactions to avoid random amplification biases, (ii) lowering the number of PCR cycles to reduce the artificial shifting of taxa abundances and (iii) redesigning primers if primer mismatches are expected for certain taxa.

References

Duarte, S., Bärlocher, F., Trabulo, J., Cássio, F., & Pascoal, C. (2015). Stream-dwelling fungal decomposer communities along a gradient of eutrophication unraveled by 454 pyrosequencing. *Fungal Diversity, 70*, 127–148.

Duarte, S., Cássio, F., Pascoal, C., & Bärlocher, F. (2017). Taxa-area relationship of aquatic fungi on deciduous leaves. *PLoS One, 12*(7), e0181545.

D'Amore, R., Ijaz, U. Z., Schirmer, M., Kenny, J. G., Gregory, R., Darby, A. C., Shakya, M., Podar, M., Quince, C., & Hall, N. (2016). A comprehensive benchmarking study of protocols and sequencing platforms for 16S rRNA community profiling. *BMC Genomics, 17*, 55.

Fernandes, F., Pereira, A., Trabulo, J., Pascoal, C., Cássio, F., & Duarte, S. (2015). Microscopy- or DNA-based analyses: Which methodology gives a truer picture of stream-dwelling decomposer fungal diversity? *Fungal Ecology, 18*, 130–134.

Hebert, P. D. N., Cywinska, A., Ball, S. L., & de Waard, J. R. (2003). Biological identifications through DNA barcodes. *Proceedings of the Royal Society of London B. Biological Sciences, 270*, 313–321.

Heino, J., Tolkkinen, M., Pirttilä, A. M., Aisala, H., & Mykrä, H. (2014). Microbial diversity and community–environment relationships in boreal streams. *Journal of Biogeography, 41*, 2234–2244.

Mykrä, H., Tolkkinen, M., Markkola, A. M., Pirttilä, A.-M., & Muotka, T. (2016). Phylogenetic clustering of fungal communities in human-disturbed streams. *Ecosphere, 7*(3), e01316.

Mykrä, H., Tolkkinen, M., & Heino, J. (2017). Environmental degradation results in contrasting changes in the assembly processes of stream bacterial and fungal communities. *Oikos, 126*, 1291–1298.

Roesch, L. F., Fulthorpe, R. R., Riva, A., Casella, G., Hadwin, A. K., Kent, A. D., Daroub, S. H., Camargo, F. A., Farmerie, W. G., & Tripplet, E. W. (2007). Pyrosequencing enumerates and contrasts soil microbial diversity. *The ISME Journal, 1*, 283.

Salter, S. J., Cox, M. J., Turek, E. M., Calus, S. T., Cookson, W. O., Moffatt, M. F., Turner, P., Parkhill, J., Loman, N. J., & Walker, A. W. (2014). Reagent and laboratory contamination can critically impact sequence-based microbiome analyses. *BMC Biology, 12*, 87.

Seena, S., Wynberg, N., & Bärlocher, F. (2008). Fungal diversity during leaf decomposition in a stream assessed through clone libraries. *Fungal Diversity, 30*, 1–14.

Suberkropp, K. (1998). Microorganisms and organic matter decomposition. In R. J. Naiman & R. E. Bilby (Eds.), *River ecology and management: Lessons from the Pacific coastal ecoregion* (pp. 120–143). Springer: New York.

Suberkropp, K., & Klug, M. J. (1976). Fungi and bacteria associated with leaves during processing in a woodland stream. *Ecology, 57*, 707–719.

Tolkkinen, M., Mykrä, H., Markkola, A.-M., Aisala, H., Vuori, K.-M., Lumme, J., Pirttilä, A. M., & Muotka, T. (2013). Decomposer communities in human-impacted streams: Species dominance rather than richness affects leaf decomposition. *Journal of Applied Ecology, 50*, 1142–1151.

Tolkkinen, M., Mykrä, H., Markkola, A.-M., & Muotka, T. (2015a). Human disturbance increases functional but not structural variability of stream fungal communities. *Functional Ecology, 29*, 1569–1577.

Tolkkinen, M., Mykrä, H., Annala, M., Markkola, A.-M., Vuori, K.-M., & Muotka, T. (2015b). Multi-stressor impacts on fungal diversity and ecosystem functions in streams: Natural vs. anthropogenic stress. *Ecology, 96*, 672–683.

van Dijk, E. L., Auger, H., Jaszczyszyn, J., & Thermes, C. (2014). Ten years of next-generation sequencing technology. *Trends in Genetics, 30*, 418–426.

Wurzbacher, C., Grimmett, I. J., & Bärlocher, F. (2016a). Metabarcoding-based fungal diversity on coarse and fine particulate organic matter in a first-order stream in Nova Scotia, Canada. *F1000Research, 4*, 1378.

Wurzbacher, C., Warthmann, N., Bourne, E., Attermeyer, K., Allgaier, M., Powell, J. R., Detering, H., Mbedi, S., Grossart, H.-P., & Monaghan, M. T. (2016b). High habitat-specificity in fungal communities in oligo-mesotrophic, temperate Lake Stechlin (North-East Germany). *MycoKeys, 16*, 17–44.

Chapter 38
Identifying Active Members of Litter Fungal Communities by Precursor rRNA

Martina Šturšová and Petr Baldrian

Keywords DNA/RNA ratio · Fungal community analysis · Fungal transcriptional activity · ITS · Microbial metabolic activity · Next-generation sequencing · Nucleic acid extraction · PCR · rRNA · Transcription

1 Introduction

Litter decomposition is a dynamic process in both terrestrial and aquatic environments (Bärlocher 2016; Baldrian 2017). This is reflected by successional changes in fungal communities during the course of decomposition, which is paralleled by chemical and physical changes of the litter (Voříšková and Baldrian 2013; Peršoh 2015), corresponding to changes in nutrient availability and interspecific competition among fungi (van der Wal et al. 2013). Frequently, the DNA of fungal endophytes or early colonizers can be recovered from plant litter, including dead wood, even at very late stages of decomposition (Rajala et al. 2011; Voříšková and Baldrian 2013). As a result, these fungi have often been regarded as decomposers. However, recovering DNA from a substrate does not reveal whether the corresponding taxa are metabolically active or simply present as inactive hyphae or spores. For example, in a study on decaying wood, only a small fraction of fungal community DNA was shown to be transcriptionally active, and species turnover within the active community was much faster than for the whole community (Rajala et al. 2011). Thus, there is both a need and an opportunity to distinguish fungi identified by DNA from metabolically active fungi – those transcribing RNA molecules.

To produce proteins, fungi need ribosomes with their structural backbone, the rRNA molecules. These molecules are present as a cassette that is transcribed as a

M. Šturšová (✉) · P. Baldrian
Laboratory of Environmental Microbiology, Institute of Microbiology of the Czech Academy of Sciences, Praha, Czech Republic
e-mail: stursova@biomed.cas.cz; baldrian@biomed.cas.cz

F. Bärlocher et al. (eds.), *Methods to Study Litter Decomposition*,
https://doi.org/10.1007/978-3-030-30515-4_38

precursor RNA molecule, including the regions of internal transcribed spacers (ITS) 1 and 2. These ITS regions are later degraded during rRNA production. Importantly, the ITS regions serve as excellent barcodes of fungal taxonomic identity (Lindahl et al. 2013), suggesting that identifying the ITS regions of RNA makes it possible to identify fungal taxa that are transcriptionally active. Furthermore, it is possible to compare the relative abundance of individual fungal taxa in the RNA and DNA pools to estimate their relative activity (Baldrian et al. 2012). This method of specifically identifying active fungal taxa is outlined below.

The approach was successfully used to identify active fungi in both terrestrial and aquatic environments (Grimmett et al. 2013; Freedman et al. 2015; Žifčáková et al. 2016). While the activity of microbiota can also be estimated by following total transcription (Žifčáková et al. 2016) or metaproteome composition (Schneider et al. 2012), these alternative methods are not suitable for fine-grained taxonomic classification of a broad range of fungi. Active fungi can be also identified using stable isotope probing (Štursová et al. 2012); however, using artificial substrates is likely a crude indicator of the actual activity in decomposing litter.

Comparing RNA and DNA pools can also be applied to bacteria, using 16S rRNA and DNA molecules (Blazewicz et al. 2013) to compare genome and ribosome pools of bacteria. Even though the ribosome/genome ratio generally increases with bacterial activity, there are many exceptions. For example, dormant spores of some bacteria are loaded with ribosomes (Blazewicz et al. 2013). Consequently, the general application of RNA/DNA ratios as a measure of bacterial activity is rather limited. More convincingly, the RNA/DNA ratio can be used as a measure of transcription of individual genes in litter, such as cellulases or β-glucosidases (Baldrian et al. 2012; Pathan et al. 2017).

2 Equipment, Chemicals, and Solutions

2.1 Equipment and Material

- Deciduous leaves collected from a stream or undisturbed forest topsoil including the litter layer
- Soil corer (optional)
- Scissors
- Spatulas
- Beakers
- Sterile polypropylene tubes (1.5 ml, DNase- and RNase-free)
- Containers for liquid nitrogen and dry ice
- Paper towels
- Laboratory balance (±10 mg precision)
- −80 °C freezer

- RNA PowerSoil Total RNA Isolation Kit (MoBio Laboratories, Qiagen, Carlsbad, CA, USA)
- RNA PowerSoil DNA Elution Accessory Kit (MoBio Laboratories, Qiagen, Carlsbad, CA, USA)
- OneStep PCR Inhibitor Removal Kit (Zymo Research, Irvine, CA, USA)
- MinElute PCR Purification Kit (Qiagen, Carlsbad, CA, USA
- Vortex with adaptor for 15-ml tubes
- Centrifuge for 15-ml tubes and cooling to 4 °C
- Centrifuge for 1.5-ml tubes
- Automatic pipettes (ranges 0.5–10, 10–100, 100–1000 μl)
- Pipette tips, DNase- and RNase-free with filter (10, 100, 1000 μl)
- Centrifugal evaporator (e.g., SpeedVac)
- Mortar and pestle
- Crushed ice
- Qubit including quantitation assays (Thermo Fisher, Waltham, MA, USA) or other RNA/DNA quantitation equipment
- PCR thermocycler

2.2 *Chemicals*

- Ethanol
- Liquid nitrogen
- Dry ice
- DNase I
- SuperScript III Reverse Transcriptase
- Random hexamer primers
- Diethyl pyrocarbonate-treated water
- Bovine serum albumin (BSA; 10 mg/ml)
- PCR nucleotide mix (10 mM)
- Primers for the PCR reactions, specifically gITS7 (5′-GTGAATCATCGAATCTTTG-3′; Ihrmark et al. 2012) and ITS4 (5′-TCCTCCGCTTATTGATATGC-3′; White et al. 1990) with attached barcode for each sample.

2.3 *Solutions*

- Phenol/chloroform/isoamyl alcohol (25:24:1)
- Pfu DNA polymerase/DyNAzyme II DNA polymerase (1:24) or other polymerase with proofreading activity

3 Experimental Procedures

3.1 Sample Preparation

1. Select representative field sites and samples. For soils or sediments covered by a litter layer, use a 5-cm-diameter corer to collect eight litter subsamples around the circumference of a 4-m-diameter circle. Alternatively, collect natural litter in the field, ensuring a mixture of leaves of different species and in different stages of decomposition. Or, incubate leaves in a stream (or forest) as described in Chap. 6 and sample on specific dates for analysis.
2. Separate litter material from any soil or sediment material directly in the field. Remove larger twigs, roots, or stones, and cut the litter into 0.5-cm pieces with sterile scissors. Combine the litter subsamples, and mix well.
3. Prepare at least four subsamples (each 0.5–3.0 g) for DNA/RNA co-extraction in sterile tubes. Freeze the subsamples immediately in liquid nitrogen, and store on dry ice. Upon arrival to the laboratory, store the frozen samples at −80 °C for no more than 6 months.

3.2 RNA/DNA Co-Extraction

1. Co-extract RNA and DNA in independent triplicates from each sample using the RNA PowerSoil Total RNA Isolation Kit combined with the OneStep PCR Inhibitor Removal Kit and the DNA Elution Accessory Kit. To increase the yield of DNA and RNA, grind samples in liquid N_2 with a mortar and pestle. RNA extraction is extremely sensitive to contamination and nucleases. Therefore, be sure to use nuclease-free water and plasticware throughout.
2. For RNA extraction, follow steps 1–8 of the RNA PowerSoil Total RNA Isolation Kit instructions.
3. Before proceeding to step 9 (addition of Solution SR4), use the OneStep PCR Inhibitor Removal Kit columns to clean the supernatant according to the manufacturer's instructions. Use as many columns as needed to clean the whole supernatant.
4. Collect the cleaned supernatant in a new 15-ml collection tube from the RNA PowerSoil Total RNA Isolation Kit. Follow steps 9–20 of the RNA PowerSoil Total RNA Isolation Kit instructions.
5. To co-extract DNA, keep the columns from step 16 of the RNA PowerSoil Total RNA Isolation Kit, and use the DNA Elution Accessory Kit to elute DNA according to instructions.
6. Store the recovered DNA at −20 °C and RNA at −80 °C. RNA can be stored for up to 6 months.

3.3 RNA Reverse Transcription

1. Quantify RNA content using Qubit with RNA quantitation assay, or use other RNA quantitation equipment.
2. Treat 50 ng to 1 μg of each RNA sample with DNase I, and perform reverse transcription using SuperScript III Reverse Transcriptase and random hexamer primers according to the manufacturer's protocol to obtain single-stranded cDNA. Store the cDNA at −20 °C. Single-stranded cDNA can be stored for up to 6 months.

3.4 cDNA and DNA Amplification

1. Pool the aliquots of DNA or cDNA originating from the same sample before PCR amplification.
2. Perform PCR amplification of each DNA or cDNA sample with the primers gITS7 and ITS4 containing sample-specific barcodes. Set up the PCR reactions in at least three independent 25-μl reactions per sample containing the following: 5 μl of 5× Q5 reaction buffer, 1.5 μl of BSA (10 mg/ml), 1 μl of each primer (0.01 mM), 0.5 μl of PCR nucleotide mix (10 mM each), 0.25 μl Q5 High-Fidelity DNA polymerase, 5 μl of 5× Q5 High GC enhancer, and 1 μl of template DNA or cDNA (approx. 5–50 ng/μl). As an alternative, a DNA polymerase with proofreading activity can be used to decrease the error level during the PCR reaction.
3. Run the PCR with the following cycling specifications: 94 °C for 5 min and 30 cycles of 94 °C for 30 s, 56 °C for 30 s, and 72 °C for 30 s, followed by 72 °C for 7 min. Use heated cover option (103–105 °C).

3.5 Library Preparation and Sequencing

1. Pool the replicate PCR products, and clean them using the MinElute Kit. Elute DNA with 20 μl of sterile water. Measure the dsDNA concentrations in samples using Qubit, and combine the barcoded samples so that the same amount of DNA is included from each sample.
2. Measure the dsDNA concentrations in composite samples using Qubit, and use them to prepare a library for high-throughput sequencing.
3. Perform the high-throughput sequencing, and sequence analysis as described in Chaps. 36 and 62 or elsewhere (Větrovský and Baldrian 2013; Lindahl et al. 2013).
4. Combine DNA sequences originating from RNA and those originating from DNA, and perform clustering at desired similarity level. For each operational

taxonomic unit (OTU), record the relative abundance of sequences in the DNA- and RNA-derived sequence pools. The sequences contained in the DNA-derived pool represent the ITS sequences present in the genomic DNA of fungi (all taxa including inactive ones). Sequences derived from the RNA-derived pool represent transcribed precursor rRNA molecules before ITS excision and thus capture members of the fungal community that are transcriptionally active at the time of sampling (Baldrian et al. 2012).

4 Final Remarks

The presented method is able to identify and quantify rRNA genes present in fungal genomes and transcripts of these genes. Theoretically, the ratio of sequence abundances of fungal taxa in the original RNA and DNA pools should broadly reflect metabolic activity as indicated by synthesis of novel ribosomes (Anderson and Parkin 2007). However, large uncertainties are to be expected because rRNA cassettes have multiple copies in fungal genomes, and their amount per genome as well as per ng of DNA varies considerably (Baldrian et al. 2013; Větrovský et al. 2016). There is also little reliable information on the lifetime of RNA transcripts, which again can differ among taxa. Therefore, the rates of activity of individual taxa in a single sample may be difficult to compare. Nevertheless, comparisons of RNA/DNA ratios may serve as reasonable proxies of the activity of individual taxa and how the individual activities may vary among samples, for example, among seasons within a year (Žifčáková et al. 2016).

References

Anderson, I. C., & Parkin, P. I. (2007). Detection of active soil fungi by RT-PCR amplification of precursor rRNA molecules. *Journal of Microbiological Methods, 68*, 248–253.

Baldrian, P. (2017). Forest microbiome: Diversity, complexity and dynamics. *FEMS Microbiology Reviews, 41*, 109–130.

Baldrian, P., Kolařík, M., Štursová, M., Kopecký, J., Valášková, V., Větrovský, T., Žifčáková, L., Šnajdr, J., Rídl, J., Vlček, Č., & Voříšková, J. (2012). Active and total microbial communities in forest soil are largely different and highly stratified during decomposition. *The ISME Journal, 6*, 248–258.

Baldrian, P., Větrovský, T., Cajthaml, T., Dobiášová, P., Petránková, M., Šnajdr, J., & Eichlerová, I. (2013). Estimation of fungal biomass in forest litter and soil. *Fungal Ecology, 6*, 1–11.

Bärlocher, F. (2016). Aquatic hyphomycetes in a changing environment. *Fungal Ecology, 19*, 14–27.

Blazewicz, S. J., Barnard, R. L., Daly, R. A., & Firestone, M. K. (2013). Evaluating rRNA as an indicator of microbial activity in environmental communities: Limitations and uses. *The ISME Journal, 7*, 2061–2068.

Freedman, Z. B., Romanowicz, K. J., Upchurch, R. A., & Zak, D. R. (2015). Differential responses of total and active soil microbial communities to long-term experimental N deposition. *Soil Biology and Biochemistry, 90*, 275–282.

Grimmett, I. J., Shipp, K. N., Macneil, A., & Bärlocher, F. (2013). Does the growth rate hypothesis apply to aquatic hyphomycetes? *Fungal Ecology, 6*, 493–500.

Ihrmark, K., Bödeker, I. T., Cruz-Martinez, K., Friberg, H., Kubartova, A., Schenck, J., Strid, Y., Stenlid, J., Brandström-Durling, M., Clemmensen, K. E., & Lindahl, B. D. (2012). New primers to amplify the fungal ITS2 region – Evaluation by 454-sequencing of artificial and natural communities. *FEMS Microbiology Ecology, 82*, 666–677.

Lindahl, B. D., Nilsson, R. H., Tedersoo, L., Abarenkov, K., Carlsen, T., Kjøller, R., Kõljalg, U., Pennanen, T., Rosendahl, S., Stenlid, J., & Kauserud, H. (2013). Fungal community analysis by high throughput sequencing of amplified markers – A user's guide. *New Phytologist, 199*, 288–299.

Pathan, S. I., Žifčáková, L., Ceccherini, M. T., Pantani, O. L., Větrovský, T., & Baldrian, P. (2017). Seasonal variation and distribution of total and active microbial community of β-glucosidase encoding genes in coniferous forest soil. *Soil Biology and Biochemistry, 105*, 71–80.

Peršoh, D. (2015). Plant-associated fungal communities in the light of meta'omics. *Fungal Diversity, 75*, 1–25.

Rajala, T., Peltoniemi, M., Hantula, J., Mäkipää, R., & Pennanen, T. (2011). RNA reveals a succession of active fungi during the decay of Norway spruce logs. *Fungal Ecology, 4*, 437–448.

Schneider, T., Keiblinger, K. M., Schmid, E., Sterflinger-Gleixner, K., Ellersdorfer, G., Roschitzki, B., Richter, A., Eberl, L., Zechmeister-Boltenstern, S., & Riedel, K. (2012). Who is who in litter decomposition? Metaproteomics reveals major microbial players and their biogeochemical functions. *The ISME Journal, 6*, 1749–1762.

Šturšová, M., Žifčáková, L., Leigh, M. B., Burgess, R., & Baldrian, P. (2012). Cellulose utilization in forest litter and soil: Identification of bacterial and fungal decomposers. *FEMS Microbiology Ecology, 80*, 735–746.

van der Wal, A., Geydan, T. D., Kuyper, T. W., & de Boer, W. (2013). A thready affair: Linking fungal diversity and community dynamics to terrestrial decomposition processes. *FEMS Microbiology Reviews, 37*, 477–494.

Větrovský, T., & Baldrian, P. (2013). Analysis of soil fungal communities by amplicon pyrosequencing: Current approaches to data analysis and the introduction of the pipeline SEED. *Biology and Fertility of Soils, 49*, 1027–1037.

Větrovský, T., Kolařík, M., Žifčáková, L., Zelenka, T., & Baldrian, P. (2016). The *rpb2* gene represents a viable alternative molecular marker for the analysis of environmental fungal communities. *Molecular Ecology Resources, 16*, 388–401.

Voříšková, J., & Baldrian, P. (2013). Fungal community on decomposing leaf litter undergoes rapid successional changes. *The ISME Journal, 7*, 477–486.

White, T. J., Bruns, T., Lee, S., & Taylor, J. W. (1990). Amplification and direct sequencing of fungal ribosomal RNA genes for phylogenetics. In M. A. Innis, D. H. Gelfand, J. J. Sninsky, & T. J. White (Eds.), *PCR protocols: A guide to methods and applications* (pp. 315–322). New York: Academic.

Žifčáková, L., Větrovský, T., Howe, A., & Baldrian, P. (2016). Microbial activity in forest soil reflects the changes in ecosystem properties between summer and winter. *Environmental Microbiology, 18*, 288–301.

Chapter 39
Gene Expression Analysis of Litter-Associated Fungi Using RNA-Seq

Elizabeth C. Bourne, Paul R. Johnston, Elisabeth Funke, and Michael T. Monaghan

Keywords Fungal community analysis · Fungal transcriptional activity · mRNA · Next-generation sequencing · RNA extraction · RNA sequencing · RNA-Seq counts · Sequence database · Transcript-level quantification · Transcriptome

1 Introduction

Many stream food webs are fuelled predominantly by allochthonous organic matter in the form of leaves, needles and other plant material (Allan and Castillo 2007). The scarcity of *in situ* primary production suggests few opportunities for fungal lifestyles (endophytic, mycorrhizal or pathogenic) based on interactions with living

---

Electronic Supplementary Material The online version of this chapter (https://doi.org/10.1007/978-3-030-30515-4_39) contains supplementary material, which is available to authorized users.

E. C. Bourne · M. T. Monaghan (✉)
Department of Ecosystem Research, Leibniz Institute of Freshwater Ecology and Inland Fisheries (IGB), Berlin, Germany

Berlin Center for Genomics in Biodiversity Research, Berlin, Germany
e-mail: monaghan@igb-berlin.de

P. R. Johnston
Department of Ecosystem Research, Leibniz Institute of Freshwater Ecology and Inland Fisheries (IGB), Berlin, Germany

Berlin Center for Genomics in Biodiversity Research, Berlin, Germany

Institut für Biologie, Freie Universität Berlin, Berlin, Germany
e-mail: paul.johnston@igb-berlin.de

E. Funke
Department of Ecosystem Research, Leibniz Institute of Freshwater Ecology and Inland Fisheries (IGB), Berlin, Germany
e-mail: funke@igb-berlin.de

F. Bärlocher et al. (eds.), *Methods to Study Litter Decomposition*,
https://doi.org/10.1007/978-3-030-30515-4_39

vascular plants (Bärlocher and Boddy 2016). Nonetheless, fungi play an ecologically important role in freshwaters as decomposers of organic carbon, including in the turnover of cellulose and lignin of plant origin (Buck et al. 2008; Floudas et al. 2012). A range of molecular genetic methods are used to study the functional roles of fungi and the association of communities with specific environmental conditions (Chaps. 34, 35, 36, 37, 38, 40, and 62). In particular, gene expression can be studied directly through direct targeting of the transcriptome.

The study of gene expression has the potential to provide an in-depth understanding of the molecular basis of leaf litter decomposition. Compared to studies of DNA, which examine the genetic potential of organisms, the study of RNA expression levels allows us to evaluate which genes and organisms are active in a particular sample, associated with a particular environmental condition or stage. RNA sequencing (RNA-seq) is therefore increasingly used to investigate fungal transcriptional activity, with the aim to identify genes or pathways that are differentially expressed across environmental conditions or experimental treatments. To this end, RNA is extracted, and the extracted rRNA is depleted to specifically yield mRNA, which is then sequenced on a suitable high-throughput sequencing platform. The read output is finally assessed through bioinformatic analysis (see also Chap. 62).

There are many methods for the quantification of gene expression and subsequent statistical analyses. Quantification generally involves either aligning reads to a genome and counting those that overlap features (specified by an accompanying annotation) (e.g. Liao et al. 2013; Anders et al. 2015) or aligning to a transcriptome (derived from an annotated genome or a *de novo* RNA-seq assembly) and aggregating counts at the gene level (Bray et al. 2016; Soneson et al. 2016; Patro et al. 2017). The latter approach is preferable due to the control of false discovery rates for genes that show only differential usage of isoforms of different length (Soneson et al. 2016). Modern transcript-level quantification methods use highly efficient pseudo-alignment (Bray et al. 2016; Patro et al. 2017), which is comparable in accuracy to traditional and more computationally intensive alignment-based methods (Li and Dewey 2011; Bray et al. 2016). This eliminates the need for specialized hardware and allows the analysis to be performed on a standard laptop computer (Bray et al. 2016). Furthermore, the use of transcript-level quantification allows for both model and non-model organisms to be analysed using the same framework.

RNA-seq counts are typically fitted to a negative binomial distribution (Robinson et al. 2010; Love et al. 2014) or, less often, by using linear modelling of transformed counts with observation-level weights (Law et al. 2014). Both the DESeq2 (Love et al. 2014) and edgeR (Robinson et al. 2010) packages for R (R Core Team 2018) accommodate gene-level estimated counts that are derived from transcript-level quantification. This is facilitated by the tximport package (Soneson et al. 2016) for R, which provides functions to read and aggregate the count data for subsequent modelling.

The methods introduced in this chapter describe RNA extraction and preparation for next-generation sequencing as well as the analysis of sequencing data. RNA extraction procedures have been modified from Johnson et al. (2012). The protocol begins at the point of harvesting fungal cultures. We do not include detailed instructions for next-generation sequencing or installation instructions for the software needed to conduct bioinformatic analyses but encourage users to consult the software 'readme' and 'installation' files. We also assume that users have a working knowledge of Linux and command-line interface.

2 Equipment, Chemicals and Solutions

2.1 Equipment

- Water bath or heat block (up to 70 °C)
- Vortex
- Microcentrifuge (13,000 *g*)
- Freezer (−80 °C)
- Fume hood
- Fluorometer (e.g. Quantus™, Promega Corp., Madison, WI, USA)
- Bioanalyzer 2100 (Agilent Genomics, Santa Clara, CA, USA)
- Microvolume spectrophotometer (e.g. NanoDrop™, Thermo Fisher Scientific, Waltham, MA, USA)
- Next-generation sequencing platform (e.g. NextSeq 500, Illumina Inc., San Diego, CA, USA)

2.2 Materials for RNA Extraction

- Fungal cultures (preserved in liquid N)
- Porcelain pestle and mortar (one per sample, or cleaned between samples with ethanol and water containing diethyl pyrocarbonate (DEPC))
- Racks for 2-ml tubes
- Stainless steel spatulas
- Forceps
- Glass beakers (500 ml), flasks (250 ml) and graduated cylinders (50, 100 and 250 ml)
- Zirconium beads (0.1 mm diameter)
- Dewar flask for liquid nitrogen
- Ice box
- Pipettes (0.5–10, 2–20 and 20–1000 µl)
- Sterile, RNase-free disposable pipette tips with barrier filters (0.1–20, 2–200, 100–1000 µl)
- Sterile, RNase-free disposable microcentrifuge tubes (2 ml, 6 per sample; 1.5 ml, 2 per sample)

2.3 Chemicals

- Reagent kits:

  RNeasy® Mini Kit (Qiagen, Hilden, Germany)
  RNase-Free DNase Set (Qiagen)
  QuantiFluor®RNASystem (Promega Corp.)
  RNA 6000 Nano kit (Agilent Genomics)
  TruSeq mRNA library prep kit (Illumina Inc.)

- Liquid nitrogen
- Ice
- Chloroform/isoamyl alcohol (24:1)
- Phenol/chloroform (5:1, pH 4.5; acid-equilibrated phenol to maximize RNA recovery, not Tris-buffered)
- Absolute ethanol
- RNase-free water
- Diethyl pyrocarbonate (DEPC) for DEPC-treated water
- Surface decontaminant (e.g. RNase AWAY®, Thermo Fisher Scientific)
- Sodium hydroxide (NaOH), optional
- Extraction buffer components:

  Hexadecyltrimethylammonium bromide (CTAB), analytical grade
  Polyvinylpyrrolidone (PVP-40)
  1 M Tris hydrochloride solution (Tris-HCl, pH 8.0), molecular-biology grade solution, RNasc free
  0.5 M ethylenediaminetetraacetic acid (EDTA, pH 8.0), molecular-biology grade solution, RNase free
  Sodium chloride (NaCl)
  Beta-mercaptoethanol (β-ME) – toxic, use fume hood!
  Diethyl pyrocarbonate (DEPC) – carcinogenic, use accordingly with caution

2.4 Solutions

- Extraction buffer (CTAB-PVP buffer), heat to 65 °C to dissolve:
  - CTAB (2% w/v)
  - PVP-40 (2% w/v)
  - 100 mM Tris-HCl (pH 8.0)
  - 25 mM EDTA
  - 2 M NaCl (warmed to 65 °C in a water bath to dissolve)
  - Add β-ME to final concentration of 2% and add immediately before use.

- DEPC-treated water:
  - In a 1-litre glass Duran bottle, add DEPC to milli-Q water to give a 0.1% solution.
  - Close the container, shake carefully to mix fully and incubate overnight at room temperature.
  - Autoclave to remove DEPC traces (15–30 min); note that Tris-containing solutions cannot be treated with DEPC.
- NaOH solution (optional):
  - To make a 0.1 N NaOH solution, add 40.0 g of NaOH to 1 litre of RNase-free water (e.g. DEPC-treated water).
  - Rinse mortars and other containers of residual RNA with saturated NaOH solution, before rinsing again in DEPC-treated water.

3 Experimental Procedures for RNA Extraction

3.1 Sample Preparation

1. At harvest time, empty flasks containing cultures into 50-ml Falcon tubes, centrifuge and decant the supernatant.
2. Flash-freeze the fungal or solid-medium pellet in liquid N and store immediately at −80 °C.

3.2 RNA Extraction

Workspace Preparation

1. Designate an RNA-only working area, since RNase-free work space and aseptic laboratory practices are critical for successful RNA extraction.
2. Use RNase AWAY® (or a similar product) to clean the work space to inactivate RNases; to this end, wipe reusable equipment with RNase AWAY® and rinse with DEPC-treated water before use and ensure that the rotor of the microcentrifuge is clean.
3. Designate an additional area that can be used for grinding the sample material (step 2, below).
4. Prepare equipment and reagents beforehand to allow a seamless workflow (e.g. place pre-aliquot extraction buffers into pre-labelled tubes where possible; see below), use filter tips for toxic work, and generally work quickly but carefully.

Homogenization and Lysis

1. In a fume hood, prepare enough extraction buffer for the day; aliquots of the stock of extraction buffer can be stored in smaller containers (e.g. Falcon tubes).
2. Add 2% (v/v) of β-mercaptoethanol (β-ME) on the day of use; β-ME will inactivate any RNases present in the sample.
3. Fully mix the extraction buffer by vortexing, and then pipette aliquots into pre-labelled 2-ml tubes.
4. Preheat the prepared extraction buffer to 65 °C.
5. In the RNA-only area outside the fume hood, use a pre-cleaned and pre-cooled pestle and mortar to grind tissue into a fine powder using liquid N. For additional friction, 0.1-mm zirconium beads can be added.
6. Using a clean spatula, add ca. 500 mg of ground, frozen sample into 1.4 ml of pre-heated (65 °C) extraction buffer in a 2-ml microcentrifuge tube.
7. Vortex the tube until the tissue is evenly distributed throughout the buffer. To facilitate mixing, invert the tube on the vortex or heat briefly in a 65 °C water bath or heat block.
8. Incubate the tube at 65 °C for 10–15 min, vortexing briefly (15 s) twice during the incubation.
9. Spin the tube at 13,000 *g* for 3 min in a microcentrifuge to ensure all insoluble matter forms a pellet at the bottom of the tube.
10. Transfer the supernatant into a new 2-ml tube.

Solvent Extractions (Use Fume Hood)

1. Add enough 24:1 chloroform/isoamyl alcohol to fill the tube (approx. 1 ml).
2. Vortex the tube for 15 s or until the phases mix and appear cloudy.
3. Spin the tube at 13,000 *g* for 3 min in a microcentrifuge.
4. Transfer the upper, aqueous phase to a new 2-ml tube, using a disposable pipette; avoid transferring any of the material (often a white precipitate of denatured proteins and DNA) from the boundary between the phases.
5. Add at least 900 μl of 24:1 chloroform/isoamyl alcohol to the tube containing the aqueous phase.
6. Vortex, spin and transfer the upper, aqueous phase to a new 2-ml tube, as above.
7. Add 1 ml of 5:1 phenol/chloroform (pH 4.5) to the tube containing the aqueous phase.
8. Vortex, spin and transfer the upper, aqueous phase to a new 2-ml tube, as above.
9. Add at least 900 μl of 24:1 chloroform/isoamyl alcohol to the tube containing the aqueous phase.
10. Vortex, spin and transfer the upper, aqueous phase to a new 2-ml tube, as above.

RNA Binding and Purification

1. Estimate the volume of the aqueous solution, add at least 0.5 volumes of solution RLT (RNeasy® Mini Kit, Qiagen) and mix by briefly shaking.
2. Estimate the new total volume in the tube, add 0.5 volumes of 100% ethanol and mix by briefly shaking.
3. Transfer the contents of the tube into a Qiagen miniRNA spin column (pink), adding no more than 700 μl.
4. Cap the tube and spin for 15 s at 8000 *g*. The column should be empty at the end of this spin.
5. Discard the flow-through from the collection tube.
6. Repeat the previous two steps with the same miniRNA spin column until all the liquid in the tube(s) has been passed through the column; the nucleic acid is now bound to the silica membrane in the spin column.
7. Apply 350 μl of solution RW1 to the spin column, then cap the tube and spin for 15 s at 8000 *g*.
8. Discard the flow-through from the collection tube.
9. Apply 80 μl of DNase digestion solution (RNase-Free DNase Set, Qiagen) to the membrane of the spin column and incubate at room temperature for 15 min.
10. Apply 350 μl of solution RW1 to the spin column, then cap the tube and spin for 15 s at 8000 *g*.
11. Discard the flow-through from the collection tube.
12. Apply 500 μl of solution RPE to the spin column, then cap the tube and spin for 15 s at 8000 *g*.
13. Discard the flow-through from the collection tube.
14. Apply 500 μl of solution RPE to the spin column, then cap the tube and spin for 15 s at 8000 *g*.
15. Discard the flow-through from the collection tube.
16. Transfer the spin column to a new collection tube, and spin at 13,000 *g* for 2 min to remove remaining liquid from the silica membrane.
17. Transfer the spin column to a new 1.5-ml microcentrifuge tube with a conical bottom.
18. Add 30 μl of RNase-free water/elution buffer to the column (apply directly on the membrane), spin at 13,000 *g* for 1 min to elute and then place on ice.
19. Divide the volume into three aliquots, one for quantification and basic quality checks (e.g. 4 μl; see below), one for quality checking using the Bioanalyzer (e.g. 2 μl) and the rest for eventual sequencing.
20. Flash-freeze samples immediately in liquid nitrogen and store at −80 °C.

RNA Quality Check

1. Depending upon the requirements of sequencing and library preparation, ensure that up to 3 μg of high-quality RNA is available (if necessary by pooling RNA from multiple extractions) to assess the concentration, quality and integrity of RNA.
2. Check the DNA concentration at this stage, and incorporate a second DNase digest step if necessary.

Concentration and Purity

1. Defrost on ice the aliquot allocated to checking the RNA concentration and quality.
2. Determine RNA and DNA concentrations using a dye-based method (e.g. QuantiFluor®RNA System or QuantiFluor®DNA System with the Quantus™ Fluorometer).
3. Perform a preliminary quality check of RNA on the NanoDrop system by determining the OD_{260nm}/OD_{280nm} ratio; if this ratio deviates much from the value of ~2.0 for pure RNA, protein, phenol or other contaminants may be present.

RNA Integrity Check

1. Prepare the reagents in the Agilent RNA 6000 Nano kit, following the manufacturer's instructions.
2. Defrost samples on ice that are set aside to be checked with the Agilent 2100 Bioanalyzer.
3. On a pre-heated block, denature the samples at 72 °C for 2 min and immediately place them on ice; this heat denaturation ensures that any secondary compounds in the RNA are reduced, thus permitting accurate analyses on the Bioanalyzer system.
4. Measure RNA quality by using the Agilent RNA 6000 Nano kit and following the manufacturer's instructions; intact RNA will yield distinct 18S and 28S fluorescence peaks on an electropherogram (Fig. 39.1a); partially degraded RNA will result in smaller peaks below the 18S peak (Fig. 39.1b) but is likely to yield good RNA-seq results; degraded RNA will result in no clear peaks (Fig. 39.1c) and is not suitable for RNA-seq.
5. Optionally, use the RNA integrity number (RIN) to assess RNA quality (Schroeder et al. 2006); the RIN value is based on the distribution of RNA in the 18S and 28S peaks compared to the full RNA profile (for intact RNA, most occurs in these two peaks) and partially upon the height of the 28S peak, which degrades more quickly than the 18S peak; the closer the RIN value is to 10, the more intact the RNA; a value >7 indicates some degradation, but suitability of the RNA for sequencing; note that the RIN value, while a useful overall indicator of sample quality, may not accurately represent the integrity of mRNA, which itself degrades at a different rate.

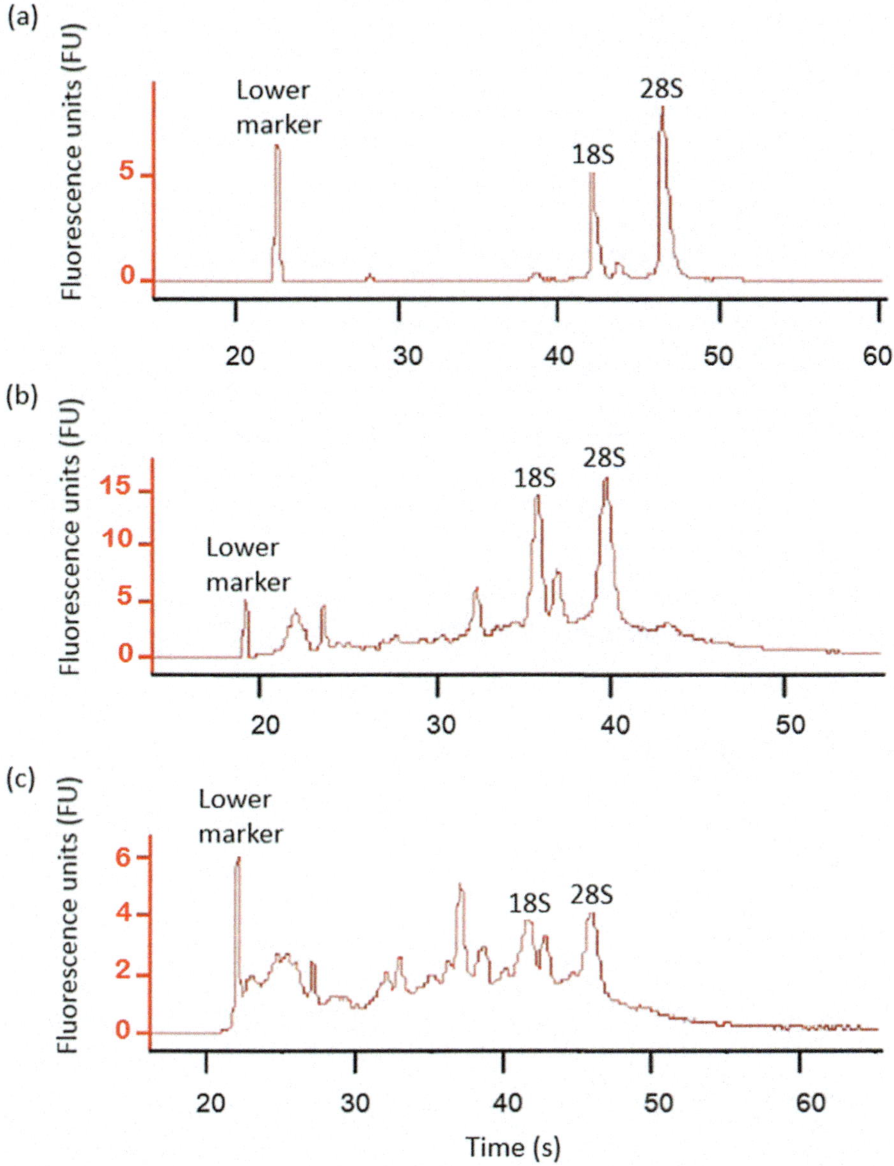

Fig. 39.1 Example electropherograms of RNA extracts from fungal cultures using a Bioanalyzer (Agilent Genomics). (**a**) Intact RNA: *Clavariopsis aquatica*, an aquatic hyphomycete; (**b**) partially degraded RNA (RIN = 7): *Helicodendron triglitziense*, an aero-aquatic hyphomycete; (**c**) degraded RNA (RIN = 5): *Clavariopsis aquatica* (RIN = 4); unlabelled peaks in panels b and c represent the presence of degraded rRNA

Further Steps

1. Ensure suitable shipping to the sequencing facility; sending total RNA frozen on dry ice is an excellent option in our opinion, although Johnson et al. (2012) found shipping of dehydrated total RNA to be more successful.

4 Experimental Procedures for RNA Sequencing

4.1 Sequencing Library Preparation and Next-Generation Sequencing

1. Select one of the commercially available kits that provide all reagents and instructions for performing mRNA selection, cDNA synthesis and fragmentation to a size suitable for the sequencing machine being used (e.g. 150-bp paired-end sequencing using a TruSeq mRNA library prep kit; Illumina Inc., San Diego, CA, USA).
2. Perform sequencing on a next-generation sequencing platform (e.g. Illumina NextSeq 500) following manufacturer's instructions.

5 RNA Sequence Data Analysis

Follow all steps as described in detail in supplemental code file Suppl. 39.1.

5.1 Software

1. Use regularly updated software, which is freely available; specifically, the following programs and versions (or higher) must be installed prior to carrying out the steps described below:

   - R 3.4.3
   - salmon 0.9.1
   - readr 1.1.0
   - tximport 1.4.0
   - DESeq2 1.16.1
   - tidyr 0.6.1
   - ggplot2 2.2.1
   - trinity 2.6.5 (optional)
   - bowtie2 2.3.4.1 (optional)

5.2 De Novo *Assembly of a Reference Transcriptome*

1. When a reference transcriptome (usually derived from a genome annotation) is unavailable, assemble the RNA-seq reads using 'Trinity' with the raw reads in fastq format as input.
2. Take care to provide the appropriate arguments to '--max_memory' and '--CPU' to control resource usage.
3. Generate a file relating transcripts to genes.

5.3 *Quantify Transcript Abundances*

1. Index the transcripts using 'salmon index' and as input a multi-FASTA file containing one sequence per transcript; perform this step in a new directory containing only the transcripts file.
2. Quantify each sample using 'salmon quant'.
3. Name the output directories according to the variables associated with the samples.
4. Adjust the '--threads' argument to match the number of available threads/CPUs on the computer.

5.4 *Differential Expression Analysis in R*

1. Begin a new R session and load and attach the required libraries; the supplemental code Suppl. 39.1 assumes that the working directory of the R process contains the 'salmon' output.
2. Import transcript counts from the salmon output subdirectories using the R package 'tximport'; the names provided by 'names(files)' will be used to denote the sample names and are inherited by the columns of the counts matrix.
3. Construct a data frame to associate sample names with the levels of variables which feature in the experimental design.
4. Create a DESeqDataSet object containing the transcript counts, sample information data frame and design formula.
5. Run DESeq to estimate size factors and dispersions.
6. Use the 'test = "LRT"' argument in conjunction with an appropriate reduced model if all levels of a variable are to be tested.
7. Extract the desired results; use the 'contrast' argument of the 'results' function if the results of a specific contrast of two levels of a variable are to be extracted; also take care to use the 'lfcThreshold' argument to specify a threshold, rather than performing *post hoc* filtering of thresholds (see supplemental code Suppl. 39.1 for details).
8. Remove the mean-variance dependence by regularized log transformation before visualization using multivariate statistical analysis.

6 Final Remarks

The presented method has been successfully used to extract RNA from two aquatic fungi, *Clavariopsis aquatica* and *Helicodendron triglitziense*, cultivated in liquid media containing milled alder leaves, milled straw or malt extract (Heeger 2018).

All solutions need to be made up in RNase-free containers and by using RNase-free components. Use sterile, disposable plasticware or treat non-disposable glassware and plasticware before use to remove potential RNases. Options for cleaning glassware include baking in a muffle furnace at 450 °C or rinsing in DEPC-treated water.

Of the four organic extraction steps following the CTAB extraction, the first two chloroform extractions act to prevent phase inversion during the acid phenol extraction, and the final chloroform extraction step removes the phenol. Efficient homogenization and lysis are critical for ensuring a good yield of RNA. In addition, large volumes of material may need to be processed to obtain sufficient RNA for subsequent steps and sequencing. It may often be necessary to extract multiple replicates and pool the RNA to obtain sufficient quantities for sequencing (ca. 3 μg).

Given the variation and multitude of factors which may interfere with a clear pattern of differential gene expression, RNA-seq studies should be replicated and based on a well-designed set-up.

References

Allan, J. D., & Castillo, M. M. (2007) *Stream ecology: Structure and function of running waters* (436 pp.). Dordrecht: Springer.

Anders, S., Pyl, P. T., & Huber, W. (2015). HTSeq – A python framework to work with high-throughput sequencing data. *Bioinformatics, 311*, 166–169.

Bärlocher, F., & Boddy, L. (2016). Aquatic fungal ecology – How does it differ from terrestrial? *Fungal Ecology, 19*, 5–13.

Bray, N. L., Pimentel, H., Melsted, P., & Pachter, L. (2016). Near-optimal probabilistic RNA-seq quantification. *Nature Biotechnology, 34*, 525.

Buck, U., Babenzien, H.-D., & Zwirnmann, E. (2008). Extracellular peroxidase activity in an experimentally divided lake (Große Fuchskuhle, northern Germany). *Aquatic Microbial Ecology, 51*, 97–103.

Floudas, D., Binder, M., Riley, R., Barry, K., Blanchette, R. A., Henrissat, B., Martínez, A. T., Otillar, R., Spatafora, J. W., Yadav, J. S., Aerts, A., Benoit, I., Boyd, A., Carlson, A., Copeland, A., Coutinho, P. M., de Vries, R. P., Ferreira, P., Findley, K., Foster, B., Gaskell, J., Glotzer, D., Górecki, P., Heitman, J., Hesse, C., Hori, C., Igarashi, K., Jurgens, J. A., Kallen, N., Kersten, P., Kohler, A., Kües, U., Kumar, T. K. A., Kuo, A., LaButti, K., Larrondo, L. F., Lindquist, E., Ling, A., Lombard, V., Lucas, S., Lundell, T., Martin, R., McLaughlin, D. J., Morgenstern, I., Morin, E., Murat, C., Nagy, L. G., Nolan, M., Ohm, R. A., Patyshakuliyeva, A., Rokas, A., Ruiz-Dueñas, F. J., Sabat, G., Salamov, A., Samejima, M., Schmutz, J., Slot, J. C., St. John, F., Stenlid, J., Sun, H., Sun, S., Syed, K., Tsang, A., Wiebenga, A., Young, D., Pisabarro, A., Eastwood, D. C., Martin, F., Cullen, D., Grigoriev, I. V., & Hibbett, D. S. (2012). The Paleozoic origin of enzymatic lignin decomposition reconstructed from 31 fungal genomes. *Science, 336*, 1715–1719.

Heeger, F. (2018). *Genomics approaches to the study of diversity and function of aquatic fungi*. Doctoral thesis, Freie Universität Berlin, Berlin, Germany (https://refubium.fu-berlin.de/handle/fub188/23679)

Johnson, M. T. J., Carpenter, E. J., Tian, Z., Bruskiewich, R., Burris, J. N., Carrigan, C. T., Chase, M. W., Clarke, N. D., Covshoff, S., de Pamphilis, C. W., Edger, P. P., Goh, F., Graham, S., Greiner, S., Hibberd, J. M., Jordon-Thaden, I., Kutchan, T. M., Leebens-Mack, J., Melkonian, M., Miles, N., Myburg, H., Patterson, J., Pires, J. C., Ralph, P., Rolf, M., Sage, R. F., Soltis, D., Soltis, P., Stevenson, D., Stewart, C. N., Jr., Surek, B., Thomsen, C. J. M., Villarreal, J. C., Wu, X., Zhang, Y., Deyholos, M. K., & Wong, G. K.-S. (2012). Evaluating methods for isolating total RNA and predicting the success of sequencing phylogenetically diverse plant transcriptomes. *PLoS One, 7*, e50226.

Law, C. W., Chen, Y., Shi, W., & Smyth, G. K. (2014). voom: Precision weights unlock linear model analysis tools for RNA-seq read counts. *Genome Biology, 15*, R29.

Li, B., & Dewey, C. N. (2011). RSEM: Accurate transcript quantification from RNA-Seq data with or without a reference genome. *BMC Bioinformatics, 12*, 323.

Liao, Y., Smyth, G. K., & Shi, W. (2013). featureCounts: An efficient general purpose program for assigning sequence reads to genomic features. *Bioinformatics, 30*, 923–930.

Love, M. I., Huber, W., & Anders, S. (2014). Moderated estimation of fold change and dispersion for RNA-seq data with DESeq2. *Genome Biology, 15*, 550.

Patro, R., Duggal, G., Love, M. I., Irizarry, R. A., & Kingsford, C. (2017). Salmon provides fast and bias-aware quantification of transcript expression. *Nature Methods, 14*, 417.

R Core Team (2018). R: A language and environment for statistical computing. R foundation for statistical computing. Vienna, Austria (https://www.R-project.org)

Robinson, M. D., McCarthy, D. J., & Smyth, G. K. (2010). edgeR: A Bioconductor package for differential expression analysis of digital gene expression data. *Bioinformatics, 26*, 139–140.

Schroeder, A., Mueller, O., Stocker, S., Salowsky, R., Leiber, M., Gassmann, M., Lightfoot, S., Menzel, W., Granzow, M., & Ragg, T. (2006). The RIN: An RNA integrity number for assigning integrity values to RNA measurements. *BMC Molecular Biology, 7*, 3.

Soneson, C., Love, M. I., & Robinson, M. D. (2016). Differential analyses for RNA-seq: Transcript-level estimates improve gene-level inferences. *F1000Research, 4*, 1521.

Chapter 40
Metaproteomics of Litter-Associated Fungi

Katharina M. Keiblinger and Katharina Riedel

Keywords Bioinformatics · Fungal community · Fungal functional capacities · Gel electrophoresis · LC-MS/MS · Protein extraction · Protein fractionation · Shotgun proteomics · Visualization software · Tandem mass spectrometry

1 Introduction

The degradation of plant litter constituents is catalyzed by microbial enzymes that cleave the polymeric compounds into small molecules that can be readily assimilated. Traditionally, activities – or potential activities – of extracellular microbial enzymes involved in litter decomposition (e.g., cellulase, protease, phenol oxidase, phosphatase; Chaps. 41, 42, 43, 44, 45, 46, 47, and 48) have been determined by enzymatic assays (Sinsabaugh et al. 2008; Burns et al. 2013). However, although widely used, simple to conduct, and useful to assess microbial carbon and nutrient acquisition, enzyme assays do not provide information on the identity and phylogenetic origin of the organisms producing the enzymes (Nannipieri et al. 2003). This is a major limitation because linking microbial community structure to enzymatic capacities is crucial to identify the microbial agents of litter decomposition and understand their responses to changing environmental conditions (Becher et al. 2013). This limitation can be overcome by shot-gun metaproteomics, which has evolved as a powerful approach allowing qualitative and quantitative assessments of entire suites of enzymes (and other proteins) not only in single organisms but also

K. M. Keiblinger (✉)
Department of Forest and Soil Sciences, Institute of Soil Research,
University of Natural Resources and Life Sciences Vienna, Vienna, Austria
e-mail: katharina.keiblinger@boku.ac.at

K. Riedel
Department of Microbial Physiology and Molecular Biology, Institute of Microbiology,
University of Greifswald, Greifswald, Germany
e-mail: riedela@uni-greifswald.de

F. Bärlocher et al. (eds.), *Methods to Study Litter Decomposition*,
https://doi.org/10.1007/978-3-030-30515-4_40

in complex communities (Schneider and Riedel 2010; Becher et al. 2013; Keiblinger et al. 2016; Starke et al. 2016). Thus, metaproteomics is useful to elucidate relationships between microbial community structure and functions in decomposing plant litter (Keiblinger et al. 2012a; Schneider et al. 2010, 2012; Becher et al. 2013; Starke et al. 2016).

Litter decomposition in terrestrial environments is carried out by a suite of fungi and other microbes that undergo successions related to changes in carbon availability and related biochemistry. The initial phase is dominated by ascomycetes, which inhabit the phyllosphere of living leaves and are strongly cellulolytic (Schneider et al. 2012; Purahong et al. 2016) but largely lack efficient lignin-degrading enzymes (Osono 2011). Later, ascomycetes are replaced by basidiomycetes (Schneider et al. 2012) capable of degrading lignin, particularly white-rot fungi (Baldrian 2008; Wal et al. 2013) that produce both laccases and manganese peroxidases (Purahong et al. 2016). In freshwaters, successional changes of fungal communities in litter are less proncounced. However, community changes correlated with the progress of litter decomposition have been described (Bärlocher 1992; Gessner et al. 1993), suggesting that functional changes occur as well.

The metaproteomics approach allows identifying drivers of key environmental processes by assigning functions as well as taxonomic origins to specific proteins. While metatranspriptomics, based on active mRNA, also provides insights into the functions of microbes active in decomposition (Chap. 39), these functions are mediated by the main players of cell physiology, the proteins. Hence, metaproteomics constitutes the most direct estimator of the activity during decomposition, including actually produced and available extracellular enzymes.

Technical challenges of metaproteomics applied to decomposing litter include interference by a broad range of substances (e.g., phenolics, complex carbohydrates, and lipids affecting protein extraction), high decomposer diversity, and low quantities of the proteins of greatest interest (e.g., extracellular cellulases, xylanases, proteases) (Keiblinger et al. 2016). However, these problems can be addressed by improved procedures for protein extraction. Furthermore, the challenge of precise peptide and protein identification has become increasingly manageable as metagenomic databases have grown, mass spectrometry has further advanced, and robust bioinformatics are being developed.

A typical metaproteomic analysis comprises the following steps (Fig. 40.1): protein extraction, pre-fractionation of proteins and peptides, mass spectrometry (MS), database search, and finally data analysis, integration, and visualization (Keiblinger & Riedel 2018). Pre-fractionation is either accomplished by a gel-free approach or one-dimensional gel electrophoresis (Keiblinger et al. 2016). The latter separates proteins based on molecular weight and eliminates co-extracted humic substances that often interfere with the analysis. However, the 'gold standard' in metaproteomics is considered the gel-free approach, because it enables fully automated processing of the extracted proteins and thus provides a fast and easily reproducible workflow.

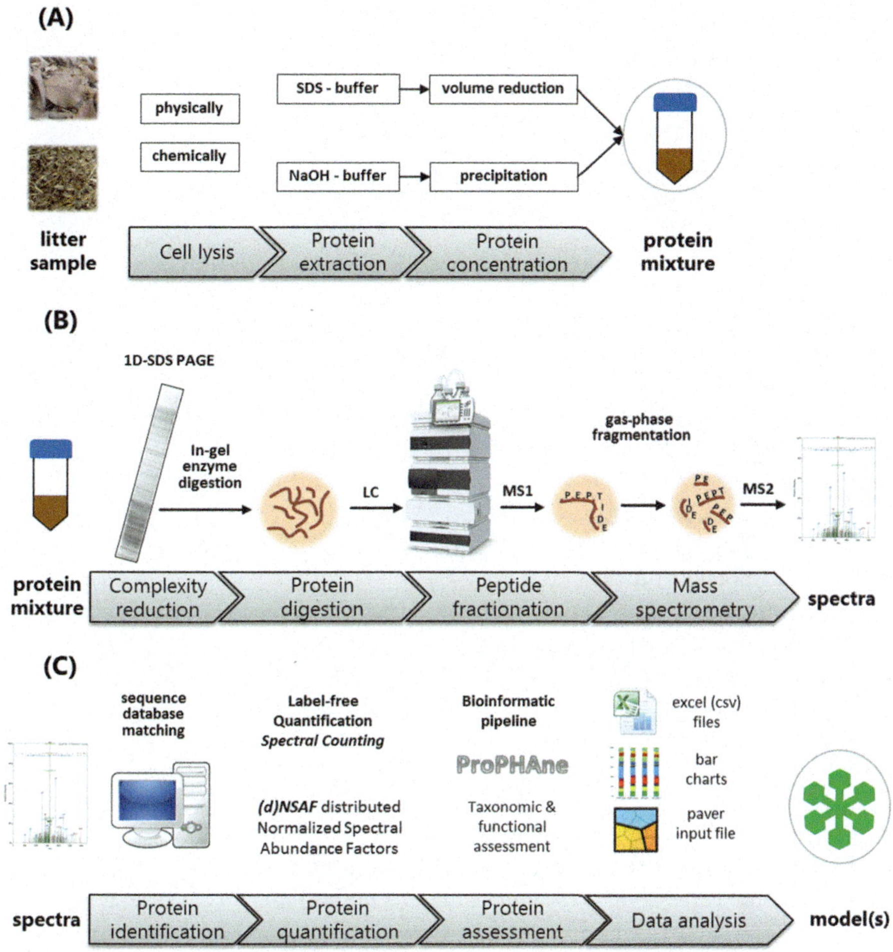

Fig. 40.1 Overview of workflow for metaproteomic analyses of decomposing plant litter. (**A**) Protein isolation, (**B**) shotgun proteomics and (**C**) protein identification, qualification, and data analysis

Comprehensive and reliable identification of proteins in leaf litter depends on (i) quantitative protein extraction, (ii) separation of proteins from the sample matrix, and (iii) removal of interfering chemical substances without altering the proteome composition. Below we describe procedures for protein extraction and sample preparation, including protein separation by in-gel digestion, as well as purification of the resulting peptide mixture before mass spectrometric analysis. The presented protocols have been successfully used to characterize enzymes and microbial communities associated with decomposing beech leaves (e.g., Keiblinger et al. 2012a; Schneider et al. 2010, 2012) and with coniferous forest floors (K. M. Keiblinger et al., unpublished data).

2 Equipment, Chemicals, and Solutions

2.1 Equipment and Material

- Autumn-shed leaves, air-dried
- Litter bags (10 × 10 cm; 10, 1, or 0.5 mm mesh size)
- Mill (e.g., Retsch Grindomax GM200) or scissors or scalpel, to homogenize sample
- Mortar and pestle
- Dewar bottle with liquid N_2
- Freezer (−80 and –20 °C)
- Cryo gloves
- Automatic pipettes: 2–20, 20–200, and 100–1000 μl ranges, 10 ml
- Pipette tips: 0.1–20, 2–200, and 100–1000 μl, 10 ml
- Pipette aid
- Disposable graduated pipettes (10 and 25 ml)
- Sterile plastic centrifuge tubes (50 ml)
- LoBind Eppendorf tubes (1.5 ml)
- Ultrasonic device with a conic probe (e.g., BANDELIN electronic GmbH & Co. KG, Berlin, Germany)
- Chambers for vertical polyacrylamide gels
- Autoclave
- Vortex
- Water bath
- Ice bath
- Centrifuge with rotor for 50 ml centrifuge tubes
- Speed-vacuum centrifuge (e.g., Eppendorf Vacuum Concentrator Plus)
- Gel cassette Mini to run gels, consisting of a buffer tank, electrode assembly, lid, and power cables (e.g., PROTEAN Tetra System, Bio-Rad Laboratories Inc., Hercules, CA, USA)
- Constant power supply (e.g., PowerPac Universal, Bio-Rad Laboratories Inc., Hercules, CA, USA)
- Plastic tray
- Scalpel
- Reaction tubes (e.g., Eppendorf, 1.5 ml)
- Thermomixer/thermoshaker
- Incubator
- Table-top centrifuge (e.g., 5430/5430R, Eppendorf, Hamburg, Germany) with rotor for 1.5 ml tubes
- Ultrasonic bath
- ZipTip μC18 pipette tips (10 μl; e.g., Merck Millipore, Burlington, MS, USA)
- LoBind Eppendorf tubes 1.5 ml (coated to prevent protein absorption)
- Aluminum heating block or water bath
- Orbital shaker

- Gel scanner to quantify proteins on gels stained with Coomassie blue over a large dynamic range (0–3.0 OD) and enable background correction for quality control (e.g., GS-800 Calibrated Densitometer, Bio-Rad Laboratories Inc., Hercules, CA, USA)

2.2 Chemicals

- Milli-Q water (ultrapure water, 18.2 MΩ; TOC ≤ 5 ppb; sterile)
- Liquid nitrogen
- Polyvinylpolypyrrolidone (PVPP), cross-linked pharmaceutical grade, non-dissolvable, to adsorb polyphenols
- Ammonium peroxodisulfate (APS)
- N,N,N,N'-tetramethylethylenediamine (TEMED), store at 4 °C
- Sodium dodecyl sulphate (SDS)
- Tris(hydroxymethyl)aminomethane (Tris), analytical grade
- Ethylenediaminetetraacetic acid (EDTA), analytical grade
- Phenol solution, equilibrated with 10 mM Tris HCl, pH 8.0, 1 mM EDTA, (e.g., commercially available as BioReagent for molecular biology, CAS Number 108–95-2, Sigma Aldrich), store at 4 °C; use caution since the reagent is toxic
- Acrylamide/bis-acrylamide: 30% (v/v) solution (electrophoresis grade), store at 4 °C; use caution since the reagent is toxic
- Pure (96%) ethanol (EtOH)
- Tris SDS buffer: 50 mM Tris, 1% (w/v) SDS, pH 7.0, autoclaved
- Sodium hydroxide (NaOH)
- Pre-chilled acetone, stored at −20 °C
- Tris-EDTA (TE) buffer: 0.5 ml 1 M Tris-HCl (pH 8.0), 1.0 ml 0.5 M EDTA, 48.5 ml Milli-Q water (18.2 MΩ; TOC ≤ 5 ppb; autoclaved)
- Methanol (MeOH)
- Ammonium acetate (NH_4CH_3COO)
- Acrylamide (6% for sequencing gels; e.g., Visible Genetics Inc., Suwanee, GA, USA), which must be compatible with the type of automated sequencer used
- Bromophenol blue
- Glycerol
- Protein ladder 10–250 kDa (e.g., PageRuler™ Plus Prestained Protein Ladder, Thermo Scientific, USA)
- Dimethyl sulfoxide (DMSO), analytical grade
- Acetonitrile (ACN)
- Ammonium sulfate ($(NH_4)_2SO_4$)
- Ortho-phosphoric acid (H_3PO_4)
- Coomassie Brilliant Blue G-250 dye
- Sequencing-grade trypsin (Promega, Mannheim, Germany)

2.3 Solutions

- Extraction buffer 1: 1% SDS, 50 mM Tris/KOH, pH 7.0 (Schneider et al. 2012)
- Extraction buffer 2: 0.1 M NaOH (Benndorf et al. 2007; Keiblinger et al. 2012b)
- Resolving gel buffer (RGB): 1.5 M Tris-HCl, pH 8.8, 1% (w/v) SDS in Milli-Q water
- Resolving gel mixture (RGM): pipette successively 4 ml Milli-Q water, 3 ml RGB, 5 ml acrylamide solution; add 80 μl 10% APS and 8 μl TEMED before use
- Stacking gel (SGB) buffer: 0.5 M Tris-HCl, pH 6.8, 1% (w/v) SDS in Milli-Q water
- Stacking gel mixture (SGM): pipette successively 4.75 ml Milli-Q water, 2 ml SGB, 1.25 ml acrylamide solution; add 80 μl APS and 8 μl TEMED before use
- Loading buffer 4x, pH 6.8: 7.5 ml glycerol, 2.5 ml β-mercaptoethanol, 1.2 g SDS, 200 μl 1% (w/v) bromophenol blue, 0.4 g Tris, 50 ml Milli-Q water
- Electrophoresis buffer 10x: Tris 30.3 g, glycine 144.1 g, 10 g SDS, 1000 ml Milli-Q water
- Fixing solution: 40% (v/v) ethanol, 10% (v/v) acetic acid, 50% (v/v) Milli-Q water
- Staining solution: 100 g $(NH_4)_2SO_4$,100 ml H_3PO_4, 1.2 g Coomassie Brilliant Blue G-250, filled up to 1000 ml with Milli-Q water and stirred overnight
- Sodium azide (NaN_3) solution for gel storage: 3 mM NaN_3 solution in Milli-Q water; use caution since the reagent is toxic
- Washing solution: 200 mM $(NH_4)HCO_3$, 30% (v/v) ACN (HPLC grade)
- Trypsin solution (1:10): dissolve 20 μg lyophilized trypsin (MS grade) in 1 ml Milli-Q water, prepare 50 μl aliquots of the solution, add 450 μl Milli-Q water, and store at −20 °C
- Wetting solution: 70% (v/v) ACN in Milli-Q water
- Equilibration solution: 3% (v/v) ACN and 0.1% (v/v) acetic acid in Milli-Q water
- Washing solution: 0.1% (v/v) acetic acid in Milli-Q water
- Elution solution: 60% (v/v) ACN and 0.1% (v/v) acetic acid in Milli-Q water

3 Experimental Procedures

3.1 Sampling, Sample Homogenization, and Storage

1. Expose leaves in litter bags in the field and collect as described in Chap. 6.
2. Shred the leaves into small pieces (e.g., with a Retsch Mill at 7000 rpm for 1 min), repeat until leaves pass through a 10-mm mesh screen.
3. Alternatively, sample forest floor, with a sufficient amount of independent and spatially separated samples spread over the experimental area to create represen-

tative composite samples (Quinn and Keough 2002; Boeddinghaus et al. 2015). For forest floor sampling, remove the upper L layer, and sieve the forest floor through a 10-mm mesh screen to obtain a homogenous sample.

4. Store samples that are not immediately processed at −80 °C.

3.2 Protein Extraction from Leaf Litter (Fig. 40.2)

1. Place liquid N_2 into a mortar until filled to pre-chill the mortar; add >5 g of fresh or frozen litter and ~0.5 g of PVPP.
2. Gently grind sample with a mortar and pestle to obtain a homogenous powder.

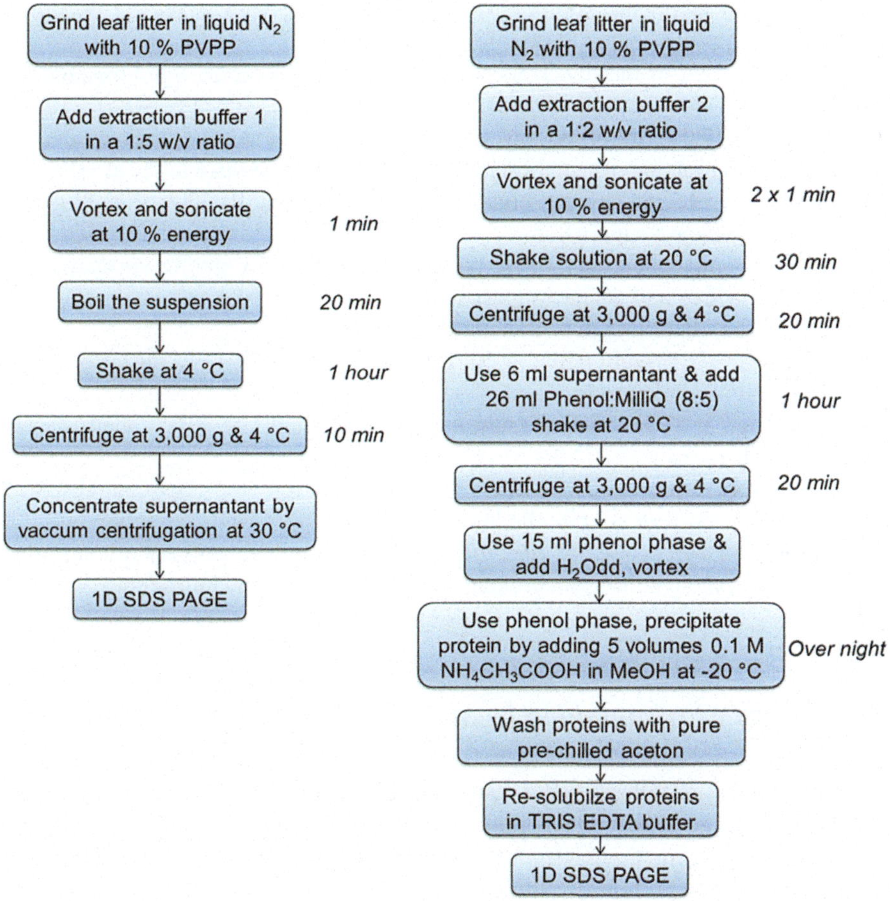

Fig. 40.2 Workflow to extract proteins from deciduous leaf litter (left) and coniferous forest floors (right)

3. Transfer ~5.5 g of the litter/PVPP mixture to a 50-ml centrifuge tube.
4. Extract proteins with 25-ml extraction buffer 1 (deciduous litter) or with 10-ml extraction buffer 2 (coniferous litter) and vortex.
5. Sonicate with an ultrasonic probe (90% pulsing, 10% energy, 2 × 1 min, cool on ice in between for 2 min).
6. For deciduous leaves (Fig. 40.2, left), boil the extract for 20 min in a water bath, centrifuge at 3000 *g* for 10 min at 4 °C, transfer supernatant to new tube, and concentrate to one-fifth by speed vacuum centrifugation at 30 °C (final volume ~5 ml). *Note*: Make sure that the tubes are large enough for the boiling procedure to avoid splashing or boiling retardation. Keep the lids open; formation of foam is possible!
7. For conifer needles (Fig. 40.2, right), shake for 30 min at 20 °C, centrifuge at 3000 *g* for 20 min at 4 °C, and then follow the protocol below.
8. Pipette 6 ml of the supernatant into a new centrifuge tube.
9. Add 26 ml of phenol solution/Milli-Q water, (8:5, v:v), shake for 60 min at 20 °C, and centrifuge at 3000 *g* for 20 min at 4 °C.
10. Transfer 15 ml of the lower phenol phase into a new centrifuge tube, and add 10 ml Milli-Q water and vortex.
11. Again transfer 15 ml of the lower phenol phase into a new centrifuge tube.
12. Precipitate proteins overnight with 0.1 M NH_4CH_3COO in MeOH at −18 °C.
13. Collect the precipitated proteins by centrifugation at 10,000 *g* for 20 min at 4 °C, and discard supernatant.
14. Wash protein pellet with 100% pre-chilled acetone by gentle vortexing, and then centrifuge and discard supernatant.
15. Air-dry the remaining protein pellet.
16. Resuspend the protein pellet in 20–100 μl of Tris-EDTA (TE) buffer by gentle mixing at 4 °C overnight.
17. Centrifuge at 18,000 *g* for 3 min at 4 °C and retain the supernatant (protein solution).

Note: To improve extraction yields, a second extraction step ('sequential extraction') can be performed. Use the initial pellet and apply fresh extraction buffer as described in the protocols, and then continue with the procedures described above. Pool the resulting supernatants prior to protein concentration.

3.3 Protein Fractionation by SDS PAGE (Fig. 40.1b)

1. Use commercial Minigels (Bio-Rad), or prepare gels with 12% resolving gel and 5% stacking gel mixture by following the manufacturer's instruction.
2. Before protein fractionation on the 1D gel, determine protein concentration. A number of total protein quantification assays are available (e.g., Chap. 12). For this specific purpose, protein concentration is determined after acid hydrolysis (Chap. 13), ideally via the ninhydrin assay (Chap. 12).

Note: The ninhydrin assay is the only assay at hand that works sequence independently, as the Bradford assay binds to positively charged and hydrophobic amino acids discriminating proteins with negative charges. In addition, the Bradford assay yields colour reactions with SDS and humic substances. Other assays may work as well, though some of the most popular ones have similar problems as Bradford.

3. Clamp the gels into the Bio-Rad Minigel electrophoresis chamber.
4. Fill gel chambers with electrophoresis buffer, and carefully remove comb.
5. Load protein ladder onto the gel according to the manufacturer's instructions (e.g., 5 μl of PageRuler™, Thermo Scientific, USA).
6. Mix protein solution with loading buffer in a 4:1 ratio (total of ~5 μg protein).
7. Denature protein samples in loading buffer for 5 min at 100 °C in an aluminum heating block.
8. Pipette ~20 μl of protein solution mixed with loading buffer into the gel slot.
9. Start electrophoresis and run at constant current of 30 mA until the bromophenol blue front reaches the edge of the gel.

3.4 Gel Staining and Destaining

1. To check the concentration and quality of the protein fractionation, first fix the proteins in the gel by shaking in 25-ml fixing solution for 30 min.
2. Wash gel in Milli-Q water to remove residual acetic acid, repeat at least once.
3. Stain gel with 20-ml staining solution and 5-ml EtOH by shaking overnight.
4. Destain the gel by shaking in Milli-Q water until the background of the gel is colorless.
5. Record results by scanning on a gel scanner.
6. Seal gel into a plastic bag (together with some droplets of 3 mM NaN_3, to prevent microbial contamination and growth during storage) and store at 4 °C.

3.5 Excision of Proteins and In-Gel Digestion

1. Transfer gel to a plastic tray with Milli-Q water and remove NaN_3 by gently shaking for 30 min at room temperature.
2. Transfer gel to a smooth and solid clean surface such as a glass plate.
3. Cut gel with a scalpel into 6–20 slices, and transfer slices to Eppendorf tubes and store at −20 °C if needed.
4. Destain gel slices in 600-μl washing solution by shaking on a Thermomixer for 15 min at 37 °C.
5. Discard washing solution and repeat the washing step to ensure the gel slices are thoroughly destained.

6. Dry gel slices in a speed vacuum centrifuge for 30–40 min at 30 °C.
7. Digest proteins by covering the gel slices with 50–100 μl of 1:10 trypsin solution for 15 min at room temperature.
8. Remove any excess trypsin solution and incubate the swollen gel slices overnight at 37 °C, with the Eppendorf tubes turned upside down.
9. Spin the Eppendorf tubes at 10,000 *g* for 30 s.
10. Add Milli-Q water until gel slices are completely covered and incubate in an ultrasonic bath for 15 min.
11. Centrifuge the Eppendorf tubes at 10,000 *g* for 30 s and transfer the total supernatant to a LoBind Eppendorf tube.
12. Concentrate the peptide containing supernatant as required (to ~10 μl) in a speed vacuum centrifuge at 30 °C.

3.6 ZipTip Desalting

1. To wet the ZipTips, pipette twice 10-μl wetting solution and discard the solution.
2. To equilibrate the C18 columns in the tips, pipette twice 10 μl of equilibration solution and discard the solution.
3. Load 10–15-μl sample on the ZipTip by aspirating 30 times with the pipette, and return the remaining solution to the LoBind tube.
4. Wash ZipTip with 10 μl of washing solution and discard the solution.
5. For elution, use 10 μl of elution solution; to ensure sufficient elution, aspirate three times and transfer the peptide-containing solution into a glass vial containing a 100-μl conical insert.
6. Concentrate the solution by drying in the speed vacuum centrifuge at 30 °C to remove acetonitrile (ACN).
7. Store samples at –20 °C.
8. Add 10 μl of Milli-Q water to the dried peptide pellets before separating the peptides by liquid chromatography with detection by electrospray ionization mass spectrometry (LC-ESI MS).

3.7 High-Pressure Liquid Chromatography Coupled with Tandem Mass Spectrometry (LC-MS/MS)

1. Separate ZipTip-treated proteolytic peptides by reversed-phase high-pressure liquid chromatography (RP-HPLC) or ultrahigh-pressure (nano) HPLC systems on a hydrophobic C18 column. For peptide elution from the column, an increasing mobile phase gradient starting from a more aqueous and continuously changing to a more organic solvent is used.

2. Chromatographic separation of peptides is followed by mass detection with a mass spectrometer (MS). Most commonly, separated peptides are introduced into the mass spectrometer via electron spray ionization (ESI) and fragmented by collision-induced dissociation (CID). CID produces a set of defined fragments originating from a specific peptide ion.
3. At each single time point of the HPLC gradient, the eluted peptides generate MS/MS spectra based on their signal of fragmented ion masses, generated from the parent ions (Becher et al. 2013).

3.8 Protein Identification, Quantification, Interpretation, and Visualization (Fig. 40.1c)

1. Compare MS/MS spectra resulting from LC-MS/MS with those of theoretic peptides from a given protein database (protein database or target database; ideally containing data of a matching full metagenome or metatranscriptome). Note that protein sequences can only be identified if represented in the target database.
2. Assign mass spectrometric data to proteins in the database, by feeding them into software packages such as MASCOT, SEQUEST, X-tandem, or MyriMatch (Becher et al. 2013).
3. Identify proteins using, for instance, Scaffold software (Searle 2010) or Proteome Discoverer (Thermo Scientific, USA). Set quality criteria for peptide and protein probability (ideally >99%) as well as for the number of unique peptides. Note that Scaffold software identifies proteins based on the peptides that match a protein with the highest probability using Scaffold. Proteome Discoverer assigns peptides to all possible proteins that match the quality criteria as well as a combination of database searches.
4. Quantify proteins using the amount of spectral counts. Normalize them by the protein length and sample-to-sample variation (Zybailov et al. 2006), resulting in so-called normalized spectral abundance factors (NSAF).

3.9 Taxonomic and Functional Assignment and Data Visualization

1. Check protein groups for homology, and exclude heterogeneous groups from further bioinformatic analysis. The PROteomics result Pruning & Homology group ANotation Engine (PROPHANE; Schneider et al. 2010), web service can be used (www.prophane.de). Check homology using MAFFT (Katoh et al. 2002) in PROPHANE.
2. Assign homologous protein hits obtained from the database search to phylogenetic and functional groups. PROPHANE provides an automated workflow for both

(i) functional analyses using various resources (COG/KOG, TIGRFAMs, and PFAMs) and (ii) lowest common anchor (LCA)-based taxonomic assessments.

3. Visualize the litter metaproteomic dataset(s), for example, by Voronoi tree maps (Eymann et al. 2017). Such tree maps offer the possibility for combining and visualizing hierarchical data structures with two additional dimensions such as area encoding (abundance data) and colour encoding (further categorical or expression change data). This is illustrated for a semiquantitative taxonomic

Fig. 40.3 Tree map visualizing the relative importance of major eukaryotic taxa derived from a proteomic analysis of decomposing beech leaves (Schneider et al., 2012). Cell sizes depict average quantities of protein, and fungi are coloured dark brown, plants dark green, and metazoa dark green-blue. Illustration kindly provided by J. Bernhardt, University of Greifswald, Germany

distribution of proteins obtained from a beech leaf litter sample from a nutrient-rich forest in Austria (Fig. 40.3).

4 Final Remarks

Protein pellets can retain a brownish color after protein concentration or precipitation, even when PVPP is used during grinding and phenol purification. When protein quantification suggests that extraction results in low yields, the use of a weakly alkaline buffer (pH 7.5–8) can be advantageous (Keiblinger et al. 2016), which increases the co-extraction of humic substances. Humic substances can be separated from proteins by SDS-PAGE, because the more highly charged humic substances move faster through the gel than proteins when an electric voltage is applied.

SDS gels stained with Coomassie blue often show a protein smear rather than distinct bands, but this does not hamper further analyses. Reversed-phase HPLC gradient elution chromatography offers high resolution and efficiency and hence is perfectly suited for one- or two-dimensional separation of peptides and proteins.

A significant improvement in sensitivity and resolution on the chromatographic side is provided by ultrahigh-pressure LC as well as modified stationary phases (Becher et al. 2013). Chromatographic separation is usually coupled with fast scanning mass spectrometers (i.e., Orbitrap Velos, LTQ Orbitrap XL) which are able to acquire higher numbers of MS/MS spectra in a given time period. High mass accuracy allows precise peptide identification, which is essential for highly complex samples such as litter (Becher et al. 2013). Detailed settings of HPLC and MS detection strongly rely on the devices that are used. These analyses are often provided by core facilities due to the know-how for operation and utilization of expensive large-scale equipment and the complex infrastructure.

The amount of identified proteins in plant litter by shotgun metaproteomics is rather low compared to the number of genes identified by metagenomics ranging from hundreds to thousands. In part, this is because only a minor part of genes is simultanously expressed, but in part also because metaproteomics still identifies only the most abundant proteins (Myrold et al. 2014) in complex protein mixtures (Keiblinger et al. 2016). Another limitation of metaproteomics is the typically limited taxonomic resolution. Often analyses are only possible to the level of phylum, class, or, at best, order (Bastida et al. 2016a; Bastida et al. 2016b; Starke et al. 2016). This is because of the currently limited availability of relevant complex protein databases. Taxonomic resolution can be improved, however, when a corresponding metagenome (or metatranscriptome) is available to derive an extended protein database. Protein abundances can provide valuable information. For example, Schneider et al. (2012) documented the greatest abundance of fungal exoenzymes that degrade plant polymers at sites where litter decomposition was fastest. This is in agreement with the idea that fungal enzyme production limits the rate of litter decomposition (Schimel and Weintraub 2003).

References

Baldrian, P. (2008). Wood-inhabiting ligninolytic basidiomycetes in soils: Ecology and constraints for applicability in bioremediation. *Fungal Ecology, 1*, 4–12.

Bärlocher, F. (1992). Community organization. In F. Bärlocher (Ed.), *The ecology of aquatic hyphomycetes* (pp. 38–76). Berlin: Springer.

Bastida, F., Jehmlich, N., Lima, K., Morris, B., Richnow, H. H., Hernandez, T., von Bergen, M., & Garcia, C. (2016a). The ecological and physiological responses of the microbial community from a semiarid soil to hydrocarbon contamination and its bioremediation using compost amendment. *Journal of Proteomics, 135*, 162–169.

Bastida, F., Torres, I. F., Moreno, J. L., Baldrian, P., Ondoño, S., Ruiz-Navarro, A., Hernández, T., Richnow, H. H., Starke, R., Garcia, C., & Jehmlich, N. (2016b). The active microbial diversity drives ecosystem multifunctionality and is physiologically related to carbon availability in Mediterranean semi-arid soils. *Molecular Ecology, 25*, 4660–4673.

Becher, D., Bernhardt, J., Fuchs, S., & Riedel, K. (2013). Metaproteomics to unravel major microbial players in leaf litter and soil environments: Challenges and perspectives. *Proteomics, 13*, 2895–2909.

Benndorf, D., Balcke, G. U., Harms, H., & Von Bergen, M. (2007). Functional metaproteome analysis of protein extracts from contaminated soil and groundwater. *ISME Journal, 1*, 224–234.

Boeddinghaus, R. S., Nunan, N., Berner, D., Marhan, S., & Kandeler, E. (2015). Do general spatial relationships for microbial biomass and soil enzyme activities exist in temperate grassland soils? *Soil Biology and Biochemistry, 88*, 430–440.

Burns, R. G., DeForest, J. L., Marxsen, J., Sinsabaugh, R. L., Stromberger, M. E., Wallenstein, M. D., Weintraub, M. N., & Zoppini, A. (2013). Soil enzymes in a changing environment: Current knowledge and future directions. *Soil Biology and Biochemistry, 58*, 216–234.

Eymann, C., Lassek, C., Wegner, U., Bernhardt, J., Fritsch, O. A., Fuchs, S., Otto, A., Albrecht, D., Schiefelebein, U., Cernava, T., Aschenbrenner, I., Berg, G., Grube, M., & Riedel, K. (2017). Symbiotic interplay of fungi, algae, and bacteria within the lung lichen *Lobaria pulmonaria* L. Hoffm. As assessed by state-of-the-art metaproteomics. *Journal of Proteome Research, 16*, 2160–2173.

Gessner, M. O., Thomas, M., Jean-Louis, A.-M., & Chauvet, E. (1993). Stable successional patterns of aquatic hyphomycetes on leaves decaying in a summer cool stream. *Mycological Research, 97*, 163–172.

Katoh, K., Misawa, K., Kuma, K., & Miyata, T. (2002). MAFFT: A novel method for rapid multiple sequence alignment based on fast Fourier transform. *Nucleic Acids Research, 30*, 3059–3066.

Keiblinger K., Masse J., Prescott C., Grayston S. (unpublished). *Microbial communities and their metabolic functions in coniferous tree species – a metaproteomics survey.*

Keiblinger, K. M., & Riedel, K. (2018). Sample preparation for metaproteome analyses of soil and leaf litter. In D. Becher (Ed.), *Microbial proteomics, methods in molecular biology* (Vol. 1841, pp. 303–318). New York: Humana Press.

Keiblinger, K. M., Schneider, T., Roschitzki, B., Schmid, E., Eberl, L., Hämmerle, I., Leitner, S., Richter, A., Wanek, W., Riedel, K., & Zechmeister-Boltenstern, S. (2012a). Effects of stoichiometry and temperature perturbations on beech leaf litter decomposition, enzyme activities and protein expression. *Biogeosciences, 9*, 4537–4551.

Keiblinger, K. M., Wilhartitz, I. C., Schneider, T., Roschitzki, B., Schmid, E., Eberl, L., Riedel, K., & Zechmeister-Boltenstern, S. (2012b). Soil metaproteomics – Comparative evaluation of protein extraction protocols. *Soil Biology and Biochemistry, 54*, 14–24.

Keiblinger, K. M., Fuchs, S., Zechmeister-Boltenstern, S., & Riedel, K. (2016). Soil and leaf litter metaproteomics—A brief guideline from sampling to understanding. *FEMS Microbiology Ecology, 92*, fiw180.

Myrold, D. D., Zeglin, L. H., & Jansson, J. K. (2014). The potential of metagenomic approaches for understanding soil microbial processes. *Soil Science Society of America Journal, 78*, 3–10.

Nannipieri, P., Ascher, J., Ceccherini, M., Landi, L., Pietramellara, G., & Renella, G. (2003). Microbial diversity and soil functions. *European Journal of Soil Science, 54*, 655–670.

Osono, T. (2011). Diversity and functioning of fungi associated with leaf litter decomposition in Asian forests of different climatic regions. *Fungal Ecology, 4*, 375–385.

Purahong, W., Wubet, T., Lentendu, G., Schloter, M., Pecyna, M. J., Kapturska, D., Hofrichter, M., Krüger, D., & Buscot, F. (2016). Life in leaf litter: Novel insights into community dynamics of bacteria and fungi during litter decomposition. *Molecular Ecology, 25*, 4059–4074.

Quinn, G. P., & Keough, M. J. (2002). *Experimental design and data analysis for biologists*. Cambridge: Cambridge University Press.

Schimel, J. P., & Weintraub, M. N. (2003). The implications of exoenzyme activity on microbial carbon and nitrogen limitation in soil: A theoretical model. *Soil Biology and Biochemistry, 35*, 549–563.

Schneider, T., & Riedel, K. (2010). Environmental proteomics: Analysis of structure and function of microbial communities. *Proteomics, 10*, 785–798.

Schneider, T., Gerrits, B., Gassmann, R., Schmid, E., Gessner, M. O., Richter, A., Battin, T., Eberl, L., & Riedel, K. (2010). Proteome analysis of fungal and bacterial involvement in leaf litter decomposition. *Proteomics, 10*, 1819–1830.

Schneider, T., Keiblinger, K. M., Schmid, E., Sterflinger-Gleixner, K., Ellersdorfer, G., Roschitzki, B., Richter, A., Eberl, L., Zechmeister-Boltenstern, S., & Riedel, K. (2012). Who is who in litter decomposition? Metaproteomics reveals major microbial players and their biogeochemical functions. *ISME Journal, 6*, 1749–1762.

Searle, B. C. (2010). Scaffold: A bioinformatic tool for validating MS/MS-based proteomic studies. *Proteomics, 10*, 1265–1269.

Sinsabaugh, R. L., Lauber, C. L., Weintraub, M. N., Ahmed, B., Allison, S. D., Crenshaw, C., Contosta, A. R., Cusack, D., Frey, S., Gallo, M. E., Gartner, T. B., Hobbie, S. E., Holland, K., Keeler, B. L., Powers, J. S., Stursova, M., Takacs-Vesbach, C., Waldrop, M. P., Wallenstein, M. D., Zak, D. R., & Zeglin, L. H. (2008). Stoichiometry of soil enzyme activity at global scale. *Ecology Letters, 11*, 1252–1264.

Starke, R., Kermer, R., Ullmann-Zeunert, L., Baldwin, I. T., Seifert, J., Bastida, F., von Bergen, M., & Jehmlich, N. (2016). Bacteria dominate the short-term assimilation of plant-derived N in soil. *Soil Biology and Biochemistry, 96*, 30–38.

Wal, A., Geydan, T. D., Kuyper, T. W., & Boer, W. (2013). A thready affair: Linking fungal diversity and community dynamics to terrestrial decomposition processes. *FEMS Microbiology Reviews, 37*, 477–494.

Zybailov, B., Mosley, A. L., Sardiu, M. E., Coleman, M. K., Florens, L., & Washburn, M. P. (2006). Statistical analysis of membrane proteome expression changes in *Saccharomyces cerevisiae*. *Journal of Proteome Research, 5*, 2339–2347.

Part V
Enzymatic Capabilities

Chapter 41
Extracellular Fungal Hydrolytic Enzyme Activity

Shawn D. Mansfield

Keywords Carbohydrates · Cellulases · Cellulose · Enzymatic potential · Hemicellulases · Hydrolysis · Litter degradation · Plant cell walls · β-Glucosidase

1 Introduction

Plant polysaccharides are the most abundant organic polymers in the biosphere. Microorganisms produce a battery of extracellular hydrolytic and oxidative enzymes to depolymerize them into smaller, more readily useable compounds. These breakdown products are then assimilated and serve as energy sources or precursors in cell biosynthesis.

The most abundant biopolymer on earth is cellulose, which is found primarily as a structural component in the cell wall of plants and marine algae. Cellulose consists of long, unbranched homopolymers of D-glucose units linked by β-1,4-glycosidic bonds to form a linear chain of over 10,000 glucose residues (Joshi and Mansfield 2007; Hon and Shiraishi 1991). Individual glucan chains adhere to each other by hydrogen bonding and van der Waals forces, and form insoluble networks. The secondary and tertiary structures of native cellulose are complex and may vary significantly depending on the source and biosynthetic machinery that produces the polymer (e.g. vascular plant or marine algae). Furthermore, the cellulose polymers of higher plants are intricately associated with lignin and hemicellulose moieties, resulting in even more complex morphologies.

Primary cellulose degradation results from either chemical or enzymatic hydrolysis of the polymer into oligomeric and monomeric soluble sugars. Due to the inherent insolubility and physical complexity of polymeric cellulose, several different enzymes are needed for complete solubilization (Mansfield et al. 1999). The

S. D. Mansfield (✉)
Department of Wood Science, University of British Columbia, Vancouver, BC, Canada
e-mail: shawn.mansfield@ubc.ca

F. Bärlocher et al. (eds.), *Methods to Study Litter Decomposition*,
https://doi.org/10.1007/978-3-030-30515-4_41

current understanding is that enzyme-mediated hydrolysis of native cellulose results primarily from the interaction of extracellular β-1,4-endoglucanases and β-1,4-exoglucanases (cellobiohydrolases) to yield cello-oligosaccharides such as cellobiose, which are subsequently cleaved to glucose by β-glucosidase. The activities of endo- and exoglucanases are synergistic (Mansfield et al. 1999). The general mechanism suggests that the endoglucanases produce free chain ends on the cellulose surface for the cellobiohydrolases to act upon. However, synergy has also been observed between different types of cellobiohydrolases (Nidetzky et al. 1993), as well as between two endoglucanases (Gübitz et al. 1998; Mansfield et al. 1998). Although all cellulolytic enzymes have similar bond specificities (β-1,4), important functional differences are found in their mode of action. More recently, it has been shown that the polysaccharide monooxygenase enzyme AA9 (formerly known as GH61) interacts synergistically with cellulases to enhance the enzymatic hydrolysis of a range of "commercially relevant" pretreated and "model" cellulosic substrates (Hu et al. 2014; Hu et al. 2016).

Generally, the activity of endoglucanases is assayed with a water-soluble substrate, such as carboxymethylcellulose (CMC) or phosphoric acid-swollen cellulose. The assay quantifies the amount of reducing sugars released from the substrate by the interaction with the enzyme (Ghose 1987). In contrast, exoglucanases differ substantially in their substrate specificity and are capable of solubilizing crystalline cellulose substrates, such as Avicel, filter paper, or cotton. Their activity is also usually measured by the amount of reducing ends that are generated (Ghose 1987). An alternative method uses a chromophoric disaccharide derivative and a homologous series of 4-methylumbelliferyl glycosides of cello-oligosaccharides (van Tilbeurgh et al. 1982, Chapter 44).

β-Glucosidases catalyze the hydrolysis of terminal, nonreducing β-D-glucose residues from β-D-glucosides, including cellobiose and cello-oligosaccharides. In some cases, mixed oligosaccharides consisting of mannose and glucose serve as substrates. In the enzymatic conversion of cellulose, it is important that the level of β-glucosidase is in excess, as cellobiose has an inhibitory effect on the cellobiohydrolases (Mansfield et al. 1999).

In addition to cellulose, various hemicelluloses are important polysaccharides in nature. These are low-molecular-weight heteropolymeric polysaccharides composed of a number of different residues, the most common of which are D-xylose, D-mannose, D-galactose, D-glucose, L-arabinose, D-rhamnose, D-galacturonic acid, D-glucuronic acid, and 4-O-methyl-D-glucuonic acid (Fengel and Wegener 1983; Sjöström 1993). The complexity and chemical nature of the hemicelluloses vary both between cell types and species.

The main xylan-derived hemicelluloses are polysaccharides with a backbone of 1,4-linked β-D-xylopyranosyl units, substituted at the carbon 2 and 3 positions. The extent of substitution is dependent on origin (Sjöström 1993): xylans of deciduous trees and conifers carry 4-O-methylglucuronic acid and L-arabinofuranosyl side groups, respectively, while xylans from annual plants may contain only the latter or both side groups. Furthermore, xylans derived from deciduous trees are acetylated, whereas xylans of conifers are not (Sjöström 1993). Xylans from annual plants may, in addition to acetyl groups, carry esterified phenolic hydroxycinnamic acids such as feruloyl and *p*-coumaroyl moieties (Grabber et al. 2000). The crucial enzyme for xylan depolymerization is endo-β-1,4-xylanase, which preferentially attacks the main xylan chain, generating non-substituted and branched or esterified oligosaccharides. The branching substituents are liberated by corresponding glycosidases or esterases (debranching or accessory enzymes): α-L-arabinofuranosidases and α-glucuronidase. Finally, acetic acid, ferulic acid, and *p*-coumaric acid residues can be liberated from the xylan by corresponding xylan esterases. β-Xylosidase liberates D-xylose from the nonreducing end of xylo-oligosaccharides.

Mannan-based hemicelluloses are substituted heteropolysaccharides that are widespread in both deciduous and coniferous trees. Their designation is largely dependent on the constituent monomers comprising the backbone and the side chains and can be divided into (1) pure mannans, (2) glucomannans, (3) galactomannans, and (4) galactoglucomannans. The biodegradation of β-mannans occurs by the synergistic action of endo-1,4-β-mannanases, β-D-mannosidases, β-D-glucosidases, α-D-galactosidases, and acetyl mannan esterases (Tenkanen et al. 1993). Endo-β-mannanases cleave polymeric mannans as well as mannooligosaccharides, usually with a degree of polymerization greater than three. Some endomannanases also cleave β-1,4 linkages between mannose and glucose in glucomannans. The degree of substitution and the distribution of the side groups significantly influence the overall capacity for endomannanase to catalyze the degradation of β-1,4 linkages. Thus, the combined actions of endomannanases and accessory enzymes such as α-galactosidase and acetyl esterase are required for total degradation of galactoglucomannan (Tenkanen et al. 1993). β-Mannosidase catalyzes the hydrolysis of terminal, nonreducing β-D-mannose residues in mannans, heteromannans, and mannooligosaccharides. Some β-mannosidases also cleave the 1,4-β-mannose-glucose linkages in glucomannans. β-Mannosidases occur in a wide range of plant and animal tissues and in many microorganisms (Gübitz et al. 1996).

The use of purified enzymes is essential to determine substrate specificities of individual enzymes and to elucidate molecular mechanisms of catalysis. However, simplified assays exist to determine each of the general classes of extracellular cellulolytic and hemicellulolytic enzymes secreted by microorganisms. This chapter presents three types of methods to quantify the major hydrolytic enzymes in fungal cultures, but does not include procedures to determine the debranching enzymes required for total cell wall carbohydrate degradation: (1) The determination of β-1,4-endoglucanases, β-1,4-endoxylanase, and β-1,4-endomannanase (on any mannan-based substrate) follows a variation of Bailey et al. (1992), with appropriate substitution for substrates and corresponding standards. For example, β-1,4-endoglucanases activity is analyzed on carboxymethylcellulose using glucose as a standard. (2) Filter paper activity is a good measure of total cellulase activity. Since exoglucanases are required for the solubilization of crystalline cellulose, this method is also a relatively good indicator of the presence of cellobiohydrolase; however, it does not specifically quantify exoglucanases (Ghose 1987). (3) Finally, β-glucosidase, β-xylosidase, and β-mannosidase are quantified based on an assay by Ghose (1987), with appropriate substitution for substrates and corresponding standards. For example, β-glucosidase activity is determined by using *p*-nitrophenyl-β-glucoside and glucose.

2 Equipment, Chemicals, and Solutions

2.1 Equipment and Material

- Analytical balance
- Cooled centrifuge (4 °C, 20,000 *g*)
- Shaking incubator (20 °C)
- Boiling water bath
- Water bath (50 °C)
- Water bath (20 °C)
- pH meter
- Magnetic stirrer
- Spectrophotometer
- Vortex
- Laboratory timer or stopwatch
- Adjustable micropipettors (0.2–1.8 ml)
- Petri plates
- Erlenmeyer flasks
- Centrifuge tubes
- Cuvettes (disposables are suitable)
- Test tubes (15 ml)
- Test tube rack
- Filter paper (Whatman No. 1)
- Fungal isolates maintained on 1% malt agar plates at 15–20 °C (to isolate aquatic hyphomycetes, see Chap. 23)
- Sterilized leaf discs

2.2 Chemicals

- Agar
- Malt extract
- Yeast extract
- KH_2PO_4
- $MgSO_4 \cdot 7H_2O$
- NaCl
- K_2HPO_4
- KNO_3

- KCl
- $(NH_4)_2SO_4$
- NaOH
- 3,5-Dinitrosalicylic acid (DNS)
- $KNaC_4H_4O_6 \cdot 4H_2O$ (Na K tartrate or Rochelle salt)
- $Na_2S_2O_5$ (Na metabisulfite)
- Phenol (melt at 50 °C)
- Deionized water
- Glycine
- Glucose
- Xylose
- Mannose
- Sodium citrate buffer (1 M, pH 4.5)
- Carboxymethylcellulose (2% w:v in 50 mM sodium citrate buffer)
- *p*-Nitrophenyl-β-glucoside
- *p*-Nitrophenyl-β-xyloside
- Birchwood xylan (1% w:v in 50-mM sodium citrate buffer)
- Mannan (1% w:v in 50-mM sodium citrate buffer)
- Ivory nut mannan (pure mannan)
- Konjac mannan (glucomannan)
- Softwood galactoglucomannan
- Locust bean mannan or guar gum (galactomannan)
- *p*-Nitrophenyl-β-mannoside

2.3 Solutions

- Mineral solution for fungal growth: 10-mM KNO_3, 2.5-mM KH_2PO_4, 2.5-mM K_2HPO_4, 3-mN NaCl, 1-mM $MgSO_4$; adjust to pH 7 before autoclaving
- Solution 1: Dinitrosalicylic acid (DNS) reagent – measure 801.0 ml of deionized water; dissolve 11.2-g NaOH and 6.0-g DNS in about 400 ml of the water in a 1000-ml container; use a powder funnel to add 173.2-g NaK tartrate and 4.7-g Na metabisulfite; use the remaining water to wash all reagents into the 1000-ml container; add 4.3-g phenol melted at 50 °C and stir to dissolve
- Solution 2: 0.4-M glycine buffer – dissolve 60 g of glycine in 1500 ml of deionized water, add 50% (v:v) NaOH solution until the pH is 10.8, and dilute to exactly 2000 ml

3 Experimental Procedures

3.1 *Fungal Growth in Liquid Culture*

1. Dispense 150 ml of mineral solution to 500-ml Erlenmeyer flask.
2. Add 1–2 g sterile leaf discs to flask.
3. Inoculate aseptically with a 5-mm plug from a 7–14-day-old culture.
4. Grow isolate as shake flask culture (ca. 140 rpm) at 15–20 °C for 7–21 days.
5. Transfer content of flask to centrifuge tube and recover culture supernatant by centrifugation (20,000 *g*) for 10 min at 4 °C.

3.2 *Endohydrolase Activity*

1. Select appropriate standards and substrate.
2. Make up stock solution of standards (i.e., 10 mM).
3. Dilute standard stock to give dilution standards (2, 4, 6, 8, and 10 μM).
4. Dispense 1.5 ml of each standard and unknown substrate in separate 15-ml test tubes.
5. Condition substrate in water bath at 50 °C for at least 5 min.
6. At 1-min intervals, add 0.5 ml of culture filtrate, standards, or buffer (blank) to each test tube containing substrate.
7. Vortex tube and return to water bath.
8. Incubate for exactly 5 min.
9. Stop reaction by adding 3.0-ml DNS reagent.
10. Vortex tube and place directly in boiling water bath for exactly 5 min.
11. Prepare enzyme blank by adding 3.0 ml of DNS to a tube containing substrate, and then add 0.5 ml of culture filtrate and place in the boiling water bath for exactly 5 min.
12. Cool tube down in water bath (20 °C).
13. Zero spectrophotometer with reaction blank at 540 nm.
14. Read absorbance of sample at room temperature at 540 nm.
15. Generate linear standard curve by plotting sugar concentration (μmol ml^{-1}) versus absorbance at 540 nm (correct for dilution of standard concentrations) by forcing line through zero. Obtain slope, intercept, and r^2 value, which should be >0.98.
16. Determine net absorbance by subtracting appropriate culture filtrate blanks from hydrolysis samples (averaged value).
17. Determine sugar concentration (μmol ml^{-1}) liberated by hydrolysis from calibration curve.
18. Calculate enzyme activity, determined as nkat ml^{-1} culture filtrate, by the following equation:

$$\frac{nkat}{ml} = \frac{\mu mol \cdot 1000}{ml\ filtrate \cdot 300s} \tag{41.1}$$

3.3 Filter Paper Activity (Total Cellulase Activity)

1. Prepare glucose standards (6.7, 5.0, 3.3, and 2.0 mg ml^{-1}).
2. Place 50 mg of Whatman No. 1 (1 cm × 6 cm) filter paper into 25-ml test tube.
3. Add 1 ml of 50 mM sodium citrate buffer (pH 4.5).
4. Condition substrate in water bath at 50 °C for at least 5 min.
5. Add 0.5 ml of culture filtrate, sugar stock solutions, and blank (buffer) to individual test tubes and mix.
6. Incubate at 50 °C for exactly 60 min.
7. Terminate reaction by adding 3.0 ml of DNS reagent.
8. Vortex tube and place in boiling water bath for exactly 5 min.
9. Cool tube down in water bath (20 °C).
10. Add 10 ml of deionized water.
11. Zero spectrophotometer with buffer blank at 540 nm.
12. Determine absorbance of sample at 540 nm.
13. Generate linear standard curve by plotting absolute amount of sugar (mg 0.5 ml^{-1}) versus absorbance at 540 nm by forcing line through zero. Obtain slope, intercept, and r^2 value, which should be >0.98.
14. Determine net absorbance by subtracting culture filtrate blanks from absorbance of hydrolysis samples.
15. Determine concentration of glucose liberated during the reaction from calibration curve.
16. Calculate total cellulase activity, determined as units ml^{-1} of culture filtrate, by the following equation:

$$\frac{Units}{ml} = \frac{mg\,Unknown \cdot 0.5 \cdot 60}{0.18} \tag{41.2}$$

3.4 β-Glucosidase Activity

1. Add 1 ml of 5 mM *p*-nitrophenyl-β-glucoside in 50-mM sodium acetate buffer (pH 4.8) to a 10-ml test tube.
2. Add 1.8 ml of 100-mM acetate buffer.
3. Condition substrate in water bath at 50 °C for at least 5 min.
4. To the substrate add 200-μl culture filtrate, and then vortex vigorously.
5. Place in water bath at 50 °C for exactly 30 min.

6. Terminate reaction by adding 4 ml 0.4-M glycine buffer.
7. Cool tube down in water bath (20 °C).
8. Use blank of 200-μl fresh culture media to zero spectrophotometer at 430 nm.
9. Determine absorbance of sample at 430 nm.
10. Generate linear standard curve by plotting sugar concentration (μmol ml^{-1}) versus absorbance at 430 nm (correct for dilution of standard concentrations) by forcing line through zero. Obtain slope, intercept, and r^2 value, which should be >0.99.
11. Average duplicate measurements.
12. Determine net absorbance by subtracting appropriate culture filtrate blanks from hydrolysis samples (averaged value).
13. Determine sugar concentration (μmol ml^{-1}) liberated by hydrolysis from calibration curve.
14. Use the following equation to calculate enzyme activity as nkat ml^{-1} of culture filtrate, where 1 nkat is the activity that releases 1 nmol of *p*-nitrophenol equivalent per second during the assay:

$$\frac{nkat}{ml} = \frac{\mu mol \cdot 1000}{ml\, Unknown \cdot 1800s} \tag{41.3}$$

4 Final Remarks

Instead of culture filtrates, extracts from stream-exposed leaves can be used. However, this generally requires tests to ensure that there is measurable activity.

Activities of hydrolytic enzymes, which are thermostable, have traditionally been measured at 50 °C, even though this is generally far higher than the temperature experienced by microorganisms in the field. These assays thus measure enzymatic potentials. If actual release of sugars under natural conditions is of interest, incubation at ambient stream temperatures is required. This may involve much longer incubation periods, and precautions will have to be taken to prevent bacterial contamination.

In the endocellulase assay, each set of assays has a reagent blank and a set of standards where buffer and standards are added to the reaction instead of culture filtrate. Additionally, each assay has an enzyme blank where DNS is added to the substrate before the enzyme so that enzyme activity is prevented and the reducing sugars in the culture filtrate can be determined.

Enzyme activity is generally expressed in nkat ml^{-1} of culture filtrate, where 1 nkat is the activity that releases 1 nmol of product (i.e., reducing sugar or *p*-nitrophenol equivalent) per second during the assay. The international unit (IU) is also often used. It represents the release of 1 μmol of product per minute.

1 katal = 1 mol s^{-1}

1 nkat = 1 nmol s^{-1}
1 IU = 1 μmol min^{-1} = 16.7 nkat

References

Bailey, M. J., Biely, P., & Poutanen, K. (1992). Interlaboratory testing of methods for assay of xylanase activity. *Journal of Biotechnology, 23*, 257–270.

Fengel, D., & Wegener, G. (1983). *Wood: Chemistry, ultrastructure, reactions*. New York: Walter de Gruyter.

Ghose, T. K. (1987). Measurement of cellulase activities. *Pure and Applied Chemistry, 59*, 257–268.

Grabber, J. H., Ralph, J., & Hatfield, R. D. (2000). Cross-linking of maize walls by ferulate dimerization and incorporation into lignin. *Journal of Agriculture and Food Chemistry, 48*, 6106–6113.

Gübitz, G. M., Hayn, M., Sommerauer, M., & Steiner, W. (1996). Mannan-degrading enzymes from *Sclerotium rolfsii*: Characterisation and synergism of two endo β-mannanases and a β-mannosidase. *Bioresource Technology, 58*, 127–135.

Gübitz, G. M., Mansfield, S. D., Böhm, D., & Saddler, J. N. (1998). Effect of endoglucanases and hemicellulases in magnetic and flotation deinking of xerographic and laser-printed papers. *Journal of Biotechnology, 65*, 209–215.

Hon, D. N.-S., & Shiraishi, N. (1991). *Wood and cellulosic chemistry*. New York: Marcel Dekker.

Hu, J., Arantes, V., Pribowo, A., Gourlay, K., & Saddler, J. N. (2014). Substrate factors that influence the synergistic interaction of AA9 and cellulases during the enzymatic hydrolysis of biomass. *Energy & Environmental Science, 7*, 2308–2315.

Hu, J., Pribowo, A., & Saddler, J. N. (2016). Oxidative cleavage of some cellulosic substrates by auxiliary activity (AA) family 9 enzymes influences the adsorption/desorption of hydrolytic cellulase enzymes. *Green Chemistry, 18*, 6329–6336.

Joshi, C. P., & Mansfield, S. D. (2007). The cellulose paradox – simple molecule, complex biosynthesis. *Current Opinion in Plant Biology, 10*, 220–226.

Mansfield, S. D., Saddler, J. N., & Gübitz, G. M. (1998). Characterization of endoglucanases from the brown rot fungi *Gloeophyllum sepiarium* and *Gloeophyllum trabeum*. *Enzyme and Microbial Technology, 23*, 133–140.

Mansfield, S. D., Mooney, C., & Saddler, J. N. (1999). Substrate and enzyme characteristics that limit cellulose hydrolysis. *Biotechnology Progress, 15*, 804–816.

Nidetzky, B., Hayn, M., Macarron, R., & Steiner, W. (1993). Synergism of *Trichoderma reesei* cellulases while degrading different celluloses. *Biotechnology Letters, 15*, 71–76.

Sjöström, E. (1993). *Wood chemistry: Fundamentals and applications* (2nd ed.). London: Academic.

Tenkanen, M., Puls, J., Rättö, M., & Viikari, L. (1993). Enzymatic deacetylation of galactoglucmannans. *Applied Microbiology and Biotechnology, 39*, 159–165.

van Tilbeurgh, H., Claeyssens, M., & de Bruyne, C. K. (1982). The use of 4-methylumbelliferyl and other chromophoric glycosides in the study of cellulolytic enzymes. *FEMS Letters, 149*, 152–156.

Chapter 42
Cellulases

Martin Zimmer

Keywords Animal endogenous cellulases · Cellobiases · Cellulases · Cellulose hydrolysis · Depolymerization · Endoglucanase · Enzyme activity · Exoglucanase · Extracellular microbial enzymes · Gut enyzmes · Gut symbionts · Lignocellulose degradation · Plant cell walls

1 Introduction

Cellulose is an unbranched chain of several thousand D-glucose monomers formed by β-1,4-glycosidic bonds. In situ, numerous parallel poly-glucose chains form insoluble complexes of crystalline cellulose fibres through hydrogen bonds and van der Waals forces. These fibres, together with the surrounding lignin matrix, form lignocellulose, the major component of plant litter, which effectively resists degradation (e.g. Royer and Minshall 2001).

Along with the ever-growing interest in industrially utilizing plant biomass, the study of mechanisms of cellulose degradation has gained increased attention (Cragg et al. 2015). Despite the variety of mechanisms discovered over the last decades (for review, see Cragg et al. 2015), some general patterns have emerged, most notably that it always requires the synergistic action of several enzyme classes to degrade native crystalline cellulose to its glucose units: In fungi (Ljungdahl and Eriksson 1985) and flagellate gut symbionts of termites (Yamin and Trager 1979), endo-β-1,4-glucanases (endocellulase, C_X-cellulase; EC 3.2.1.4) cleave inner β-1,4-glycosidic bonds and thus generate oligosaccharides. Cellobiohydrolases (exo-β-1,4-glucanase, exocellulase, C1-cellulase; EC 3.2.1.91) split off cellobiose,

M. Zimmer (✉)
Leibniz Centre for Tropical Marine Research (ZMT), University of Bremen, Bremen, Germany
e-mail: martin.zimmer@leibniz-zmt.de

F. Bärlocher et al. (eds.), *Methods to Study Litter Decomposition*,
https://doi.org/10.1007/978-3-030-30515-4_42

a glucose dimer, from the non-reducing end of the oligosaccharide chain (Wood and Garcia-Campayo 1990). While amorphous and soluble cellulose may be degraded through the action of endocellulases alone, the degradation of crystalline cellulose requires the activity of an exocellulase (Wood and Garcia-Campayo 1990), at least in the case of common extracellular fungal cellulases. Cellobiase (β-glucosidase, EC 3.2.1.21) cleaves cellobiose into two glucose moieties (Ljungdahl and Eriksson 1985). Possibly, glucohydrolase (EC 3.2.1.74) is also involved in cellulose degradation as a component of exoglucanase (Goyal et al. 1991), splitting off glucose – instead of cellobiose – from the non-reducing end of poly- and oligosaccharides. Alternatively, cellobiose can be degraded oxidatively through the action of a cellobiose-quinone-oxidoreductase (EC 1.1.5.1; Evans et al. 1991), which also reduces quinones and phenoxy radicals generated during phenol oxidation and converts cellobiose to cellobionic acid (Ljungdahl and Eriksson 1985).

In aerobic bacteria, cellulose hydrolysis has been attributed to two types of enzymes adhering to the cell wall and acting like the fungal endocellulases and cellobiases (Ljungdahl and Eriksson 1985; Wood and Garcia-Campayo 1990). Exocellulases have been found in only a few bacteria (Rapp and Beermann 1991). The cleavage of crystalline cellulose by organisms lacking exocellulase appears to be due to an intramolecular synergism of bacterial endocellulases (Din et al. 1994). Cellulolysis by anaerobic bacteria results from the action of "cellulosomes" (Leschine 1995), which are multi-protein complexes containing endo- or exocellulases as well as xylanases (EC 3.2.1.37; see Chap. 41) and several other proteins with structural or substrate-binding functions (Wood and Garcia-Campayo 1990).

Most detritivorous invertebrates lack large digestive gut chambers to cultivate microbial gut symbionts. This has favoured the evolution of endogenous cellulases in these animals. Genes of diverse cellulase genes have been documented in Insecta (Shelomi et al. 2014), Gastropoda (Tsuji et al. 2013), Crustacea (King et al. 2010; Kostanjsek et al. 2010; Bui and Lee 2015) and Annelida (Nozaki et al. 2009). They possibly act through similar mechanisms as those of aerobic bacteria (Rouland et al. 1988) and can degrade cellulose without microbial assistance. An alternative mechanism of cellulose degradation by enzymes of animal origin has recently been described for detritivorous land crabs (Allardyce et al. 2010). Upon removal of cross-linkages between cellulose chains by hemicellulases, endo-β-1,4-glucanases cleave glycosidic bonds within amorphous cellulose regions, possibly releasing short oligomers into solution. A glucohydrolase attacks these short oligomers, and may also attack the chain ends on the surface of the cellulose fibre that have been exposed by the endo-β-1,4-glucanases, eventually resulting in the release of glucose. Nevertheless, cellulases of animal origin are often complemented by microbial enzymes acquired from gut symbionts (e.g. Koenig et al. 2013) or ingested along with the food (e.g. Martin 1984).

Numerous approaches to measuring cellulase activity have been published over the last decades. They can be grouped as (1) determining the change in physical properties of the substrate, (2) monitoring the loss of substrate and (3) quantifying the accumulation of hydrolysis products. A commonly used approach is the IUPAC (International Union of Pure and Applied Chemistry) method of gravimetrically determining the loss of cellulose filter paper upon exposure to enzyme preparations or environmental samples (e.g. Yu et al. 2016). When quantifying the products of cellulose hydrolysis, endoglucanase activity can be assayed with a water-soluble cellulose derivate, such as carboxymethylcellulose (CMC) or phosphoric acid-swollen cellulose (Chap. 43), but exocellulase activity can only be estimated by measuring the release of glucose from crystalline cellulose such as Avicel, filter paper or cotton. However, the common method of determining the release of reducing groups from cellulose (Ghose 1987; see Chap. 41) is not specific to glucose and is prone to interference by a number of substances in environmental samples (Skambracks and Zimmer 1998). The approach described here uses a more specific, enzyme-based technique to quantify the amount of glucose released from crystalline cellulose by the combined activities of several enzymes (Skambracks and Zimmer 1998).

2 Equipment and Materials

2.1 Equipment

- Homogenizer (e.g. electronic disperser or mortar and pestle for leaf litter; rotation grinder or ultrasonic disintegrator for gut and faeces samples)
- Incubation tubes; e.g., glass tubes with screw caps (15–20 ml) for leaf litter; plastic reaction tubes (1.5 ml) for gut and faeces samples
- Analytical balance
- Shaker
- Centrifuge
- Pipettes (100–1000 μl; 10–100 μl)
- Plastic cuvettes
- Spectrophotometer

2.2 Material

- Field-collected leaf litter
- Dissected guts of detritivores having fed on leaf litter (gut epithelium best removed)
- Faeces of detritivores having fed on leaf litter

2.3 Chemicals

- α-Cellulose
- KH_2PO_4
- Na_2HPO_4
- Citric acid monohydrate (citrate)
- NaN_3
- Commercially available kit for the determination of glucose and fructose in food; if the required solutions are prepared in the laboratory:
  - Tri-ethanolamine-HCl
  - MgSO4. 7H2O
  - NaOH, 5 mol l–1
  - NADP-Na2
  - ATP-Na_2H_2
  - $NaHCO_3$
- Hexokinase (130–250 U mg^{-1})
- Glucose-6-phosphate-dehydrogenase (100 U mg^{-1})
- Double-distilled water (H_2O)

2.4 Solutions

- Solution 1 (contained in kit): 0.75 M tri-ethanolamine buffer (pH 7.6), including 10 mM NADP, 80 mM ATP.
- Solution 2 (contained in kit): 2 mg ml^{-1} hexokinase and 1 mg ml^{-1} glucose-6-phosphate-dehydrogenase.
- Solution 3: 0.05 M K-Na-phosphate buffer: 415 ml 0.1 M KH_2PO_4 + 85 ml 0.1 M Na_2HPO_4 + 500 ml double-distilled water, pH 6.2; if prepared accurately, pH does not need to be adjusted.
- Solution 4: Citrate-phosphate buffer: 400 ml 0.1 M citrate (citric acid monohydrate) + 600 ml 0.2 M Na_2HPO_4, pH 5.8; if prepared accurately, pH does not need to be adjusted.
- Solution 5: Citrate-phosphate buffer, pH 5.8, with 0.05% NaN_3.

3 Experimental Procedures

3.1 Extraction of Microbial Enzymes

1. Weigh samples of leaf litter (corresponding to 50–100 mg dry mass), dissected guts (5–10 mg) or faeces (5–10 mg).

2. Determine dry mass/fresh mass ratios to estimate dry mass of samples from fresh mass.
3. The appropriate method of enzyme extraction depends on the source of enzymes; with extracellular enzymes, thoroughly chopping up sample material with a homogenizer is sufficient for accurate measurement of enzyme activity; with cell-bound enzymes, additional sonication is recommended.
4. Homogenize litter samples with the appropriate method in 10 ml of 0.05 M phosphate buffer or gut or faeces samples in 1 ml of 0.05 M phosphate buffer; although cellulases are quite stable, place samples on ice during homogenization to avoid thermal denaturation of enzymes.
5. Homogenates can be stored frozen (−20 °C) until used for assays.
6. Centrifuge suspensions (5 min; ca. 10,000 *g*, 4 °C).
7. Use supernatants for extracellular cellulase activity; pellets can be used for estimating cellular enzyme activities (such as dehydrogenases).

3.2 Determination of Cellulase Activity

1. Add 20 mg α-cellulose to 200 μl aliquots of the supernatant.
2. Add 200 μl citrate-phosphate buffer with 0.05% NaN_3.
3. Incubate on a shaker for 18–24 h at 20 °C.
4. Centrifuge for 5 min at ca. 10,000 *g* and 4 °C.
5. Add 50 μl of the supernatant to 450 μl of Solution 1.
6. Add 1000 μl of double-distilled water.
7. Measure absorbance (A_0) at 340 nm.
8. Add 10 μl of Solution 2.
9. Incubate for 30 min at room temperature.
10. Measure absorbance (A_{30}) at 340 nm.
11. Calculate ΔA as $A_{30}–A_0$.
12. Calculate glucose concentration, *c*, of the sample in mg ml^{-1}:

$$c = \frac{test\ volume \cdot molecular\ weight\ of\ glucose}{extinction\ coefficient\ of\ NADPH \cdot cuvette\ length \cdot sample\ volume \cdot 1000} \cdot \Delta A$$
$$= \frac{1.51 \cdot 180.16}{6.3 \cdot 1.0 \cdot 0.05 \cdot 1000} \left(\frac{mL \cdot \frac{g}{mol}}{\frac{L}{mmol \cdot cm} \cdot cm \cdot mL \cdot 1000} \right) \tag{42.1}$$

13. Run controls without adding α-cellulose to the incubation, and subtract *c* values from those of samples before calculating cellulase activity.
14. Calculate cellulase activity (μg glucose mg^{-1} h^{-1}):

$$cellulase activity = \frac{c \cdot dilution\,factor \cdot 1000}{sampledrymass \cdot incubation\ \ time} \quad \left(\frac{\frac{mg}{mL} \cdot mL \cdot 1000}{mg \cdot h} \right) \tag{42.2}$$

where the dilution factor = 50 for litter and 5 for guts and faeces, sample dry mass = 50–100 mg for litter and 5–10 mg for guts and faeces and incubation time = 18–24 h.

4 Final Remarks

If no kit is used, Solution 1 can be prepared in the laboratory:

- Solution 1a: dissolve 7.0 g tri-ethanolamine-HCl and 0.125 g $MgSO_4$. $7H_2O$ in 40 ml H_2O; add ca. 2 ml NaOH (5 M) to adjust pH to 7.6; add H_2O to 50 ml.
- Solution 1b: dissolve 25 mg NADP-Na_2 in 2.5 ml H_2O.
- Solution 1c: dissolve 125 mg ATP-Na_2H_2 and 124 mg $NaHCO_3$ in 2.5 ml H_2O.
- Mix 50 ml of Solution 1a with 2.5 ml of Solution 1b and 2.5 ml of Solution 1c; store at 4 °C for up to 4 weeks.

References

Allardyce, B. J., Linton, S. M., & Saborowski, R. (2010). The last piece in the cellulase puzzle: The characterisation of β-glucosidase from the herbivorous gecarcinid land crab *Gecarcoidea natalis*. *The Journal of Experimental Biology, 213*, 2950–2957.

Bui, T. H. H., & Lee, S. Y. (2015). Endogenous cellulase production in the leaf litter foraging mangrove crab *Parasesarma erythodactyla*. *Comparative Biochemistry and Physiology Part B: Biochemistry and Molecular Biology, 179*, 27–36.

Cragg, S. M., Beckham, G. T., Bruce, N. C., Bugg, T. D. H., Distel, D. L., Dupree, P., Green Etxabe, A., Goodell, B. S., Jellison, J., McGeehan, J. E., McQueen-Mason, S. J., Schnorr, K., Walton, P. H., Watts, J. E. M., & Zimmer, M. (2015). Lignocellulose degradation mechanisms across the tree of life. *Current Opinion in Chemical Biology, 29*, 108–119.

Din, N., Damude, H. G., Gilkes, N. R., Miller, R. C., Warren, A. J., & Kilburn, D. G. (1994). C1-Cx revisited: Intramolecular synergism in a cellulase. *Proceedings of the National Academy of Sciences of the USA, 91*, 11383–11387.

Evans, C. S., Gallagher, I. M., Atkey, P. T., & Wood, D. A. (1991). Localisation of degradative enzymes in white-rot decay of lignocellulose. *Biodegradation, 2*, 93–106.

Ghose, T. K. (1987). Measurement of cellulase activities. *Pure and Applied Chemistry, 59*, 257–268.

Goyal, A., Ghosh, B., & Eveleigh, D. E. (1991). Characteristics of fungal cellulases. *Bioresource Technology, 36*, 37–50.

King, A. J., Cragg, S. M., Li, Y., Dymond, J., Guille, M. J., Bowles, D. J., Bruce, N. C., Graham, I. A., & McQueen-Mason, S. J. (2010). Molecular insight into lignocellulose digestion by a marine isopod in the absence of gut microbes. *Proceedings of the National Academy of Sciences of the USA, 107*, 5345–5350.

Koenig, H., Li, L., & Froehlich, J. (2013). The cellulolytic system of the termite gut. *Applied Microbiology and Biotechnology, 97*, 7943–7962.

Kostanjsek, R., Milatovic, M., & Srus, J. (2010). Endogenous origin of endo-beta-1,4-glucanase in common woodlouse *Porcellio scaber* (Crustacea Isopoda). *Journal of Comparative Physiology B, 180*, 1143–1153.

Leschine, S. B. (1995). Cellulose degradation in anaerobic environments. *Annual Review of Microbiology, 49*, 399–426.

Ljungdahl, L. G., & Eriksson, K.-E. (1985). Ecology of microbial cellulose degradation. *Advances in Microbial Ecology, 8*, 237–299.

Martin, M. M. (1984). The role of ingested enzymes in the digestive processes of insects. In J. M. Anderson, A. D. M. Rayner, & D. W. H. Walton (Eds.), *Invertebrate-microbial interactions* (pp. 155–172). Cambridge: Cambridge University Press.

Nozaki, M., Miura, C., Tozawa, Y., & Miura, T. (2009). The contribution of endogenous cellulase to the cellulose digestion in the gut of earthworm (*Pheretima hilgendorfi*: Megascolecidae). *Soil Biology and Biochemistry, 41*, 762–769.

Rapp, P., & Beermann, A. (1991). Bacterial cellulases. In C. H. Haigler & P. J. Weimer (Eds.), *Biosynthesis and biodegradation of cellulose* (pp. 269–276). New York: Marcel Dekker.

Rouland, C., Civas, A., Renoux, J., & Petek, F. (1988). Purification and properties of cellulases from the termite *Macrotermes mülleri* (Termitidae, Macrotermitinae) and its symbiotic fungus *Termitomyces* sp. *Comparative Biochemistry and Physiology, 91B*, 449–458.

Royer, T. V., & Minshall, G. W. (2001). Effects of nutrient enrichment and leaf quality on the breakdown of leaves in a hardwater stream. *Freshwater Biology, 46*, 603–610.

Shelomi, M., Watanabe, H., & Arakawa, G. (2014). Endogenous cellulase enzymes in the stick insect (Phasmatodea) gut. *Journal of Insect Physiology, 60*, 25–30.

Skambracks, D., & Zimmer, M. (1998). Combined methods for the determination of microbial activity in leaf litter. *European Journal of Soil Biology, 34*, 105–110.

Tsuji, A., Tominaga, K., Nishiyama, N., & Yuasa, K. (2013). Comprehensive enzymatic analysis of the cellulolytic system in digestive fluid of the sea hare *Aplysia kurodai*. Efficient glucose release from sea lettuce by synergistic action of 45 kDa endoglucanase and 210 kDa β-glucosidase. *PLoS One, 8*, e65418.

Wood, T. M., & Garcia-Campayo, V. (1990). Enzymology of cellulose degradation. *Biodegradation, 1*, 147–161.

Yamin, M. A., & Trager, W. (1979). Cellulolytic activity of an axenically-cultivated termite flagellate, *Trichomitopsis termopsidis*. *Journal of General Microbiology, 13*, 417–420.

Yu, X., Liu, Y., Cui, Y., Cheng, Q., Zhang, Z., Lu, J. H., Meng, Q., Teng, L., & Ren, X. (2016). Measurement of filter paper activities of cellulose with microplate-based assay. *Saudi Journal of Biological Sciences, 23*, S93–S98.

Chapter 43
Viscosimetric Determination of Endocellulase Activity

Björn Hendel and Jürgen Marxsen

Keywords Carboxymethylcellulose · Cellobiase · Chromogenic substrates · Cellulose degradation · Cellulase · Enzyme activity · Exocellulase · Fluorogenic substrates · Leaf litter · Native cellulose

1 Introduction

Cellulose is the most common organic polymer on earth (Klemm et al. 2005) and provides an important carbon source for food webs of both aquatic and terrestrial ecosystems that receive or produce organic matter of vascular plant origin. Heterotrophic microorganisms such as fungi and bacteria cleave the high-molecular weight compounds into smaller fragments, part of which they incorporate into their own biomass, thus making the carbon available to food web members unable to hydrolyze native cellulose.

The enzymatic degradation of cellulose involves up to three types of extracellular enzymes working synergistically to transform the polymer to glucose monomers (Robson & Chambliss 1989; Gilbert and Hazlewood 1993; Chaps. 42 and 44): (1) Endocellulases (endo-β-1,4-glucanases) cleave internal β-1,4-glycosidic bonds randomly within the native chain of cellulose; (2) exocellulases (exo-β-1,4-glucanases, mainly cellobiohydrolases) release cellobiose (or glucose) from the nonreducing ends of cellulose; and (3) β-glucosidases (cellobiases) hydrolyse cellobiose into two glucose units.

The activity of β-glucosidases and exocellulases can be determined by means of fluorogenic (or chromogenic) model substrates (Chap. 44), whereas endocellulase activity can be estimated by monitoring the viscosity of a standard cellulose solution. The viscosity declines in parallel to the average molecular weight of the

B. Hendel · J. Marxsen (✉)
iFZ Research Centre for Biosystems, Land Use and Nutrition, Department of Animal Ecology and Systematics, Justus Liebig University, Giessen, Germany
e-mail: hendel@chemie-bio.uni-siegen.de; jmar@zo.jlug.de

F. Bärlocher et al. (eds.), *Methods to Study Litter Decomposition*,
https://doi.org/10.1007/978-3-030-30515-4_43

dissolved molecules; in the presence of endocellulase activity, the large cellulose chains are cut into smaller fragments. Naturally occurring cellulose is not soluble in water; therefore, it is replaced in the assay by water soluble carboxymethylcellulose, where numerous hydroxyl groups are substituted by carboxymethyl groups. Carboxymethylcellulose does, however, provide the same β-1,4-glycosidic bonds as natural cellulose, which are the target of cellulases (Hulme 1988).

The reduction of viscosity is monitored in vertical glass capillary tubes (Micro-Ubbelohde viscometer) containing a solution of carboxymethylcellulose and enzymes. The efflux time of the solution in the viscometer is recorded as a measure of viscosity, and standardized enzyme units are calculated from these data. The procedure presented here has been adopted from Almin and Eriksson (1967) and Hendel (1999). Only a few published values appear to be available for endocellulase activity associated with particulate organic matter in aquatic and soil ecosystems. They are summarized in Table 43.1.

Table 43.1 Endocellulase activity in decomposing leaves and wood from aquatic and soil environments determined by measuring changes in the viscosity of a carboxymethylcellulose solution

Plant material	Environment	Geographic region	Range of activity (IEU g^{-1} AFDM)	References
White oak (*Quercus alba*) leaves	Woodland stream	James River Basin, VA, USA	300–8800	1, 2
Red maple (*Acer rubrum*) leaves	Woodland stream	James River Basin, VA, USA	300–5200	1, 2
Flowering dogwood (*Cornus florida*) leaves	Woodland stream	James River Basin, VA, USA	300–2000	1, 2
Chestnut oak (*Quercus prinus*) leaves	Woodland soil	Brush Mountain, VA, USA	600–2500	2
Red maple (*Acer rubrum*) leaves	Woodland soil	Brush Mountain, VA, USA	300–4200	2
Flowering dogwood (*Cornus florida*) leaves	Woodland soil	Brush Mountain, VA, USA	300–8600	2
Black alder (*Alnus glutinosa*) leaves	Open grassland stream (Breitenbach)	Central Germany	31–987	3
European beech (*Fagus sylvatica*) leaves	Open grassland stream (Breitenbach)	Central Germany	25–746	3
Black alder (*Alnus glutinosa*) wood	Open grassland stream (Breitenbach)	Central Germany	61–558	3
European beech (*Fagus sylvatica*) wood	Open grassland stream (Breitenbach)	Central Germany	43–213	3

All values are given in international enzyme units (IEU) per gram of ash-free dry mass (AFDM). The activity values of Sinsabaugh et al. (1981) and Linkins et al. (1990) were derived from the published figures
1 = Sinsabaugh et al. (1981), 2 = Linkins et al. (1990), 3 = Hendel (1999).

2 Equipment, Chemicals, and Solutions

2.1 Equipment

- Sharp knife or scalpel
- Homogenizer (e.g., Polytron)
- Homogenization vessels, 100 ml
- Water bath (5 °C)
- Bench centrifuge
- Micro-Ubbelohde viscometer: 5 ml volume, capillary diameter approximately 0.5 mm (Fig. 43.1)
- Viscometer tripod
- Automatic viscosity system or stopwatch with water bath set at 25.0 °C. An automatic system (e.g., ViscoSystem AVS 350, Xylem Analytics, Schott, Mainz, Germany) replaces manual measurement. It consists of a water bath with tem-

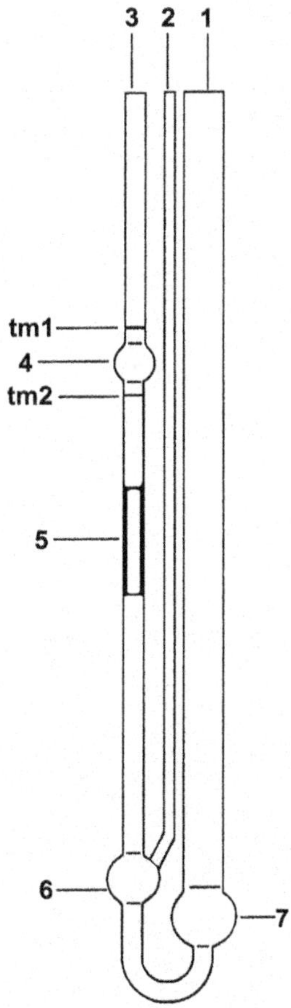

Fig. 43.1 Ubbelohde micro-viscometer. 1: Filling tube, 2: Venting tube, 3: Capillary tube, 4: Measuring sphere, 5: Capillary, 6: Reference level vessel, 7: Reservoir, tm1: Upper timing mark, tm2: Lower timing mark

perature control. A light barrier measures the time needed for the surface of the solution to fall from mark tm1 to tm2 (Fig. 43.1).
- Gooch crucibles (20 ml, porosity 4 with an approximate pore size of 10–16 μm)
- Eppendorf-type pipettes (2.5 and 5 ml)

2.2 *Chemicals*

- Carboxymethylcellulose, substitution grade 0.5–0.7
- Acetic acid: 100%, analytical grade
- Sodium hydroxide (NaOH), analytical grade
- Autoclaved water from sampling site

2.3 *Solutions*

- Acetate buffer: 50 mmol l^{-1} (3.0025 g acetic acid 100% per liter), adjusted to pH 5.0 with NaOH
- Carboxymethylcellulose stock solution: 50 g l^{-1} in acetate buffer. Dissolve pellets or powder at ambient temperature on a magnetic stirrer overnight. Filter stock solution through a Gooch crucible and immediately freeze (−18 °C) aliquots of 100–200 ml, depending on the number of samples to be processed per day. Thaw aliquots just before use in assays. Use the same stock solution for all samples.

3 Experimental Procedures

3.1 *Sample Preparation*

1. Collect leaves, wood, or other types of organic matter and transport to the laboratory in a cooled, insulated container.
2. Process samples as soon as possible, but no later than 6 h after collection.
3. Remove any adhering debris and macroinvertebrates.
4. With a knife or scalpel cut samples in pieces. The final size depends on the homogenizer used; it must be able to homogenize them to a point where fragments are no longer visible.
5. Transfer precut pieces of organic matter (fresh mass ≥ 1 g) into a homogenization vessel containing unfiltered, autoclaved water from the sampling site (e.g., 80 ml for a 100 ml vessel).

6. Place homogenization vessel in a water bath at 5 °C, and homogenize samples for 2 min while preventing the sample temperature from rising above 18 °C.
7. Use the homogenate for enzyme assays.

3.2 *Enzyme Analysis*

1. Use a wide-mouth pipette to remove 5 ml of the homogenate from each homogenization vessel and transfer to a clean crucible.
2. Filter each sample and collect the filtrate containing the enzymes in a centrifugation tube.
3. Centrifuge for 8 min at ca. 3000 *g* to separate any remaining solids from the fluid.
4. Mix 1.5 ml of enzyme solution with 5 ml of carboxymethylcellulose stock solution.
5. Immediately transfer 4 ml of the mixture to the filling tube of the viscometer.
6. Measure efflux time of the solution, which is the time needed to fall from the upper mark (tm1) to the lower mark (tm2) at 25.00 ± 0.05 °C (Fig. 43.1).
7. Repeat the measurement four times at 5-min intervals.

3.3 *Calculation*

Calculate enzyme activity using the following equation (Hulme 1988):

$$A = 1.27\frac{dx}{dt}\left(\frac{1}{\eta_{sp}}\right)_{t=0} \tag{43.1}$$

where A = enzyme activity and η_{sp} = specific viscosity of the sample solution. The specific viscosity can be calculated as:

$$\eta_{sp} = \frac{t_s}{t_0} - 1 \tag{43.2}$$

where t_s = efflux time of enzyme solution and t_0 = efflux time of the acetate buffer. The derivative *dx/dt* of Eq. 43.1 is calculated from the plot of $1/\eta_{sp}$ against elapsed time. At least the first three, and generally all five, measurements lie on a straight line, whose slope is the required value. The intercept has no meaning.

The result appears as international enzyme units (IEU). One IEU corresponds to the amount of enzyme that catalyzes the hydrolysis of one microequivalent of β-1,4-glucosidic bonds per min at defined conditions of pH and temperature (International Union of Biochemistry 1984).

4 Final Remarks

The use of carboxymethylcellulose with an exactly defined substitution grade is critically important (Eriksson and Hollmark 1969), as is strict maintenance of the temperature during the experiment within an interval of ±0.05 °C around 25 °C.

Filtration of the homogenized solution can be replaced by centrifugation at 38,000 *g* for 20 min. Processing one sample in the viscometer requires at least 40 min. When a single viscometer is available, about 12 samples can be processed during a normal work day.

Best results are obtained with an automatic viscosity system. Such a system minimizes measurement error when determining the time taken by the liquid to fall from tm1 and tm2 (Fig. 43.1), due to automatic light barriers and exact temperature adjustment. Nevertheless, a standard capillary with a defined diameter of approximately 0.5 mm in a water bath with a defined temperature can yield accurate and sufficiently precise values, although errors tend to be significantly larger.

References

Almin, K. E., & Eriksson, K.-E. (1967). Enzymic degradation of polymers. I. Viscosimetric method for the determination of enzymic activity. *Biochimica et Biophysica Acta, 139*, 238–247.

Eriksson, K.-E., & Hollmark, B. H. (1969). Kinetic studies of the action of cellulase upon sodium carboxymethyl cellulose. *Archives of Biochemistry and Biophysics, 133*, 233–237.

Gilbert, H. J., & Hazlewood, G. P. (1993). Bacterial cellulases and xylanases. *Journal of General Microbiology, 139*, 187–194.

Hendel, B. (1999). *Der mikrobielle Abbau von Holz und Laub im Breitenbach unter besonderer Berücksichtigung der Bedeutung extrazellulärer Enzyme*. Doctoral Thesis. University of Giessen, Germany.

Hulme, M. A. (1988). Viscosimetric determination of carboxymethylcellulase activity. In W. A. Wood & S. T. Kellogg (Eds.), *Methods in enzymology vol. 160, biomass, part a, cellulose and hemicellulose* (pp. 130–135). New York: Academic.

International Union of Biochemistry. (1984). *Enzyme nomenclature: Recommendations of the commission on biochemical nomenclature*. New York: Academic.

Klemm, D., Heublein, B., Fink, H.-P., & Bohn, A. (2005). Cellulose: Fascinating biopolymer and sustainable raw material. *Angewandte Chemie International Edition, 44*, 3358–3393.

Linkins, A. E., Sinsabaugh, R. L., McClaugherty, C. A., & Melillo, J. M. (1990). Comparison of cellulase activity on decomposing leaves in a hardwood forest and woodland stream. *Soil Biology and Biochemistry, 22*, 423–425.

Robson, L. M., & Chambliss, G. H. (1989). Cellulases of bacterial origin. *Enzyme and Microbial Technology, 11*, 626–644.

Sinsabaugh, R. L., III, Benfield, E. F., & Linkins, A. E., III. (1981). Cellulase activity associated with the decomposition of leaf litter in a woodland stream. *Oikos, 36*, 184–190.

Chapter 44
Fluorometric Determination of the Activity of β-Glucosidase and Other Extracellular Hydrolytic Enzymes

Björn Hendel and Jürgen Marxsen

Keywords Cellulose · Hemicellulases · Proteinases · Phosphatase · Litter degradation · MUF · Hydrolytic enzymes · Leaf litter · Enzyme extraction · Fluorometry

1 Introduction

Cellulose is the major structural polysaccharide of vascular plants. It confers tensile strength (Chap. 22) and cannot normally be cleaved by animals (Chap. 41). Like other polymers, cellulose needs to be cleaved into smaller subunits before microbial cells can assimilate this carbon source. Cleavage is achieved by extracellular enzymes produced mainly by bacteria and fungi (Marxsen 2011). These enzymes are either bound to cell surfaces or are released into the environment (Wetzel 1991). Thus, cleavage of cellulose and other macromolecules by extracellular enzymes is a crucial initial step in the microbial degradation of plant litter (Marxsen and Fiebig 1993).

Activities of polysaccharide-degrading and many other hydrolytic extracellular enzymes can be precisely measured by means of fluorogenic model substrates, which are available for a suite of natural compounds (Hoppe 1983). These model substrates most commonly consist of 4-methylumbelliferone (MUF) linked to compounds such as glucose, phosphate, or amino acids. The bonds between these molecules are cleaved by the enzymes in a way similar to the natural oligomeric or polymeric substances (Fig. 44.1). Since the MUF released by hydrolysis is fluorescent, it can be quantified fluorometrically to indicate the level of extracellular enzyme activity in a given sample. Another widely used fluorogenic molecule is

B. Hendel · J. Marxsen (✉)
iFZ Research Centre for Biosystems, Land Use and Nutrition, Department of Animal Ecology and Systematics, Justus Liebig University, Giessen, Germany
e-mail: hendel@chemie-bio.uni-siegen.de; jmar@zo.jlug.de

F. Bärlocher et al. (eds.), *Methods to Study Litter Decomposition*,
https://doi.org/10.1007/978-3-030-30515-4_44

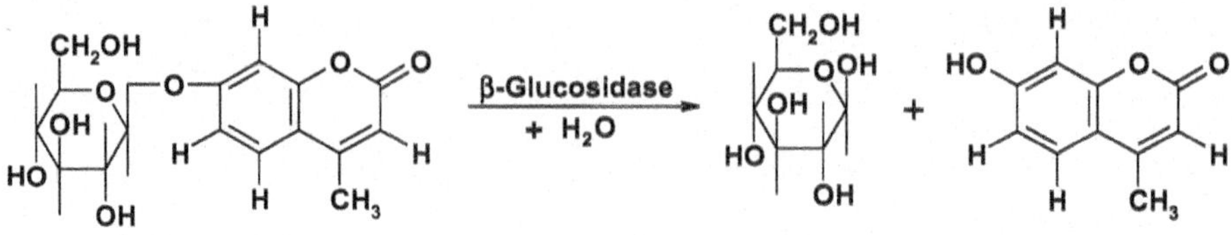

Fig. 44.1 Example of a MUF-compound and its cleavage: hydrolysis of nonfluorescent MUF-β-glucoside into glucose and fluorescent MUF

Table 44.1 Main model substrates with 4-methylumbelliferone (MUF) or 7-amido-4-methyl-coumarin (MCA) as fluorogenic compound, the corresponding natural substrates and the enzymes cleaving these molecules (Marxsen 2011)

Model substrate	Natural substrates	Enzymes
MUF-α-glucoside	Maltose, starch	α-Glucosidases
MUF-β-glucoside	Cellobiose, cellulose	β-Glucosidases
MUF-β-xyloside	Hemicellulose, xylans	β-Xylosidases
MUF-β-N-acetyl-glucosaminide	Chitobiose, chitin	β-N-acetyl-glucosaminidases
Leucine-MCA	Peptides, proteins	Peptidases
MUF-laurate	Lipids	Lipases
MUF-phosphate	Polyphosphates	Phosphatases
MUF-sulfate	Sulfate esters	Sulfatases

7-amido-4-methyl-coumarin (MCA); it can be used in enzyme assays in the same way as MUF.

Free methylumbelliferone exhibits its maximum fluorescence at an excitation wavelength between 355 and 380 nm at a pH above 10. The emission wavelength is between 440 and 460 nm. The corresponding values for free MCA are 370–390 nm (emission) and 430–465 nm (excitation). Humic compounds occurring in many samples from aquatic and terrestrial ecosystems may shift both the excitation and emission wavelengths. Consequently, the excitation and emission wavelengths specific to a given type of sample need to be determined to enhance the reliability and sensitivity of measurements. A main advantage of the technique is indeed its high sensitivity, which is several orders of magnitudes above that of methods based on chromogenic substances like pNP(*p*-nitrophenyl)-compounds (Tank et al. 1998). This facilitates short incubation periods (60 min) and incubation at low in situ temperatures.

MUF model compounds are available for several natural substances (Table 44.1). The method for β-glucosidase described below can thus be modified for the detection of a broad range of extracellular enzymes. These include – in addition to carbohydrate-degrading enzymes like β-xylosidase, cellobiohydrolase, and chitinase – enzymes that degrade other major biopolymers such as proteins and lipids as well as enzymes involved in nutrient remineralization, such as phosphatases and sulfatases (Hoppe 1993; Marxsen et al. 1998; Marxsen 2011). The procedure

Table 44.2 Hydrolytic extracellular enzyme activities in decomposing leaves and wood from aquatic and soil environments determined by measuring the release of methylumbelliferone from model substrates

Plant material	Environment	Geographic region	Enzyme	Range of activity ($\mu mol\ g^{-1}$ AFDM h^{-1})	References
White birch (*Betula papyfera*), wood	Woodland soil, riparian and lotic sites	Northern New York, USA	β-glucosidase	5–25	1
White birch (*Betula papyfera*), wood	Woodland soil, riparian and lotic sites	Northern New York, USA	β-xylosidase	1.3–6.5	1
Black alder (*Alnus glutinosa*), leaves	Grassland stream	Central Germany	β-glucosidase	120–4300	2
European beech (*Fagus sylvatica*), leaves	Grassland stream	Central Germany	β-glucosidase	100–2100	2
Black alder (*Alnus glutinosa*), leaves	Grassland stream	Central Germany	Cellobio-hydrolase	2–720	2
European beech (*Fagus sylvatica*), leaves	Grassland stream	Central Germany	Cellobio-hydrolase	1–520	2
Black alder (*Alnus glutinosa*), leaves	Grassland stream	Central Germany	β-xylosidase	40–750	2
European beech (*Fagus sylvatica*), leaves	Grassland stream	Central Germany	β-xylosidase	20–320	2
Black alder (*Alnus glutinosa*), wood	Grassland stream	Central Germany	β-glucosidase	4–2000	2, 3
European beech (*Fagus sylvatica*), wood	Grassland stream	Central Germany	β-glucosidase	1–890	2, 3
Black alder (*Alnus glutinosa*), wood	Grassland stream	Central Germany	Cellobio-hydrolase	2–140	2, 3
European beech (*Fagus sylvatica*), wood	Grassland stream	Central Germany	Cellobio-hydrolase	1–70	2, 3
Black alder (*Alnus glutinosa*), wood	Grassland stream	Central Germany	β-xylosidase	3–280	2, 3
European beech (*Fagus sylvatica*), wood	Grassland stream	Central Germany	β-xylosidase	4–210	2, 3

1 = Sinsabaugh et al. (1992), 2 = Hendel (1999), 3 = Hendel and Marxsen (2000)

presented here for enzyme activities associated with plant litter has been adopted from Hendel and Marxsen (2000).

Activities of extracellular polysaccharide-degrading enzymes associated with particulate organic matter in aquatic and soil environments are presented in Table 44.2. Tank et al. (1998) applied chromogenic *p*-nitrophenyl substrates for measuring extracellular β-glucosidase and β-xylosidase activities associated with yellow poplar wood in streams and found activities similar to those of Sinsabaugh et al. (1992) with white birch wood (Table 44.2).

2 Equipment, Chemicals, and Solutions

2.1 Equipment and Materials

- Sharp knife or scalpel
- Homogenizing device (e.g., Polytron)
- Homogenization vessels (100 ml)
- Glassware for preparing solutions
- Shaking water baths (5 and 10 °C)
- Erlenmeyer flasks (25 ml) as incubation vessels
- Fluorometer with fluorescence quartz glass cuvettes
- Pipettes
- Centrifuge tubes (10–12 ml)
- Centrifuge (for 10–12 ml tubes, 3000 g)

2.2 Chemicals

- Methylumbelliferone (MUF), analytical grade
- Methylumbelliferyl-β-D-glucopyranoside (MUF-Glc), analytical grade
- 2-Methoxyethanol (methyl cellosolve, MCS), analytical grade
- Glycine, analytical grade
- Ammonia (25%), analytical grade
- Sodium hydroxide (NaOH) pellets, analytical grade
- Autoclaved water from sampling site

2.3 Solutions

- MUF stock solution, 300 μmol l^{-1}
- MUF-Glc stock solution, 5 mmol l^{-1}

- Ammonium glycine buffer (pH 10.5): dissolve 3.75 g glycine in 14.8 ml ammonia (25%), make up to 1000 ml with deionized water, and adjust pH with sodium hydroxide solution to 10.5.

3 Experimental Procedures

3.1 *Sample Preparation*

1. Collect plant litter or other organic matter, and transport to the laboratory in a cooled, insulated container.
2. Process sample as soon as possible, but no later than 6 h after collection.
3. Remove any adhering debris and macroinvertebrates.
4. Use a knife or scalpel to cut samples in pieces, the final size of which depends on the homogenizer used (see below).
5. Transfer a representative sample (fresh mass ≥ 1 g) into a homogenization vessel containing unfiltered, autoclaved water from the sampling site (e.g., 80 ml for a 100 ml vessel).
6. Place homogenization vessel in a water bath at about 5 °C.
7. Homogenize samples for 2 min so that litter fragments are no longer visible by the naked eye, while preventing sample temperatures from rising above 18 °C.
8. Use the homogenate for all enzyme assays.

3.2 *Enzyme Analysis*

1. Determine the wavelength of maximum excitation and emission of a water sample with the fluorimeter.

Incubation

1. Prepare fresh stock solutions each time, adding 1% (v/v) MCS to enhance solubility of MUF substrates, although some substrates are easily soluble without adding MCS.
2. Prepare the incubation vessels (Erlenmeyer flasks) and an appropriate number of calibration standards (≥3) and blanks (≥2).
3. Transfer 200 μl of leaf or wood homogenate into each Erlenmeyer flask (samples, calibration standards and blanks).
4. For samples, add 1 ml of MUF-Glc solution to the homogenate; for calibration standards, add 660 μl of MUF solution.
5. Make up the final volume in each flask to 20 ml.

6. Place each flask in a shaking water bath at 10 °C, the purpose of shaking being to reduce cell adhesion to the flask walls.

Measurement of Enzyme Activity

1. After 5 min, remove 4 ml from each flask and transfer to a centrifuge tube.
2. Add 400 μl of ammonium glycine buffer to each flask to raise pH and thus stop enzyme activity and convert MUF to its anionic form, which amplifies fluorescence.
3. Shake flask gently.
4. Centrifuge for 1 min at ~3000 g.
5. Calibrate the fluorimeter, measure fluorescence of the sample, and correct for naturally occurring fluorescence (blanks).
6. Repeat the procedure for measuring enzyme activity after 60 min.

Calculation of Enzyme Activity

1. Calculate the enzyme activity in the samples based on the difference between MUF concentrations after 5 and 65 min, assuming a linear increase of fluorescence.

4 Final Remarks

The procedure described here is adjusted for application with a single substrate concentration, which is normally in the saturation range of the enzyme. This means that this approach quantifies potential enzyme activities. If complete enzyme kinetics are to be determined, concentrations are best chosen to range from one order of magnitude below the Michaelis-Menten constant of the enzyme (K_m) to one order of magnitude above. If substrates other than MUF-Glc are used, the substrate concentration required for saturation needs to be determined before running analyses.

After appropriate adjustment, the procedure can be used to quantify extracellular enzyme activity in a variety of sample types. In addition to leaves and wood (Hendel and Marxsen 2000; Frossard et al. 2013; Tlili et al. 2017), MUF enzyme assays have been successfully applied to sediments (e.g., Marxsen and Fiebig 1993; Marxsen et al. 1998; Frossard et al. 2012), biofilms (e.g., Freeman et al. 1990; Romaní 2000; Pohlon et al. 2010), and water (Hoppe 1983, 1993), including water from springs or groundwater where enzyme activities are typically very low (Hendel and Marxsen 1997).

Sample throughput can be markedly increased if the assay is downsized for microplate assays (e.g., Sinsabaugh et al. 2005; Stursova et al. 2006; Frossard et al.

2013; Tlili et al. 2017). An excellent online source of information is the website of the Enzymes in the Environment Research Coordination Network (https://enzymes.nrel.colostate.edu/enzymes-home.html), where detailed protocols for microplate assays are presented, which were originally designed for soil analyses (McMahon and Steinweg 2009, last update 2012, and Sinsabaugh 2009).

References

Freeman, C., Lock, M. A., Marxsen, J., & Jones, S. (1990). Inhibitory effects of high molecular weight dissolved organic matter upon metabolic processes in biofilms from contrasted rivers and streams. *Freshwater Biology, 24*, 159–166.

Frossard, A., Gerull, L., Mutz, M., & Gessner, M. O. (2012). Disconnect of microbial structure and function: Enzyme activities and bacterial communities in nascent stream corridors. *The ISME Journal, 6*, 680.

Frossard, A., Gerull, L., Mutz, M., & Gessner, M. O. (2013). Shifts in microbial community structure and function in stream sediments during experimentally simulated riparian succession. *FEMS Microbiology Ecology, 84*, 398–410.

Hendel, B. (1999). *Der mikrobielle Abbau von Holz und Laub im Breitenbach unter besonderer Berücksichtigung der Bedeutung extrazellulärer Enzyme*. Doctoral thesis. University of Giessen, Germany.

Hendel, B., & Marxsen, J. (1997). Measurement of low level extracellular enzyme activity in natural waters using fluorigenic model substrates. *Acta Hydrochimica et Hydrobiologica, 25*, 253–258.

Hendel, B., & Marxsen, J. (2000). Extracellular enzyme activity associated with degradation of beech wood in a central European stream. *International Review of Hydrobiology, 85*, 95–105.

Hoppe, H.-G. (1983). Significance of exoenzymatic activities in the ecology of brackish water: Measurements by means of methylumbelliferyl-substrates. *Marine Ecology – Progress Series, 11*, 299–308.

Hoppe, H.-G. (1993). Use of fluorogenic model substrates for extracellular enzyme activity (EEA) measurement of bacteria. In P. F. Kemp, B. F. Sherr, E. B. Sherr, & J. J. Cole (Eds.), *Handbook of methods in aquatic microbial ecology* (pp. 423–431). Boca Raton: Lewis Publishers.

Marxsen, J. (2011). Bacteria and fungi. In R. Wagner, J. Marxsen, P. Zwick, & E. J. Cox (Eds.), *Central European stream ecosystems. The long term study of the Breitenbach* (pp. 131–194). Weinheim: Wiley-VCH.

Marxsen, J., & Fiebig, D. M. (1993). Use of perfused cores for evaluating extracellular enzyme activity in stream-bed sediments. *FEMS Microbiology Ecology, 13*, 1–11.

Marxsen, J., Tippmann, P., Heininger, H., Preuß, G., & Remde, A. (1998). Enzymaktivität. In Vereinigung für Allgemeine und Angewandte Mikrobiologie (Ed.), *Mikrobiologische Charakterisierung aquatischer Sedimente – Methodensammlung* (pp. 87–114). Munich: Oldenbourg

McMahon, S., & Steinweg, M. (2009). Microplate enzyme assay using fluorescence-based MUB & MUC-linked substrates. https://enzymes.nrel.colostate.edu/enzymes-experimental-protocols.html

Pohlon, E., Marxsen, J., & Küsel, K. (2010). Pioneering bacterial and algal communities and extracellular enzyme activities of stream biofilms. *FEMS Microbiology Ecology, 71*, 364–373.

Romaní, A. M. (2000). Characterization of extracellular enzyme kinetics in two Mediterranean streams. *Archiv für Hydrobiologie, 148*, 99–117.

Sinsabaugh, R. L. (2009). Fluorescence-based protocol using MUB & MUC-linked substrates. https://enzymes.nrel.colostate.edu/enzymes-experimental-protocols.html

Sinsabaugh, R. L., Antibus, R. K., Linkins, A. E., McClaugherty, C. A., Rayburn, L., Repert, D., & Weiland, T. (1992). Wood decomposition over a first-order watershed: Mass loss as a function of lignocellulose activity. *Soil Biology and Biochemistry, 24*, 743–749.

Sinsabaugh, R. L., Gallo, M. E., Lauber, C., Waldrop, M. P., & Zak, D. R. (2005). Extracellular enzyme activities and soil organic matter dynamics for northern hardwood forests receiving simulated nitrogen deposition. *Biogeochemistry, 75*, 201–215.

Stursova, M., Crenshaw, C. L., & Sinsabaugh, R. L. (2006). Microbial responses to long-term N deposition in a semiarid grassland. *Microbial Ecology, 51*, 90–98.

Tank, J. L., Webster, J. R., Benfield, E. F., & Sinsabaugh, R. L. (1998). Effect of leaf litter exclusion on microbial enzyme activity associated with wood biofilms in streams. *Journal of the North American Benthological Society, 17*, 95–103.

Tlili, A., Jabiol, J., Behra, R., Gil-Allué, C., & Gessner, M. O. (2017). Chronic exposure effects of silver nanoparticles on stream microbial decomposer communities and ecosystem functions. *Environmental Science and Technology, 51*, 2447–2455.

Wetzel, R. G. (1991). Extracellular enzymatic interactions: Storage, redistribution, and interspecific communication. In R. J. Chróst (Ed.), *Microbial enzymes in aquatic environments* (pp. 6–28). Berlin: Springer.

Chapter 45
Pectin-Degrading Enzymes: Polygalacturonase and Pectin Lyase

Keller Suberkropp

Keywords Enzyme activity · Leaf litter · Litter degradation · Litter softening · pH optimum · Plant litter maceration · Pectin lyase · Polygalacturonase · Spectrophotometry · Streams

1 Introduction

Pectic substances, or pectins, are most abundant in the middle lamellae of primary cell walls of vascular plants, where their main function is to cement cells together into tissues. Pectins are the initial polysaccharide substrates encountered by decomposers in nonlignified or weakly lignified plant tissue, and their removal exposes other polysaccharides such as xylans, mannans, and cellulose to microbial degradation (Chamier and Dixon 1982). Pectin-degrading enzymes produced by plant pathogenic fungi have been implicated in the maceration of living plant tissue (Bateman and Basham 1976; Friend 1977). Macerating activity has also been noted in leaf litter exposed to aquatic hyphomycetes (Fig. 45.1), suggesting that this process is important in the decomposition of plant litter in streams (Suberkropp and Klug 1980; Chamier and Dixon 1982, 1983).

Pectin is a polymer of galacturonic acid in which various percentages of the carboxyl residues have been methylated. There are three major classes of enzymes that degrade pectin. These include (1) hydrolytic enzymes, such as polygalacturonases, which hydrolyze the glycosidic bonds between the galacturonic acid residues; (2) enzymes, such as pectin lyase, that cleave the glycosidic bonds between the galacturonic residues by β-elimination; and (3) esterases, such as pectin methyl esterases, that cleave the methyl group from the galacturonic acid residues (Rexová-Benková and Markovič 1976).

K. Suberkropp (✉)
Department of Biological Sciences, University of Alabama, Tuscaloosa, AL, USA
e-mail: kellersuber@gmail.com

F. Bärlocher et al. (eds.), *Methods to Study Litter Decomposition*,
https://doi.org/10.1007/978-3-030-30515-4_45

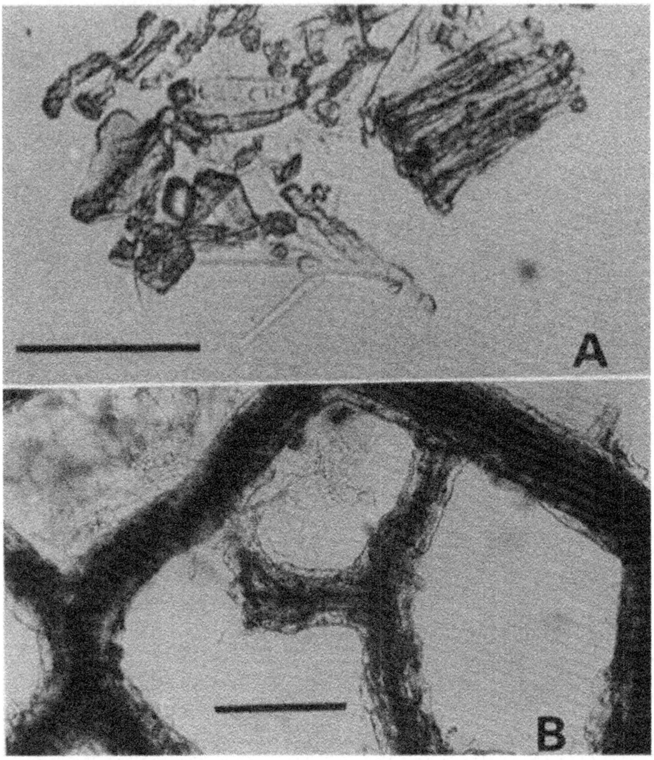

Fig. 45.1 Remains of hickory (*Carya glabra*) leaf discs colonized by *Tetracladium marchalianum* at 10 °C for 6 weeks. (**A**) The fine particulate fraction contains leaf parenchyma cells and fungal spores. Scale bar represents 50 μm. (**B**) The coarse particulate fraction contains skeletonized leaf veins. Scale bar represents 100 μm (from Suberkropp and Klug 1980)

Some pectinases cleave the glycosidic bonds between galacturonic acid residues randomly within the polysaccharides (endopectinases), whereas others cleave bonds and release subunits from the ends of the polysaccharides (exopectinases). Endopectinases cause a reduction in the viscosity of pectin solutions. Consequently, a useful assay for endopectinases involves following decreases in the viscosity of the reaction mixture using viscometers (Chamier and Dixon 1982).

Assays for the hydrolytic enzymes and lyases which depolymerize pectic polysaccharides are presented below. The proposed protocol is based on Suberkropp et al. (1983) and Jenkins and Suberkropp (1995). Polygalacturonases typically have a pH optimum around 5, pectin lyases around 8. This is reflected in greater activities for pectin lyases in hardwater streams and greater activities for polygalacturonases in softwater streams (Table 45.1).

Table 45.1 Maximum activity (± SE) of pectin lyase and polygalacturonase associated with tulip tree (*Liriodendron tulipifera*) leaves decomposing in two streams differing in pH (from Jenkins and Suberkropp 1995)

Stream type	pH	Pectin lyase (units g^{-1} h^{-1})	Polygalacturonase (mol g^{-1} h^{-1})
Hardwater	8.2	14.0 ± 1.6	15.2 ± 2.0
Softwater	6.3	3.1 ± 0.2	64.2 ± 20.4

2 Equipment, Chemicals, and Solutions

2.1 Equipment and Materials

- Spectrophotometer (550 and 540 nm)
- pH meter
- Magnetic stirrer and hot plate
- Water or dry bath (100 °C)
- Test tubes
- Glassware for preparing and storing solutions
- Adjustable pipette (1 ml)
- Dialysis tubing
- Bench centrifuge (9000 g)
- Analytical balance (0.1 mg precision)
- Drying oven (50 °C)
- Muffle furnace (500 °C)

2.2 Chemicals

- Filtered stream water
- Water, distilled or deionized
- Pectin
- Polygalacturonic acid
- Potassium acetate
- Bicine
- Calcium chloride ($CaCl_2 \cdot 2H_2O$)
- Thiobarbituric acid
- Dinitrosalicylic acid
- Thimerasol

- Potassium hydroxide (KOH)
- Hydrochloric acid (HCl)

2.3 Solutions

- Solution 1 – 1% pectin: add 1 g pectin very slowly to 80 ml of water with constant stirring. When dissolved, dialyze overnight, centrifuge (9000 *g*, 20 min), add 1 ml of 1% thimerosal (to inhibit microbial growth), and adjust final volume to 100 ml. Storage at 4–5 °C is possible for an extended period.
- Solution 2 – 0.5% polygalacturonic acid: add 0.5 g polygalacturonic acid to 80 ml of water. Adjust pH of solution to 5.0 and add 1 ml of 1% thimerosal. Adjust final volume to 100 ml. Storage at 4–5 °C is possible for an extended period.
- Solution 3 – 0.2 M bicine with 0.03 M $CaCl_2$: add 3.26 g bicine and 0.44 g $CaCl_2 \cdot 2H_2O$ to ca. 70 ml water, adjust pH to 8.0 with 0.1 M KOH solution, add 1 ml of 1% thimerosal. Adjust final volume to 100 ml. Storage at 4–5 °C is possible for an extended period.
- Solution 4 – 0.2 M potassium acetate: add 1.96 g potassium acetate to ca. 70 ml water, adjust pH to 5.0 with 0.1 M HCl; add 1 ml of 1% thimerosal. Adjust final volume to 100 ml. Storage at 4–5 °C is possible for an extended period.
- Solution 5 – 0.04 M thiobarbituric acid: add 0.58 g thiobarbituric acid to water and stir. Adjust final volume to 100 ml (Ayers et al. 1976).
- Solution 6 – Dinitrosalicylic acid reagent: dissolve 1 g 3,5-dinitrosalicylic acid in 20 ml of 2 M NaOH, and 50 ml water. Add 30 g NaK tartrate and adjust final volume to 100 ml with water. Store in stoppered bottle to protect from CO_2 (Bernfield 1955).
- Solution 7 – Galacturonic acid standards: prepare galacturonic acid solution (500 μg ml^{-1}) by dissolving 50 mg galacturonic acid in 100 ml water. Store aliquots at −20 °C and thaw when needed.

3 Experimental Procedures

3.1 Sample Preparation

1. Collect decomposing leaves and transport to laboratory in an ice chest.
2. Wash leaves in stream water and cut 4 leaf discs (ca. 12 mm diameter) for each replicate (2 for the assay and 2 for controls).
3. Place leaf discs in filtered stream water.
4. Boil the leaf discs to be used for controls for 10 min.

3.2 *Pectin Lyase*

1. Pipette 2 ml bicine plus $CaCl_2$ (Solution 3) and 1 ml pectin (Solution 1) into reaction tubes and incubate in a water or dry bath set at ambient temperature or at a fixed temperature, depending on the specific question of the study.
2. Add leaf discs to reaction mixtures at 15 s intervals. Incubate for 180 min.
3. Remove 0.5 ml reaction mixture and add it to 2.5 ml each of 0.04 thiobarbituric acid (Solution 5) and 0.1 M HCl. Add 5 ml water and boil for 30 min. Cool tubes to room temperature and measure absorbance at 550 nm (A_{550}).
4. Rinse leaf discs from each tube with water, dry to constant weight at 50 °C, weigh to the nearest 0.1 mg, ash at 500 °C for 4 h to determine leaf ash-free dry mass (AFDM).
5. Because there is no commercially available standard for the product containing the double-bond, express enzyme activity as A_{550} g^{-1} leaf material h^{-1} from differences between absorbance readings from experimental and boiled control discs.

3.3 *Polygalacturonase*

1. Pipette 2 ml acetate (Solution 4) and 1 ml polygalacturonic acid (Solution 2) into reaction tubes and incubate in water bath set at ambient temperature or at a fixed temperature, depending on the specific question of the study.
2. Add leaf discs to reaction mixtures at 15 s intervals. Incubate for 180 min.
3. Remove 1.0 ml of the reaction mixture and add it to 1.0 ml of dinitrosalicylic acid reagent (Solution 6), place in boiling water for 5 min, cool in running tap water, add 20 ml water, mix and measure absorbance at 540 nm.
4. Rinse leaf discs from each tube with water, dry at 50 °C, weigh, and ash at 500 °C for 4 h and reweigh to determine leaf AFDM.
5. Determine activity as g galacturonic acid produced g^{-1} leaf litter h^{-1} by comparing absorbance with standard curves prepared with known concentrations of galacturonic acid (0–500 g ml^{-1}; Solution 7) and subtracting values obtained from the boiled controls.

4 Final Remarks

The extracellular enzymes considered above are typically stable and active over a wide range of temperatures. If compatible with the goal of the study, enzyme reactions can be carried out at higher temperatures (e.g. 30 °C) to increase the rate of catalysis and the amount of product when enzyme activity is very low or not detectable at ambient temperatures.

References

Ayers, W. A., Papavizas, G. C., & Diem, A. F. (1976). Polygalacturonase production by *Rhizoctonia solani*. *Phytopathology, 67*, 1250–1258.

Bateman, D. F., & Basham, H. G. (1976). Degradation of plant cell walls and membranes by microbial enzymes. In R. Heitefuss & P. U. Williams (Eds.), *Physiological plant pathology, encyclopedia of plant physiology* (New series) (Vol. 4, pp. 316–355). Berlin: Springer.

Bernfield, P. (1955). Amylases, α and β. *Methods in Enzymology, 1*, 149–150.

Chamier, A.-C., & Dixon, P. A. (1982). Pectinases in leaf degradation by aquatic hyphomycetes: The enzymes and leaf maceration. *Journal of General Microbiology, 128*, 2469–2483.

Chamier, A.-C., & Dixon, P. A. (1983). Effects of calcium-ion concentrations on leaf maceration by *Tetrachaetum elegans*. *Transactions of the British Mycological Society, 81*, 415–418.

Friend, J. (1977). Biochemistry of plant pathogens. In D. H. Northcote (Ed.), *Plant biochemistry II* (Vol. 13, pp. 141–182). Baltimore: University Park Press.

Jenkins, C. C., & Suberkropp, K. (1995). The influence of water chemistry on the enzymatic degradation of leaves in streams. *Freshwater Biology, 33*, 245–253.

Rexová-Benková, L., & Markovič, O. (1976). Pectic enzymes. *Advances in Carbohydrate Chemistry and Biochemistry, 33*, 323–385.

Suberkropp, K., & Klug, M. J. (1980). The maceration of deciduous leaf litter by aquatic hyphomycetes. *Canadian Journal of Botany, 58*, 1025–1031.

Suberkropp, K., Arsuffi, T. L., & Anderson, J. P. (1983). Comparison of degradative ability, enzymatic activity, and palatability of aquatic hyphomycetes grown on leaf litter. *Applied and Environmental Microbiology, 46*, 237–244.

Chapter 46
Lignin-Degrading Enzymes: Phenoloxidase and Peroxidase

Björn Hendel, Robert L. Sinsabaugh, and Jürgen Marxsen

Keywords Electron-donating substrate · Enzymatic potential · Extracellular oxidative enzymes · L-DOPA · Leaf litter · Lignin degradation · Litter homogenization · Microplate analyses · Polyphenolics · Spectrophotometry

1 Introduction

Lignin is a principal constituent of vascular plants and, after cellulose, the second most abundant naturally occurring compound. Because of its intimate association with cellulose fibers in plants, lignin is important in regulating the flow of carbon into decomposer food webs and, more generally, through ecosystems. Lignin consists primarily of phenylpropane units, which are randomly polymerized into three-dimensional macromolecules. Degradation of the lignin molecule is an oxidative process that may extend over long periods. The enzymic equipment for depolymerizing lignin can be found in fungi and bacteria. Several types of enzymes involved in lignin degradation have been described (Baldrian 2006; Hoegger et al. 2006; Sinsabaugh 2010; Bach et al. 2013). These include monooxygenases (phenoloxidases, laccases), dioxygenases, and peroxidases. Measuring the activity of these enzymes in environmental samples is constrained by the solubility of available substrates and complicated by competition with organic and mineral components in the sample matrix that can act as alternate electron donors and acceptors.

Assays of oxidative enzyme activity involve a substrate that serves as an electron donor, generating a product that can be quantified spectrophotometrically (Mason

B. Hendel · J. Marxsen (✉)
iFZ Research Centre for Biosystems, Land Use and Nutrition, Department of Animal Ecology and Systematics, Justus Liebig University, Giessen, Germany
e-mail: hendel@chemie-bio.uni-siegen.de; jmar@zo.jlug.de

R. L. Sinsabaugh
Department of Biology, University of New Mexico, Albuquerque, NM, USA
e-mail: rlsinsab@unm.edu

F. Bärlocher et al. (eds.), *Methods to Study Litter Decomposition*,
https://doi.org/10.1007/978-3-030-30515-4_46

1948). The procedure described here is the most common and uses L-3,4-dihydroxyphenylalanine (L-DOPA) as the electron-donating substrate for the detection of phenoloxidase activity. L-DOPA is a useful substrate to determine enzyme activities in environmental samples, because it is soluble in water and readily oxidized at circumneutral pH. One of the products of DOPA oxidation has a red tint and can be quantified by measuring its absorbance at a wavelength of 460 nm. When hydrogen peroxide (H_2O_2) is added to the sample, the activity of peroxidases can also be estimated using the same approach. Other substrates, notably pyrogallol (1,2,3-trihydroxybenzene) and ABTS (2,20-azino-bis(3-ethylbenzthiazoline-6-sulfonic acid), can also be used. Bach et al. (2013) compared phenol oxidase and peroxidase assays using these substrates, as well as L-DOPA, for a wide variety of soils under varying pH conditions. Their findings and recommendations are valuable when selecting substrates.

The L-DOPA method can be used for all types of organic matter occurring in aquatic environments, including plant litter (Table 46.1), and it is fast and accurate (Sinsabaugh and Linkins 1990). However, activity associated with particulate organic matter is more easily detected than activity in the water column. Activity also tends to be higher in humic systems and when associated with fine as opposed to coarse organic particles.

There are several important caveats. In particular, the ecological significance of extracellular oxidative enzyme activity can be difficult to interpret (Münster and De Haan 1998; Sinsabaugh and Foreman 2003; Bach et al. 2013), because these enzymes are involved in both the formation and degradation of polyphenols (Stevenson 1994) and, in some organisms, they may play a role in mitigating the potentially inhibitory effects of reactive phenols (Freeman et al. 2001). Furthermore, the fact that L-DOPA is readily oxidized means that different classes of oxidative enzymes may contribute variously to colour formation during assays. These contributions may vary with assay conditions and sample composition. Another potential problem is that L-DOPA can be oxidized non-enzymatically under some conditions (Bach et al. 2013). In our experience, samples containing reduced Mn are especially problematic. Thus, appropriate controls and attention to reaction kinetics are important. During assays, a fixed incubation time, generally the minimum time needed to detect a signal 2–3 times greater than the negative controls is critical for comparing activities among samples. For water column samples, which typically have activities one or more orders of magnitude lower than particulate organic matter or sediments, several hours of incubation may be needed to detect activity relative to controls.

The specific procedure presented here has been adopted from Sinsabaugh and Linkins (1990), Hendel (1999), and Hendel and Marxsen (2000).

Table 46.1 Phenoloxidase and peroxidase extracellular enzyme activities in decomposing leaves and wood from aquatic and soil environments determined by oxidation of DOPA

Enzyme	Plant material	Plant species	Environment	Geographic region	Enzyme activity (μmol g^{-1} AFDM h^{-1})	References
Phenoloxidase	Leaf litter	Black alder (*Alnus glutinosa*)	Grassland stream	Central Germany	8–66	1
Phenoloxidase	Leaf litter	European beech (*Fagus sylvatica*)	Grassland stream	Central Germany	3–16	1
Phenoloxidase	Leaf litter	Black poplar (*Populus nigra*)	Forested stream	Northeast Spain	<1–7	2
Phenoloxidase	Leaf litter	London plane (*Platanus acerifolia*)	Forested stream	Northeast Spain	<1–16	2
Phenoloxidase	Leaf litter	Cotton wood (*Populus deltoides*)	Riparian zone and river channel	Rio Grande, NM, USA	28–34	3[a]
Phenoloxidase	Leaf litter	Russian olive (*Elaeagnus angustifolia*)	Riparian and river channel	Rio Grande, NM, USA	20–33	3[a]
Phenoloxidase	Needle litter	Ponderosa pine (*Pinus ponderosa*)	Mixed conifer forest	Sierra Nevada, CA, USA	1–12	4
Phenoloxidase	Benthic CPOM	Mixed	Shallow ponds	Southwest coast of Spain	1–450	5
Phenoloxidase	Wood	Black alder (*Alnus glutinosa*)	Grassland stream	Central Germany	0.6–5.1	1, 6
Phenoloxidase	Wood	European beech (*Fagus sylvatica*)	Grassland stream	Central Germany	0.4–7.0	1, 6
Phenoloxidase	Wood	Yellow poplar (*Liriodendron tulipifera*)	Woodland stream	North Carolina, USA	3–40	7
Phenoloxidase	Wood	White birch (*Betula papyfera*)	Woodland stream, riparian zone and soil	Northern New York, USA	2–30	8
Peroxidase	Leaf litter	Black alder(*Alnus glutinosa*)	Grassland stream	Central Germany	5–12	1
Peroxidase	Leaf litter	European beech (*Fagus sylvatica*)	Grassland stream	Central Germany	<1–13	1
Peroxidase	Leaf litter	Cotton wood (*Populus deltoides*)	Riparian zone and river channel	Rio Grande, NM, USA	14–36[a]	3
Peroxidase	Leaf litter	Russian olive (*Elaeagnus angustifolia*)	Riparian zone and river channel	Rio Grande, NM, USA	11–37[a]	3
Peroxidase	Wood	Black alder (*Alnus glutinosa*)	Grassland stream	Central Germany	<1–15	1, 6

Table 46.1 (continued)

Enzyme	Plant material	Plant species	Environment	Geographic region	Enzyme activity (μmol g^{-1} AFDM h^{-1})	References
Peroxidase	Wood	European beech (*Fagus sylvatica*)	Grassland stream	Central Germany	<1–5.8	1, 6
Peroxidase	Wood	Yellow poplar (*Liriodendron tulipifera*)	Woodland stream	North Carolina, USA	0–25	7

1 = Hendel (1999), 2 = Artigas et al. (2011), 3 = Harner et al. (2009), 4 = Waldrop et al. (2003), 5 = Alvarez and Guerrero (2000), 6 = Hendel and Marxsen (2000), 7 = Tank et al. (1998), 8 = Sinsabaugh et al. (1992)

[a]Range of average values from different environments

2 Equipment, Chemicals, and Solutions

2.1 *Equipment*

- Sharp knife or scalpel
- Homogenizing device (e.g., Polytron)
- Homogenization vessels (100 ml)
- Water baths (5 and 10 °C)
- Shaking incubator (20 °C)
- Bench centrifuge (3000 g)
- Spectrophotometer (460 nm)
- Pipettes, Eppendorf type or equivalent (e.g., 1000 and 2500 μl)

2.2 *Chemicals*

- Acetic acid (100%), analytical grade
- Sodium hydroxide (NaOH), analytical grade
- L-3,4-Dihydroxyphenylalanine (L-DOPA), analytical grade
- Hydrogen peroxide (H_2O_2, 30%), analytical grade

2.3 *Solutions*

- Acetate buffer: 50 mmol l^{-1} (3.0025 g 100% acetic acid per liter, pH adjusted with NaOH to 5.0)
- L-DOPA stock solution: 5 mmol l^{-1}(0.98575 g l^{-1}) in 50 mM acetate buffer; as this solution is unstable, it needs to be prepared immediately before running assays
- Hydrogen peroxide solution: 0.3% (v/v)

3 Experimental Procedures

3.1 *Sample Preparation*

1. Collect leaf litter or wood in the field (fresh mass preferably ≥1.0 g) and transport to the laboratory in a cooled, insulated container.
2. Preferably process collected samples immediately.
3. Remove any adhering debris and macroinvertebrates.
4. Chop litter with a knife or scalpel.
5. Transfer the chopped wood or leaf litter into a homogenization vessel containing autoclaved water from the sampling site (e.g., 80 ml for a 100 ml vessel).
6. Place the homogenization vessel in a water bath set at about 5 °C, and homogenize sample for at least 2 min, while preventing the sample temperature from rising above 18 °C.
7. Use the homogenate for enzyme assays.

3.2 *Enzyme Analysis*

1. Phenoloxidase: mix 2 ml of homogenate with 2 ml of DOPA stock solution; run at least 4 analytical replicates.
2. As a control, mix 2 ml of the homogenate with 2 ml of acetate buffer.
3. Peroxidase: mix 2 ml of the homogenate with 2 ml of DOPA stock solution; run at least 4 analytical replicates.
4. Add 200 µl of hydrogen peroxide solution and mix gently.
5. As a control, mix 2 ml of the homogenate with 2 ml acetate buffer and 200 µl hydrogen peroxide solution.
6. Incubate for exactly 60 min at 20 °C in a shaking incubator.
7. Centrifuge for 1 min at approximately 3000 g.
8. Transfer supernatant to cuvette (size: 1 cm).
9. Measure absorbance immediately at 460 nm in spectrophotometer.
10. Calculate enzyme activity according to:

$$A = \mathrm{Abs}_{460} \times k^{-1} \times T^{-1} \times OMC^{-1} \quad (46.1)$$

where

A = enzyme activity in µmol h^{-1} g^{-1}
Abs_{460} = sample absorbance at 460 nm minus control absorbance; for peroxidase also minus absorbance for phenoloxidase
k = extinction coefficient, which is 1.66 l $mmol^{-1}$(1.66 ml $µmol^{-1}$) for DOPA under the conditions of this assay
T = incubation time, h
OMC = organic matter concentration (g organic matter per ml of sample homogenate).

4 Final Remarks

The pH optimum of the reaction is about 8. Assays should consequently be run at this pH, if the goal is to estimate maximum potential activities. In contrast, to assess activities occurring naturally in the environment (e.g., within leaf litter or soils), acetate buffer at pH 5, as used in the protocol here, is often preferable because 5 is the typical pH of litter and soils. However, in some aquatic environments, such as hardwater streams, the intrinsic pH of litter might deviate from 5 and should hence be determined before choosing a pH for carrying out the enzyme assays.

The assay can be adapted for 96-well microplates (Saiya-Cork et al. 2002; Gallo et al. 2004). In this procedure, 200 µl aliquots of sample homogenate are dispensed into replicate wells. For phenoloxidase, 50 µl of 25 mM DOPA is added to each sample well. Peroxidase assays receive 50 µl of 25 mM DOPA plus 10 µl of 0.3% H_2O_2. Negative control wells for phenoloxidase contain 200 µl of acetate buffer and 50 µl of DOPA solution; blank control wells contain 200 µl of sample suspension and 50 µl of acetate buffer. For peroxidase, negative and blank control wells also receive 10 µl of H_2O_2. There are 16 replicate sample wells for each assay and 8 replicate wells for blanks and controls. The microplates are incubated in the dark at 20 °C for up to 18 h. Activity is quantified by measuring absorbance at 460 nm using a microplate spectrophotometer. Results are expressed in units of nmol h^{-1} g^{-1} organic matter using a micromolar extinction coefficient of 7.9. As an excellent online source of information, the website of the Enzymes in the Environment Research Coordination Network (https://enzymes.nrel.colostate.edu/enzymes-home.html) is recommended, where detailed microplate protocols for litter, soil, and surface waters are presented.

References

Alvarez, S., & Guerrero, M. C. (2000). Enzymatic activities associated with decomposition of particulate organic matter in two shallow ponds. *Soil Biology and Biochemistry, 32*, 1941–1951.

Artigas, J., Gaudes, A., Muñoz, I., Romaní, A. M., & Sabater, S. (2011). Fungal and bacterial colonization of submerged leaf litter in a Mediterranean stream. *International Review of Hydrobiology, 96*, 221–234.

Bach, C. E., Warnock, D. D., Van Horn, D. J., Weintraub, M. N., Sinsabaugh, R. L., Allison, S. D., & German, D. P. (2013). Phenol oxidase and peroxidase activities in relation to substrate, assay conditions and soil type. *Soil Biology and Biochemistry, 67*, 183–191.

Baldrian, P. (2006). Fungal laccases – Occurrence and properties. *FEMS Microbiology Reviews, 30*, 215–242.

Freeman, C., Ostle, N., & Kang, H. (2001). An 'enzymatic latch' on a global carbon store. *Nature, 409*, 149.

Gallo, M., Amonette, R., Lauber, C., Sinsabaugh, R. L., & Zak, D. R. (2004). Microbial community structure and oxidative enzyme activity in nitrogen-amended north temperate forest soils. *Microbial Ecology, 48*, 218–229.

Harner, M. J., Crenshaw, C. L., Abelho, M., Stursova, M., Follstad Shah, J. J., & Sinsabaugh, R. L. (2009). Decomposition of leaf litter from a native tree and an actinorhizal invasive across riparian habitats. *Ecological Applications, 19*, 1135–1146.

Hendel, B. (1999). *Der mikrobielleAbbau von Holz und Laub im Breitenbach unter besonderer Berücksichtigung der Bedeutung extrazellulärer Enzyme*. Doctoral thesis. University of Giessen, Germany.

Hendel, B., & Marxsen, J. (2000). Extracellular enzyme activity associated with degradation of beech wood in a Central European stream. *International Review of Hydrobiology, 85*, 95–105.

Hoegger, P. J., Kilaru, S., James, T. Y., Thacker, J. R., & Kues, U. (2006). Phylogenetic comparison and classification of laccase and related multicopper oxidase protein sequences. *FEBS Journal, 273*, 2308e2326.

Mason, H. S. (1948). The chemistry of melanin: III. Mechanism of the oxidation of dihydroxyphenylalanine by tyrosinase. *Journal of Biological Chemistry, 172*, 83–99.

Münster, U., & De Haan, H. (1998). The role of microbial extracellular enzymes in the transformation of dissolved organic matter in humic waters. In D. O. Hessen & L. J. Tranvik (Eds.), *Aquatic humic substances* (pp. 199–257). Berlin: Springer.

Saiya-Cork, K. R., Sinsabaugh, R. L., & Zak, D. R. (2002). Effects of long term nitrogen deposition on extracellular enzyme activity in an *Acer saccharum* forest soil. *Soil Biology and Biochemistry, 34*, 1309–1315.

Sinsabaugh, R. L. (2010). Phenol oxidase, peroxidase and organic matter dynamics of soil. *Soil Biology and Biochemistry, 42*, 391–404.

Sinsabaugh, R. L., & Foreman, C. M. (2003). Integrating DOM metabolism and microbial diversity. In S. Findlay & R. L. Sinsabaugh (Eds.), *Aquatic ecosystems: The interactivity of dissolved organic matter* (pp. 426–454). San Diego: Academic.

Sinsabaugh, R. L., & Linkins, A. E. (1990). Enzymic and chemical analysis of particulate organic matter from a boreal river. *Freshwater Biology, 23*, 301–309.

Sinsabaugh, R. L., Antibus, R. K., Linkins, A. E., McClaugherty, C. A., Rayburn, L., Repert, D., & Weiland, T. (1992). Wood decomposition over a first-order watershed: Mass loss as a function of lignocellulose activity. *Soil Biology and Biochemistry, 24*, 743–749.

Stevenson, F. J. (1994). Biochemistry of the formation of humic substances. In F. J. Stevenson (Ed.), *Humus chemistry* (pp. 188–211). New York: Wiley.

Tank, J. L., Webster, J. R., Benfield, E. F., & Sinsabaugh, R. L. (1998). Effect of leaf litter exclusion on microbial enzyme activity associated with wood biofilms in streams. *Journal of the North American Benthological Society, 17*, 95–103.

Waldrop, M. P., McColl, J. G., & Powers, R. F. (2003). Effects of forest postharvest management practices on enzyme activities in decomposing litter. *Soil Science Society of America Journal, 67*, 1250–1256.

Chapter 47
Phenol Oxidation

Martin Zimmer

Keywords Brown-rot fungi · Catecholases · Laccases · Lignin · Phenol oxidases · Polyphenolics · Quinones · Soft-rot fungi · Tannins · Tyrosinases · White-rot fungi

1 Introduction

Lignins are major structural components of plant cell walls and therefore of plant litter (Sarkanen and Ludwig 1971). They are complex polymers of a small number of methoxylated phenolic compounds such as coumaryl alcohol, sinapyl alcohol, and coniferyl alcohol (Boerjan et al. 2003). Due to strong C-C linkages and alkyl-aryl ether bonds, lignins effectively resist chemical and enzymatic attack (Hagerman and Butler 1991). Therefore, lignin degradation requires phenol oxidation (Breznak and Brune 1994). Other important phenolic litter constituents include condensed tannins (Harrison 1971; Savoie and Gourbiére 1989; see Chap. 19), which are regularly structured polymers of flavan-3-ols and flavan-3,4-diols that are linked through C-C bonds between the monomers (Swain 1979; Hagerman and Butler 1991). As with lignins, the degradation of condensed tannins begins with oxidation. Hydrolyzable tannins, in contrast, are glucose esters of gallic acid or ellagic acid units and are hence subject to hydrolysis by esterases.

The degradation of the lignin moiety of lignocellulose (see Chap. 46) is strongly dependent on microbial activity (Breznak and Brune 1994). However, not every microbial species involved in decomposition is capable of degrading lignocellulose (Ljungdahl and Eriksson 1985). In contrast to brown- and white-rot fungi, which are

M. Zimmer (✉)
Leibniz Centre for Tropical Marine Research (ZMT), University of Bremen, Bremen, Germany
e-mail: martin.zimmer@leibniz-zmt.de

F. Bärlocher et al. (eds.), *Methods to Study Litter Decomposition*,
https://doi.org/10.1007/978-3-030-30515-4_47

primarily terrestrial, the litter-degrading soft-rot and other fungi (mostly Ascomycetes and Deuteromycetes) prominent in aquatic environments are only weakly adapted to degrading lignin (Rabinovich et al. 2004).

Numerous enzymes are involved in phenol oxidation (Sinsabaugh 2010). Laccases (EC 1.10.3.2) have been found in plants and many fungi (Mayer 1987), but only in one bacterium (Faure et al. 1995). Tyrosinases (EC 1.14.18.1) are known from fungi (Wood 1980; Claus and Filip 1990), actinomycetes (Claus and Filip 1990), and plants (Summers and Felton 1994). Both laccases and tyrosinases may be important in wood and leaf-litter decomposition (Wood 1980; Thurston 1994). Catechol oxidase (EC 1.10.3.1) has a similar function as laccases in many plants and fungi (Mayer 1987), and is involved in the oxidative polymerization of phenolics. Phenol-oxidizing enzymes of invertebrate origin are mostly considered to be involved in moulting, wound-healing, and immune response, but there is evidence suggesting that hemocyanin of crustaceans can be activated into a phenoloxidase-like enzyme (Jaenicke et al. 2009) that might serve in the oxidative breakdown of ingested lignins and other phenolic compounds (Cragg et al. 2015). Along this line of argument, Yoruk and Marshall (2003) suggest adding SDS (up to 2 mM) to assays of phenol oxidase activity to activate these enzymes. It should be kept in mind, however, that the outcome of such measurements will not reflect the natural conditions in the environment.

Determining the activity of phenol-oxidizing enzymes can be based on (1) the decrease in the concentration of a particular substrate, or (2) the increase in oxidation products, which potentially covers the combined activities of several enzymes with different substrate affinities when using a common substrate. Much more frequently than in decomposition studies, phenol oxidation is measured in the context of invertebrate immune response (e.g., Laughton and Siva-Jothy 2011). The corresponding techniques cannot be transferred directly to decomposition studies, but the basic approach, relying on the brownish color of oxidation products of most phenolic compounds, is the same. It is virtually impossible to determine individual activities of any of the various phenol-oxidizing enzymes in an environmental sample.

Since the oxidation products of phenolic compounds (called quinones) are not clearly defined, no specific extinction coefficient can be determined. The method therefore yields only relative phenol oxidation capacity (ΔA mg^{-1} h^{-1}), and, thus, comparison of the data is confined to using the same substrate. These quinones are prone to further cross-reaction and polymerization, eventually resulting in undefined melanins. To prevent the quinones formed upon oxidation of, for example, catechol as an often-used substrate to determine phenol oxidase activity, from further chemical change, Perucci et al. (2000) suggest adding L-proline to the assay to stabilize quinone and the corresponding red-brownish coloration of the assay.

The method by Zimmer and Topp (1998) described here provides an estimate of the overall phenol oxidation capacity based on the amount of oxidation products. To this end, a suitable phenolic substrate (see Faure et al. 1995) is mixed with the sample, and the change in absorbance resulting from the release of colored (brownish) oxidation products is followed over time. However, different substrates may be indicative of different types of phenol oxidases, such as *o*-diphenol oxidase (catecholase), oxidizing, for example, catechol or L-3,4-dihydroxyphenylalanine (L-DOPA); monophenol oxidase (tyrosinase), oxidizing, for example, tyrosine to L-DOPA; laccase, oxidizing, for example, syringaldazine or the artificial substrate 2,2′-azino-bis(3-ethylbenzothiazoline-6-sulphonic acid) (ABTS) that is often used for determining phenol oxidase activity but is highly sensitive to the pH of the assay.

2 Equipment and Materials

2.1 Equipment

- Homogenizer (e.g., electronic disperser or mortar and pestle for leaf litter; rotation grinder, or ultrasonic disintegrator for gut and feces samples)
- Incubation tubes: glass tubes with screw caps (15–20 ml) for leaf litter and plastic reaction tubes (1.5 ml) for gut and feces samples
- Analytical balance
- Shaker
- Centrifuge (10,000 g).
- Micropipettes (100–1000 μl; 10–100 μl)
- Plastic cuvettes
- Spectrophotometer (340 nm)

2.2 Materials

- Leaf litter collected in the field
- Dissected guts of detritivores having fed on leaf litter (gut epithelium preferably removed)
- Feces of detritivores having fed on leaf litter

2.3 Chemicals and Solutions

- 0.05 M phosphate buffer: 415 ml 0.1 M KH_2PO_4 + 85 ml 0.1 M Na_2HPO_4 + 500 ml distilled water, pH 6.2; if prepared accurately, the pH does not need to be adjusted
- 50 mM catechol and 50 mM L-proline in 0.05 M phosphate buffer, pH 6.2
- Depending on the phenol oxidase investigated, phenolic substrates other than catechol may be more appropriate (e.g., gallic acid, L-DOPA, tyrosine, syringaldazine). The wavelength for photometric determination of oxidation products may have to be adjusted accordingly (Faure et al. 1995; Zimmer and Topp 1998); for the 4-(*N*-proline)-*o*-benzoquinone product of proline and the catechol-derived quinone, a wavelength of 525 nm is most appropriate (Perucci et al. 2000)

- Depending on the phenolic substrate, the addition of up to 20% ethanol may be necessary to dissolve the substrate (Faure et al. 1995; Zimmer and Topp 1998)

3 Experimental Procedures

3.1 *Extraction of Microbial Enzymes*

1. Weigh samples of leaf litter (corresponding to 50–100 mg dry mass), dissected guts (5–10 mg), or feces (5–10 mg).
2. Determine dry mass/fresh mass ratios to estimate dry mass of samples from fresh mass.
3. The appropriate method of enzyme extraction depends on the source of enzymes; with extracellular enzymes, thoroughly chopping up samples with a homogenizer is sufficient for accurate measurement of enzyme activity; with cell-bound enzymes, additional sonication is recommended to release enzymes into the supernatant.
4. Homogenize litter samples in 10 ml of 0.05 M phosphate buffer or gut or feces samples in 1 ml of 0.05 M phosphate buffer. Homogenization must be done on ice to avoid thermal denaturation of enzymes.
5. Homogenates may be stored frozen (−20 °C) until used for assays.
6. Centrifuge suspensions (5 min; ca. 10,000 g, depending on the available centrifuge and reaction tubes; 4 °C).

3.2 *Determination of Phenol Oxidase Activity*

1. Add 100 µl of the supernatant to 900 µl of 50 mM catechol solution and mix thoroughly.
2. Follow change in absorbance (ΔA) at 340 nm at 1-min intervals for the first 10 min.
3. Determine relative catechol oxidation as mean ΔA per min by linear regression analysis.
4. Calculate relative phenol oxidase activity (ΔA mg^{-1} h^{-1}) as:

$$\text{phenol oxidase activity} = \frac{\Delta A \times 60 \times \text{dilution factor}}{\text{sample dry mass}} \qquad (47.1)$$

where the dilution factor = 100 for litter and 10 for guts and feces.

References

Boerjan, W., Ralph, J., & Baucher, M. (2003). Lignin biosynthesis. *Annual Review of Plant Biology, 54*, 519–546.

Breznak, J. A., & Brune, A. (1994). Role of microorganisms in the digestion of lignocellulose by termites. *Annual Review of Entomology, 39*, 453–487.

Claus, H., & Filip, Z. (1990). Effects of clays and other solids on the activity of phenoloxidases produced by some fungi and actinomycetes. *Soil Biology and Biochemistry, 22*, 483–488.

Cragg, S. M., Beckham, G. T., Bruce, N. C., Bugg, T. D. H., Distel, D. L., Dupree, P., Green Etxabe, A., Goodell, B. S., Jellison, J., McGeehan, J. E., McQueen-Mason, S. J., Schnorr, K., Walton, P. H., Watts, J. E. M., & Zimmer, M. (2015). Lignocellulose degradation mechanisms across the tree of life. *Current Opinion in Chemical Biology, 29*, 108–119.

Faure, D., Bouillant, M.-L., & Bally, R. (1995). Comparative study of substrates and inhibitors of *Azospirillum lipoferum* and *Pyricularia oryzae* laccases. *Applied and Environmental Microbiology, 61*, 1144–1146.

Hagerman, A. E., & Butler, L. G. (1991). Tannins and lignins. In G. A. Rosenthal & M. R. Berenbaum (Eds.), *Herbivores: Their interactions with secondary plant metabolites – I: The chemical participants* (pp. 355–388). New York: Academic.

Harrison, A. F. (1971). The inhibitory effect of oak leaf litter tannins on the growth of fungi, in relation to litter decomposition. *Soil Biology and Biochemistry, 3*, 167–172.

Jaenicke, E., Fraune, S., Irmak, P., Maym, S., Augustin, R., & Zimmer, M. (2009). Is activated hemocyanin instead of phenoloxidase involved in immune response in woodlice? *Developmental and Comparative Immunology, 33*, 1055–1063.

Kubitzki, K., & Gottlieb, O. R. (1984). Phytochemical aspects of angiosperm origin and evolution. *Acta Botanica Neerlandica, 33*, 457–468.

Laughton, A., & Siva-Jothy, M. T. (2011). A standardised protocol for measuring phenoloxidase and prophenoloxidase in the honey bee, *Apis mellifera*. *Apidologie, 42*, 140–149.

Ljungdahl, L. G., & Eriksson, K. E. (1985). Ecology of microbial cellulose degradation. *Advances in Microbial Ecology, 8*, 237–299.

Mayer, A. M. (1987). Polyphenol oxidases in plants – Recent progress. *Phytochemistry, 26*, 11–20.

Perucci, P., Casucci, C., & Dumontet, S. (2000). An improved method to evaluate the *o*-diphenol oxidase activity of soil. *Soil Biology and Biochemistry, 32*, 1927–1933.

Rabinovich, M. L., Bolobova, A. V., & Vasil'chenko, L. G. (2004). Fungal decomposition of natural aromatic structures and xenobiotics: A review. *Applied Biochemistry and Microbiology, 40*, 1–17.

Sarkanen, K. V., & Ludwig, C. H. (1971). *Lignins – Occurrence, formation, structure and reactions*. New York: Wiley.

Savoie, J. M., & Gourbière, F. (1989). Decomposition of cellulose by the species of the fungal succession degrading *Abies alba* needles. *FEMS Microbial Ecology, 62*, 307–314.

Sinsabaugh, R. L. (2010). Phenol oxidase, peroxidase and organic matter dynamics of soil. *Soil Biology and Biochemistry, 42*, 391–404.

Summers, C. B., & Felton, G. W. (1994). Prooxidant effects of phenolic acids on the generalist herbivore *Helicoverpa zea* (Lepidoptera: Noctuidae): Potential mode of action for phenolic compounds in plant anti-herbivore chemistry. *Insect Biochemistry and Molecular Biology, 24*, 943–953.

Swain, T. (1979). Tannins and lignins. In G. A. Rosenthal & D. H. Janzen (Eds.), *Herbivores: Their interactions with secondary plant metabolites – I: The chemical participants* (pp. 657–718). San Diego: Academic.

Thurston, C. F. (1994). The structure and function of fungal laccases. *Microbiology, 140*, 19–26.

Wood, D. A. (1980). Production, purification and properties of extracellular laccase of *Agaricus bisporus*. *Journal of General Microbiology, 117*, 327–338.

Yoruk, R., & Marshall, M. R. (2003). Physicochemical properties and function of plant phenol oxidase: A review. *Journal of Food Biochemistry, 27*, 361–422.

Zimmer, M., & Topp, W. (1998). Nutritional biology of terrestrial isopods (Isopoda: Oniscidea): Copper revisited. *Israel Journal of Zoology, 44*, 453–462.

Chapter 48
Proteinase Activity: Azocoll and Thin-Layer Enzyme Assay

Manuel A. S. Graça and Felix Bärlocher

Keywords Enzyme activity · Nitrogen acquisition · Polyphenolics · Protein immobilization · Proteinase inhibitors · Proteinase pH optimum · Proteolytic activity · Shredders · Spectrophotometry · Tipula gut

1 Introduction

Senescent leaves are rich in structural polysaccharides. Invertebrates can gain access to this energy source by (a) ingesting leaves partially digested by fungi, (b) relying on gut endosymbionts, and/or (c) using fungal enzymes in their own gut (Bärlocher 1982; Graça 1993). While carbon is thus readily available in senescent leaves, nitrogen tends to be scarce, even relative to the much lower demand compared to carbon (Klug and Kotarski 1980; Bernays 1981). In addition, nitrogen accessibility to invertebrates is lowered by the presence of plant polyphenolics that remain active after senescence and complex proteins, where most of the cellular nitrogen is located (Chaps. 18, 19, and 20; MacManus et al. 1985; Waterman and Mole 1994; Zhang et al. 2008). Leaves with low nitrogen and high polyphenolic content are therefore a low-quality food resource for detritivores.

Invertebrate detritivores with low mobility such as tipulid larvae cannot afford to reject low-quality food. Natural selection should have favoured adaptations allowing them to overcome the protein-masking effects of polyphenolics and use N resources more efficiently than highly mobile detritivores (Bärlocher and Porter 1986). This seems to have been achieved by alkaline protein digestion; the pH of the hindgut of

M. A. S. Graça (✉)
MARE – Marine and Environmental Sciences Centre & Department of Life Sciences, University of Coimbra, Coimbra, Portugal
e-mail: mgraca@ci.uc.pt

F. Bärlocher
Department of Biology, Mount Allison University, Sackville, NB, Canada
e-mail: fbaerloc@mta.ca

F. Bärlocher et al. (eds.), *Methods to Study Litter Decomposition*,
https://doi.org/10.1007/978-3-030-30515-4_48

some immature tipulid shredders can reach or exceed 10.5–11.0 (Martin et al. 1980; Sinsabaugh et al. 1985; Bärlocher and Porter 1986; Canhoto and Graça 2006). Since phenol–protein complexes are less stable under alkaline conditions (Swain 1979; Bärlocher et al. 1989), the pH regime of the midgut allows these tipulids to digest protein in the presence of polyphenolics. However, this is only possible if proteinases remain active at high pH. This has in fact been demonstrated for several tipulid proteinases (Martin et al. 1980; Sinsabaugh et al. 1985; Bärlocher and Porter 1986; Graça and Bärlocher 1998). Given that enzymes active against polysaccharides reach their maximum activity at acidic to circum-neutral conditions, it is not surprising that polysaccharide and protein digestion in the tipulid gut occur at separate locations characterized by different pH values (Martin et al. 1980; Bärlocher and Porter 1986; Fig. 48.1).

Here, we describe two simple methods to estimate generalized proteolytic activity. Proteinases can be subdivided on the basis of the peptide linkage they attack (e.g., serine proteinases and cysteine proteinases). The two assays measure the combined effects of all these proteinases.

The first assay involves the use of azocoll. Azocoll is a suspension of powdered cowhide to which a bright-red dye is attached (azo dye-bound collagen) (Oakley et al. 1946). The cowhide contains the usual assortment of peptide linkages characteristic of all proteins. When a proteolytic enzyme breaks one of these linkages, the bound dye is released into solution. The rate of this release (determined by measuring absorbance of the filtrate at 520 nm) is used to measure overall proteolytic activity. The described protocol follows Martin et al. (1980).

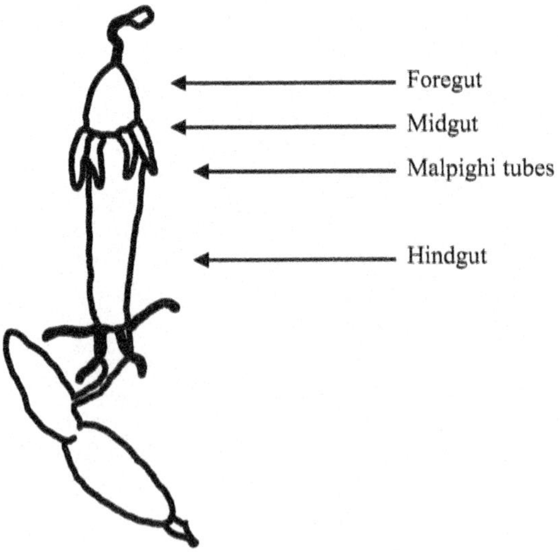

Fig. 48.1 Tipulid gut. (Adapted from Martin et al. 1980)

Wilkström et al. (1981, 1982) described another method to measure the activity of proteinases, known as the "thin-layer enzyme assay". It takes advantage of the property of proteins to adsorb to hydrophobic solid surfaces as thin layers, increasing wettability of these surfaces. Polystyrene Petri dishes are hydrophobic surfaces which can be coated with a protein solution. Application of a gut extract containing proteinases will cause protein digestion, and the consequent decrease in wettability can be visualized as a decrease in the condensation of water vapour. The magnitude of protein digestion is proportional to the zone of reduced wettability, expressed as D^2 (squared diameter) of the affected area.

2 Equipment, Chemicals, and Solutions

2.1 Equipment and Material

- Immature specimens of *Tipula* sp.
- Homogenizer or a manual tissue grinder
- Centrifuge (14,000 g)
- Sephadex G-25 M column
- Spectrophotometer (520 nm)
- Polystyrene Petri dishes
- Microsyringe (10 μl)
- Cork borer
- Fibreglass filter or membrane filter
- Oven (37 °C)
- Pan with hot water (50 °C)

2.2 Chemicals and Solutions

- Ethanol (95%)
- Sterile deionized water
- Bovine Serum Albumin (BSA)
- Agar
- Azocoll
- HCl (0.01 M)
- Buffers: 0.1 M acetate (pH 5.1, 5.6), 0.1 M phosphate (pH 6.0, 6.5, 7.0, 7.5), 0.1 M tris(hydroxymethyl) aminomethane (Tris; 8.0, 8.5, 9.0), 0.1 M carbonate (pH 9.5, 10.0, 10.5, 11.0), 0.1 M phosphate (11.5, 12.0, 12.5)

3 Experimental Procedures

3.1 Sample Preparation

1. Take live immature specimens of a tipulid shredder. Place in freezer for a few minutes. Open the specimens, and remove the entire gut from the anterior to the posterior end.
2. Homogenize the desired gut section (e.g., midgut, hindgut; combine sections of 3–10 individuals) in 2.5 ml of sterile deionized water.
3. Centrifuge the homogenized gut at 14,000 g for 4 min.
4. Desalt the supernatant by passing it through a Sephadex G-25 M column and store at 4 °C until needed.
5. Optionally, determine protein content of the extract (see Chap. 12) to express results per mg protein.

3.2 Azocoll

1. Mix 0.5 ml of gut extract with 0.5 ml of buffer and add 25 mg azocoll.
2. Incubate at 37 °C for 15–90 min, depending on enzyme activity.
3. Terminate reaction by adding 3.0 ml of 0.01 M HCl.
4. Pass through fibreglass filter.
5. Measure absorbance at 520 nm.
6. Express results as amount of enzyme required to bring about a change of absorbance at 520 nm under the conditions of the assay (e.g., number of 0.001 absorbance units per minute per gut section at pH = 8 and incubation at 37 °C).

3.3 Thin-Layer Enzyme Assay

1. Clean the internal surface of polystyrene Petri dishes with 95% ethanol and dry in an oven at 37 °C.
2. Coat the internal surface of the Petri dishes with protein (3 ml of BSA, 1 mg ml^{-1} for 30 min). Discard the excess protein solution and gently rinse with sterile deionized water. Dry at room temperature or in an oven at 37 °C.
3. Prepare an agar solution (2%) and apply 10 ml over the coated Petri dishes. Allow to gel at room temperature.
4. Cut 3–4 mm diameter wells in the agar and apply 10 μl of desalted gut extract.
5. Incubate the Petri dishes for 18 h at 37 °C.
6. Remove the agar manually, gently wash with distilled water and dry.

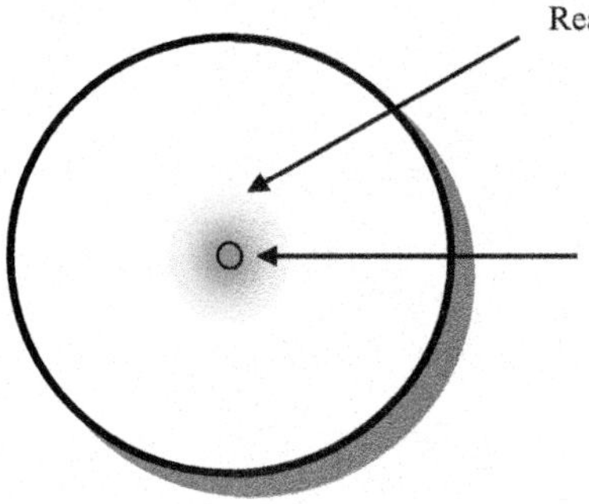

Fig. 48.2 Decreased wettability of the internal surface of a polystyrene Petri dish, caused by digestion of proteins

7. Expose the Petri dish to water vapour from a hot (50 °C) water pan. Digestion of protein is visible as condensation of water vapour due to reduced surface wettability (Fig. 48.2).
8. Determine the size of the reactive area or an equivalent measure (e.g. squared diameter).
9. To estimate proteolytic activity as a function of pH, buffers can be used to adjust the pH of the agar to values similar to the ones in the gut or environment.

References

Bärlocher, F. (1982). The contribution of fungal enzymes to the digestion of leaves by *Gammarus fossarum* Koch (Amphipoda). *Oecologia, 52*, 1–4.

Bärlocher, F., & Porter, C. W. (1986). Digestive enzymes and feeding strategies of three stream invertebrates. *Journal of the North American Benthological Society, 5*, 58–66.

Bärlocher, F., Tibbo, P. G., & Christie, S. H. (1989). Formation of phenol-protein complexes and their use by two stream invertebrates. *Hydrobiologia, 173*, 243–249.

Bernays, E. A. (1981). Plant tannins and insect herbivores: An appraisal. *Ecological Entomology, 6*, 353–360.

Canhoto, C., & Graça, M. A. S. (2006). Digestive tract and leaf processing capacity of the stream invertebrate *Tipula lateralis*. *Canadian Journal of Zoology-Revue Canadienne de Zoologie, 84*, 1087–1095.

Graça, M. A. S. (1993). Patterns and processes in detritus-based stream systems. *Limnologica, 23*, 107–114.

Graça, M. A. S., & Bärlocher, F. (1998). Proteolytic gut enzymes in *Tipula caloptera* – Interaction with phenolics. *Aquatic Insects, 21*, 11–18.

Klug, M. J., & Kotarski, S. F. (1980). Ecology of the microbial microbiota in the posterior hind gut of larval stages of the crane fly *Tipula abdominalis*. *Applied and Environmental Microbiology, 40*, 408–416.

MacManus, J. P., Davis, K. G., Beart, J. E., Goffney, S. H., Lilley, T. H., & Haslam, E. (1985). Polyphenol interaction. Part 1. Introduction: Some observations on the reversible complexation of polyphenols with proteins and polysaccharides. *Journal of the Chemical Society, Perkin Transactions, II*, 1429–1438.

Martin, M. M., Martin, J. S., Kukor, J. J., & Merrit, R. W. (1980). The digestion of protein and carbohydrate by the stream detritivore *Tipula abdominalis* (Diptera, Tipulidae). *Oecologia, 46*, 360–364.

Oakley, C. L., Warrack, G. H., & van Heyningen, W. E. (1946). The collagenase (K Toxin) of *Cl. welchii* type A. *Journal of Pathology, 58*, 229–235.

Sinsabaugh, R. L., Linkins, A. E., & Benfield, E. F. (1985). Cellulose digestion and assimilation by three leaf-shredding aquatic insects. *Ecology, 66*, 1464–1471.

Swain, T. (1979). Tannins and lignins. In G. A. Rosenthal & D. H. Janzen (Eds.), *Herbivores: Their interactions with secondary plant metabolites* (pp. 657–682). New York: Academic.

Waterman, P. G., & Mole, S. (1994). *Analysis of phenolic plant metabolites*. London: Blackwell Scientific Publications.

Wikström, M., Elwing, H., & Linde, A. (1981). Determination of proteolytic activity: A sensitive and simple assay utilizing substrate adsorbed to a plastic surface and radial diffusion in gel. *Analytical Biochemistry, 118*, 240–246.

Wikström, M. B., Elwing, H., & Möller, A. J. R. (1982). Proteins adsorbed to a hydrophobic surface used for determination of proteolytic activity. *Enzyme and Microbial Technology, 4*, 265–268.

Zhang, L. H., Lin, Y. M., Ye, G. F., Liu, X. W., & Lin, G. H. (2008). Changes in the N and P concentrations, N:P ratios, and tannin content in *Casuarina equisetifolia* branchlets during development and senescence. *Journal of Forest Research, 13*, 302–311.

Part VI
Litter Consumers

Chapter 49
Processing of Aquatic Invertebrates Colonizing Decomposing Litter

John S. Richardson

Keywords Community structure · Detritivores · Invertebrate identification · Invertebrate sorting · Leaf litter · Litter bags · Litter colonization · Shredders · Shredder biomass · Streams

1 Introduction

Invertebrates may substantially accelerate the breakdown of leaf litter (e.g. Hieber and Gessner 2002; Kominoski et al. 2011), although that is not always the case in all freshwaters (e.g. Boyero et al. 2011). This prompts questions such as which invertebrate species are consuming litter (Chap. 50), how they interact with each other, which litter types they select (Chap. 52), and the stage of breakdown at which they colonize leaf litter. Many invertebrate species contribute to litter breakdown through consumption and fragmentation, but some species use litter mostly as a habitat (Richardson 1992).

There are many reasons to determine which species and size classes of invertebrates are involved in leaf litter breakdown. We may want to test hypotheses about the succession of species or the roles of species at different stages of breakdown. For instance, do particular invertebrate species colonize litter early when microbes are still beginning to grow, when most of the leaf tissue has been colonized by microbes, or when most of the non-lignified tissue is gone? We may also want to compare the types, numbers, and biomass of invertebrates on different types of leaf litter. Estimation of biomass of invertebrates can be linked to resource (litter) abundance, and maybe useful for comparing the relative contributions to litter breakdown by invertebrate species of different size (e.g. Hieber and Gessner 2002; Tonin et al. 2018). Or, we might be interested in comparing different

J. S. Richardson (✉)
Department of Forest and Conservation Sciences, University of British Columbia, Vancouver, BC, Canada
e-mail: john.richardson@ubc.ca

F. Bärlocher et al. (eds.), *Methods to Study Litter Decomposition*,
https://doi.org/10.1007/978-3-030-30515-4_49

freshwaters that differ in nutrient status or other characteristics that might affect invertebrate composition and abundances on leaf litter.

Litter bags of different mesh sizes are commonly used in litter breakdown studies (Chap. 6), with different types of mesh influencing the size range of the consumers found in litter bags. Typically, fine-mesh litter bags have a mesh size of 250 or 500 µm (Chap. 6), which are supposed to keep out most macroinvertebrates (i.e. specimens >1 mm in length). However, even fine-mesh litter bags may be colonized by a lot of meiofauna, i.e. multicellular organisms as small as 30 µm (e.g. Madji et al. 2017), which may affect breakdown rates (Ribblett et al. 2005). In addition, depending on the length of time the litter bags spend in water, small early instars of macroinvertebrates might colonize fine-mesh bags and subsequently grow inside to become large individuals (Baldy and Gessner 1997). Coarse-mesh bags that allow access by macroinvertebrates are typically of 10-mm mesh, but some studies use 5 mm. Thus, the size of mesh used for making litter bags needs to be appropriate to the questions asked about the relations of invertebrates to leaf litter colonization and breakdown.

The choice of sieve sizes used in the laboratory for sorting invertebrate samples depends on the lower limit of sizes one is willing to identify. A common lower size for catching and enumerating macroinvertebrates is 500 µm. Although the preferred mesh size varies among studies, this cut-off corresponds nicely to the fine-mesh litter bags described above. This choice represents a trade-off between the time required to process samples and the characterization of the invertebrate community. Often the 500-µm size is considered equivalent to macroinvertebrates, although the earliest stages of many macroinvertebrates are much smaller.

Ideally, litter bags retrieved from the field are processed fresh, so that invertebrates are still alive and can be picked in this condition. Alternatively, leaves might be frozen or preserved in other ways before processing the samples, in which case invertebrates will be killed. One caution to take is that live invertebrates can actively make their way through the meshes of sieves, which would be unlikely for dead invertebrates that are curled up and unable to squeeze through a mesh. This suggests that if any freshly processed samples are being compared in a given study to samples that have been preserved, a step needs to be inserted to kill any live invertebrates (e.g. with hot water or ethanol) before passing the sample through the sieve. This is to ensure that behaviour does not add systematic variation to the data. The same reasoning applies to data that will be compared with other studies; protocols need to be consistent for meaningful comparisons.

2 Equipment and Materials

- Resealable plastic (e.g. Zip Lock®) bags
- Ice chest
- Sieves (500 µm, 250 µm, or smaller mesh size, depending on the specific questions)
- 2 wash bottles (one for water, one for ethanol)

- Preservative, preferably 70% ethanol, because formalin is considerably more toxic
- Spatula or knife for scraping sample from sieve into the storage jar
- Jars (one per sample, ~200 mL) to store invertebrates and any retained debris prior to sorting
- White enamel or plastic tray or, for small samples, Petri dishes placed on a white surface
- Forceps (fine, needle-nosed forceps) or pipettes
- Dissecting microscope
- Taxonomic keys
- Vials large enough to store sorted invertebrates (usually at least 20 mL capacity), with well-sealing caps and a good label, preferably printed on both sides
- Analytical balance (± 0.001 mg or greater precision), if invertebrate biomass is to be determined by weighing

3 Experimental Procedures

3.1 General

1. Place litter bags retrieved from the field individually in resealable plastic bags to prevent invertebrates from moving between litter bags.
2. Unless invertebrates can be separated from litter directly in the field, transport the litter to the laboratory in ice chests.
3. As a rule, process leaf litter freshly or after storage for <24 h in a refrigerator. Alternatively, freeze litter bags and the associated invertebrates for later processing, particularly if there are a large number of samples to be processed in a short period. However, freezing makes most invertebrates soft and poorly preserved once thawed. Preservation with ethanol or formaldehyde has even more unwanted impacts, with ethanol preservation resulting in significant leaching of additional components from leaves, and formaldehyde being a rather harmful chemical.

3.2 Collecting Invertebrates from Leaf Samples

1. Carefully remove leaves from a litter bag and place them in a white tray with water. This can be done in the field or laboratory.
2. Gently separate leaves while avoiding breakage of leaves and rinse invertebrates off in the water (Fig. 49.1a).
3. Place rinsed leaves into another tray of clean water and rinse again. Repeat as necessary until there are no more invertebrates on the leaves. It can be useful to use warm water (no more than ~30 °C to avoid killing the invertebrates) to encourage invertebrates in tubes on or in the leaves to dislodge.

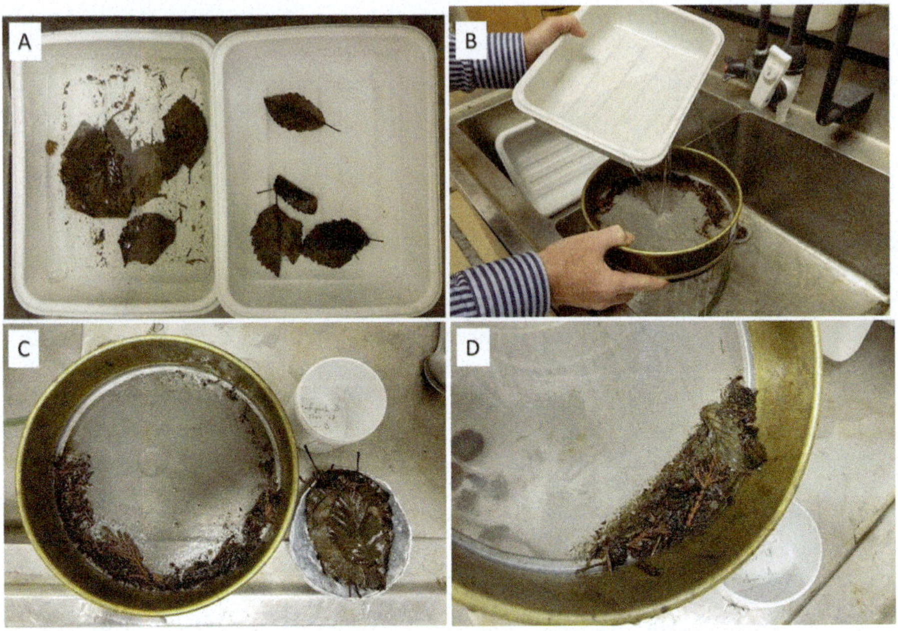

Fig. 49.1 Steps in separating invertebrates from leaf litter. (**A**) Rinse leaves in water to remove invertebrates and place rinsed leaves into a tray with clean water. This step may need repeating. (**B**) Carefully pour water containing invertebrates and any retained debris through a sieve with appropriate mesh size (e.g. 500 or 250 μm). (**C**) Sieve with invertebrates and the retained debris; rinsed leaves to the bottom right of the photo, separated and ready to dry. (**D**) Invertebrates and retained debris from the litter bags ready for picking live invertebrates and preservation

4. Pour the water from the tray with the invertebrates (and retained debris) through the selected sieve (Fig. 49.1b).
5. Once all the remaining water has been passed through the sieve and the tray has been rinsed, wash the contents of the sieve to one end (Fig. 49.1c, d).
6. If picking invertebrates alive, rinse the contents of the sieve into a large white tray.
7. Invert the sieve and carefully spray water from the wash bottle from the underside of the sieve ("back washing").
8. If the invertebrates are to be preserved and sorted later, rinse the contents of the sieve into a labelled storage jar (e.g. 200 mL) with a preservative such as 70% ethanol. Add enough ethanol to approximately double the volume of the sample.
9. Catch invertebrates clinging to the sieve with forceps or pipettes and place into the storage jar.
10. Place an informative label inside the jar (e.g. sampling site, sample reference number, date collected, person doing the study), using a pencil or laser printer for the label.
11. Store the sample for later identification and enumeration.

12. If samples are to be stored for an extended period, ensure that the jar or vial is well sealed to avoid evaporation, which can be done with an extra layer of parafilm on the jar or vial (and under the lid) or by using jars that have lids with well-sealing inserts.

3.3 Sorting Invertebrates

1. For living specimens, use forceps or a pipette to pick live invertebrates from the white tray and place them into a vial with preservative for later identification and counting.
2. For dead invertebrates, place sample into a Petri dish (e.g. 10 cm diameter by 10 mm high) with enough ethanol to cover the sample.
3. If a sample is large, subdivide into aliquots and examine them separately.
4. Looking through a microscope with at least 10× power, remove the invertebrates from other particles using fine, needle-nosed forceps, and place into the storage vial with ethanol.

3.4 Identifying and Counting Invertebrates

1. Count specimens after identification, preferably to the genus or species level. Use taxonomic keys appropriate to the region where samples were taken, if such keys are available (e.g. Fernández and Dominguez 2001; Yule and Sen 2004; Merritt et al. 2019; Mugnai et al. 2010; Tachet et al. 2010; Thorp and Covich 2009). The required resolution of identifications depends foremost on the scientific question, although the lowest practical level of identification can also be influenced by the available keys, skill at identification, and time that can be afforded. For poorly characterized communities, identification may be limited to morphotypes, which are organisms that look similar, except possibly for size, but cannot be assigned to a particular taxonomic group.
2. If biomass is to be estimated, one approach is to measure sizes of invertebrates and calculate biomass from existing size–biomass relations (e.g. Burgherr and Meyer 1997; Benke et al. 1999). This method is most accurate when an eyepiece or stage micrometer is used with a microscope, but graph paper can also yield sufficiently precise results.
3. As an alternative approach, dry and weigh invertebrates directly, which typically works well for larger invertebrates (dry mass > 5 mg). For small invertebrates it will be necessary to pool a large enough number of specimens for weighing, depending on the sensitivity of the balance.

3.5 Data Analysis

1. Express the data as total numbers of invertebrates per litter bag, numbers per g ash-free dry mass (AFDM) of leaf litter, or numbers or biomass of each invertebrate taxon, depending on the specific hypotheses to be tested.
2. Analyse data based on the study design, for example, ANOVA designs. Repeated sampling events may make it necessary to use generalized linear mixed models with time as a random factor, which works well for contrasting total numbers or numbers of particular species.
3. If differences in community composition are to be documented, use some form of multivariate comparison, typically ordination such as principal components analysis, non-metric multidimensional scaling, redundancy analysis, etc.
4. Analyse diversity of the litter-colonizing invertebrate communities as described, for instance, in Chap. 62.

4 Final Remarks

The level of resolution of identification does not always need to be the same for each group. Analyses of taxon richness can be meaningful even if some groups are identified only to order, while other groups are identified to lower levels such as genus (e.g. Buss and Vitorino 2010). Although studies on population dynamics require species-level information, identification to the level of genus, family, or even morphotype is often sufficient in the context of biomonitoring. What is more, as the costs of DNA sequencing are rapidly declining, it is likely that a point will be reached where it becomes effective to use genomics methods of identification.

It is common to express numbers or biomass of invertebrates either as numbers per litter bag or as number per g dry mass or AFDM of leaf litter remaining. The latter approach assumes that consumer interactions with the litter are related to the amount of leaf resource available, as well as to resource quality. This means of expression fits with a large body of literature on consumer-resource dynamics (e.g. Murdoch et al. 2003), and where the interest is in leaves as a basal resource in food webs, the numbers or biomass per unit of resource are a suitable measure.

Often leaves enclosed in fine-mesh litter bags (e.g. mesh size of 500 μm) are used to exclude macroinvertebrates. Breakdown within such bags is then attributed to microbial processes. However, that mesh size does not exclude small invertebrates, which can be very numerous and include meiofauna as well as the early instars of macroinvertebrates, the latter of which might grow to be relatively large within fine-mesh bags (Baldy and Gessner 1997).

Litter bags can also be used to enclose macroinvertebrates with leaf litter, rather than exclude specimens that are larger than the mesh size. Thus, the invertebrates that colonize leaf litter in coarse-mesh bags can be experimentally manipulated by subsequently enclosing leaves and invertebrates in fine mesh, and potentially altering densities, species composition, or size structure, among other possibilities; for examples, see McKie et al. (2009) and Madji et al. (2015).

References

Baldy, V., & Gessner, M. O. (1997). Towards a budget of leaf litter decomposition in a first-order woodland stream. *Comptes Rendus de l'Académie des Sciences, Series III, 320*, 747–758.

Benke, A. C., Huryn, A. D., Smock, L. A., & Wallace, J. B. (1999). Length-mass relationships for freshwater macroinvertebrates in North America with particular reference to the southeastern United States. *Journal of the North American Benthological Society, 18*, 308–343.

Boyero, L., Pearson, R. G., Gessner, M. O., Barmuta, L. A., Ferreira, V., Graça, M. A. S., Dudgeon, D., Boulton, A. J., Callisto, M., Chauvet, E., Helson, J. E., Bruder, A., Albariño, R. J., Yule, C. M., Arunachalam, M., Davies, J. N., Figueroa, R., Flecker, A. S., Ramírez, A., Death, R. G., Iwata, T., Mathooko, J. M., Mathuriau, C., Gonçalves, J. F. Jr., Moretti, M. S., Jinggut, T., Lamothe, S., M'Erimba, C., Ratnarajah, L., Schindler, M. H., Castela, J., Buria, L. M., Cornejo, A., Villanueva, V. D., & West, D. C. (2011). A global experiment suggests climate warming will not accelerate litter decomposition in streams but might reduce carbon sequestration. *Ecology Letters, 14*, 289–294.

Burgherr, P., & Meyer, E. I. (1997). Regression analysis of linear body dimensions *vs* dry mass in stream macroinvertebrates. *Archiv für Hydrobiologie, 139*, 101–112.

Buss, D. F., & Vitorino, A. S. (2010). Rapid bioassessment protocols using benthic macroinvertebrates in Brazil: Evaluation of taxonomic sufficiency. *Journal of the North American Benthological Society, 29*, 562–571.

Fernández, H. R. & Dominguez, E. (2001). *Guía para la determinación de los artrópodos bentónicos sudamericanos*. San Miguel de Tucumán, Tucumán, Argentina: Universidad Nacional de Tucumán, Facultad de Ciencias Naturales e Instituto Miguel Lillo, 282 pp.

Hieber, M., & Gessner, M. O. (2002). Contribution of stream detrivores, fungi, and bacteria to leaf breakdown based on biomass estimates. *Ecology, 83*, 1026–1038.

Kominoski, J. S., Marczak, L. B., & Richardson, J. S. (2011). Riparian forest composition affects stream litter decomposition despite similar microbial and invertebrate communities. *Ecology, 92*, 151–159.

Majdi, N., Traunspurger, W., Richardson, J. S., & Lecerf, A. (2015). Small stonefly predators affect micro- and meiobenthic communities in stream leaf packs. *Freshwater Biology, 60*, 1930–1943.

Majdi, N., Threis, I., & Traunspurger, W. (2017). It's the little things that count: Meiofaunal density and production in the sediment of two headwater streams. *Limnology and Oceanography, 62*, 151–163.

McKie, B. G., Schindler, M., Gessner, M. O., & Malmqvist, B. (2009). Placing biodiversity and ecosystem functioning in context: Environmental perturbations and the effects of species richness in a stream field experiment. *Oecologia, 160*, 757–770.

Merritt, R. W., Cummins, K. W., & Berg, M. B. (2019). *An introduction to the aquatic insects of North America* (5th ed.). Dubuque: Kendall Hunt Publishing, 1480 pp.

Mugnai, R., Nessimian, J. L., & Baptista, D. F. (2010). *Manual de Identificação de Macroinvertebrados Aquáticos do Estado do Rio de Janeiro*. Rio de Janeiro: Technical Books, 176 pp.

Murdoch, W. W., Briggs, C. J., & Nisbet, R. M. (2003). *Consumer-resource dynamics* (Monographs in population biology) (Vol. 36). Princeton: Princeton University Press, 464 pp.

Ribblett, S. G., Palmer, M. A., & Coats, D. W. (2005). The importance of bacterivorous protists in the decomposition of stream leaf litter. *Freshwater Biology, 50*, 516–526.

Richardson, J. S. (1992). Food, microhabitat, or both? Macroinvertebrate use of leaf accumulations in a montane stream. *Freshwater Biology, 27*, 169–176.

Tachet, H., Richoux, P., Bournaud, M., & Usseglio-Polatera, P. (2010). *Invertébrés d'eau douce : systématique, biologie, écologie*. Paris: CNRS Editions, 607 pp.
Thorp, J. H., & Covich, A. P. (2009). *Ecology and classification of North American freshwater invertebrates* (3rd ed.). New York: Academic, 1021 pp.
Tonin, A. M., Pozo, J., Monroy, S., Basaguren, S., Pérez, J., Gonçalves, J. F., Jr., Pearson, R., Cardinale, B. J., & Boyero, L. (2018). Interactions between large and small detritivores influence how biodiversity impacts litter decomposition. *Journal of Animal Ecology, 87*, 1465–1474.
Yule, C. M., & Sen, Y.-H. (2004). *Freshwater invertebrates of the Malaysian region*. Kuala Lumpur: Akademi Sains Malaysia, 861 pp.

Chapter 50
Identifying Stream Invertebrates as Plant Litter Consumers

Luz Boyero, Richard G. Pearson, Ricardo J. Albariño, Marcos Callisto, Francisco Correa-Araneda, Andrea C. Encalada, Frank Masese, Marcelo S. Moretti, Alonso Ramírez, April E. Sparkman, Christopher M. Swan, Catherine M. Yule, and Manuel A. S. Graça

Keywords Leaf litter · Gut content analysis · Detritivores · Shredders · Leaf scrapers · Leaf miners · Wood borers · Invertebrate feeding

1 Introduction

Invertebrates play a key role in the decomposition of plant litter (i.e. leaf litter and wood) in streams (Graça 2001) through their feeding, case-making and burrowing activities. Animals in the shredder functional feeding group (Cummins 1973; Cummins and Klug 1979), which have mouth parts capable of cutting and chewing

Electronic Supplementary Material The online version of this chapter (https://doi.org/10.1007/978-3-030-30515-4_50) contains supplementary material, which is available to authorized users.

L. Boyero (✉)
Department of Plant Biology and Ecology, University of the Basque Country, Leioa, Spain & Ikerbasque, Bilbao, Spain
e-mail: luz.boyero@ehu.eus

R. G. Pearson
College of Science and Engineering, TropWater, James Cook University, Townsville, Australia
e-mail: richard.pearson@jcu.edu.au

R. J. Albariño
INstituto de Investigaciones en BIOdiversidad y MedioAmbiente (INIBIOMA), Laboratorio de Fotobiología, National University of Comahue – CONICET, Quintral, Argentina
e-mail: ralbarin@comahue-conicet.gob.ar

M. Callisto
Instituto de Ciencias Biologicas, Federal University of Minas Gerais, Belo Horizonte, Brazil
e-mail: callisto@mono.icb.ufmg.br

F. Bärlocher et al. (eds.), *Methods to Study Litter Decomposition*,
https://doi.org/10.1007/978-3-030-30515-4_50

pieces of litter (Ramírez and Gutiérrez-Fonseca 2014), make the greatest invertebrate contribution to litter decomposition. Other functional feeding groups can also contribute by scraping leaf surfaces (scrapers) or by making tunnels in leaf mesophyll (miners) or wood (borers). Here we focus on all invertebrates that consume plant litter as a substantial proportion of their diets at some time in their life cycle. Hence, rather than adopting a functional feeding group perspective – which is based on feeding mode and relies on mouthpart morphology and feeding behaviour (Ramírez and Gutiérrez-Fonseca 2014) – we focus on the food items consumed, usually determined through gut content analysis (e.g. Cheshire et al. 2005; Chará-Serna et al. 2012). We thus use the term litter-consuming invertebrates to include invertebrates specialized in leaf shredding, scraping or mining and wood shredding or boring, but also more generalist consumers that feed on a range of other materials such as fine particulate organic matter (FPOM) or periphyton while including a substantial proportion (usually ≥40%) of litter in their diets (Cheshire et al. 2005).

We describe the methods for gut content analysis, as this is the most straightforward way to determine whether and to what extent an invertebrate feeds on litter.

F. Correa-Araneda
Unidad de Cambio Climático y Medio Ambiente, Instituto de Estudios de Hábitat (IEH), Facultad de Arquitectura y Construcción, Universidad Autónoma de Chile, Temuco, Chile
e-mail: francisco.correa@uautonoma.cl

A. C. Encalada
Colegio de Ciencias Biológicas y Ambientales, University of San Francisco de Quito, Quito, Ecuador
e-mail: aencalada@usfq.edu.ec

F. Masese
Department of Fisheries & Aquatic Sciences, University of Eldoret, Eldoret, Kenya
e-mail: fmasese@uoeld.ac.ke

M. S. Moretti
Ecology of Aquatic Insects Lab, University of Vila Velha, Vila Velha, Brazil
e-mail: marcelo.moretti@uvv.br

A. Ramírez
Department of Biological Sciences, North Carolina State University, Raleigh, NC, USA

University of Puerto Rico, San Juan, PR, USA
e-mail: alonso.ramirez@ncsu.edu

A. E. Sparkman · C. M. Swan
Department of Geography & Environmental Systems, University of Maryland, Baltimore, MD, USA
e-mail: asparkm1@umbc.edu; cmswan@umbc.edu

C. M. Yule
School of Science and Engineering, University of the Sunshine Coast, Queensland, Australia
e-mail: cyule@usc.edu.au

M. A. S. Graça
MARE – Marine and Environmental Sciences Centre & Department of Life Sciences, University of Coimbra, Coimbra, Portugal
e-mail: mgraca@ci.uc.pt

However, on some occasions, material in the gut may not be readily identifiable, and mouthpart examination or behavioural observations may be required (e.g. Albariño and Díaz-Villanueva 2006; Mayer et al. 2008). Another useful approach to identify the origin of assimilated food (plant vs. animal, allochthonous vs. autochthonous) is stable isotope analysis (e.g. Mantel et al. 2004), although this method does not differentiate litter from FPOM and it does not identify material that is rapidly metabolized but not assimilated (Schmidt et al. 2017). DNA-based methods have proved successful in determining the diets of other invertebrates, but have not yet been used to identify benthic freshwater invertebrates as litter consumers (e.g. Blankenship and Yayanos 2005; Hardy et al. 2010; Carreon-Martínz and Heath 2010).

2 Equipment and Materials

- Analytical balance (± 0.1 mg)
- Dissecting microscope (at least 50× and preferably higher magnification)
- Compound microscope
- Scalpel
- Forceps
- Dissection needles (insect mounting pins mounted in a cork or a glass or wooden rod)
- Glass slides and coverslips
- Polyvinyl alcohol-lactophenol or an alternative mountant

3 Procedures

3.1 Invertebrate Collection

1. Locate litter packs in the stream bed and capture all litter-consuming invertebrates by scooping the litter with a dip net. If the relative abundance of litter-consuming invertebrates is important for the study, collect with regular methods of sampling benthic invertebrates (e.g. Surber or Hess sampler).
2. Very large and mobile litter consumers such as crayfish or freshwater crabs may be missed with these methods. Therefore, target such litter consumers separately by using electro-shocking procedures (not efficient for burrowing crayfish) or baited traps (in which case the gut contents tend to include the bait!).
3. If particular invertebrate species are to be collected, inspect individual leaves in litter packs retained in different areas of the stream bed and collect the invertebrates of interest with forceps or a small soft brush. For example, cased caddisflies are often located in depositional areas, while large tipulids are more common in riffle areas.
4. Although the sample size and level of replication will be dictated by the particular purpose of each study, ideally observe no fewer than 20 specimens.

3.2 Invertebrate Abundance and Biomass

1. To determine invertebrate abundance and biomass, pick all individuals from each litter sample, and separate them into species or morphospecies (hereafter called "species"). This can be done with live or frozen samples (see also Chap. 49).
2. Divide invertebrates of each species into size classes (e.g. small, medium, large), at least for initial analysis, to determine whether there is an ontogenetic shift in diet. If all sizes are found to have similar diets, this requirement may be dispensed with, or particular size groups may be targeted.
3. For each species, select a group of at least 5 specimens from each size class in each sample, ideally with full guts. This will result in at least 20 individuals per species per site and time. This could include all size groups if all have the same diet.
4. Remove cased caddisflies from their cases. Surface dry individuals on tissue paper, then weigh them collectively to the nearest 0.1 mg. Note the mass and the number of individuals weighed to give a mean mass per individual for each species, sample unit and, where appropriate, size class.
5. Oven-dry the litter sample (50 °C, 72 h) and weigh to the nearest 0.01 g.
6. Incinerate the litter sample in a muffle furnace (550 °C, 4 h) and reweigh to obtain an estimate of litter ash-free dry mass to express densities or biomass of animals per unit of litter dry mass.

3.3 Gut Mounting

1. Examine each individual animal under a dissecting microscope.
2. Where possible, remove the gut, place it on a microscope slide, and squeeze out the contents.
3. For small individuals where guts cannot be removed, detach the head and squash the whole animal to eject the gut contents on the slide. For small animals without a sclerotized or calcareous exoskeleton covering the body, such as chironomids, entire animals may be mounted to assist with identifying the gut contents. Note that retaining the heads can help with identification (e.g. for chironomids).
4. Add a drop of polyvinyl alcohol-lactophenol to a slide, then add the gut contents and finally a coverslip. This mountant will allow slides to be usable for some days or weeks, but is not permanent.

3.4 Gut Content Analysis

1. Examine the slide under 100× magnification, using a graticule with a 10 × 10 grid. Discard slides where guts appear empty.

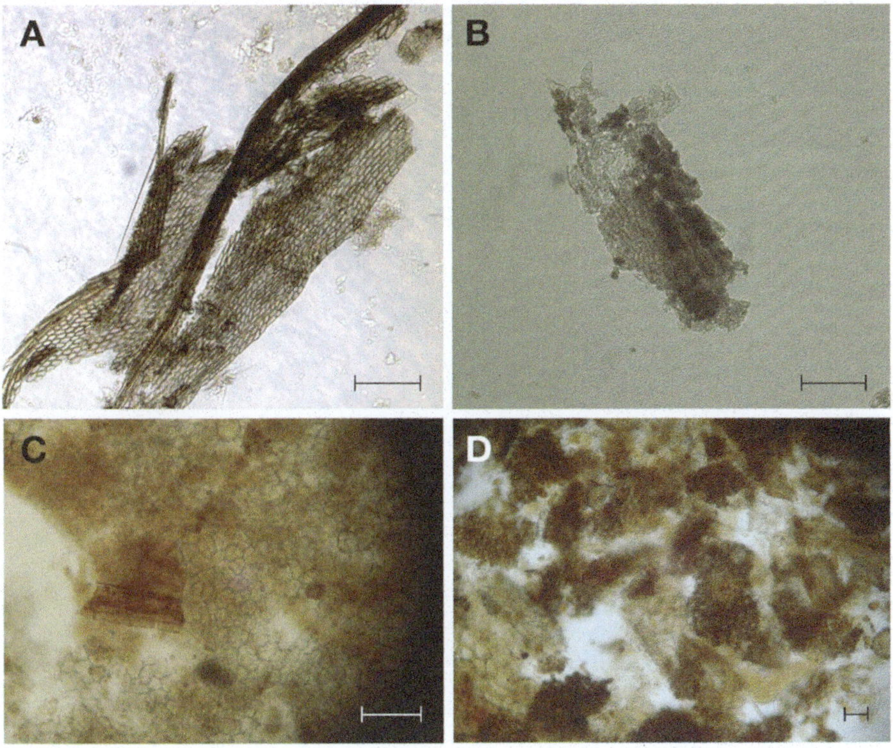

Fig. 50.1 Slides photographed under a compound microscope showing gut contents of litter-consuming invertebrates. Plant cell walls are clearly distinguished in some cases (**a**, **c**) but not others (**b**, **d**). Bar = 100 μm. (Photos by Ana Eguiguren (**a**, **b**) and Richard Pearson (**c**, **d**))

2. Visually estimate the proportion (%) of grid cells containing vascular plant tissue, which is identified by the presence of cell walls (Fig. 50.1). Use averages for species and size classes.
3. Assign feeding categories as follows: when ≥40% of gut contents consist of vascular plant tissue, classify the specimen as a litter-consuming invertebrate; consider animals with 40–70% of vascular plant tissue in their guts as generalists; and regard those in which vascular plant tissue represents >70% of their gut contents as specialists.

4 Final Remarks

Table 50.1 presents a summary of families recorded as litter-consuming invertebrates in studies from across the globe. The list includes not only families in which most or many species are litter consumers but also families where this type of diet

Table 50.1 List of litter-feeding invertebrate families from different biogeographic regions (Nea, Nearctic; Neo, Neotropical; Pal, Palearctic; Afr, Afrotropical; Ind, Indomalayan; Aus, Australasian)

Order/Family	Nea	Neo	Pal	Afr	Ind	Aus
Gastropoda						
Ampullaridae		x				
Arctiidae		x				
Hydrobiidae						x
Lymnaeidae			x			
Melanopsidae			x			
Pachychilidae					x	
Planorbidae		x	x			x
Tateinae			x			
Thiaridae		x				
Amphipoda						
Corophiidae			x			
Crangonyctidae			x			
Gammaridae[a]			x		x	
Hyalellidae		x				
Hyalidae						x
Paracalliopiidae						x
Paramelitidae						x
Perthiidae						x
Pontogeneiidae						x
Talitridae			x			
Isopoda						
Amphisopodidae						x
Asellidae			x			
Cirolanidae					x	
Janiridae		x				x
Oniscidae						x
Decapoda						
Aeglidae		x				
Astacidae	x		x			
Atyidae[b]		x				x
Deckeniidae				x		
Gecarcinucidae				x	x	x
Grapsidae			x			
Palaemonidae		x		x		
Paramelitidae						x
Parastacidae		x				x
Parathelphusidae					x	
Potamidae					x	
Potamonautidae				x		
Pseudothelphusidae		x				

(continued)

Table 50.1 (continued)

Order/Family	Nea	Neo	Pal	Afr	Ind	Aus
Sundathelphusidae					x	
Trichodactylidae		x				
Xiphocaridae		x				
Ephemeroptera						
Baetidae		x		x		
Caenidae		x				
Ephemerellidae	x		x		x	
Euthyplociidae		x				
Leptohyphidae		x				
Leptophlebiidae		x			x	x
Melanemerellidae		x				
Oligoneuriidae		x				
Polymitarcyidae		x				
Potamanthidae	x					
Siphlonuridae	x					
Xiphocentronidae		x				
Blattodea						
Blaberidae		x			x	
Plecoptera						
Austroperlidae		x				x
Brachypterainae	x					
Capniidae[a]	x		x		x	
Chloroperlidae			x			
Diamphipnoidae		x				
Gripopterygidae		x				x
Leuctridae[a]	x		x		x	
Nemouridae[a]	x		x		x	
Notonemouridae		x				x
Peltoperlidae	x				x	
Perlidae[b]		x				
Perlodidae			x			
Pteronarcyidae	x				x	
Scopuridae					x	
Taeniopterygidae[a]	x		x		x	
Heteroptera						
Corixidae[b]		x	x			
Coleoptera						
Chrysomelidae		x	x			x
Curculionidae	x	x	x	x	x	x
Dryopidae		x	x		x	
Elmidae	x	x	x	x	x	x
Eulichadidae	x				x	

(continued)

Table 50.1 (continued)

Order/Family	Nea	Neo	Pal	Afr	Ind	Aus
Haliplidae	x	x	x		x	
Helodidae					x	x
Helophoridae		x	x			
Hydraenidae		x	x			x
Hydrochidae		x	x			
Hydrophilidae	x	x	x			x
Lutrochidae		x				
Psephenidae		x				
Ptilodactylidae	x	x			x	x
Scirtidae	x	x	x		x	x
Staphylinidae		x				
Diptera						
Axymyiidae	x					
Ceratopogonidae		x				
Cylindrotomidae			x			
Chironomidae[b]	x	x	x		x	x
Cylindrotominae	x					
Dixidae		x				
Dolichopodidae					x	
Ephydridae	x	x	x		x	
Hydrelliinae	x					
Pelecorhynchidae	x					
Psychodidae			x			
Ptychopteridae	x		x		x	
Scathophagidae	x		x			
Stratiomyidae		x	x			
Tanyderidae						x
Tipulidae[1]	x	x	x	x	x	x
Lepidoptera						
Coleophoridae	x					
Cosmopterigidae	x					
Crambidae	x	x	x	x	x	
Epipyropidae		x				
Musotiminae	x					
Nepticulidae	x					
Noctuidae	x	x			x	
Opostegidae		x				
Pyralidae[a]		x			x	x
Pyraustinae	x					
Schoenobiinae	x					
Spilomelinae	x					
Tortricidae	x	x				

(continued)

Table 50.1 (continued)

Order/Family	Nea	Neo	Pal	Afr	Ind	Aus
Trichoptera						
Anomalopsychidae		x				
Apataniidae	x		x			
Beraeidae			x			
Brachycentridae	x		x			
Calamoceratidae[a]	x	x	x	x	x	x
Calocidae						x
Conoesucidae						x
Ecnomidae		x		x		x
Glossosomatidae		x				
Helicopsychidae		x				
Hydropsychidae		x				
Hydroptilidae	x	x			x	
Lepidostomatidae[a]	x	x	x	x	x	
Leptoceridae[a]	x	x	x	x	x	x
Limnephilidae[a]	x	x	x		x	x
Limnocentropodidae					x	
Odontoceridae	x	x	x		x	
Oeconesidae						x
Parasericostomatidae[a]		x				
Phryganeidae	x		x		x	
Pisuliidae				x		
Polycentropodidae	x	x			x	
Philopotamidae		x				
Psychomyiidae			x			
Ptilocolepidae	x		x			
Rhyacophilidae	x		x			
Rossianidae	x					
Sericostomatidae[a]	x	x	x	x	x	
Tasiimidae		x				x
Theliopsychinae	x					
Xiphocentronidae		x				
N° of known families	51	70	50	15	43	38

[1]Includes the subfamily *Limoniinae*

[a]Families in which most species are litter consumers; [b]Families in which litter consumers are exceptional

is occasional or exceptional. Here we focus on the family level of taxonomic resolution because of space limitations, but a table of known genera and species is provided online, together with feeding modes (leaf shredder, leaf scraper, leaf miner, wood shredder/borer, or generalist) and a full reference list.

References

Albariño, R. J., & Díaz-Villanueva, V. (2006). Feeding ecology of two plecopterans in low order Andean-Patagonian streams. *International Review of Hydrobiology, 91*, 122–135.

Blankenship, L. E., & Yayanos, A. A. (2005). Universal primers and PCR of gut contents to study marine invertebrate diets. *Molecular Ecology, 14*, 891–899.

Carreon-Martínz, L., & Heath, D. D. (2010). Revolution in food web analysis and trophic ecology: Diet analysis by DNA and stable isotope analysis. *Molecular Ecology, 19*, 25–27.

Chará-Serna, A. M., Chará, J. D., Zúñiga, M. D. C., Pearson, R. G., & Boyero, L. (2012). Diets of leaf litter-associated invertebrates in three tropical streams. *International Journal of Limnology, 48*, 139–144.

Cheshire, K., Boyero, L., & Pearson, R. G. (2005). Food webs in tropical Australian streams: Shredders are not scarce. *Freshwater Biology, 50*, 748–769.

Cummins, K. W. (1973). The utilization of leaf litter by stream detritivores. *Ecology, 54*, 336–345.

Cummins, K. W., & Klug, M. J. (1979). Feeding ecology of stream invertebrates. *Annual Review of Ecology and Systematics, 10*, 147–172.

Graça, M. A. S. (2001). The role of invertebrates on leaf litter decomposition in streams – a review. *International Review of Hydrobiology, 86*, 383–393.

Hardy, C. M., Kurull, E. S., Hartley, D. M., & Oliver, R. L. (2010). Carbon source accounting for fish using combined DNA and stable isotope analyses in a regulated lowland river weir pool. *Molecular Ecology, 19*, 197–212.

Mantel, S. M. K., Salas, M., & Dudgeon, D. (2004). Foodweb structure in a tropical Asian forest stream. *Journal of the North American Benthological Society, 23*, 728–755.

Mayer, G., Maier, G., Maas, A., & Waloszek, D. (2008). Mouthparts of the Ponto-Caspian invader *Dikerogammarus villosus* (Amphipoda: Pontogammaridae). *Journal of Crustacean Biology, 28*, 1–15.

Ramírez, A., & Gutiérrez-Fonseca, P. E. (2014). Functional feeding groups of aquatic insect families in Latin America: A critical analysis and review of existing literature. *Revista de Biología Tropical, 62*, 155–167.

Schmidt, K., Blanchette, M., Pearson, R. G., Alford, R. A., & Davis, A. M. (2017). Trophic roles of tadpoles in tropical Australian streams. *Freshwater Biology, 62*, 1929–1941.

Chapter 51
Shredder Feeding and Growth Rates

Manuel A. S. Graça and José M. González

Keywords Detritivores · Leaf conditioning · Ingestion rate · Food consumption · Field assay · Leaf litter · Invertebrate feeding · Invertebrate growth

1 Introduction

Low-order streams running through forests receive large amounts of coarse particulate organic matter, mainly in the form of leaves. In a pioneering paper, Fisher and Likens (1973) calculated that ~99% of the annual energy supply to an undisturbed second-order stream was allochthonous (for a review, see Abelho 2001), providing an essential resource to microorganisms and invertebrate detritivores. Detritivores capable of feeding on coarse particulate organic matter, the functional feeding group designated shredders (Cummins and Klug 1979), discriminate among litter types and preferentially consume microbially colonized litter (Bärlocher and Kendrick 1973; Graça 2001; Lopez-Rojo et al. 2018).

Shredders also tend to maximize their food intake and grow faster when fed on litter with high nutrient concentrations and low levels of chemical and physical plant defences (Chaps. 11, 12, 13, 14, 15, 16, 17, 18, 19, 20, 21 and 22). However, in some cases, shredders increase their ingestion rates when feeding on a low-quality diet. This phenomenon has been interpreted as a compensatory mechanism allowing sufficient assimilation of resources despite low concentrations of labile carbon, nutrients or both in the food (Iversen 1974; Graça et al. 1993; Flores et al. 2014). Information on growth and ingestion rates can be important in studies aiming to understand how global warming or stressful conditions such as exposure to

M. A. S. Graça (✉)
MARE – Marine and Environmental Sciences Centre & Department of Life Sciences, University of Coimbra, Coimbra, Portugal
e-mail: mgraca@ci.uc.pt

J. M. González
Departamento de Biología y Geología, Física y Química inorgánica, Universidad Rey Juan Carlos, Madrid, Spain
e-mail: jose.gonzalez@urjc.es

F. Bärlocher et al. (eds.), *Methods to Study Litter Decomposition*,
https://doi.org/10.1007/978-3-030-30515-4_51

Table 51.1 Feeding rates of selected invertebrates expressed in terms of mg of ingested leaf dry mass per mg of invertebrate dry mass per day

Consumer	Ingestion rate (mg leaf mg^{-1} animal day^{-1})	Reference
Gammarus fossarum (Amhipoda)	0.002–0.037	1
Micropternasequax (Trichoptera)	0.004–0.40	2
Limnephilus atlanticus (Trichoptera)	0.08–0.44	3
11 shredder species	0.11–1.03	4
Gammarus pulex (Amphipoda)	~0.12–0.42	5
Gammarus fossarum (Amphipoda)	0.16–0.57	6
Gammarus fossarum (Amphipoda)	~0.22	7
Schizopelex festiva (Trichoptera)	≤0.23	8
Klapopteryx kuscheli (Plecoptera)	≤0.37	9
Lepidostoma complicatum (Trichoptera)	0.41–0.97	10
Stenophylax sp. (Trichoptera)	~0.50–0.90	11
Triplectides gracilis (Trichoptera)	0.47–2.34	12

1 = Lecerf et al. (2005); 2 = Rumbos et al. (2010); 3 = Balibrea et al. (2017); 4 = Santatoja et al. (2018); 5 = Foucreau et al. (2016); 6 = Jabiol and Chauvet (2012); 7 = Zubrod et al. (2010); 8 = Graça and Poquet (2014); 9 = Albariño and Balseiro (2001); 10 = Kochi and Kagaya (2005); 11 = Mas-Martí et al. (2017); 12 = Kiffer et al. (2018)

Table 51.2 Growth rates of selected shredders feeding on leaf litter, expressed in terms of dry mass gained per shredder dry biomass and time in days based on linear (Eq. 51.5) or exponential (Eq. 51.6) growth models

Invertebrate species	Model	Growth rate (μg mg^{-1} day^{-1})	Reference
Limnephilus atlanticus (Trichoptera)	Linear	~0.2–11.8	1
Allogamus ligonifer (Trichoptera)	Linear	0.49	2
Sericostoma personatum (Trichoptera)	Linear	1.5–8.5	3
Sericostoma personatum (Trichoptera)	Exponential	1.4–7.4	3
Klapopteryx kuscheli (Plecoptera)	Linear	2–4	4
Sericostoma vittatum (Trichoptera)	Linear	5–16	5

1 = Balibrea et al. (2017); 2 = Pradhan et al. (2012); 3 = Friberg and Jacobsen (1999; values calculated from raw data); 4 = Albariño and Balseiro (2001); 5 = Hepp et al. (2017)

pesticides may affect shredders (Zubrod et al. 2015; Sweeney et al. 2018) or litter decomposition (Rumbos et al. 2010; Rodrigues et al. 2018).

The aim of the procedures described in this chapter is to estimate feeding and growth rates of shredders. Estimates can be made under laboratory conditions where the diet, water chemistry, temperature and photoperiod can be controlled. The latter is particularly important for some nocturnal caddisfly shredders (Wargner 1990; Feio and Graça 2000). Estimates can also be obtained under field conditions; however, large variability in ingestion and growth rates is expected in this case. Ingestion is calculated as the decrease in leaf litter mass exposed to detritivores for a defined period, assuming that the difference is due to feeding. Ingestion is expressed in terms of litter dry mass consumed per shredder body mass and time or in terms of litter mass consumed per individual and time (Table 51.1). Growth rate is measured as mass increase per unit of consumer body dry mass over time by assuming either linear or exponential growth (Table 51.2).

2 Equipment and Materials

- Oven (50 °C)
- Freezer (−20 °C), optional
- Analytical balance (0.01 mg precision)
- Vacuum pump and filtering system
- Beakers (approx. 10 cm high, 8 cm diameter)
- Aeration system (plastic tube, pipette tips, compressor)
- Cork borer
- Large litter bags (10 × 10 cm; 0.5 mm mesh size)
- Small litter bags (2 × 2 cm; 0.5 mm mesh size)
- Clips to fix fine-mesh bags
- Culture racks for biological samples (e.g. Nunc®, 5 × 4 chambers rack) or aluminium trays
- Aluminium foil
- Stream water
- Stream sand, sieved (2 mm mesh size), ignited (500 °C) and washed with distilled water
- Fibre glass, membrane or paper filters
- Leaf litter conditioned in a stream in mesh bags (e.g. for 2 weeks)
- Dissecting microscope
- Graduated eyepiece graticule or digital camera attached to the dissecting microscope as well as a computer and software for measurements (e.g. freeware ImageJ, downloadable at https://imagej.nih.gov/ij)
- Enclosures for field incubations
- Shredders

3 Experimental Procedures

3.1 General

1. Place ~4 g of leaves in 0.5 mm mesh bags, and incubate in a stream for 2–3 weeks to facilitate microbial colonization.
2. Retrieve the mesh bags from the stream, and place them in an ice chest for transport to the laboratory
3. In the laboratory, remove the leaves from the mesh bags; clean them with filtered stream water, distilled water or running tap water, preferably unchlorinated (if tap water is chlorinated, strongly aerate it for 2 days prior to use).
4. Cover the bottom of beakers to a depth of ~5 mm with sand collected in a stream; this is particularly important for case-building caddisflies.
5. Add filtered stream water (~200 ml) and aerate (Fig. 51.1).

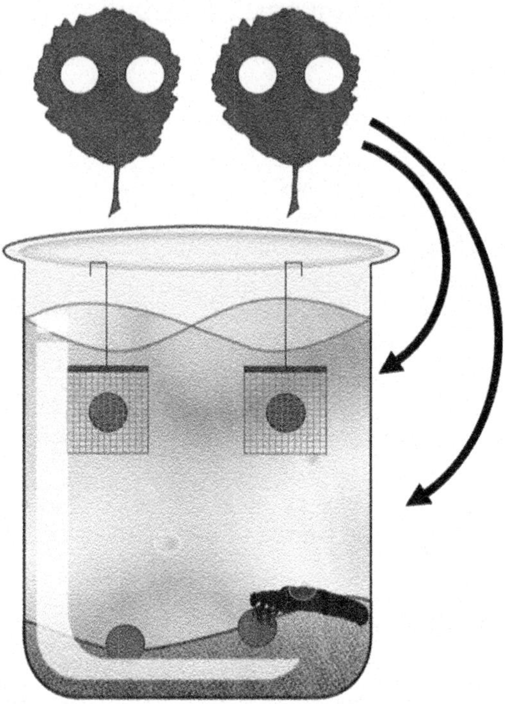

Fig. 51.1 Consumer in a beaker with two leaf disks and matching disks protected from the consumer in fine mesh bags

3.2 *Feeding Rates by the Paired-Leaf-Disk Method*

1. Cut pairs of disks from the leaves with a cork borer, preferably from each side of the main vein, which are assumed to have the same mass.
2. Expose one disk of at least two pairs to the invertebrate, and place the other in a small litter bag attached to the rim of the beaker with a clip.
3. Place invertebrates individually in the beakers (≥10 replicates), and incubate until the area of the leaf disks has been reduced to about half of its original size.
4. Prepare labelled aluminium pans; weigh and store in a rack box.
5. After the feeding period, retrieve the control disks and the remains of the exposed disks, and place them in the individually labelled aluminium pans in the racks.
6. Dry the leaf disks to constant mass at 50 °C (generally 48 h).
7. Transfer the animals (case of caddisflies removed) to weighed aluminium pans, and dry at 50 °C. Prior to drying in the oven, specimens may be killed by exposure in a freezer for ~20 min.
8. Weigh control and exposed leaf disks and animals to the nearest 0.01 mg.
9. Compute ingestion (I) as the difference in dry mass (mg) between the control (M_{cont}) and exposed leaf disks (M_{exp}) divided by shredder dry body mass (M_s) and time (t) in days:

$$I = \frac{M_{cont} - M_{\exp}}{M_s \times t} \tag{51.1}$$

10. Alternatively, although less frequently reported, express ingestion rates in terms of mass of ingested food per animal per day.

3.3 Feeding Rate by the Single-Leaf-Disk Method

1. If pairs of disks cannot be obtained, cut individual leaf disks with a cork borer or use leaf pieces of similar size.
2. Place each leaf disk between two pieces of filter paper for a constant time (e.g. 5 s) to eliminate excess water, and weigh the disk to determine its initial fresh mass (M_{fresh}).
3. Proceed as described above by placing the shredders and leaf disks inside the beakers, and incubate until the leaf disks have been reduced to approximately half of their initial area.
4. Retrieve the disks, oven dry, and weigh them.
5. Calculate the initial dry mass of the exposed disks based on the fresh mass (M_{fresh}) of ~20 leaf disks that have been oven-dried and reweighed to determine their dry mass (M_{dry}).
6. Apply the mean ratio of the dry to fresh mass of the 20 leaf disks as a correction factor (c_1) to relate the initial fresh mass of all samples to dry mass (M_{dry})

$$c_1 = \frac{M_{dry}}{M_{fresh}} \tag{51.2}$$

7. Compute a second correction factor (c_2) to account for changes in mass not related to ingestion (e.g. microbial decomposition) by using an estimate of the initial (M_i^p) and final (M_f^p) dry mass of the disks protected (p) from feeding in the small 0.5-mm mesh bag suspended in the beakers:

$$c_2 = \frac{M_f^p}{M_i^p} \tag{51.3}$$

8. Compute ingestion (I) as the difference between the final dry mass of the leaf disks (M_f) and the estimated initial dry mass derived from the initial fresh mass (M_i) corrected for water content and non-consumptive mass loss; then divide by shredded dry mass (M_s) and elapsed time in days (t):

$$I = \frac{(M_i \times c1 \times c2) - M_f}{M_s \times t} \tag{51.4}$$

3.4 Growth Rate

1. Prepare leaves and individual beakers as described in Sect. 51.3.1.
2. Collect a sufficient number of shredders in the field.
3. Allocate specimens to one of two sets, the first comprising >20 individuals that cover the whole size range of the species to establish body size-dry mass relationships by regression analysis and, the second, to be used in experiments, consisting of specimens of similar size, ideally ~1/3 of their maximum size.
4. Measure the size of each specimen under a dissecting microscope, aided with an eyepiece graticule, or take a picture with a digital camera to measure body size on the photo by using appropriate software.
5. Typically measure a standard length of the specimens, but if that is not possible (e.g. for several cased caddisflies), consider other measures such as head capsule or pronotum width, cephalothorax, femur or tibia length (Kiffer et al. 2018) or even case width, if the case is regular as in the caddisfly genera *Sericostoma* or *Nectopsyche*.
6. If the specimens are robust (e.g. Asellidae, Tipulidae), they can be quickly blotted dry on filter paper for about 5 s, weighed and then returned to the water.
7. Place the shredders in labelled aluminium pans (caddisflies without cases), and dry to constant mass at 50 °C, generally for 48 h. Prior to placing them in the oven for drying, specimens may be killed by freezing(~20 min).
8. Calculate a regression to derive body dry mass from the body size measurements.
9. For specimens used in experiments, obtain the initial mass as in steps 4 and 5 and the above regression equations or as in step 6 above.
10. Place shredders in aerated beakers with a sand-covered bottom and provide food ad libitum.
11. Replace water, sand and food every 3–7 days, and measure body size of the shredders each time.
12. Run the experiment until appreciable changes in size are recorded, which typically requires 4–10 weeks for species from temperate streams.
13. If mass increases linearly, calculate relative growth rate (G) as the difference between the final (M_f) and initial (M_i) animal dry mass, divided by the initial specimen mass and elapsed time in days (t):

$$G = \frac{M_f - M_i}{M_i \times t} \tag{51.5}$$

14. If size is measured at timed intervals, relative growth rates can also be estimated by regressing size against time. The slope indicates mass increase per animal per day. Dividing slope by initial mass gives results comparable to those of Eq. 51.5.
15. Since growth is exponential in most cases, the following equation is more appropriate:

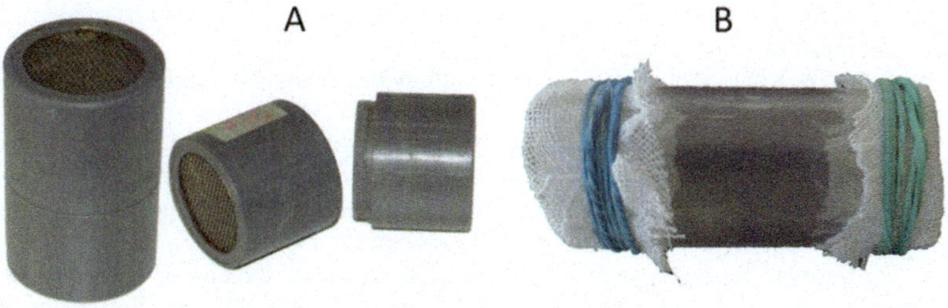

Fig. 51.2 Small enclosures made of PCV tubes to deploy invertebrates supplied with food to measure ingestion and growth rates in field conditions. Mesh screens of a defined size cover both ends of the tubes. Note the thread to unscrew the tube in type A and rubber bands to hold the loose mesh screens in place in type B

$$G = \frac{\ln(M_f) - \ln(M_i)}{M_i \times t} \tag{51.6}$$

16. Again, comparable results are obtained by regressing size against time, with the slope being an estimate of growth.

4 Final Remarks

Growth and feeding rates can also be measured in field conditions by using small enclosures (e.g. Hutchens et al. 1997; McKie et al. 2009; Hines et al. 2016). Enclosures can be made from PVC pipes, for example, 4–6 cm in diameter and 8–9 cm long and covered with a net at both ends that may be unscrewed (Fig. 51.2a) or, simpler, with the two ends covered by a net attached with rubber bands (Fig. 51.2b).

Note that large specimens approaching their final size may exhibit slow or no growth or, in the case of holometabolous insects, stop growing altogether to initiate pupation.

References

Abelho, M. (2001). From litterfall to breakdown in streams: A review. *Scientific World Journal, 1*, 656–680.

Albariño, R., & Balseiro, E. G. (2001). Food quality, larval consumption, and growth of *Klapopteryx kuscheli* (Plecoptera Austroperlidae) from a South Andes stream. *Journal of Freshwater Ecology, 16*, 517–526.

Balibrea, A., Ferreira, V., Gonçalves, V., & Raposeiro, P. M. (2017). Consumption, growth and survival of the endemic shredder *Limnephilus atlanticus* (Trichoptera, Limnephilidae) fed with distinct leaf species. *Limnologica, 64*, 31–37.

Bärlocher, F., & Kendrick, B. (1973). Fungi and food preferences of *Gammarus pseudolimnaeus*. *Archiv für Hydrobiologie, 72*, 501–516.

Cummins, K. W., & Klug, M. J. (1979). Feding ecology of stream invertebrates. *Annual Review of Ecology and Systematics, 10*, 47–172.

Feio, M. J., & Graça, M. A. S. (2000). Food consumption by the larvae of *Sericostoma vittatum* (Trichoptera), an endemic species from the Iberian Peninsula. *Hydrobiologia, 439*, 7–11.

Fisher, S. G., & Likens, G. E. (1973). Energy flow in bear brook, New Hampshire: An alternative approach to stream metabolism. *Ecological Monographs, 43*, 421–439.

Flores, L., Larrañaga, A., & Elosegi, A. (2014). Compensatory feeding of a stream detritivore alleviates the effects of poor food quality when enough food is supplied. *Freshwater Science, 33*, 134–141.

Foucreau, N., Piscart, C., Puijalon, S., & Hervant, F. (2016). Effects of rising temperature on a functional process: Consumption and digestion of leaf litter by a freshwater shredder. *Fundamental and Applied Limnology, 187*, 295–306.

Friberg, N., & Jacobsen, D. (1999). Variation in growth of the detritivoreshredder *Sericostoma personatum* (Trichoptera). *Freshwater Biology, 42*, 625–635.

Graça, M. A. S. (2001). The role of invertebrates on leaf litter decomposition in streams – A review. *International Review of Hydrobiology, 86*, 383–393.

Graça, M. A., & Poquet, J. M. (2014). Do climate and soil influence phenotypic variability in leaf litter, microbial decomposition and shredder consumption? *Oecologia, 174*, 1021–1032.

Graça, M. A. S., Maltby, L., & Calow, P. (1993). Importance of fungi in the diet of *Gammarus pulex* and *Asellus aquaticus*. I: Feeding strategies. *Oecologia, 93*, 139–144.

Hepp, L. U., Pratas, J., & Graça, M. A. S. (2017). Arsenic in stream waters is bioaccumulated but neither biomagnified through food webs nor biodispersed to land. *Ecotoxicology and Environmental Safety, 139*, 132–138.

Hines, J., Reyes, M., & Gessner, M. O. (2016). Density constrains cascading consequences of warming and nitrogen from invertebrate growth to litter decomposition. *Ecology, 97*, 1635–1642.

Hutchens, J. J., Benfield, E. F., & Webster, J. R. (1997). Diet and growth of a leaf-shredding caddisfly in southern Appalachian streams of contrasting disturbance history. *Hydrobiologia, 346*, 193–201.

Iversen, T. M. (1974). Ingestion and growth in *Sericostoma personatum* (Trichoptera) in relation to nitrogen content of ingested leaves. *Oikos, 25*, 278–282.

Jabiol, J., & Chauvet, E. (2012). Fungi are involved in the effects of litter mixtures on consumption by shredders. *Freshwater Biology, 57*, 1667–1677.

Kiffer, W. P., Jr., Mendes, F., Casotti, C. G., Costa, L. C., & Moretti, M. S. (2018). Exotic *Eucalyptus* leaves are preferred over tougher native species but affect the growth and survival of shredders in an Atlantic Forest stream (Brazil). *PLoS One, 13*, e0190743.

Kochi, K., & Kagaya, T. (2005). Green leaves enhance the growth and development of a stream macroinvertebrate shredder when senescent leaves are available. *Freshwater Biology, 50*, 656–667.

Lecerf, A., Dobson, M., Dang, C. K., & Chauvet, E. (2005). Riparian plant species loss alters trophic dynamics in detritus-based stream ecosystems. *Oecologia, 146*, 432–442.

Lopez-Rojo, N., Martinez, A., Perez, J., Basaguren, A., Pozo, J., & Boyero, L. (2018). Leaf traits drive plant diversity effects on litter decomposition and FPOM production in streams. *PLoS One, 13*, e0198243.

Mas-Marti, E., Sanpera-Calbet, I., & Munoz, I. (2017). Bottom-up effects of streambed drying on consumer performance through changes in resource quality. *Aquatic Sciences, 79*, 719–773.

McKie, B. G., Schindler, M., Gessner, M. O., & Malmqvist, B. (2009). Placing biodiversity and ecosystem functioning in context: Environmental perturbations and the effects of species richness in a stream field experiment. *Oecologia, 160*, 757–770.

Pradhan, A., Seena, S., Pascoal, C., & Cassio, F. (2012). Copper oxide nanoparticles can induce toxicity to the freshwater shredder *Allogamus ligonifer*. *Chemosphere, 89*, 1142–1150.

Rodrigues, A. C. M., Bordalo, M. D., Golovko, O., Koba, O., Barata, C., Soares, A., & Pestana, J. L. T. (2018). Combined effects of insecticide exposure and predation risk on freshwater detritivores. *Ecotoxicology, 27*, 794–802.

Rumbos, C. I., Stamopoulos, D., Georgoulas, G., & Nikolopoulou, E. (2010). Factors affecting leaf litter decomposition by *Micropterna sequax* (Trichoptera: Limnephilidae). *International Review of Hydrobiology, 95*, 383–394.

Santonja, M., Pellan, L., & Piscart, C. (2018). Macroinvertebrate identity mediates the effects of litter quality and microbial conditioning on leaf litter recycling in temperate streams. *Ecology and Evolution, 8*, 2542–2553.

Sweeney, B. W., Funk, D. H., Camp, A. A., Buchwalter, D. B., & Jackson, J. K. (2018). Why adult mayflies of *Cloeon dipterum* (Ephemeroptera:Baetidae) become smaller as temperature warms. *Freshwater Science, 37*, 64–81.

Wagner, R. (1990). A laboratory study on the life-cycle of *Sericostoma personatum* (Kirby and Spence), and light dark-dependent food-consumption. *Hydrobiologia, 208*, 201–212.

Zubrod, J. P., Bundschuh, M., & Schulz, R. (2010). Effects of subchronic fungicide exposure on the energy processing of *Gammarus fossarum* (Crustacea; Amphipoda). *Ecotoxicology and Environmental Safety, 73*, 1674–1680.

Zubrod, J. P., Englert, D., Wolfram, J., Wallace, D., Schnetzer, N., Baudy, P., Konschak, M., Schulz, R., & Bundschuh, M. (2015). Waterborne toxicity and diet-related effects of fungicides in the key leaf shredder *Gammarus fossarum* (Crustacea: Amphipoda). *Aquatic Toxicology, 169*, 105–112.

Chapter 52
Feeding Preferences

Cristina Canhoto, Manuel A. S. Graça, and Felix Bärlocher

Keywords Food choice tests · Leaf conditioning · Food selection · Litter quality · Food quality · Shredders · Detritivores · Invertebrate feeding behaviour · Litter consumption · Analysis of food choice data

1 Introduction

When given the choice, consumers often preferentially feed on some food items while ignoring others. This can profoundly influence species diversity at lower trophic levels and may give important clues to co-evolution between predator and prey. Not surprisingly, there is a vast body of literature on food choices in freshwater (e.g. Graça et al. 1993; Bastian et al. 2007), marine (Steinberg 1985; Pennings et al. 2000) and terrestrial (Bonkowsky et al. 2000; Ihnen and Zimmer 2008) consumers. The general design of most of these studies is straightforward: in a container (representing one replicate, or a block), a consumer is given a choice among several food types. The amount of the various items consumed is then estimated, generally

Electronic Supplementary Material The online version of this chapter (https://doi.org/10.1007/978-3-030-30515-4_52) contains supplementary material, which is available to authorized users.

C. Canhoto (✉)
Centre for Functional Ecology, Department of Life Sciences, University of Coimbra, Coimbra, Portugal
e-mail: ccanhoto@ci.uc.pt

M. A. S. Graça
MARE – Marine and Environmental Sciences Centre & Department of Life Sciences, University of Coimbra, Coimbra, Portugal
e-mail: mgraca@ci.uc.pt

F. Bärlocher
Department of Biology, Mount Allison University, Sackville, NB, Canada
e-mail: fbaerloc@mta.ca

F. Bärlocher et al. (eds.), *Methods to Study Litter Decomposition*,
https://doi.org/10.1007/978-3-030-30515-4_52

by comparing area or mass before and after exposure to consumers. The difference is assumed to be due to feeding. When mass loss can occur independently from the feeders (i.e. microbial-mediated degradation, leaf physical breakdown), it must be estimated with separate control replicates without consumers (e.g. Mas-Martí et al. 2015), or, maintaining invertebrate-inaccessible food items in the same container (e.g. Díaz Villanueva et al. 2012) (Fig. 52.1).

Statistical analysis of food choice experiments presents several problems (Bärlocher 1999). To begin with, mass losses in controls have been used to calculate 'correction factors', i.e. average losses occurring in the absence of consumers were subtracted from changes found in the presence of the consumers. The resulting values were assumed to represent consumption and were used in statistical analyses. This approach ignores variability between control replicates: when two independent variables are added or subtracted, the total variance corresponds to the sum of the variances of the two variables. The consequences will be negligible when the variance among control replicates is much smaller than the variance among treatment replicates. If the variance in controls is sizeable, one can maintain overall variance by forming random pairs between control and experimental data, and determine the differences. To avoid random pairing, a test can be based on the comparison of the

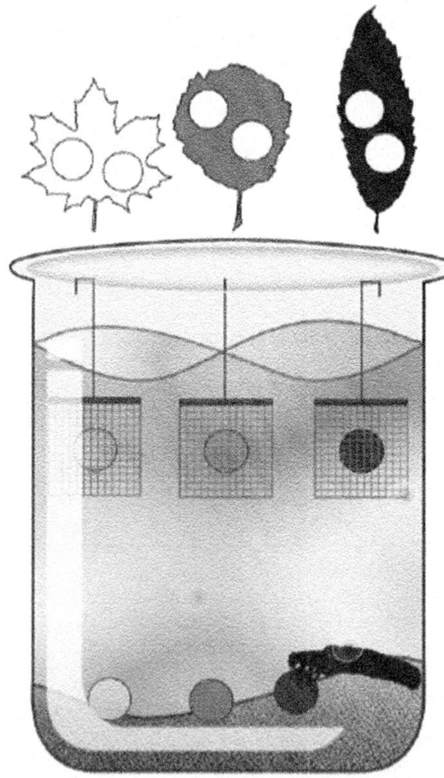

Fig. 52.1 Feeding chamber with 3 pairs of disks from different leaf types exposed to a consumer in a food choice experiment

means of two multivariate vectors represented by controls and treatments (Manly 1986, 1993; Krebs 1999).

The second problem of food choice experiments concerns ‘independence of variables’. When two or more food items are presented in the same container, consumption of one cannot be assumed to be independent of consumption of the other(s). Conventional statistical tests are therefore inappropriate. As alternative, one can use Friedman’s test. It is a modification of Fisher’s permutation test: actual measurements are first converted to ranks (to maintain interdependencies, food items within each feeding container are ranked separately). This conversion to ranks was necessary at the time when the test was developed because of a lack of powerful computers. It results in a loss of power of the test, making it less likely that significant differences can be identified.

An alternative approach to the multiple choice experiments is to run tests with only two food items presented in each experimental replicate, and prepare replicates with all pair-wise combinations of tested foods. This approach was first suggested by Petersen and Renaud (1989) and used by Friberg and Jacobsen (1994) and Graça et al. (2001). This alternative allows the ranking of food items in terms of preference. However, the problem with this approach is the high number of trials needed; for instance, if one wants to evaluate preference of an invertebrate for 7 food types, 21 pair trials must be run. In addition, the sequence of preferences based on pair-wise choices may not correspond to the sequence when all foods are offered simultaneously (Manly 1993).

Detritivores are selective feeders and their species-specific preferences have been related with leaf senescence (Yeates and Barmuta 1999), leaf structure and nutrient contents (Dray et al. 2014; Balibrea et al. 2017), presence of unpalatable or indigestible compounds (Canhoto and Graça 1999), microbial conditioning (Bärlocher and Kendrick 1973; Arsuffi and Suberkropp 1989). There is a general consensus that tougher, thicker, cuticle-protected leaves with high C:N:P ratios and rich in refractory (i.e. lignin) and secondary compounds (e.g. polyphenols) are preferentially consumed by most invertebrates over less recalcitrant leaf litter (e.g. Bärlocher and Kendrick 1973; Canhoto and Graça 1995). *Quercus robur* and *Alnus glutinosa* are examples of more and less recalcitrant leaves, respectively.

In this chapter, we describe an approach to assess feeding preferences of shredders from streams; it can easily be adapted for terrestrial, marine or other freshwater consumers.

2 Equipment and Materials

- Oven (50 °C)
- Analytical balance (± 0.01 mg)
- Vacuum pump and filtering system
- Plastic cups (approx. 10 cm high, 8 cm diameter)
- Aeration system (plastic tube, pipette tips, compressor)

- Cork borer
- Small litter bags (2 × 2 cm; 0.5 mm mesh size)
- Clips
- Pins with coloured heads
- Culture racks for biological samples (e.g. Nunc®, 5 × 4 chambers rack)
- Aluminium foil
- Stream water
- Stream sand devoid of organic matter (ignited at 500 °C for 8 h)
- Fibre glass, membrane or paper filters
- Food items of 2 or more categories. (e.g. leaf species).
- Shredders (e.g. *Sericostoma* sp., *Tipula* sp., *Gammarus* sp.)

3 Experimental Procedures

3.1 General

1. Place ~4 g of leaves in litter bags and incubate in a stream for an appropriate time (e.g. 3 weeks) for conditioning.
2. After incubation, remove the leaves from the litter bags and rinse under running tap water.
3. Cut pairs of disks (one from each side of the main vein) with a cork borer (Ø = 1 cm). One disk of each pair will be used as control, and the other will be offered to the invertebrates (Fig. 52.1). Prepare material for ca. 20 replicates.
4. Filter the stream water and add ~200 ml to each cup.
5. Cover the bottom of containers with stream sand (~5 mm).
6. Prepare the aeration system as shown in Fig. 52.2

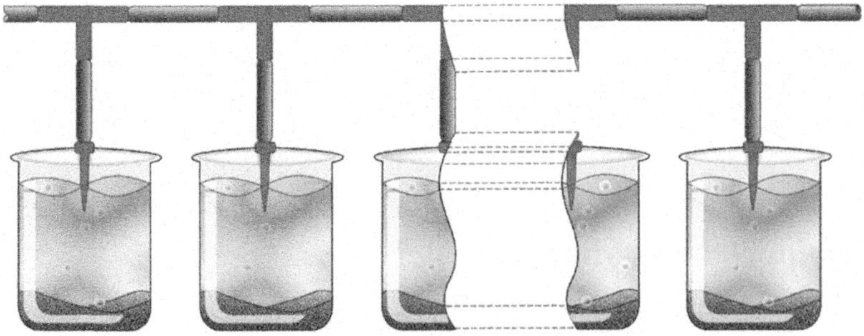

Fig. 52.2 Series of chambers for feeding experiments

3.2 Selection Among Several Food Items

1. Mark the leaf disks of each leaf type with coloured pins (e.g. sp. 1 – white head; sp. 2 – blue head; sp. 3 – green head).
2. Place one control disk of each pair inside a small litter bag that will be attached to the cup with a clip.
3. Place the other identified disks inside the cup as shown in Fig. 52.1.
4. Allocate one invertebrate shredder to each cup and allow feeding until one of the leaf disks is reduced to about half of its initial size.
5. Make small aluminium pans with bottom of a pencil, label and weigh them.
6. After the feeding period, retrieve the leaf material (disks exposed to shredders and control disks) and place them individually in the aluminium pans.
7. Dry the disks in the oven for 48 h and weigh to the nearest 0.01 mg.
8. Proceed in the same way with the shredders.
9. Estimate individual consumption for each leaf type (mg) as the difference between the mass of the control leaf disk (M_{cont}) (whose initial mass is assumed to be similar to its corresponding disk from the same leaf) and the mass of the corresponding leaf disk exposed to shredders (M_{exp}). Results can be expressed as mg dry mass consumed per individual dry mass of shredder (M_s) over the feeding time in days (t):

$$C = \frac{M_{cont} - M_{exp}}{M_s \cdot t} \tag{52.1}$$

3.3 Statistical Analysis

1. Evaluate food preferences by Friedman's test, which is based on ranks.
2. Alternatively, evaluate actual consumption values by a permutation test, using Resampling Stats or a similar program (see Chap. 59).
   (a) Define a test statistic *S* that measures differences in consumption rates of the various food items, e.g. the sum of squared deviations from average consumption. Calculate the value of *S* for the original data.
   (b) Within each container, randomly assign measured consumption values to the available food items. For each of these permutations, determine the new value of *S*. Repeat this 10,000 times; this gives the distribution of all possible values of *S*.
   (c) How extreme is the original value of *S* compared to the entire distribution? Determine the proportion of *S* values that is at least as large as the original value of *S*; this proportion corresponds to the *p*-value of traditional statistical methods.

4 Final Remarks

The paired design described above avoids underestimates of the variability of consumption rates. If this design is not possible, individual leaf disks can be dried and weighed. Before the experiment, the disks should be rehydrated and exposed to shredders. A series of leaf disks without shredders should be used as controls. In this case, the initial mass of each disk is known. Consumption can be estimated as:

$$C = \frac{Mi - (Mf \cdot F)}{I \cdot t} \tag{52.2}$$

where C = consumption, F = correction factor given by the ratio of initial to final mass of a set of control disks, I = invertebrate dry mass, t = duration of feeding trial in days, M_i = initial leaf disk mass, and M_f = final leaf disk mass. The use of an average correction factor will underestimate the variance of C. If the ratio of initial to final mass in control disks is highly variable, a comparison of two multinomial vectors (mass losses in control and in feeding containers) may be more appropriate (Manly 1993).

For terrestrial consumers such as isopods, feeding experiments can be carried out in 5.5 mm diameter Petri dishes. High humidity can be achieved by applying wet filter paper disks to the lid of the Petri dish.

References

Arsuffi, T. L., & Suberkropp, K. (1989). Selective feeding by shredders on leaf-colonizing stream fungi: Comparison of macroinvertebrate taxa. *Oecologia, 79*, 30–37.

Balibrea, A., Ferreira, V., Gonçalves, V., & Raposeiro, P. M. (2017). Consumption, growth and survival of the endemic shredder *Limnephilus atlanticus* (Trichoptera, Limnephilidae) fed with distinct leaf species. *Limnologica, 64*, 31–37.

Bärlocher, F. (1999). *Statistical analysis of feeding experiments: A resampling approach*. https://tinyurl.com/yb33mwez

Bärlocher, F., & Kendrick, B. (1973). Fungi and food preferences of *Gammarus pseudolimnaeus*. *Archiv für Hydrobiologie, 72*, 501–516.

Bastian, M., Boyero, L., Jackes, B. R., & Pearson, R. G. (2007). Leaf litter diversity and shredder preferences in an Australian tropical rain-forest stream. *Journal of Tropical Ecology, 23*, 219–229.

Bonkowsky, M., Griffiths, B. S., & Ritz, K. (2000). Food preferences of earthworms for soil fungi. *Pedobiologia, 44*, 666–676.

Canhoto, C., & Graça, M. A. S. (1995). Food value of introduced eucalypt leaves for a Mediterranean stream detritivore: *Tipula lateralis*. *Freshwater Biology, 34*, 209–214.

Canhoto, C. M., & Graça, M. A. S. (1999). Leaf barriers to fungal colonization and shredders (*Tipula lateralis*) consumption of decomposing *Eucalyptus globulus*. *Microbial Ecology, 37*, 163–172.

Dray, M. W., Crowther, T. W., Thomas, S. M., A'Bear, A. D., Godbold, D. L., Ormerod, S. J., Hartley, S. E., & Jones, H. (2014). Effects of elevated CO_2 on litter chemistry and subsequent invertebrate detritivore feeding responses. *PLoS One, 9*(1), e86246.

Friberg, N., & Jacobsen, D. (1994). Feeding plasticity of two detritivore-shredders. *Freshwater Biology, 32*, 133–142.

Graça, M. A. S., Maltby, L., & Calow, P. (1993). Importance of fungi in the diet of *Gammarus pulex* and *Asellus aquaticus*. I: Feeding strategies. *Oecologia, 93*, 139–144.

Graça, M. A. S., Cressa, C., Gessner, M. O., Feio, M. J., Callies, K. A., & Barrios, C. (2001). Food quality, feeding preferences, survival and growth of shredders from temperate and tropical streams. *Freshwater Biology, 46*, 947–957.

Ihnen, K., & Zimmer, M. (2008). Selective consumption and digestion of litter microbes by *Porcellio scaber* (Isopoda: Oniscidea). *Pedobiologia, 51*, 335–342.

Krebs, C. J. (1999). *Ecological methodology*. Menlo Park: Benjamin/Cummings.

Manly, B. F. J. (1986). *Multivariate statistical methods*. London: Chapman & Hall.

Manly, B. F. J. (1993). Comments on design and analysis of multiple-choice feeding-preference experiments. *Oecologia, 93*, 149–152.

Mas-Martí, E., Muñoz, I., Oliva, F., & Canhoto, C. (2015). Effects of increased water temperature on leaf litter quality and detritivore performance: A whole-reach manipulative experiment. *Freshwater Biology, 60*, 184–197.

Pennings, S. C., Carefoot, T. H., Zimmer, M., Danko, J. P., & Ziegler, A. (2000). Feeding preferences of supralittoral isopods and amphipods. *Canadian Journal of Zoology, 78*, 1918–1929.

Petersen, C. H., & Renaud, P. E. (1989). Analysis of feeding preference experiments. *Oecologia, 80*, 82–86.

Steinberg, P. D. (1985). Feeding preferences of *Tegula funebralis* and chemical defenses of marine brown algae. *Ecological Monographs, 55*, 333–349.

Díaz Villanueva, V., Albariño, R., & Canhoto, C. (2012). Positive effect of shredders on microbial biomass and decomposition in stream microcosms. *Freshwater Biology, 57*, 2504–2513.

Yeates, L. V., & Barmuta, L. A. (1999). The effects of willow and eucalypt leaves on feeding preference and growth of some Australian aquatic macroinvertebrates. *Australian Journal of Ecology, 24*, 593–598.

Chapter 53
Energy Budget of Shredders

Manuel A. S. Graça

Keywords Detritivores · Food assimilation · Feeding rate · Food consumption · Scope for growth · Respiration · Invertebrate growth · Faecal production · Conversion efficiency · Ingestion rate · Leaf litter · Calorimetry · Egestion

1 Introduction

Energy budgets provide information on energy allocation and physiological responses of consumers under different environmental conditions to optimize their energy gain. This can be achieved by adjusting rates of food ingestion, assimilation, respiration or growth. Stressful conditions may affect the energy balance of consumers by increasing metabolic costs associated with maintaining homeostasis. Therefore, stressful conditions affect performance, since animals reallocate energy from growth and reproduction to other body functions. For instance, increased temperature causes the amphipod shredder *Gammarus pulex* to increase food intake (Foucreau et al. 2016) and the caddisfly *Sericostoma vittatum* to decrease assimilation efficiency (Díaz Villanueva et al. 2011).

Microbial colonization of litter is also expected to affect rates of resource use by shredders. For example, *Gammarus pulex* increases its assimilation efficiency when fed litter colonized by microbes, and the isopod *Asellus aquaticus* offered such a diet increases both growth and reproductive output (Graça et al. 1993). Finally, growth of the caddisfly shredder *Psychoglypha* is affected by the identity of individual fungi colonizing the consumed leaves (Arsuffi and Suberkropp 1986).

The difference between food intake and metabolic output is known as Scope for Growth (SfG), which is an indicator of energy available for growth and reproduction. Negative values indicate that energy reserves are used for

M. A. S. Graça (✉)
MARE – Marine and Environmental Sciences Centre & Department of Life Sciences, University of Coimbra, Coimbra, Portugal
e-mail: mgraca@ci.uc.pt

F. Bärlocher et al. (eds.), *Methods to Study Litter Decomposition*,
https://doi.org/10.1007/978-3-030-30515-4_53

Table 53.1 Selected reference values for assimilation efficiency (AE) and gross growth efficiency (GGE) of shredders

Shredder	Plant material and conditions	AE (%)	GGE (%)	Reference
Triplectides gracilis	5 leaf species	20.0–60.9[a]	n.d.	1
Sericostoma vittatum	3 leaf species, 2 temperatures	24.0–62.3[a]	n.d.	2
11 shredder species	2 leaf species, conditioned and unconditioned	25.5–56.6	n.d.	3
Pycnopsyche spp	2 leaf species	26.3–59.6	1.0–1.2	4
Anabolia nervosa	2 leaf species and 1 macrophyte species	44[b]	n.d.	4
Verger cf. *limnophilus*	*Potentilla anserina* leaves	46.7[a]	3	6
Klapoptery kuscheli	7 leaf species	n.d.	0.0–34.6	7
Sericostoma vittatum	*Alnus glutinosa leaves* collected in streams	n.d.	0.6–1.2	8
Sericostoma vittatum	*Alnus glutinosa leaves* conditioned in litter bags	n.d.	2.0–2.6	8
Sericostoma vittatum	12 leaf species,10 °C	n.d.	7–22	9
Sericostoma vittatum	12 leaf species, 15 °C	n.d.	36–74	9
Schizopelex sp.	*Castanea sativa* leaves	n.d.	30–43	10
Schizopelex sp.	*Castanea sativa* leaves exposed to copper	n.d.	11–18	10
Tipula abdominalis	2 leaf species	n.m.	36–75[b]	11

1 = Kiffer et al. (2018); 2 = Díaz Villanueva et al. (2011); 3 = Santonja et al. (2018); 4 = Hutchens et al. (1997); 5 = Jacobsen and Sand-Jensen (1994); 6 = Díaz Villanueva & Trochine (2005); 7 = Albariño & Balseiro (2001); 8 = Flores et al. (2014); 9 = Landeira_Dabarca et al. (2019): 10 = Silva et al. (2018); 11 = Fuller et al. (2015)

n.d. not determined; [a]calculated from the reported data; [b]values given in terms of carbon

maintenance (Maltby 1993). The reasoning for using SfG instead of actual growth or reproduction is that SfG can be determined within days, whereas measuring growth and reproduction would take weeks to months (Naylor et al. 1989; Maltby 1993). Exposure of *Gammarus pulex* to metals, ammonia and other stressors decreases their food ingestion, assimilated energy and SfG (Maltby and Naylor 1990; Maltby et al. 1990; Maltby 1993). In other systems, SfG has also been found to be sensitive to acidification (Pedersen et al. 2014), hypoxia (Shin et al. 2014) and salinity (Normant and Lamprecht 2006; Baillieul et al. 1996).

Here we describe a methodology based on Waldbauer (1968) and Naylor et al. (1989) to calculate energy budgets of shredders feeding on leaf litter. The approach involves determining food ingestion, excretion and respiration under laboratory conditions. Rates determined in mass or carbon units can be converted to energy units. Typical rates of processes contributing to energy budgets of shredders from

Table 53.2 Selected reference energy values (J mg^{-1} animal dry mass day^{-1}) for processes contributing to energy budgets of shredders. I = ingestion; F = faeces; A = assimilation; R = respiration; SfG = scope for growth

Shredder	Plant material and conditions	I	F	A	R	SfG	Reference
Asellus aquaticus	Unconditioned *Ulmus* leaves	1.0	0.45	0.55	0.39	0.16	1
Asellus aquaticus	Conditioned *Ulmus* leaves	4.0	2.75	1.25	0.37	0.8	1
Gammarus pulex	*Alnus* leaves	n.r.	n.r.	1.88–2.53	0.20–0.30	1.55–2.22	2
Gammarus pulex	*Alnus* leaves exposed to zinc	n.r.	n.r.	1.04	0.24	0.8	2
Gammarus pulex	*Alnus* leaves exposed to DCA	n.r.	n.r.	0.72	0.36	0.36	2
Gammarus pulex	*Alnus* leaves at 50% oxygen saturation	n.r.	n.r.	0.46	0.37	0.10	2
Brotia hainanensis	*Bauhinia* leaves	2.12	n.r.	1.5	0.5	1.0	3
Brotia hainanensis	*Bauhinia* leaves exposed to cadmium	0.8	n.r.	0.5	0.5	0.0	3

1 = Graça et al. (1993); 2 = Maltby et al. (1990); 3 = Lam (1996)
n.r. not reported; DCA = 3,4-dichloroaniline

streams are summarized in Tables 53.1 and 53.2, but caution is needed when applying these values, since they strongly depend on consumer identity, food quality and environmental conditions. Data on ingestion rates (0.002–2.5; median 0.4 mg leaf dry mass mg^{-1} animal dry mass day^{-1}) and growth rates (0.2–60; median 8.5 µg mg^{-1} animal dry mass day^{-1}) are compiled in Tables 51.1 and 51.2 of Chap. 51.

2 Equipment and Materials

- Oven (50 °C)
- Desiccator
- Analytical balance (precision ±0.01 mg)
- Vacuum pump and filtering system
- Plastic beakers (approx. 10 cm high, 8 cm diameter) or nested cups with the bottom of the upper one removed and replaced by a netting
- Netting or veil (1-mm mesh size)
- Aeration system (plastic tubing, pipette tips, compressor)
- Cork borer
- Small mesh bags (2 × 2 cm; 0.5 mm mesh size) and clips to fix them at the rim of the beakers or cups (see Chap. 51)
- Aluminium caps
- Stream water
- Milli-Q water (≤0.06 µs cm^{-1})

- Fibre glass filters (Whatman GF/F, 47 mm, 0.7 μm average pore size) and paper filters
- Plastic Pasteur pipettes
- Leaf litter conditioned in mesh bags in a stream (e.g. for 2 weeks)
- Shredders
- Micro bomb calorimeter (e.g. 6725 Semimicro Calorimeter, Parr Instrument Company, Moline, IL, USA)
- Benzoic acid

3 Experimental Procedures

3.1 Collection and Preparation of Leaf Material and Shredders

1. Place ~4 g of leaves in 0.5 mm mesh bags and incubate in a stream for 2–3 weeks for microbial colonization.
2. In the laboratory, remove the leaves from the bags, and clean them with filtered stream water, distilled water or unchlorinated running tap water.
3. Cut disks from leaves with a cork borer to feed the shredders, preferably pairs of disks from contiguous areas of the same leaf.
4. Collect invertebrate shredders from a stream using a hand net (e.g. *Gammarus*) from the mineral substrate (e.g. *Sericostoma, Tipula*) or from litter packs (e.g. Calamoceratidae, Limnephilidae).

3.2 Experimental Procedures to Determine Shredder Feeding Rate, Faecal Production, Assimilation and Growth

1. Prepare two-chamber feeding arenas consisting of two nested cups, one of them with the bottom removed and inserted into another cup with a bottom, the two cups being separated by a netting or veil (Fig. 53.1).
2. Rear individual shredders in the top chamber, provided with known amounts of food (at least two leaf disks; see below), filtered stream water (~200 ml) and aeration (see Chap. 51).
3. Set a temperature similar to that of the stream where the shredders were collected.
4. Adjust the photoperiod (e.g. 12 h light:12 h dark) as it affects the activity of some shredders.
5. With a cork borer, cut pairs of disks from contiguous areas of a leaf.

Fig. 53.1 Two-chamber feeding arena to measure feeding rates and faecal production. The two cups are separated by a net or veil. The invertebrate and food (leaf disks) are placed in the upper chamber, whereas the lower chamber serves to collect faeces to avoid coprophagy. Control disks protected from feeding are placed inside fine-mesh bags

6. Offer one of the disks of each pair to the shredder and place the other one in a small litter bag fixed at the rim of the beakers or cups to avoid consumption by the shredder.
7. Run the experiment for 7–14 days, replenishing food when necessary.
8. About every 4 days pass the water containing sedimented faeces of the shredders through a pre-ashed (500 °C, 5 h), labelled and pre-weighed GF/F filter.
9. Dry the filter at 50 °C, allow them to cool in a desiccator and weigh to the nearest 0.01 mg.
10. After the feeding period, retrieve the leaf material (disks exposed to shredders and control disks) and place them individually in aluminium pans.
11. Dry the disks in the oven for 48 h, allow them to cool in a desiccator, and weigh to the nearest 0.01 mg.
12. Proceed in the same way with the shredders.

3.3 Calculations

1. Calculate ingestion rates (I) as the difference in weight between the control disks (M_{cont}) and the disks exposed to the shredders (M_{exp}) divided by shredder dry mass (M_s) and exposure time in days (t):

$$I = \frac{M_{\mathrm{cont}} - M_{\mathrm{exp}}}{M_s \times t} \tag{53.1}$$

2. Calculate faecal production (F) as the difference between the final (f_f) and initial (f_i) filter mass, corrected for changes in the filter mass resulting from the manipulations (f_c; i.e. the mean ratio of the final to the initial mass of filters used to filter water without faeces; Naylor et al. 1989), the shredders dry mass (M_s) and exposure time (t):

$$F = \frac{\left(f_f - (f_i \times f_c)\right)}{M_s \times t} \tag{53.2}$$

3. Compute assimilation (A) rate (μg mg^{-1} day^{-1}) as the difference between ingested food (I) and faecal production (F) divided by shredder dry mass (M_s) and exposure time (t):

$$A = \frac{I - F}{M_s \times t} \tag{53.3}$$

4. Compute assimilation efficiency (AE) as the ratio of assimilated (A) to ingested (I) food:

$$AE = \frac{A}{I} \times 100 \tag{53.4}$$

5. If growth rates can be obtained as changes in shredders mass over time, which generally requires at least 4 weeks of feeding (see Chap. 51), estimate gross growth efficiency (GGE) as the mass increase (G; from Eqs. 51.5 or 51.6 in Chap. 51) per ingested food (I) (Hutchens et al. 1997):

$$GGE = \frac{G}{I} \times 100 \tag{53.5}$$

where G is given by the difference in final (M_f) and initial (M_i) animal dry mass, divided by the initial animal dry mass and elapsed time (t):

$$G = \frac{M_f - M_i}{M_i \times t} \tag{53.6}$$

3.4 *Calorimetry*

1. To express results in terms of energy units, determine the energy content of oven-dried faeces and leaf disks in a micro-bomb calorimeter, using benzoic acid as a standard.

3.5 Respirometry

1. Measure respiration rates as described in Chap. 33; briefly, place shredders individually in respiration chambers and measure changes in oxygen concentration during the incubation period.
2. Express the results per shredder dry mass and exposure time.
3. Express oxygen consumption in terms of energy units by assuming that 1 mg $O_2 = 14.77$ J and the diet is essentially carbohydrates (Elliot and Davidson 1975; Naylor et al. 1989).

3.6 Scope for Growth

1. Determine ammonia excretion (*E*) by placing animals without food in beakers with Milli-Q water for 6 h and measuring ammonia concentrations (APHA 2005) to express excretion rate in terms of mg of NH_4^+ per g shredder dry mass and exposure time.
2. Convert ammonia excretion to energy units by assuming 24.85 J per mg of excreted NH_4^+ (Elliott and Davidson 1975).
3. Compute the scope for growth (*SfG*) as the difference between the assimilated energy (*A*; energy ingested minus energy in faeces) and the loss through respiration (*R*) and excretion (*E*) per mg of shredder mass and exposure time:

$$SfG = A - R - E \tag{53.7}$$

4 Final Remarks

If pairs of disk are difficult to obtain, individual leaf disks may be dried, weighed and rehydrated before exposing them to shredders (see Chap. 51).

Simple beakers might be used as an alternative to the two nested cups. However, to minimize losses of faeces by coprophagy, faecal pellets should be collected with a pipette at least twice a day. The faecal pellets are transferred to pre-weighed aluminium pans (~1 cm diameter) and dried in an oven. It is useful to pool the faecal samples over the entire exposure period before drying and weighing the aluminium pans with the faecal pellets to calculate the total faecal production during the entire feeding period.

Naylor et al. (1989) assumed that energy loss via excreta was minimal in *Gammarus pulex* and ignored this term in the equation. The reported energy values for leaf disks ranged from 14.7 to 21.6 J mg^{-1} dry mass and for *Gammarus pulex* faecal pellets, it ranged from 8.7 to 14.3 J mg^{-1} dry mass (Naylor et al. 1989; Graça et al. 1993). These values are highly dependent on leaf identity, degree of leaf conditioning and the identity of shredder.

References

Albariño, R., & Balseiro, E. G. (2001). Food quality, larval consumption, and growth of *Klapopteryx kuscheli* (Plecoptera: Austroperlidae) from a South Andes stream. *Journal of Freshwater Ecology, 16*, 517–526.

APHA American Public Health Association. (2005). *Standard methods for the examination of water and wastewater*. Washington, DC: American Public Health Association.

Arsuffi, T. L., & Suberkropp, K. (1986). Growth of two stream caddisflies (Trichoptera) on leaves colonized by different fungal species. *Journal of the North American Benthological Society, 5*, 297–305.

Baillieul, M., Selens, M., & Blust, R. (1996). Scope for growth and fitness of *Daphnia magna* in salinity-stressed conditions. *Functional Ecology, 10*, 227–233.

Díaz Villanueva, V., & Trochine, C. (2005). The role of microoganisms in the diet of *Verger cf. limnophilus* (Trichoptera: Limnephilidae) larvae in a Patagonian Andean temporary pond. *Wetlands, 25*, 473–479.

Díaz Villanueva, V., Albariño, R., & Canhoto, C. (2011). Detritivores feeding on poor quality food are more sensitive to increased temperatures. *Hydrobiologia, 678*, 155–165.

Elliot, J. M., & Davidson, W. (1975). Energy equivalents of oxygen consumption in animal energetics. *Oecologia, 19*, 195–120.

Flores, L., Larrañaga, A., & Elosegi, A. (2014). Compensatory feeding of a stream detritivore alleviates the effects of poor food quality when enough food is supplied. *Freshwater Science, 33*, 134–141.

Foucreau, N., Piscart, C., Puijalon, S., & Hervant, F. (2016). Effects of rising temperature on a functional process: Consumption and digestion of leaf litter by a freshwater shredder. *Fundamental and Applied Limnology, 187*, 295–306.

Fuller, C. L., Evans-White, M., & Entrekin, S. A. (2015). Growth and stoichiometry of a common aquatic detritivore respond to changes in resource stoichiometry. *Oecologia, 177*, 837–848.

Graça, M. A. S., Maltby, L., & Calow, P. (1993). Importance of fungi in the diet of *Gammarus pulex* and *Asellus aquaticus*. II. Effects on growth, reproduction and physiology. *Oecologia, 96*, 304–309.

Hutchens, J. J., Benfield, E., & Webster, J. R. (1997). Diet and growth of a leaf-shredding caddisfly in southern Appalachian streams of contrasting disturbance history. *Hydrobiologia, 346*, 193–201.

Jacobsen, D., & Sand-Jensen, K. (1994). Growth and energetics of a trichopteran larva feeding on fresh submerged and terrestrial plants. *Oecologia, 94*, 412–418.

Kiffer, W. P., Jr., Mendes, F., Casotti, C. G., Costa, L. C., & Moretti, M. S. (2018). Exotic *Eucalyptus* leaves are preferred over tougher native species but affect the growth and survival of shredders in an Atlantic Forest stream (Brazil). *PLoS One, 13*, e0190743.

Lam, P. K. S. (1996). Sublethal effects of cadmium on the energetics of a tropical freshwater snail, *Brotia hainanensis* (Brot, 1872). *Environmental Toxicology and Water Quality, 11*, 345–349.

Landeira-Dabarca, A., Pérez, J., Graça, M. A. S., & Boyero, L. (2019). Joint effects of temperature and litter quality on detritivore-mediated breakdown in streams. *Aquatic Sciences, 81*(1).

Maltby, L. (1993). Stress, shredders and streams: Using *Gammarus* energetics to assess water quality. In D. W. Sutcliffe (Ed.), *Water quality and stress indicators in marine and freshwater ecosystems: Linking levels of organisation (individuals, populations, communities)* (pp. 98–11). Ambleside: Freshwater Biological Association.

Maltby, L., & Naylor, C. (1990). Preliminary observation on the ecological relevance of the *Gammarus* "scope for growth" assay: Effect of zinc on reproduction. *Functional Ecology, 4*, 393–397.

Maltby, L., Naylor, C., & Calow, P. (1990). Effect of stress on a fresh-water benthic detritivore: Scope for growth in *Gammarus pulex*. *Ecotoxicology and Environmental Safety, 19*, 285–291.

Naylor, C., Maltby, L., & Calow, P. (1989). Scope for growth in *Gammarus pulex*, a freshwater benthic detritivore. *Hydrobiologia, 188/189*, 517–523.

Normant, M., & Lamprecht, I. (2006). Does scope for growth change as a result of salinity stress in the amphipod *Gammarus oceanicus*? *Journal of Experimental Marine Biology and Ecology, 334*, 158–163.

Pedersen, S. A., Hakedal, O. J., Salaberria, I., Tagliati, A., Gustavson, L. M., Jenssen, B. M., Olsen, A. J., & Altin, D. (2014). Multigenerational exposure to ocean acidification during food limitation reveals consequences for copepod scope for growth and vital rates. *Environmental Science and Technology, 48*, 12275–12284.

Santonja, M., Pellan, L., & Piscart, C. (2018). Macroinvertebrate identity mediates the effects of litter quality and microbial conditioning on leaf litter recycling in temperate streams. *Ecology & Evolution, 8*, 2542–2553.

Shin, P. K. S., Chan, C. S. K., & Cheung, S. G. (2014). Physiological energetics of the fourth instar of Chinese horseshoe crabs (*Tachypleus tridentatus*) in response to hypoxic stress and re-oxygenation. *Marine Pollution Bulletin, 85*, 522–525.

Silva, V., Marques, C. R., Campos, I., Vidal, T., Keizer, J. J., Goncalves, F., & Abrantes, N. (2018). Combined effect of copper sulfate and water temperature on key freshwater trophic levels - approaching potential climatic change scenarios. *Ecotoxicology and Environmental Safety, 148*, 384–392.

Waldbauer, G. P. (1968). The consumption and utilization of food by insects. *Advances in Insect Physiology, 5*, 229–288.

Chapter 54
The Role of Shredders in Litter Dynamics at Stream-Scale

José M. González and Manuel A. S. Graça

Keywords Detritivores · Secondary production · Gross production efficiency · Ingestion rate · Food consumption · Invertebrate growth · Leaf litter · Invertebrate feeding · Egestion · Energy budget · Temperature

1 Introduction

Shredders are invertebrates feeding on detritus particles greater than 1 mm (Cummins 1973; Graça 2001). They play a key role in leaf litter decomposition by incorporating litter energy and nutrients into secondary production and producing large amounts of fine particles of leaves and feces (Fig. 54.1). Because these transformations deliver matter and energy from coarse detritus to other ecosystem compartments, the quantification of shredder biomass, feeding rates, and secondary production provide great insight on stream functioning (Wallace and Webster 1996; Gessner et al. 1999).

In this chapter, we introduce a method for calculating litter consumed by shredders in a stream stretch. This requires both laboratory and field data on invertebrate

---

Electronic Supplementary Material The online version of this chapter (https://doi.org/10.1007/978-3-030-30515-4_54) contains supplementary material, which is available to authorized users.

J. M. González (✉)
Departamento de Biología y Geología, Física y Química inorgánica, Universidad Rey Juan Carlos, Madrid, Spain
e-mail: jose.gonzalez@urjc.es

M. A. S. Graça
MARE – Marine and Environmental Sciences Centre & Department of Life Sciences, University of Coimbra, Coimbra, Portugal
e-mail: mgraca@ci.uc.pt

F. Bärlocher et al. (eds.), *Methods to Study Litter Decomposition*,
https://doi.org/10.1007/978-3-030-30515-4_54

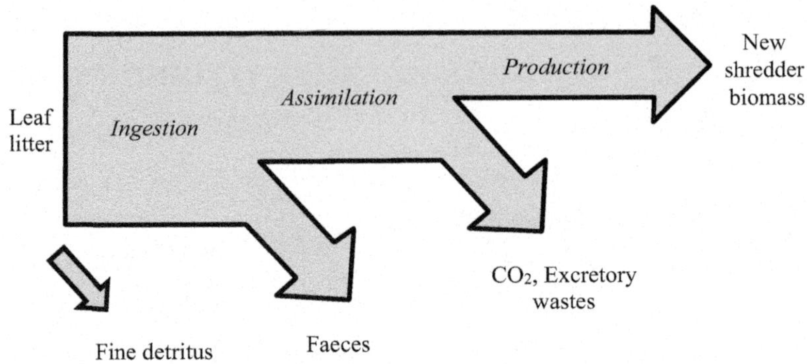

Fig. 54.1 Fate of the energy removed from leaf litter by shredders. Note that, in the infrequent case of shredders in adult stages, not all production is channeled into growth, but a proportion will be used to produce descendants

identity, individual biomass, and water temperature (e.g., Rumbos et al. 2010; Mas-Martí et al. 2015). The procedures that will be shown here are applicable to shredder populations; unless the assemblages are species-poor, proxies are used to calculate the energetics of the entire guild (e.g., Benke and Wallace 2015).

The core procedure to calculate the amount of litter consumed by a field population is to combine (a) the secondary production of shredders measured at the field with (b) their gross production efficiencies, i.e., the ratio between shredder production and ingestion rate, determined usually in the laboratory. The methods to calculate secondary production (see Benke 1993; Benke and Huryn 2006; Dolbeth et al. 2012 for reviews) can be classified in two groups: cohort and size-based methods. Cohort methods are applicable to populations with synchronous development and allow calculating both annual production and production between sampling intervals; therefore, they provide information on the temporal patterns of leaf litter use. Size-based methods are used when the cohorts are not distinguishable; they render only one general estimate of annual production and leaf litter processed.

Determination of gross production efficiency requires measuring both ingestion and growth under field-like conditions. Both rates depend on the individual biomass (e.g., Brown et al. 2007), and their calculation in this context requires working with animals of the entire size range observed in the studied assemblage. Direct measures of ingestion and growth rates in stream conditions are not frequent in the literature, due to logistic difficulties of maintaining the animals in field enclosures for weeks and supplying them with known quantities of food, but some have adopted this approach (e.g., Benke and Jacobi 1986; Mas-Martí et al. 2015). By performing these measures in the laboratory, researchers achieve controlled and constant conditions, providing information on the relationship between shredder performance and physical environment (e.g., González and Graça 2003; Galic and Forbes 2017). Laboratory conditions also allow measuring fecal and fine particle production by shredders and possibly assimilation efficiency (if fecal material can be identified and isolated).

Shredders differ greatly in their requirements to grow: caddisflies constructing mineral cases will need sand or small gravel in their laboratory or field enclosures, whereas other caddisflies cut pieces of leaf disks to construct their cases, introducing noise in the feeding rate calculations unless corrections are made. Short pilot studies are highly recommended to confirm that shredders will survive in confined conditions and to obtain a rough indication of the number of animals and the amount of leaf litter needed for the whole experiment. The methods showed here can be adapted to fit the research and the shredder requirements.

2 Site Selection and Equipment

2.1 Site Selection

Select a 100-m-long stream reach. If the stretch is heterogeneous (e.g., combinations of riffles and pools, or presence of debris dams), longer lengths and stratified sampling may be needed.

2.2 Equipment and Material

- Water temperature recorder
- Benthos sampler that provides quantitative measures of abundance (e.g., Surber and Hess samplers); mesh size depends on the shredder size (0.25–0.5 mm)
- Measuring tape
- Set of nested sieves (e.g., 10, 5, 1 and 0.5 or 0.25 mm)
- Bucket
- Plastic trays
- 75% alcohol
- Dissecting microscope (10–40×)
- Petri dishes
- Forceps
- Device for measuring invertebrates (e.g., ocular micrometer, or digital camera attached to microscope and software for measurements)
- Balance (0.01 mg precision)
- Autumn-shed leaves
- Litter bags (≥ 10 × 10 cm, mesh size ≤0.5 mm)
- Cork borer
- Centrifuge
- Drying oven
- Muffle furnace
- Plastic flasks

- Air pump, tubing, and Pasteur pipettes
- Equipment to maintain water at constant temperature in the laboratory or a temperature-controlled room

3 Experimental Procedure

3.1 Secondary Production Determinations

1. Construct size-mass relationships for the shredder species of interest in the study reach. To this end, take shredders of entire size range, and maintain them alive and refrigerated for a maximum of 6 h of transport and manipulation. In the laboratory, anesthetize the animals (iced or carbonated water), measure and transfer them to preweighed aluminum foils; they will be dried at 50 °C to constant mass (generally ≥2 days) and weighed. Biomass is very often measured in terms of dry mass, but consider determining ash-free dry mass (AFDM). Size determinations are rather idiosyncratic: body length and head capsule width are most frequently used (Benke et al. 1999), but are not mandatory (e.g., Mas-Martí et al. 2015; Martins et al. 2017). For caddisfly larvae with a regular case, the anterior case opening can be measured and regressed against mass. Size-mass regressions can be also taken from the literature, but, since they will be central to obtain accurate calculations of secondary production, growth rates, and leaf litter consumption, we strongly advise to construct ad hoc size-mass regressions (see Benke et al. 1999).
2. If the local shredders have asynchronous life histories, or if they take about ≥1 year to complete their life cycle, take quantitative benthos samples monthly. If shredders have shorter life histories, modify the sampling schedule accordingly (e.g., sample every 3 weeks or more frequently).
3. Take ~ 5 random benthos samples per sampling occasion and habitat. Adjust the number of samples to recover at least 50 individuals on each sampling occasion.
4. Transport the samples alive in ice chests to the laboratory and sort the specimens in <12 h. If this is not possible, preserve in 75% alcohol.
5. Wash the samples through a set of nested sieves. Place the portion of the sample retained by sieves of mesh ≥1 mm in white trays and sort. Sorting is much easier if the animals are alive. The material retained in the finer sieves must be preserved and sorted using a binocular microscope.
6. Measure the specimens under a binocular microscope or take photos to measure them digitally.
7. Determine the individual biomass of each animal using the size-mass regression.
8. Calculate the secondary production of the studied population. If it has synchronous development, the most common method is based on instantaneous growth;

this procedure calculates production between consecutive samples as the product of instantaneous growth rate and mean population biomass. For shredders with indistinguishable cohorts, the most common method by far is based on size-frequency (Benke and Huryn 2006).

3.2 *Ingestion, Egestion, and Fine Detritus Production*

Calculate rates of food intake, egestion, fine detritus production, and growth as indicated in Chap. 51 using animals of all the sizes found in the study reach.

3.3 *Putting the Pieces Together: How Much Litter Did Shredders Consume in the Stretch?*

1. Compute the gross production efficiency (GPE_i, biomass/leaf mass) of each shredder captured in the field as:

$$GPE_i = \frac{b_i \cdot g_i}{CR_i}, \quad (54.1)$$

where b_i is individual biomass of the shredder "*i*" and g_i and CR_i are, respectively, instantaneous growth and consumption rates calculated from the models that predict them from individual biomasses (see Chap. 51).

2. If the secondary production of the field population of shredders was calculated using cohort production methods:

   (a) Calculate the average gross production efficiency of the shredders captured at the start of each sampling interval weighted by their individual consumption rates (gpe, biomass/leaf mass) as:

$$gpe = \sum_{i}^{n} GPE_i \cdot cr_i \quad (54.2)$$

where cr_i is the consumption rate of shredder i divided by the sum of the consumption rates of all shredders captured at that sampling.

   (b) Calculate the ingestion rate of leaf litter at the stream (LLI, detritus mass) during the production interval as:

$$LLI = \frac{P}{gpe} \quad (54.3)$$

where p is the biomass produced by the shredder population during the sampling interval.

(c) Calculate the annual amount of leaf litter consumed by the shredder population adjusted by the length of each production interval.

3. If the secondary production of the field population of shredders was calculated using the size-frequency method, the procedure to calculate the amount of litter consumed by the whole population is similar to that for cohort production methods:

   (a) Use the size-mass regression to calculate the mass of all shredders captured during the study.
   (b) Calculate annual gpe as shown above.
   (c) Calculate annual LLI.

4. A similar approach may be used to calculate fecal and fine detritus mass produced by the field population of shredders:

   (a) Calculate the ratios between secondary production and fine detritus and fecal production for each shredder individual ($SP:FFP_b$, biomass/detritus mass) as:

$$SP : FFP_b = \frac{B \cdot g_b}{FP_b} \tag{54.4}$$

where FP_b is fecal and fine detritus production of a given individual shredder of b biomass.

(b) This ratio will substitute GPE in the calculations showed in points 2.a and 3.b.

4 Final Remarks

Water temperature greatly influences shredder energetics, including ingestion, egestion, and growth rates. Consider, thus, the need to perform these laboratory experiments at water temperatures covering the entire temperature range observed in the field. In this case, at least one water temperature recorder must be deployed at each habitat present in the study reach. The equations used to calculate CR, FP, g, GPE, and SP:FFP ratio should include temperature as independent variable. Average gross production efficiency should incorporate daily mean water temperature and be calculated for daily intervals.

References

Benke, A. C. (1993). Concepts and patterns of invertebrate production in running waters. *Verhandlungen – Internationale Vereinigung fur Theoretische und Angewandte Limnologie, 25*, 15–38.

Benke, A. C., & Huryn, A. D. (2006). Secondary production of macroinvertebrates. In F. R. Hauer & G. A. Lamberti (Eds.), *Methods in stream ecology*. San Diego: Academic.

Benke, A. C., & Jacobi, D. I. (1986). Growth rates of mayflies in a subtropical river and their implications for secondary production. *Journal of the North American Benthological Society, 5*, 107–114.

Benke, A. C., & Wallace, J. B. (2015). High secondary production in a coastal plain river is dominated by snag invertebrates and fueled mainly by amorphous detritus. *Freshwater Biology, 60*, 236–255.

Benke, A. C., Huryn, A. D., Smock, L. A., & Wallace, J. B. (1999). Length–mass relationships for freshwater macroinvertebrates in North American with particular reference to the southeastern United States. *Journal of the North American Benthological Society, 18*, 308–343.

Brown, J. H., Allen, A. P., & Gillooly, J. F. (2007). The metabolic theory of ecology and the role of body size in marine and freshwater ecosystems. In A. G. Hildrew, D. G. Raffaelli, & R. Edmonds-Brown (Eds.), *Body Size: The structure and function of aquatic ecosystems*. Cambridge: Cambridge University Press.

Cummins, K. W. (1973). Trophic relations of aquatic insects. *Annual Review of Entomology, 18*, 183–206.

Dolbeth, M., Cusson, M., Sousa, R., & Pardal, M. A. (2012). Secondary production as a tool for better understanding of aquatic ecosystems. *Canadian Journal of Fisheries and Aquatic Sciences, 69*, 1230–1253.

Galic, N., & Forbes, V. E. (2017). Effects of temperature on the performance of a freshwater amphipod. *Hydrobiologia, 785*, 35–46.

Gessner, M. O., Chauvet, E., & Dobson, M. (1999). A perspective on leaf litter breakdown in streams. *Oikos, 85*, 377–384.

González, J. M., & Graça, M. A. S. (2003). Conversion of leaf litter to secondary production by a shredding caddis-fly. *Freshwater Biology, 48*, 1578–1592.

Graça, M. A. S. (2001). The role of invertebrates on leaf litter decomposition in streams - a review. *International Review of Hydrobiology, 86*, 383–393.

Martins, R. T., Melo, A. S., Gonçalves, J. F., Campos, C. M., & Hamada, N. (2017). Effects of climate change on leaf breakdown by microorganisms and the shredder *Phylloicus elektoros* (Trichoptera: Calamoceratidae). *Hydrobiologia, 789*, 31–44.

Mas-Martí, E., Romaní, A. M., & Muñoz, I. (2015). Consequences of warming and resource quality on the stoichiometry and nutrient cycling of a stream shredder. *PLoS One, 10*, e0118520.

Rumbos, C. I., Stamopoulos, D., Georgoulas, G., & Nikolopoulou, E. (2010). Factors affecting leaf litter decomposition by *Micropterna sequax* (Trichoptera: Limnephilidae). *International Review of Hydrobiology, 95*, 383–394.

Wallace, J. B., & Webster, J. R. (1996). The role of macroinvertebrates in stream ecosystem function. *Annual Review of Entomology, 41*, 115–139.

Part VII
Litter Manipulations

Chapter 55
Manipulation of Leaf Litter Stoichiometry

Julio Arce-Funck, Vincent Felten, and Michael Danger

Keywords Aquatic hyphomycetes · Ecological stoichiometry · Litter quality · Nutrient immobilization · Aquatic fungi · Leaf litter · Food quality · Phosphorus · Leaf conditioning · Fungal inoculum

1 Introduction

Elemental composition (e.g. N and P contents, C/N and C/P ratios) is one of the main determinants of resource quality for consumers (Sterner and Elser 2002). Imbalances between elemental requirements of consumers and elemental composition of their resources can have large impacts on consumer life history traits (Elser et al. 2001). It can result in predictable consequences on ecosystem functioning, in particular through consumer-driven nutrient recycling (Vanni 2002). The study of elemental imbalances and their ecological consequences constitute the central tenet of the Ecological Stoichiometry framework (Sterner and Elser 2002).

To date, most studies on ecological stoichiometry have been carried out at the plant-herbivore interface, mainly in lacustrine ecosystems. Yet, terrestrial plant detritus being among the resources exhibiting the lowest N and P content (Moore et al. 2004), stoichiometric constraints are expected to be maximized for detritivorous species (Martinson et al. 2008). In lakes, phytoplankton elemental quality used for experimentally feeding herbivores is generally manipulated by applying short-term ($\leq$6 h) nutrient pulses, leading to fast nutrient intake and large increases in phytoplankton nutrient content (e.g. Elser et al. 2001). This short-term exposure to nutrients limits the risk of confounding factors, such as nutrient-induced changes in primary producer biomass or other resource quality parameters (e.g. essential fatty acid profiles, phytoplankton physical or chemical defences).

J. Arce-Funck (✉)
MARE – Marine and Environmental Sciences Centre & Department of Life Sciences, University of Coimbra, Coimbra, Portugal
e-mail: julioarcefunck@gmail.com

V. Felten · M. Danger
University of Lorraine, Metz, France
e-mail: vincent.felten@univ-lorraine.fr; michael.danger@univ-lorraine.fr

F. Bärlocher et al. (eds.), *Methods to Study Litter Decomposition*,
https://doi.org/10.1007/978-3-030-30515-4_55

Studies investigating litter resource quality for consumers generally use leaf litter incubated in situ (e.g. Cross et al. 2007; Danger et al. 2012) or in the laboratory (González et al. 2014; Halvorson et al. 2015). Thus, increases in leaf litter elemental content cannot be disconnected from other physical and chemical changes occurring during leaf litter decomposition (Suberkropp et al. 1976). These changes are generally stimulated by nutrient availability (Suberkropp and Chauvet 1995). In particular, increases in leaf litter N and P contents might concomitantly occur with drastic changes in leaf litter lipids, soluble sugars and polyphenols and lignin contents (Suberkropp et al. 1976).

The method described here, while limiting confounding factors, allows the manipulation of leaf litter elemental content, taking advantage of the largely non-homeostatic elemental composition of both fungi (Danger et al. 2013, 2016; Gulis et al. 2017) and bacteria (Chrzanowski et al. 1996). The manipulation of leaf litter elemental quality is based on two major steps. First, leaf litter is inoculated for a few weeks with aquatic fungi or natural microbial consortia in nutrient-poor conditions, either in the laboratory (Danger et al. 2013) or in situ (Rollin et al. 2018). The incubation period is adjusted based on the initial recalcitrance of the leaf species. In the second step, the microbially colonized leaves are exposed to a single nutrient pulse in the laboratory. Microbial immobilization can lead to large changes in leaf litter nutrient content. The magnitude of these changes depends on the intensity of the nutrient pulse, on the microbial development at the time of the pulse and on the nature of the leaf litter species. For example, Danger et al. (2013) obtained up to four-fold reductions of leaf litter C/P ratios in alder (*Alnus glutinosa*) and sycamore (*Acer pseudoplatanus*) 3 days after completion of the nutrient pulse. The technique was also applied by Arce-Funck et al. (2016) and Rollin et al. (2018) (see Table 55.1).

2 Equipment, Chemicals and Solutions

2.1 Equipment and Material

- Autumn-shed leaves collected at abscission, air-dried
- Litter bags (15 × 25 cm, 0.5 mm mesh size)
- Erlenmeyer flasks (300 mL)
- Orbital shaker
- Glass flasks (GL 45, 100 and 1000 mL)
- Homogenizer Ultra-Turrax
- Ball mill
- Analytical balance (±0.1 mg)
- Pipettes
- Autoclave
- Ziploc bags
- Freezer at −20 °C
- CHN analyzer

Table 55.1 Leaf litter P manipulation effects on the P content of leaf litter from two plant species

References	Leaf species	Treatment	Mean (% P)	SD	P increase factor compared to NC or LQ treatment
Danger et al. (2013)	Sycamore (*Acer pseudoplatanus*)	NC	0.020	<0.001	
		P0	0.018	0.001	0.97
		P1	0.042	0.006	1.62
		P2	0.070	<0.001	2.27
	Alder (*Alnus glutinosa*)	NC	0.034	0.002	
		P0	0.033	0.002	0.89
		P1	0.055	0.001	2.08
		P2	0.077	0.011	3.48
Arce-Funck et al. (2016)	Sycamore (*Acer pseudoplatanus*)	NC	0.015	0.003	
		P0	0.013	0.001	0.89
		P1	0.031	0.010	2.13
		P2	0.070	0.017	4.76
	Alder (*Alnus glutinosa*)	NC	0.030	0.003	
		P0	0.028	0.002	0.92
		P1	0.045	0.011	1.50
		P2	0.063	0.005	2.08
Rollin et al. (2018)	Sycamore (*Acer pseudoplatanus*)	LQ	0.036	0.008	
		HQ1	0.124	0.016	3.44

P0, no P added; P1, low P concentration; P2, high P concentration; NC, non-conditioned; LQ, low quality

- Spectrophotometer
- Cuvettes (1 cm)
- Tin capsules
- Cork borer
- Spray bottle
- Pure cultures of aquatic hyphomycetes (see Chaps. 23 and 24)
- Microbiological safety workbench
- Centrifugation tubes (50 mL)
- Dispenser (e.g. 100 mL)

2.2 *Chemicals and Solutions*

- Deionized water
- Potassium nitrate (KNO_3)
- Calcium chloride ($CaCl_2$)
- Magnesium sulfate ($MgSO_4$)
- Iron chloride ($FeCl_3$, $6H_2O$)
- Iron sulfate ($FeSO_4$, $7H_2O$)
- Boric acid (H_3BO_3)

- Manganese nitrate ($Mn(NO_3)_2$, $4H_2O$)
- Zinc sulfate ($ZnSO_4$, $7H_2O$)
- Cobalt chloride ($CoCl_2$, $6H_2O$)
- Ammonium molybdate ($(NH_4)_6Mo_7O_{24}$, $4H_2O$)
- Copper sulfate ($CuSO_4$, $5H_2O$)
- Dipotassium phosphate (K_2HPO_4)
- Ethlylene diamine tetraacetic acid (EDTA)

3 Experimental Procedures

3.1 Preparation

1. Collect senescent leaves in early autumn. In the laboratory, let the leaves dry at room temperature and store them in boxes, in the dark and at room temperature.
2. When the experiment is ready to proceed, spray the leaves with deionized water to make them less brittle. Punch out leaf discs using a cork borer, avoiding the main veins. Dry a set of 20 discs at 60 °C for elemental analyses of carbon, nitrogen and phosphorus (Chap. 11).
3. Prepare stock solutions for the experimental P-free culture medium in 100 mL-glass flasks as in Table 55.2. For long-term storage, autoclave stock solutions.
4. Prepare the experimental P-free culture medium in a 1 L-glass flask as follows:

   Add approximately 500 mL of deionized water, then 10 mL of each stock solution flask from 1 to 5 (Table 55.2), complete to 1000 mL with deionized water and autoclave.
5. Colonize leaves in the laboratory (Fig. 55.1A) or in the field (Fig. 55.1B).

3.2 Fungal Inoculation in the Laboratory

1. Cut 12 mm diameter leaf disks from leaves with a cork borer.
2. Introduce the discs in four 500-mL-Erlenmeyer flasks containing 350 mL of deionized water (~75 discs in 350 mL) and autoclave (120 °C, 15 min; Fig. 55.1A, a). Additionally, autoclave 2–3 L of deionized water to rinse the samples when needed. These four Erlenmeyer flasks will represent four groups (labelled I, II, III and IV) that will ultimately represent a gradient of P.
3. The first level (I in Fig. 55.1A) corresponds to leached leaf discs and reflects the quality of leaves before fungal colonization and nutrient immobilization. The second level (II in Fig. 55.1A) allows evaluating the effect of fungal colonization on leaf litter elemental composition: leaf litter is conditioned by fungi but receives no P. The third and fourth levels (III and IV in Fig. 55.1A) are conditioned with two different concentrations of P and reflect the effect of fungal colonization and microbial immobilization of P.

Table 55.2 Initial concentrations, calculated for a final volume of 1 l, of stock solutions for culture media

Stock solution (100 mL flasks)	Compound	Concentration stock solution (mg L^{-1})	Initial volume (mL)	Final concentration in culture medium (mg L^{-1})
Solution 1	KNO_3	1000	10	10
Solution 2	$CaCl_2$	500	10	5
	$MgSO_4$	500	10	5
Solution 3	$FeCl_3$, $6H_2O$	6.25	10	0.0625
	$FeSO_4$, $7H_2O$	6.25	10	0.0625
Solution 4	H_3BO_3	0.6	10	0.006
	$Mn(NO_3)_2$, $4H_2O$	0.6	10	0.006
	$ZnSO_4$, $7H_2O$	0.6	10	0.006
	$CoCl_2$, $6H_2O$	0.6	10	0.006
	$(NH_4)_6Mo_7O_{24}$, $4H_2O$	0.6	10	0.006
Solution 5	$CuSO_4$, $5H_2O$	0.3	10	0.003

Adapted from Gessner and Chauvet (1993) and Kilham et al. (1998)

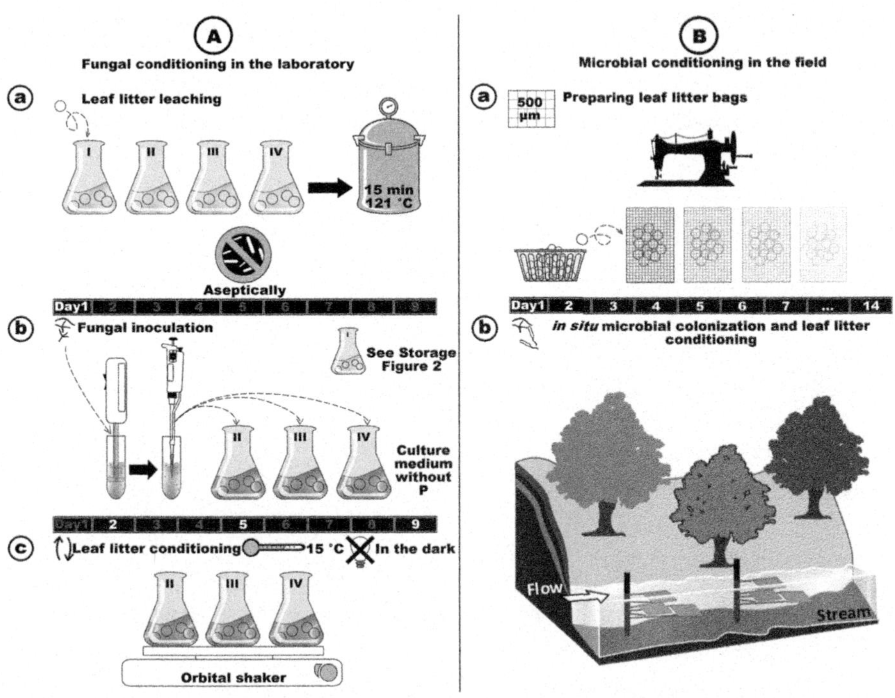

Fig. 55.1 (**A**): Three steps of leaf litter conditioning in the laboratory: (a) leaching, (b) fungal inoculation and (c) conditioning. (**B**): Microbial conditioning in the field, (a) leaf discs are placed in litter bags and (b) incubated in the stream

4. In a microbiological safety workbench, rinse the discs from one Erlenmeyer flask three times with sterile deionized water and prepare Ziploc ® bags containing 20 discs each. Store them at −20 °C. These samples, labelled "I", will be used to evaluate the leaching and sterilization effects on the elemental composition of leaves.
5. Discard the deionized water from the three remaining Erlenmeyer flasks containing sterilized leaf discs. Rinse each flask three times with sterile deionized water to eliminate leaf litter leachates and add 150 mL of the sterile culture media without P with a sterile dispenser (Fig. 55.1A, b).
6. Prepare the fungal assemblage using inoculum consisting of aquatic hyphomycete spores (Danger et al. 2013) or mycelium (Arce-Funck et al. 2016). See also Chaps. 23, 24 and 26.
7. If spores of aquatic hyphomycete cultures are used, add ~2000 spores per species to inoculate each experimental flask.
8. When the assemblage is prepared with fungal mycelium, introduce the same volume of each species in a sterile 50 mL-falcon with a given volume of sterile culture medium (e.g. 1 mL of each fungal species in 5 mL of sterile growth medium). Homogenize with a sterile Ultra-Turrax, and add 500 μL of this fungal suspension to inoculate the experimental flasks labelled II, III and IV and containing the leaf discs.
9. Inoculated flasks are incubated between 12 and 15 °C in the dark on an orbital shaker for about 2 weeks (Fig. 55.1A, c).
10. To limit the risks of nutrient limitation (other than P), culture medium can be renewed regularly (e.g. every 3–4 days) with sterile P-free culture medium.

3.3 *Fungal Inoculation in the Field*

1. Prepare leaf litter bags: Fifteen 500 μm mesh size-litter bags of 15 × 25 cm are prepared (Fig. 55.1B, a). Leaf litter discs (12 mm in diameter) are cut using a cork borer and 200 discs are put in each litter bag.
2. Incubate the litter bags in a non-polluted headwater stream for about 2 weeks (Fig. 55.1B, b).
3. In the laboratory, rinse the discs with stream water.
4. Place batches of 200 leaf discs in 500-mL-Erlenmeyer flasks containing 200 mL of sterile P-free culture medium (Table 55.2).

4 Phosphorus Immobilization and Storage Procedure

1. To manipulate elemental content, leaf litter is exposed to a pulse of P at the end of the microbial growth period. This allows microorganisms to immobilize P. The P immobilization stage must remain short to minimize changes in other leaf litter quality parameters (Fig. 55.2).

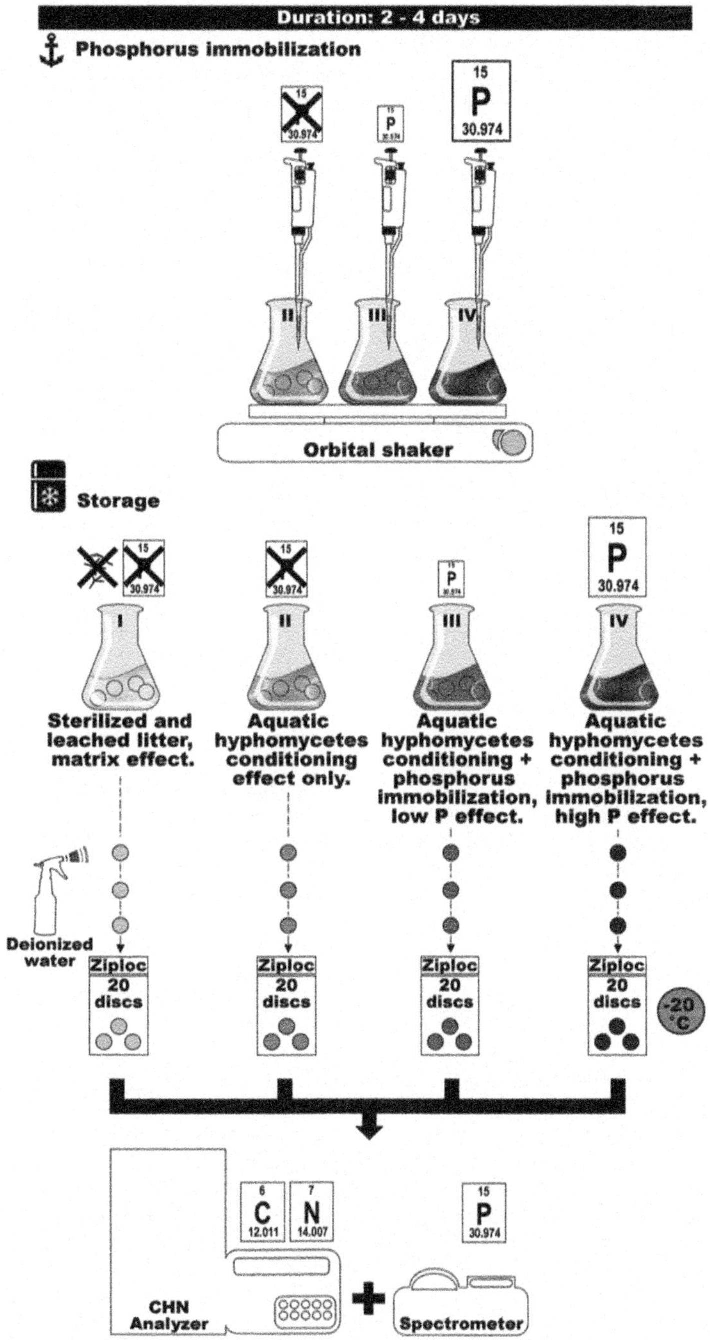

Fig. 55.2 Phosphorus immobilization and storage procedure. Phosphorus icon sizes are proportional to the quantity of P added and colour gradient of Erlenmeyer flasks corresponds to the final P gradient

2. When fungal inoculation is done in the laboratory, phosphorus is added during the last renewal of culture medium. Microorganisms are incubated for 2–4 days after the nutrient pulse to immobilize added nutrients. The concentrations depend on the objectives of the study and might vary with the chosen leaf litter species. P-additions can be calculated using the initial P-content of leaf litter. For example, Danger et al. (2013) chose to add enough P to potentially raise the initial P content of leaf litter by a factor of 2 and 10. Taking into account the initial P content of leaf discs and their total mass in each Erlenmeyer flask, 150 μL and 1 mL of a sterile K_2HPO_4 solution at 33.7 g L^{-1} were added to flasks labelled III and IV, respectively. In the case of in situ microbial inoculations, Rollin et al. (2018) used a stock solution of K_2HPO_4 at 38.2 g L^{-1} (i.e. 8.71 g P L^{-1}) and added 4.46 mL to 150 mL of growing medium, with final concentration reaching 0.26 g P L^{-1}.
3. To eliminate P not incorporated in leaf litter, rinse leaf litter discs three times with sterile deionized water.
4. Disc can be placed in Ziploc® bags (20 discs per bag) and stored at −20 °C until use. Discs A batch of 20 discs of each condition should be reserved for elemental analyses using CHN analyzer and a spectrophotometer, following the procedure described earlier.

References

Arce-Funck, J., Crenier, C., Danger, M., Cossu-Leguille, C., Guérold, F., & Felten, V. (2016). Stoichiometric constraints modulate impacts of silver contamination on stream detritivores: An example test with *Gammarus fossarum*. *Freshwater Biology, 61*, 2075–2089.

Chrzanowski, T. H., Kyle, M., Elser, J. J., & Sterner, R. W. (1996). Element ratios and growth dynamics of bacteria in an oligotrophic Canadian shield lake. *Aquatic Microbial Ecology, 11*, 119–125.

Cross, W. F., Wallace, J. B., & Rosemond, A. D. (2007). Nutrient enrichment reduces constraints on material flows in a detritus-based food web. *Ecology, 88*, 2563–2575.

Danger, M., Cornut, J., Elger, A., & Chauvet, E. (2012). Effects of burial on leaf litter quality, microbial conditioning and palatability to three shredder taxa. *Freshwater Biology, 57*, 1017–1030.

Danger, M., Arce-Funck, J., Devin, S., Heberlé, J., & Felten, V. (2013). Phosphorus content in detritus controls life history traits of a detritivore. *Functional Ecology, 27*, 807–815.

Danger, M., Gessner, M. O., & Bärlocher, F. (2016). Ecological stoichiometry of aquatic fungi: Current knowledge and perspectives. *Fungal Ecology, 19*, 100–111.

Elser, J. J., Hayakawa, K., & Urabe, J. (2001). Nutrient limitation reduces food quality for zooplankton: Daphnia response to seston phosphorus enrichment. *Ecology, 82*, 898–903.

Gessner, M. O., & Chauvet, E. (1993). Ergosterol-to-biomass conversion factors for aquatic hyphomycetes. *Applied and Environmental Microbiology, 59*, 502–507.

González, A. L., Romero, G. Q., & Srivastava, D. S. (2014). Detrital nutrient content determines growth rate and elemental composition of bromeliad-dwelling insects. *Freshwater Biology, 59*, 737–747.

Gulis, V., Kuehn, K. A., Schoettle, L. N., Leach, D., Benstead, J. P., & Rosemond, A. D. (2017). Changes in nutrient stoichiometry, elemental homeostasis and growth rate of aquatic litter-associated fungi in response to inorganic nutrient supply. *The ISME Journal, 11*, 2729–2739.

Halvorson, H. M., Scott, J. T., Sanders, A. J., & Evans-White, M. A. (2015). A stream insect detritivore violates common assumptions of threshold elemental ratio bioenergetics models. *Freshwater Science, 34*, 508–518.

Kilham, S. S., Kreeger, D. A., Lynn, S. T., Goulden, C. E., & Herrera, L. (1998). COMBO: A defined freshwater culture medium for algae and zooplankton. *Hydrobiologia, 377*, 147–159.

Martinson, H. M., Schneider, K., Gilbert, J., Hines, J. E., Hamback, P. A., & Fagan, W. F. (2008). Detritivory: Stoichiometry of a neglected trophic level. *Ecological Research, 23*, 487–491.

Moore, J. C., Berlow, E. L., Coleman, D. C., de Ruiter, P. C., Dong, Q., Hastings, A., Johnson, N. C., McCann, K. S., Melville, K., Morin, P. J., Nadelhoffer, K., Rosemond, A. D., Post, D. M., Sabo, J. L., Scow, K. M., Vanni, M. J., & Wall, D. H. (2004). Detritus, trophic dynamics and biodiversity. *Ecology Letters, 7*, 584–600.

Rollin, M., Coulaud, R., Danger, M., Sohm, B., Flayac, J., Bec, A., Chaumot, A., Geffard, O., & Felten, V. (2018). Additive effect of calcium depletion and low resource quality on Gammarus fossarum (Crustacea, Amphipoda) life history traits. *Environmental Science and Pollution Research, 12*, 11264–11280.

Sterner, R. W., & Elser, J. J. (2002). *Ecological stoichiometry: The biology of elements from molecules to the biosphere*. Princeton: Princeton University Press.

Suberkropp, K., & Chauvet, E. (1995). Regulation of leaf breakdown by fungi in streams: Influences of water chemistry. *Ecology, 76*, 1433–1445.

Suberkropp, K., Godshalk, G. L., & Klug, M. J. (1976). Changes in the chemical composition of leaves during processing in a woodland stream. *Ecology, 57*, 720–727.

Vanni, M. J. (2002). Nutrient cycling by animals in freshwater ecosystems. *Annual Review of Ecology and Systematics, 33*, 341–370.

Chapter 56
Isotopic Labelling of Leaf Litter Nitrogen

Bernd Zeller, Severine Bienaimé, and Etienne Dambrine

Keywords Stable isotope labelling · Foliar urea spraying · Stem injection technique · Leaf litter · N cycle · Isotopic tracer · ^{15}N · Tree leaves · Urea

1 Introduction

Leaf litter decomposition is an essential process to recycle nutrients in ecosystems (Colpaert and van Tichelen 1996; Gessner et al. 2010). This is especially true for nitrogen (N) in remote forest ecosystems, where little reactive N is provided by atmospheric deposition or fixation of atmospheric N_2 (Setäla et al. 1996; Nadelhoffer et al. 1999). Given the importance of N as a nutrient, net fluxes of N released from leaf litter have been studied for decades, and these investigations have established that the kinetics depends on both site-specific characteristics (climate, soil chemistry, microbial community structure) and physical and biochemical litter quality. The use of ^{15}N labelled litter has greatly facilitated studying these fluxes (Berg 1988). This has provided detailed insight into the spatial and temporal analysis of the fate of litter N in the soil-plant system at time scales ranging from months to decades (Zeller et al. 2000; Pena et al. 2013; Guo et al. 2013; Boberg et al. 2014; Blok et al. 2016). A prerequisite for such studies is leaf material in which all relevant N pools are homogeneously labelled.

Plants take up N via different pathways. Therefore, to produce isotopically enriched leaf litter, ^{15}N may be applied as a tracer to soil, stems or leaves (Sommer et al. 2017). The N assimilated by plants is rapidly metabolized in roots or leaves, primarily to glutamate, which then serves as a donor of amino groups for the

B. Zeller (✉) · S. Bienaimé
Biogeochemistry of Forest Ecosystems, INRA, Champenoux, France
e-mail: bernhard.zeller@inra.fr; severine.bienaime@inra.fr

E. Dambrine
CARRTEL, INRA, Le Bourget du Lac, France
e-mail: etienne.dambrine@inra.fr

F. Bärlocher et al. (eds.), *Methods to Study Litter Decomposition*,
https://doi.org/10.1007/978-3-030-30515-4_56

synthesis of other amino acids and proteins, which either remain at the synthesis site or are transported to other plant parts. During leaf senescence, a substantial fraction of leaf N (up to 50%) is translocated to roots and stems. Part of these stored reserves is remobilized in the next growing season and transported back to new leaf tissue. Therefore, an N label once applied will contribute to the leaf and litter N pool of a plant for several years.

Addition of ^{15}N to soil, whether as nitrate or ammonium, is not generally successful at producing highly labelled litter, because the label is strongly diluted by a large pool of non-labelled N in soil before it is taken up by plants. This approach may be used, however, when plants are grown from seeds in hydroponic or sand cultures where the applied ^{15}N cannot be diluted by soil N. An important advantage of this approach is that the entire plant is labelled at virtually the same isotope concentration, corresponding to that of the applied ^{15}N solution, since isotopic fractionation appears to be negligible. A downside is that the biochemical composition of plants grown in such cultures may not be fully representative of that of wild plants grown in the field.

Stem injection of ^{15}N is another easy way to produce ^{15}N-labelled leaf litter. The transpiration of trees induces an upward flux of xylem sap from the roots to the canopy. Therefore, a reservoir with labelling solution can be connected to a borehole in the stem that extends to the xylem to ensure that the upward sap flow induced by transpiration carries the label to the leaves (Swanston and Myrold 1997; Proe et al. 2000; Garten and Brice 2009). In principle, this method can also be applied to conifers despite the potential problem that the resin these plants produce can clog the bore hole (Augusto et al. 2011; Nair et al. 2014). Key advantages of the technique are that it works well to label big trees in their natural environment and that uncontrolled dispersion of the label to soil, surrounding trees, etc. is limited. Another advantage of the stem-injection approach is that it is suitable for multiple labelling with ^{15}N, ^{26}Mg and ^{42}Ca (Augusto et al. 2011)

Foliar application of leaves by spraying with ^{15}N labelled urea is a third method. It is particularly useful to obtain large amounts (kilograms) of homogeneously labelled leaf litter (Zeller et al. 1998). Another key advantage is, as with the stem-injection method, that the chemical and biochemical composition of leaves grown in the field is not modified. Spraying urea on the foliage of cultivated plants like cereals and fruit trees has in fact been used for decades to improve crop growth. Urea fertilizer is chosen for this purpose because it contains 46% N, the highest N content of all widely used fertilizers, and because leaf uptake and assimilation are fairly rapid. However, efficiency of urea as fertilizer depends highly on weather conditions, because the applied tracer solution may be carried away by high winds, thus significantly reducing the amount of fertilizer N available for leaf uptake. In addition, urea uptake by leaves depends on plant-specific leaf traits such as thickness of a wax layer and ecophysiological responses such as stomata closure induced by water stress.

This chapter presents two of the three complementary approaches described above to obtain ^{15}N-labelled leaf litter. The first approach involves spraying leaves with ^{15}N-labelled urea as described in Zeller et al. (1998); the second describes the stem-injection technique following Paula et al. (2015).

2 Equipment and Material

2.1 *Spraying Method*

- Hand sprayer producing a fine mist (e.g. 1.5 l; Berthoud, Belleville, France)
- Transparent plastic bags (10–1500 l)
- ^{15}N-labelled urea (10–98 atom percent ^{15}N)

2.2 *Stem-Injection Method*

- ^{15}N-labelled NH_4NO_3 (10–98 atom percent ^{15}N)
- Distilled water
- Polyethylene flasks (100 and 1000 ml)
- Drill and drill bit (6 mm)
- Polyethylene tubing (6 mm diameter) flexible enough to connect flasks
- Non-toxic mineral putty (e.g. Terostat®)

3 Experimental Procedures

3.1 *Leaf Spraying*

1. Select plants (location, species, age) and determine the labelling period (spring or summer in temperate climates, end of the rainy season in tropical climates).
2. Prepare transparent plastic bags to enclose plants into a loosely fitting cover for 24 h after spraying; alternatively, build a simple mobile greenhouse chamber to cover the plants during spraying and the following day.
3. Prepare the labelling solution (3.0 g of ^{15}N labelled urea in 1000 ml of distilled water) and fill it into the hand sprayer.
4. Choose a calm and cloudy afternoon to label the plants, since wind will disperse the mist applied with the hand sprayer and warm weather increases evaporation, thus reducing the labelling efficiency.
5. Spray the foliage with the labelling solution, ensuring that the mist covers the whole foliage while avoiding that droplets form on the leaves and fall to the ground.
6. Cover the plant or the plots with large plastic bags or a mini greenhouse for 24 h.

7. To obtain a homogeneous ^{15}N label of the leaves, repeat the procedure a few weeks later during the growing season (normally 2–4 applications) until the intended enrichment level is reached (typically 1–10 atom percent ^{15}N), as judged by measuring the N concentration and ^{15}N enrichment 3 weeks after spraying the leaves.
8. Consider labelling trees in two consecutive years to increase homogeneity of the ^{15}N label in leaves and leaf litter.

3.2 Stem Injection

1. Drill a hole (6 mm in diameter) into the stem (25 mm deep), 1 m above the ground.
2. Lubricate the drill by continuously spraying it with water to prevent air from entering xylem vessels and thus avoid cavitation.
3. Immediately after removing the drill, push the rigid end of a polyethylene tube (6 mm in diameter) into the drilled hole and connect the tube to a flask containing 100 ml of distilled water (Fig. 56.1).
4. Pack the bark around the tube with non-toxic mineral putty (e.g. Terostat®) to prevent the labelled solution from leaking.

Fig. 56.1 Flasks and tubing to deliver distilled water or an ^{15}N solution to the xylem of trees selected for ^{15}N-labelling by the stem-injection method. Photos: R.R. Paula (left) and B. Zeller (right)

5. Cover the soil around the labelled trees with a plastic sheet to avoid accidental contamination of the soil with ^{15}N.
6. Make sure that the connection is working by verifying for 15–30 min that the water level in the flask progressively decreases; otherwise, restart the procedure by drilling a new hole.
7. Connect the 100-ml flask to a 1-l flask containing 0.9 g of N as potassium or ammonium nitrate (98 atom percent of $^{15}NO_3^-$) dissolved in 400 ml of distilled water (Nair et al. 2014). Using two flasks connected to the same drill hole allows first to check whether water uptake of the tree occurs and then labelling the tree in a safe way, thus avoiding loss of time and money but also any accidental contamination of the soil by ^{15}N when inserting the tube into the drill hole.
8. Backfill the drill hole with Terostat putty as soon as the 400 ml of labelled solution has been completely absorbed, which in young acacia trees (diameter 0.06–0.25 m, height 5–15 m), for example, takes 12–36 h.
9. Consider repeated labelling of a given tree to increase homogeneity of the ^{15}N label in leaves and leaf litter, which for larger trees (stem diameter > 10 cm) requires multiple boreholes distributed at 0, 120 and 240 degrees around the stem.

3.3 *Litter Collection*

1. Manually pick senescent leaves from labelled plants just before abscission, or wrap trees in a net to collect the litter that falls to the ground.
2. Continue collecting litter for weeks, months or even years.
3. Check homogeneity and strength of the label before using the litter in decomposition experiments or for other purposes.

4 Final Remarks

The ^{15}N enrichment of the urea solution must be adapted to the possible dilution of the ^{15}N label in the plant and the number of intended applications. For many studies, a final enrichment of 1–4 atom percent ^{15}N in the leaf litter is appropriate. To reach this level of labelling, multiple sprayings by a solution containing 5–10 atom percent of ^{15}N in the urea during two subsequent years may be needed. Repeating the stem injection will improve homogeneity of the labelling as well.

Caution is needed because urea may cause damage of sprayed leaves (visible brown spots). However, a concentration of 3 g per litre of distilled water has proved appropriate for many plant species, both trees and forbs. Nevertheless, in case of doubt, any negative effects at elevated concentration can be evaluated prior to the labelling.

References

Augusto, L., Zeller, B., Midwood, A., Swanston, C., Dambrine, E., Schneider, A., & Bosc, A. (2011). Direct labelling of eight-year-old *Pinus pinaster* trees with ^{15}N, ^{26}Mg and ^{42}Ca: Mass flow transport versus exchange transport. *Annals of Forest Sciences, 68*, 169–178.

Berg, B. (1988). Dynamics of nitrogen (^{15}N) in decomposing Scots pine (*Pinus sylvestris*) needle litter. Long-term decomposition in a Scots pine forest VI. *Canadian Journal of Botany, 66*, 1539–1546.

Blok, D., Elberling, B., & Michelsen, A. (2016). Initial stages of tundra shrub litter decomposition may be accelerated by deeper winter snow but slowed down by spring warming. *Ecosystems, 19*, 155–169.

Boberg, J. B., Finlay, R. D., Stenlid, J., Ekblad, A., & Lindahl, B. D. (2014). Nitrogen and carbon reallocation in fungal mycelia during decomposition of boreal forest litter. *PLoS One, 9*, e92897.

Colpaert, J. V., & van Tichelen, K. K. (1996). Decomposition, nitrogen and phosphorus mineralization from beech leaf litter colonized with ectomycorrhizal or litter decomposing basidiomycetes. *New Phytologist, 134*, 123–132.

Garten, C., & Brice, D. (2009). Belowground fate of ^{15}N injected into sweetgum trees (*Liquidambar styraciflua*) at the ORNL FACE experiment. *Rapid Communications in Mass Spectrometry, 23*, 3094–3100.

Gessner, M. O., Swan, C. M., Dang, C. K., McKie, B. G., Bardgett, R. D., Wall, D. H., & Hättenschwiler, S. (2010). Diversity meets decomposition. *Trends in Ecology and Evolution, 25*, 372–380.

Guo, C., Simon, J., Gasche, R., Naumann, S. P., Bimüller, C., Pena, R., Polle, A., Kögel-Knabner, I., Zeller, B., Rennenberg, H., & Dannenmann, M. (2013). Minor contribution of leaf litter to N nutrition of beech (*Fagus sylvatica*) seedlings in a mountainous beech forest of Southern Germany. *Plant and Soil, 369*, 657–668.

Nadelhoffer, K. J., Downs, M. R., & Fry, B. (1999). Sinks for ^{15}N-enriched additions to an oak forest and a red pine plantation. *Ecological Applications, 9*, 72–86.

Nair, R., Weatherall, A., Perks, M., & Mencuccini, M. (2014). Stem injection of ^{15}N-NH_4NO_3 into mature Sitka spruce (*Picea sitchensis*). *Tree Physiology, 34*, 1130–1140.

Paula, R. R., Bouillet, J. P., Trivelin, P. C. O., Zeller, B., Moraes Gonçalves, J. L., Nouvellon, Y., Bouvet, J. M., Plassard, C., & Laclau, J. P. (2015). Evidence of short-term belowground transfer of nitrogen from *Acacia mangium* to *Eucalyptus grandis* trees in a tropical planted forest. *Soil Biology and Biochemistry, 91*, 99–108.

Pena, R., Tejedor, J., Zeller, B., Dannenmann, M., & Polle, A. (2013). Inter-specific temporal and spatial differences in ectomycorrhizal fungal assemblages for the acquisition of litter-derived nitrogen. *New Phytologist, 199*, 520–528.

Proe, M. F., Mead, D. J., & Byrne, D. (2000). Effect of pruning on nitrogen dynamics within crowns of *Pinus radiata*. *Tree Physiology, 20*, 653–661.

Setälä, H., Marshall, V. G., & Trofymow, J. A. (1996). Influence of body size of soil fauna on litter decomposition and ^{15}N uptake by poplar in a pot trial. *Soil Biology and Biochemistry, 28*, 1661–1675.

Sommer, J., Dippold, M., & Kuzyakov, Y. (2017). Different responses of ash and beech on nitrate versus ammonium leaf labeling. *Journal of Plant Nutrition and Soil Science, 180*, 446–453.

Swanston, C. W., & Myrold, D. D. (1997). Incorporation of nitrogen from decomposing red alder leaves into plants and soil of a recent clearcut in Oregon. *Canadian Journal of Forest Research, 27*, 1496–1502.

Zeller, B., Colin-Belgrand, M., Dambrine, E., & Martin, F. (1998). ^{15}N partitioning and production of ^{15}N-labelled litter in beech trees following [^{15}N]urea spray. *Annals of Forest Sciences, 55*, 375–383.

Zeller, B., Colin-Belgrand, M., Dambrine, E., Martin, F., & Bottner, P. (2000). Decomposition of ^{15}N-labelled litter and fate of nitrogen derived from litter in a beech forest. *Oecologia, 123*, 550–559.

Chapter 57
Decomposition and Consumption Tablets (DECOTABs)

Gea H. Van der Lee, Ellard R. Hunting, J. Arie Vonk, and Michiel H. S. Kraak

Keywords Cellulose · Food quality · Invertebrate feeding · Litter chemistry · Litter quality · Litter surrogate · Microbial decomposition · Standardized substrate

1 Introduction

Estimates of decomposition and consumption rates of plant litter generally rely on the incubation of a known quantity of leaves (e.g. alder leaf disks; Webster and Benfield 1986) or a standardized surrogate litter (e.g. cotton strip assay; Tiegs et al. 2007) in both field and laboratory experiments. The natural complexity in field settings is commonly captured by enclosing plant material in litter bags of varying mesh sizes to differentiate between microbial decomposition and invertebrate consumption (Boulton and Boon 1991; Chap. 6). The mechanisms behind decomposition and consumption rates are further studied by adding a standardized quantity of plant material under controlled conditions in microcosms (Dang et al. 2009) or mesocosms (Hines et al. 2014). These approaches provide valuable information on, for instance, effects of the physicochemical environment on decomposition rates and decomposers (Bjelke 2005; Dang et al. 2009; Hines et al. 2014), the composition of detritivore communities (González and Graça 2003) or the role of litter quality for microbial decomposition and invertebrate consumption (Suberkropp et al. 1976; Gessner 1991; Swan and Palmer 2006). However, while ideal for standardization, the use of plant litter also suffers from some limitations.

G. H. Van der Lee (✉) · J. A. Vonk · M. H. S. Kraak
Institute for Biodiversity and Ecosystem Dynamics (IBED-FAME),
University of Amsterdam, Amsterdam, The Netherlands
e-mail: g.h.vanderlee@uva.nl; j.a.vonk@uva.nl; m.h.s.kraak@uva.nl

E. R. Hunting
School of Biological Sciences, University of Bristol, Bristol, UK
e-mail: e.r.hunting@bristol.ac.uk

F. Bärlocher et al. (eds.), *Methods to Study Litter Decomposition*,
https://doi.org/10.1007/978-3-030-30515-4_57

First, the chemical composition of natural plant material (e.g. concentrations of nutrients and secondary plant metabolites) is highly variable among plant species (Webster and Benfield 1986). This hampers comparisons of results from different studies and of environmental influences on decomposition over large spatial and typically also temporal scales. Collecting leaves from the same species and at a single time reduces this variability, but chemical composition inevitably varies due to variation in litter quality both within and among both biogeographic regions and leaf species (LeRoy et al. 2006; Lecerf and Chauvet 2008; Graça and Poquet 2014). Moreover, litter quality can vary within individual trees, depending on position in the canopy and the intensity of solar irradiance the leaves were exposed to (Sariyildiz and Anderson 2003).

Another major constraint for experimental purposes is that neither natural leaf litter nor standardized litter surrogates are easily manipulated chemically, although this is not impossible (e.g. Talbot and Treseder 2012; Danger et al. 2013). This constraint limits the potential to test for effects of chemical litter composition, including food quality aspects like nutrient and polyunsaturated fatty acid (PUFA) contents (Chaps. 17 and 55), on decomposition rates (Lecerf and Chauvet 2008). Likewise, it is difficult to address the effects of specific natural and anthropogenic compounds (e.g. antibiotics, fungicides or metals) in the environment that potentially inhibit microbial decomposition or invertebrate consumption (Rader et al. 1994).

Decomposition and consumption tablets (DECOTABs; Fig. 57.1) may overcome these constraints. DECOTABs are easy-to-prepare agar-based pellets containing

Fig. 57.1 DECOTABs that are (**a**) being fed upon by *Limnephilus lunatus*, a litter-consuming caddisfly; (**b**) deployed in a laboratory setting; (**c**) retrieved from a field study; and (**d**) set up for drying in the oven

25% agar (dry mass) with the remaining 75% being organic matter at choice (Kampfraath et al. 2012). This makes them highly standardized, inexpensive and adjustable, allowing extensive manipulations of chemical composition. The standard DECOTAB has been proposed to consist of a high concentration (75%) of cellulose powder (Kampfraath et al. 2012), but cellulose can be replaced or its composition altered by adding (fine) particulate organic matter (Hunting et al. 2016), ground plant litter (Hunting et al. 2016; Vonk et al. 2016) or specific substances such as PUFAs (Vonk et al. 2016). This chemical versatility combined with the texture of DECOTABs allows using them as realistic surrogates of natural organic matter and enables testing preferences of microbes and invertebrates for specific food components (see also Chap. 52). Furthermore, it is possible to add antibiotics or fungicides to specifically inhibit bacterial or fungal activity (Kampfraath et al. 2012).

Although only recently developed (Kampfraath et al. 2012), the use of DECOTABs in field and laboratory studies has already led to important insights into the effects of agricultural practices on organic matter consumption and decomposition (Hunting et al. 2016), the importance of linoleic acid in food sources for several detritivorous invertebrates (Vonk et al. 2016) and how dissolved oxygen affects the relative importance of microbes and invertebrates to organic matter decomposition and consumption (Van der Lee et al. 2017). Overall, these studies have revealed that the mass loss of DECOTABs can vary greatly, depending on the physicochemical environment, detritivore community composition and litter quality. The following protocol for DECOTAB preparation and deployment has been adopted from Kampfraath et al. (2012).

2 Equipment and Materials

- Drying oven (e.g. 50 °C)
- Analytical balance (±0.1 mg precision)
- 2-l Erlenmeyer flask
- Magnetic stirrer with hot plate and stirring bar
- Moulds: standard multi-well polycarbonate mould 15 mm in diameter and 5 mm deep (alternative shapes and sizes possible)
- Purified agar
- Powdered cellulose
- Ascorbic acid
- Coarse-mesh litter bags (e.g. 4 mm mesh size) for field study
- Fine-mesh bags (e.g. 500 μm mesh size) for field study
- PUFA such as linoleic acid (99% purity, 0.40 ng l^{-1} agar)
- Antibiotic such as chloramphenicol (60 mg l^{-1} agar)
- Fungicide such as natamycin (21 mg l^{-1} agar)

3 Experimental Procedures

3.1 Standard DECOTAB Preparation

1. Boil 1 l of deionized water with 20 g purified agar for 3 min (100 °C). The agar solution is ready when it becomes transparent.
2. Cool solution to 60 °C while continuously stirring on a hot plate. Do not let the solution cool below 50 °C, since it will then solidify and reheating fails to redissolve the agar.
3. While stirring the solution, add 60 g of powdered cellulose and 10 mg of ascorbic acid as antioxidant.
4. Pour solution into the moulds.
5. Remove DECOTABs from the moulds after 7 min of cooling and store in a closed container at 7 °C. To prepare more DECOTABs, steps 4 and 5 can be repeated, while the solution is kept at 60 °C and continuously stirred.
6. Dry a subset of 20 DECOTABs per type at 50 °C for 2 days, and weigh on an analytical balance to determine the initial dry mass of the DECOTABs.
7. The DECOTABs can be stored in a closed container at 4 °C for up to 3 weeks without noticeable decay or dehydration. Sterilization of DECOTABs (e.g. by submersion in 70% ethanol) and storage in sterile deionized water can substantially prolong this period.

3.2 Alternative DECOTABs

The standard DECOTAB composition can be altered by substituting cellulose for (fine) particulate organic matter or plant litter, or by adding specific substances. Three examples of alternative DECOTABs are described below, but a wide variety of other types or combinations are possible:

Particulate Organic Matter

- Collect sediment, soil or other type of substrate at the study area.
- Re-suspend and allow inorganic material (e.g. quartz sand) to settle, and collect supernatant containing the (fine) particulate organic matter fraction.
- Dry at 50 °C for 2 days.
- Grind the plant material to powder to pass a 0.5–2 mm mesh size.
- Substitute the powdered cellulose in the standard DECOTABs for the prepared particulate organic matter (Hunting et al. 2016), and then proceed as for the standard DECOTABs.

Plant Litter

- Collect plant litter.
- Rinse, and air-dry for 2 days.
- Grind the plant material to powder to pass a 0.5–2 mm mesh size.
- Substitute the powdered cellulose in the standard DECOTABs for the ground plant litter (Hunting et al. 2016; Vonk et al. 2016), and then proceed as for the standard DECOTABs.

Addition of Specific (Hydrophobic) Substances

- Add PUFAs such as 0.40 g L^{-1} linoleic acid (Vonk et al. 2016) in accordance with the linoleic acid concentration of submerged macrophytes (Van Ginneken et al. 2011) or other specific plant substances at the third step of the standard DECOTAB preparation.
- Add an antibiotic such as 60 mg l^{-1} chloramphenicol (Kampfraath et al., 2012) to curb bacterial activity.
- Add a fungicide such as 21 mg l^{-1} natamycin to curb fungal activity (Pedersen 1992).
- Otherwise, proceed as for the standard DECOTABs.

3.3 Deployment and Retrieval

- Place DECOTABs in coarse-mesh or fine-mesh bags and deploy them in the field using an appropriate experimental design.
- Regularly check the mass remaining and retrieve the DECOTABs when approximately 50% of their mass appears to have been lost.
- Alternatively, place DECOTABs in mesocosms or microcosms inoculated with microorganisms or stocked with invertebrates. A deployment time of 21 days is recommended, or less when the observed consumption rates are high.
- Rinse the DECOTABs carefully after removal from the field, mesocosms or microcosms.
- Dry the DECOTABs at 50 °C for 2 days, and weigh them on an analytical balance to determine the final DECOTAB dry mass.

3.4 Data Analysis

Decomposition rates of DECOTABs (k) in the field or laboratory can be calculated as a function of time by fitting the data to an exponential decay model as described in Chap. 6 for leaf litter. Alternatively, microbial mass loss rate (MMLR) can be expressed as a function of time by assuming linear mass loss:

$$\text{Microbial mass loss rate } (\text{MMLR}) = \frac{DM_i - DM_f}{t} \tag{57.1}$$

where MMLR for each DECOTAB is the difference between the initial DECOTAB dry mass (DM_i) and the dry mass of the corresponding DECOTAB deployed in fine-mesh bags (DM_f) at the end of the exposure period, divided by the duration of the deployment time (t) (e.g. in days). Similarly, the consumption rate of DECOTABs by invertebrates (ICR) can be calculated according to:

$$\text{Invertebrate consumption rate}(ICR) = \frac{DM_i - DM_c}{t} - \text{MMLR} \tag{57.2}$$

where the invertebrate consumption rate is the difference between the initial DECOTAB dry mass (DM_i) and the dry mass of the corresponding DECOTAB deployed in coarse-mesh bags (DM_c) at the end of the exposure period, divided by the duration of the deployment time (t) (e.g. in days), minus the mass loss rate of the DECOTAB in the fine-mesh bags (MMLR). For consumption under laboratory conditions, see Chap. 51.

4 Final Remarks

DECOTABs do not capture the full complexity of natural litter. The texture, in particular, does not adequately mimic natural leaf litter. However, DECOTABs do provide a relatively coarse standardized substrate with a firm texture that can be specifically adjusted to address a range of questions requiring the manipulation of chemical composition. Consequently, DECOTABs will be particularly useful for experimental tests of hypotheses relating to the importance of the chemical composition of organic matter.

A dedicated website is available at www.ibed.uva.nl/decotab to facilitate interaction among DECOTAB users and to share ideas, information and files relevant to the use of DECOTABs.

References

Bjelke, U. (2005). Processing of leaf matter by lake-dwelling shredders at low oxygen concentrations. *Hydrobiologia, 539*, 93–98.

Boulton, A. J., & Boon, P. I. (1991). A review of methodology used to measure leaf litter decomposition in lotic environments: Time to turn over an old leaf? *Marine and Freshwater Research, 42*, 1–43.

Dang, C. K., Schindler, M., Chauvet, E., & Gessner, M. O. (2009). Temperature oscillation coupled with fungal community shifts can modulate warming effects on litter decomposition. *Ecology, 90*, 122–131.

Danger, M., Arce Funck, J., Devin, S., Heberle, J., & Felten, V. (2013). Phosphorus content in detritus controls life-history traits of a detritivore. *Functional Ecology, 27*, 807–815.

Gessner, M. O. (1991). Differences in processing dynamics of fresh and dried leaf litter in a stream ecosystem. *Freshwater Biology, 26*, 387–398.

González, J. M., & Graça, M. A. S. (2003). Conversion of leaf litter to secondary production by a shredding caddis-fly. *Freshwater Biology, 48*, 1578–1592.

Graça, M. A. S., & Poquet, J. M. (2014). Do climate and soil influence phenotypic variability in leaf litter, microbial decomposition and shredder consumption? *Oecologia, 174*, 1021–1032.

Hines, J., Reyes, M., Mozder, T. J., & Gessner, M. O. (2014). Genotypic trait variation modifies effects of climate warming and nitrogen deposition on litter mass loss and microbial respiration. *Global Change Biology, 20*, 3780–3789.

Hunting, E. R., Vonk, J. A., Musters, C. J. M., Kraak, M. H. S., & Vijver, M. G. (2016). Effects of agricultural practices on organic matter degradation in ditches. *Scientific Reports, 6*, 21474.

Kampfraath, A. A., Hunting, E. R., Mulder, C., Breure, A. M., Gessner, M. O., Kraak, M. H. S., & Admiraal, W. (2012). DECOTAB: A multipurpose standard substrate to assess effects of litter quality on microbial decomposition and invertebrate consumption. *Freshwater Science, 31*, 1156–1162.

Lecerf, A., & Chauvet, E. (2008). Intraspecific variability in leaf traits strongly affects alder leaf decomposition in a stream. *Basic and Applied Ecology, 9*, 598–605.

LeRoy, C. J., Whitham, T. G., Keim, P., & Marks, J. C. (2006). Plant genes link forests and streams. *Ecology, 87*, 255–261.

Pedersen, J. C. (1992). Natamycin as a fungicide in agar media. *Applied and Environmental Microbiology, 58*, 1064–1066.

Rader, R. B., McArthur, J. V., & Aho, J. M. (1994). Relative importance of mechanisms determining decomposition in a southeastern blackwater stream. *American Midland Naturalist, 132*, 19–31.

Sariyildiz, T., & Anderson, J. M. (2003). Decomposition of sun and shade leaves from three deciduous tree species, as affected by their chemical composition. *Biology and Fertility of Soils, 37*, 137–146.

Suberkropp, K., Godshalk, G. L., & Klug, M. J. (1976). Changes in the chemical composition of leaves during processing in a woodland stream. *Ecology, 57*, 720–727.

Swan, C. M., & Palmer, M. A. (2006). Composition of speciose leaf litter alters stream detritivore growth, feeding activity and leaf breakdown. *Oecologia, 147*, 469–478.

Talbot, J. M., & Treseder, K. K. (2012). Interactions among lignin, cellulose, and nitrogen drive litter chemistry–decay relationships. *Ecology, 93*, 345–354.

Tiegs, S. D., Langhans, S. D., Tockner, K., & Gessner, M. O. (2007). Cotton strips as a leaf surrogate to measure decomposition in river floodplain habitats. *Journal of the North American Benthological Society, 26*, 70–77.

Van der Lee, G. H., Kraak, M. H. S., Verdonschot, R. C. M., Vonk, A. J., & Verdonschot, P. F. M. (2017). Oxygen drives benthic-pelagic decomposition pathways in shallow wetlands. *Scientific Reports, 7*, 15051.

Van Ginneken, V. J. T., Helsper, J. P. F. G., De Visser, W., Van Keulen, H., & Brandenburg, W. A. (2011). Polyunsaturated fatty acids in various macroalgal species from north Atlantic and tropical seas. *Lipids in Health and Disease, 10*, 104.

Vonk, J. A., Van Kuijk, B. F., Van Beusekom, M., Hunting, E. R., & Kraak, M. H. S. (2016). The significance of linoleic acid in food sources for detritivorous benthic invertebrates. *Scientific Reports, 6*, 35785.

Webster, J. R., & Benfield, E. F. (1986). Vascular plant breakdown in freshwater ecosystems. *Annual Review of Ecology and Systematics, 17*, 567–594.

Chapter 58
Inoculation of Leaf Litter with Aquatic Hyphomycetes

Eric Chauvet

Keywords Fungal inoculum · Ingoldian fungi · Fungal spores · Conidia · Aquatic fungi · Leaf litter · Microcosm · Fungal community

1 Introduction

There is evidence that colonization of new substrates by aquatic hyphomycetes in streams occurs predominantly via conidia. This is suggested by the functional significance of the distinctive morphology of these spores (Webster 1987) and the very high proportion of fungal production allocated to their formation (Suberkropp 1991). The colonization of new substrates by mycelial contact is common in soil where litter tends to accumulate in layers that favour exploration by individual hyphae or hyphal aggregations (rhizomorphs; Bärlocher and Boddy 2016). In small forest streams, in contrast, litter tends to accumulate in front of obstacles. Hyphal colonization of litter by fungal mycelium in streams is still possible and has allowed successful *in vitro* inoculation using hyphal homogenates or sections of agar overgrown with aquatic hyphomycete colonies (e.g. Bärlocher and Corkum 2003; Suberkropp 2003; Pascoal et al. 2010). However, in addition to reflecting the typical life cycle of aquatic hyphomycetes, inoculation with conidia of these fungi offers the advantage of ensuring homogeneity of inocula impacting on the surface of the substrates to be colonized. Inoculation with conidia of aquatic hyphomycetes is therefore preferable to using hyphal homogenates, especially when the simultaneous colonization of substrates by several fungal species is intended.

The aim of this chapter is to provide a simple procedure for inoculating leaf litter with conidia of pure cultures of single or several species of aquatic hyphomycetes. The presented protocol is recommended when conducting field or laboratory

---

E. Chauvet (✉)
EcoLab, Université de Toulouse, CNRS, Toulouse, France
e-mail: eric.chauvet@univ-tlse3.fr

F. Bärlocher et al. (eds.), *Methods to Study Litter Decomposition*,
https://doi.org/10.1007/978-3-030-30515-4_58

experiments with plant material colonized by known communities of aquatic hyphomycetes. The procedures can easily be adapted to conidia from sources other than cultures grown on agar (e.g. leaves colonized by fungi in a stream) or to the nature (e.g. cotton strips) or form of the substrate (e.g. whole leaves or leaf discs). The protocol is derived from the microcosm procedure described in Suberkropp (1991). It is particularly appropriate when a large amount of inoculated material is required to cover multiple combinations such as in experiments designed to assess biodiversity effects of aquatic hyphomycetes on litter decomposition (e.g. Jabiol et al. 2013). Some alternatives to the protocol described below, also with conidia, have been used successfully. For instance, aquatic hyphomycetes may be cultured and conidial suspensions be prepared for inoculation in sterile tissue culture flasks with caps (allowing gas exchange) and placed on an orbital shaker (V. Gulis, personal communication). Finally, aquatic hyphomycete mycelia produced from conidial inocula quickly invade leaf litter in nature, often resulting in the accrual of large amounts of leaf-associated fungal biomass on decomposing leaves (Gessner et al. 2007). As a consequence, once leaves are fully colonized, it is not always necessary to maintain substrates in strictly sterile conditions during experiments as long as the sources of contamination and the experimental duration are limited.

2 Equipment, Chemicals and Solutions

2.1 Equipment and Material

- Autoclave
- Laminar flow cabinet (for aseptic manipulations)
- Microscope (10× and 40× objectives)
- Erlenmeyer flasks, 100 ml; alternatively: stream-mimicking microcosms (Suberkropp 1991)
- Erlenmeyer flasks, 1 l
- Magnetic stirrers and stirring bars (1 and 5 cm length); a multiple-position magnetic stirrer can be advantageous
- Membrane filters (5 μm pore size) and filtering apparatus
- Volumetric glass or Eppendorf pipettes (1–5 ml)
- Supply of pressurized air (e.g. aquarium pumps), tubing and cotton plugs
- Cork borer (ca. 10 mm diameter)
- Autumn-shed leaves, air-dried
- Pure culture of aquatic hyphomycete on 0.1% MEA, preferably freshly isolated and grown to maintain the capacity to sporulate (see Chap. 23 for isolation techniques and Chap. 24 for culture maintenance)

2.2 Chemicals and Solutions

- Deionized water, autoclaved
- Mineral nutrient solution, autoclaved: 10 mg KNO_3, 0.55 mg K_2HPO_4, 100 mg $CaCl_2 \cdot 2H_20$, 10 mg $MgSO_4 \cdot 7H_20$, 500 mg 3-morpholino-propanesulfonic acid (MOPS) and 1 l deionized water
- 0.1% trypan blue or cotton blue in 60% lactic acid

3 Experimental Procedures

1. Place agar plugs cut from the margins of an actively growing fungal colony in a 100 ml Erlenmeyer flask containing 50 ml of sterile water or mineral nutrient solution. Subject the suspension to forced sterile aeration and shaking or both; the set-up for producing conidial inocula is similar to that for incubation of inoculated leaf discs as detailed below, except for the size of stirring bars (1 cm; Fig. 58.1). Periodically (e.g. every 2 days) replace water (or nutrient solution) by using aseptique technique.
2. When water or medium is renewed, determine the abundance of suspended conidia by passing a small aliquot (precisely measured with a glass or Eppendorf pipette) over a membrane filter, staining the collected conidia with a few drops of trypan blue or cotton blue solution, and examining them under the micro-

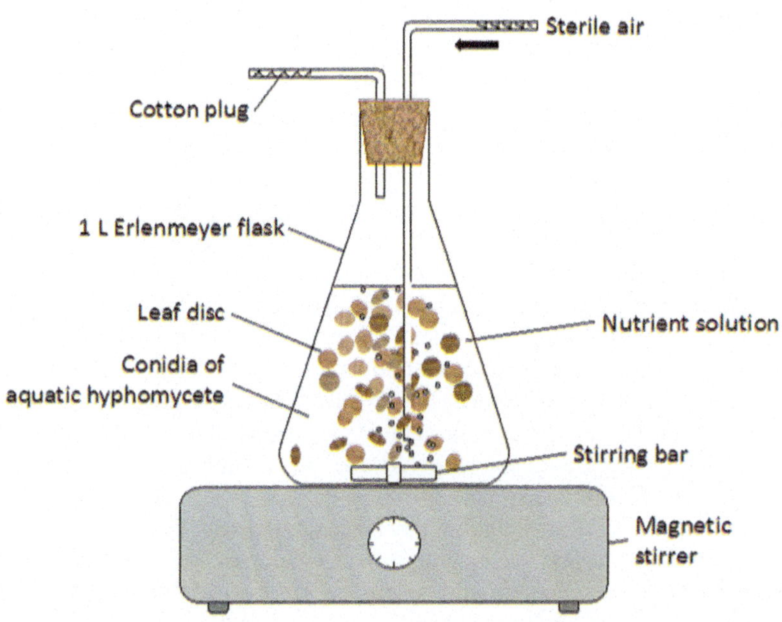

Fig. 58.1 Inoculation of leaf discs by aquatic hyphomycete conidia

scope. Conidial density can be determined by scanning the entire filter surface or a measured subsection (Chap. 26). As a very large number of conidia is generally required for inoculation, maximum sporulation is targeted, which typically occurs a few days after submersion, depending on species identity, culture age and other factors (see Chap. 23 for anamorph induction and Chap. 24 for culture maintenance). To maximize the number of conidia produced, multiple flasks or microcosms can be used.

3. Use a cork borer to cut the number of required leaf discs plus some spare ones, preferably by avoiding the main vein. Pre-wetting dried leaves during the preceding few hours normally facilitates cutting discs.
4. Autoclave 360 leaf discs in 1-l Erlenmeyer flasks containing deionized water (or nutrient solution) and a 5-cm stirring bar. Aseptically replace the leachate by 800 ml of nutrient solution. The leaf discs and solution can be set aside (in the dark and a cool place) for up to a few days until inoculation. Just before inoculation, aseptically equip the flasks with the aeration system and gently (ca. 80 rpm) stir the suspension as illustrated in Fig. 58.1.
5. When a sufficient conidial density for inoculation has been reached, collect an appropriate volume of the suspension in a 100-ml sterile flask containing a 1-cm stirring bar. Determine the exact conidial density in the flask from at least two measured aliquots aseptically collected with a pipette (under agitation of the flask) as detailed in Step 2. Calculate the corresponding volume to collect from the flask for inoculation, based on the required number of approximately 90,000 conidia for optimal colonization of 360 leaf discs (Suberkropp 1991; Treton et al. 2004). If the conidial density is too high to estimate conidial numbers reliably, dilute the suspension with sterile nutrient solution in a new 100-ml sterile flask; in this case, determine the reduced conidial density in the new flask for inoculation. When leaf discs are inoculated with a mixture of different fungal species, divide 90,000 by the number of species to determine the number of conidia required per species.
6. Aseptically add the required volume of the spore suspension, as determined in Step 5, to the leaf disc suspension kept under agitation. Initiate aeration and adjust stirring speed to distribute the conidia homogenously. Leaf discs should slowly be moving around. After 5 min, stop both agitation and aeration for 30 min to favour conidial sedimentation and impaction onto leaf surfaces (K. Suberkropp, personal communication). This step may be repeated during the very first few hours to optimize leaf colonization. The duration of incubation (generally 10–40 days) depends on temperature, growth rate of the fungal species and objectives of the experiment. To check fungal development, particularly when several species are present, and to ensure the absence of contamination, periodically collect aliquots to determine the abundance of newly released conidia as detailed in Step 2.
7. Because the timing of inoculation may be critical, particularly when several species are involved, it may be advantageous to maintain continuously sporulating batches of cultures. To this end, inoculate sterile leaf discs with a conidial suspension (as detailed above, although there is no need to determine the conidial

density in this case). When leaf discs start to release conidia in substantial amount (determined at times when nutrient solution is renewed; see Step 2), introduce fresh leaf discs into the culture; they constitute a new substrate for mycelial development followed by sporulation. Repeated input of new leaf material allows extending conidial production over several weeks.

References

Bärlocher, F., & Boddy, L. (2016). Aquatic fungal ecology – How does it differ from terrestrial? *Fungal Ecology, 19*, 5–13.

Bärlocher, F., & Corkum, M. (2003). Nutrient enrichment overwhelms diversity effects in leaf decomposition by stream fungi. *Oikos, 101*, 247–252.

Gessner, M. O., Gulis, V., Kuehn, K. A., Chauvet, E., & Suberkropp, K. (2007). Fungal decomposers of plant litter in aquatic ecosystems. In C. Kubikak & I. Druzhinina (Eds.), *The Mycota – Environmental and microbial relationships* (Vol. IV, pp. 301–324). Berlin: Springer.

Jabiol, J., McKie, B. G., Bruder, A., Bernadet, C., Gessner, M. O., & Chauvet, E. (2013). Trophic complexity enhances ecosystem functioning in an aquatic detritus-based model system. *Journal of Animal Ecology, 82*, 1042–1051.

Pascoal, C., Cássio, F., Nikolcheva, L., & Bärlocher, F. (2010). Realized fungal diversity increases functional stability of leaf litter decomposition under zinc stress. *Microbial Ecology, 59*, 84–93.

Suberkropp, K. (1991). Relationships between growth and sporulation of aquatic hyphomycetes on decomposing leaf litter. *Mycological Research, 95*, 843–850.

Suberkropp, K. (2003). Methods for examining interactions between freshwater fungi and macroinvertebrates. In C. K. M. Tsui & K. D. Hyde (Eds.), *Freshwater mycology* (pp. 159–171). Hong Kong: Fungal Diversity Press.

Treton, C., Chauvet, E., & Charcosset, J. Y. (2004). Competitive interaction between two aquatic hyphomycete species and increase in leaf litter breakdown. *Microbial Ecology, 48*, 439–446.

Webster, J. (1987). Convergent evolution and the functional significance of spore shape in aquatic and semi-aquatic fungi. In A. Rayner, C. Brasier, & D. Moore (Eds.), *Evolutionary biology of the fungi* (pp. 191–201). Cambridge: Cambridge University Press.

Part VIII
Data Analyses

Chapter 59
A Primer for Statistical Analysis

Felix Bärlocher

Keywords Descriptive statistics · Inferential statistics · Bounded rationality · Probability theory · Bayesian statistics · Regression analysis · Correlation analysis · χ^2 (chi square) test · Permutation tests · Big Data challenge · Path analysis · Normal distribution · Statistical hypotheses · Decision theory · Statistical power · Null hypothesis · Alternative hypothesis

1 Introduction

Most scientific investigations begin with the collection of data. Summarizing and representing the data is generally labelled "descriptive statistics"; conclusions, predictions, or diagnoses based on these data fall under the domain of "inferential statistics." Inferences are never completely certain and are therefore expressed as probabilities. Consequently, to use statistical methods effectively, we need at least a basic understanding of the concepts of probability.

In everyday life, we continuously make "statistical" statements: we know, for example, that men tend to be taller than women or that Scandinavians tend to have lighter skin than Egyptians. Such common-sense conclusions are generally reliable if the differences are large. Often, however, natural variability (environmental noise) is so great that it can mask the effect of factors that we investigate. Statistical evaluation is therefore essential, since our natural intuition can mislead us (Paulos 1995). For example, there is no scientifically justifiable doubt today that smoking poses a health risk. But we may still hear the argument that somebody knows friends who smoked every day and lived a healthy life into their 80s or 90s and that therefore smoking may be harmless after all. We also tend to make unwarranted connections between a chance event and a particularly memorable success or failure: an athlete may have experienced a spectacular victory while wearing a particular sweater or pair of socks. Or, we may see a black cat, and a few minutes later, we have an acci-

F. Bärlocher (✉)
Department of Biology, Mount Allison University, Sackville, NB, Canada
e-mail: fbaerloc@mta.ca

F. Bärlocher et al. (eds.), *Methods to Study Litter Decomposition*,
https://doi.org/10.1007/978-3-030-30515-4_59

dent. This tendency to interpret events in close temporal sequence as causally related can lead to superstitions, or prejudice – or, it may lead to new insights into actual mechanisms. Statistics can help us making rational decisions. It does not claim to reveal the truth. It has the more modest goal of increasing the probability that we correctly separate "noise" from "signal." It helps us avoid both ignorance (being unaware of real connections between two variables) and superstition (accepting false connections between two variables).

The way we evaluate chance and probabilities has been shaped by evolution (Pinker 2002). Attitudes that helped our ancestors survive and reproduce were favored by natural selection. They were not necessarily those that infallibly separate signals from noise. To begin with, a complete evaluation of our environment would be time-consuming and exceed the capabilities of our central nervous system: "Our minds are adapted to a world that no longer exists, prone to misunderstandings, correctable only by arduous education" (Pinker 2002). Economists and psychologists refer to this shortcoming of our intellect politely as "bounded rationality." It plays an enormous role in many everyday choices and decisions. Investigations into how we perceive probabilities were pioneered by D. Kahnemann and O. Tversky (e.g., Kahnemann et al. 1982); Kahnemann was awarded the Nobel Prize in Economics for this work.

2 Roots of Statistical Methods

The word "Statistik" was coined by G. Achenwall (Göttingen, Germany, 1719–1747). It is derived from "statista" (Italian for statesman), and refers to the knowledge that a statesman is supposed to have. Some early examples of statistical applications include population counts, estimates of harvests in a country, taxes, etc.). Early statistical societies restricted themselves to the collection of data for economical and political purposes. They often deliberately refused to draw conclusions based on their data: the motto of the Statistical Society of London was *Aliis exterendum* – let others do the threshing, i.e., the extraction of conclusions (Gigerenzer et al. 1989; Bärlocher 2008).

An important breakthrough was made when Adolphe Quetelet (1796–1874) introduced the concept of the "average man," whose thoughts and deeds coincide with those of the entire society. He also recognized the importance of large numbers. Increasingly, the interpretation of collected data became important. The deliberate connection of measurements with probabilistic statements was initiated toward the end of the nineteenth century.

The impetus for probability theory came from games of chance. Its formal beginning is usually connected to an exchange of letters in 1654 between Blaise Pascal and Pierre de Fermat, discussing a gambling problem put to them by the Chevalier de Méré. The modern basis of probability was presented by Jakob Bernoulli (1654–1705) in *Ars Conjectandi*. Other important developments were the derivation of the normal distribution by de Moivre, and its further elaboration by Karl Friedrich

Gauss. Thomas Bayes (1702–1763) introduced the important distinction between a priori and a posteriori probabilities. Bayesian Statistics, where a priori probability is often subjective and is well-established in economics and law. Its application to biology and other sciences has been increasing but is still somewhat controversial.

Francis Galton (1822–1911) is considered the founder of eugenics and biometrics. Biometrics (or biometry) is defined as the application of mathematical techniques to organisms or life processes. Today, it is generally used more narrowly to describe the use of statistical methods in biological investigations. Galton developed the basis for regression and correlation. Another important technique, the χ^2 (chi square) test, was introduced by Karl Pearson (1857–1936).

The most influential theoretician of modern statistics is undoubtedly Sir Ronald A. Fisher (1890–1962). His work on analysis of variance, significance tests, experimental design, etc., continues to dominate the practice of data analysis (Zar 1996). His approach was modified and expanded by Jerzy Neyman (1894–1981) and Egon S. Pearson (1895–1980).

Statistics is often viewed as a monolithic, internally consistent structure of universally accepted concepts and laws. This is far from being the case (Gigerenzer et al. 1989). Deep-seated philosophical differences concerning the proper analysis and interpretation of data persist to this day, and no universally accepted approach seems to be in sight (Meehl 1978; Howson and Urbach 1993). What is represented as "the" statistical method in textbooks has been called a "hybrid theory," trying to reconcile the often contradictory approaches and interpretations by Fisher, on the one side, and Neyman/Pearson, on the other side. Both differ from Bayesian statistics. A relatively new approach called model selection replaces traditional null hypothesis tests by simultaneously confronting several hypotheses by data. The enormous increase in computer power has allowed the manipulation of collected data and the production of "synthetic" data, which may provide clues to their underlying structure (Monte Carlo techniques, Bootstrap, resampling, and permutation methods; Efron and Tibshirani 1993; Manly 1997; Good 1994, 1999).

The widespread use of molecular techniques in all areas of ecology, especially in taxonomic surveys (Chap. 62), has generated huge amounts of data. Extracting knowledge from large data sets has been labelled the "Big Data Challenge." Mayer-Schönberger and Cukier (2013) claim that we are approaching a state where N = All, i.e., we essentially capture the entire population of data. This lowers our sampling error, which lessens the impact of measurement errors: "what we lose in accuracy at the micro level we gain in insight at the macro level" (Mayer-Schönberger and Cukier 2013). Another crucial, controversial suggestion is changing the emphasis away from investigating causality toward pattern recognition by correlations: "The correlations may not tell us *why* something is happening, but they alert us *that* it is happening" (Mayer-Schönberger and Cukier 2013). This seems to circumvent the widely accepted truism "Correlation does not imply causation." However, as Montello (2015) pointed out, a more appropriate statement would be something like "correlation is causality, but ambiguously so." Path analysis, a subset of structural equation modeling, is an extension of multiple regression. It evaluates hypothesized causal connections between sets of variables. Shipley (2000) provides an excellent introduction to this topic.

The development of powerful microcomputers and sophisticated statistical programs allows the application of very complex statistical models by naïve users. A taskforce of the American Psychological Association (APA, meeting on statistical inference, Newark, 14–15 December, 1996) saw this as problematic: the underlying assumptions are often ignored; little effort is made to determine whether the results are reasonable, and the precision of the analysis is often overestimated. The task force's recommendations include: making an attempt to verify the results by independent computation; more emphasis on simpler experimental designs; more emphasis on descriptive data analysis. This includes graphic representation (see Tukey 1977), calculation of averages with confidence intervals, and consideration of direction and size of effects.

3 Fisher's Approach

3.1 Assuming Normal Distribution

How do we know that something is true? A naïve empiricist might reply that if we observe an event or a series of events often enough, it must be true. The Scottish philosopher David Hume (1711–1776) correctly argued that mere repetition of an event does not necessarily imply that it will occur in the future. An often used example concerns swans: Europeans are likely to encounter only white swans and might conclude that all swans are white.

If repeated observations do not reliably reveal the truth, how do we decide which interpretation of nature is valid? The solution that has been accepted by most scientists (but see Howson and Urbach 1993; Berry 1996), and forms the basis of classical statistics, was suggested by Sir Karl Popper (1935). He agrees with Hume that our knowledge is always preliminary and based on assumptions or hypotheses. We can never verify these hypotheses. However, if a hypothesis does not represent the truth, it is vulnerable to being falsified. A useful hypothesis allows us to make predictions that are not obvious. We design an experiment to test these predictions; if they do not occur, we have falsified the hypothesis. For example, a European could propose the hypothesis that all swans are white. If he happens to visit New Zealand, he will sooner or later encounter a black swan, which falsifies his hypothesis. Or, as Thomas Huxley (1825–1895) put it: "The great tragedy of science is the slaying of a beautiful hypothesis by a nasty, ugly, little fact." Scientific research essentially is a weeding out of hypotheses that do not survive rigorous testing. Popper's reasoning was enormously influential. In economics, its basic philosophy has been expressed as follows: "The ultimate test of the validity of a theory is not conformity to the canons of formal logic, but the ability to deduce facts that have not yet been observed, that are capable of being contradicted by observation, and that subsequent observation does not contradict" (Friedman 1966). The same approach has been applied to natural selection: An organism, its organs, and behavior can be interpreted as "hypotheses" concerning the nature of the environment. If they are inappropriate, they will be "rejected" by nature, i.e., the organism dies.

Biological hypotheses rarely allow yes or no predictions. Experiments more commonly produce continuous or discrete data, whose measurement cannot be accomplished without errors. Their true value must therefore be expressed in probabilistic terms. To take this into account, Fisher used the following approach:

- Formulate a null hypothesis (H_0). For example, we propose that two groups of animals on different diets have the same final body weight.
- Define a test statistic characterizing the difference between the two groups (the most obvious number to choose is simply the difference between the two averages; more commonly, the *t*-value is used). Measure the actual value of this statistic.
- Assume that the weights of animals vary according to a defined probabilistic distribution (generally a normal distribution).
- Assuming that the two groups have in fact the same final weight (i.e., H_0 is correct), how likely is it that the test statistic will reach a value that is at least as extreme as the one actually measured (extreme is measured in terms of distance from the most probable value, which is the average)? This probability, generally determined from the assumed data distribution, is called *p*.
- If *p* falls below a pre-established critical value α (frequently 0.05 or 0.01), we reject the null hypothesis. We label the two values as significantly different.

To repeat, *p* measures the probability that our test statistic (a number measuring a discrepancy between two or more groups) reaches a value at least as high as the one actually found **IF** the null hypothesis is correct. It does not tell us anything about the probability that H_0 is correct or false. Because our measurements are always subject to random error, extreme values are possible and will occur. The value of α therefore also represents the probability that we incorrectly reject a null hypothesis that is in fact true (Table 59.1). According to Fisher, we can reject H_0, but we can never prove it to be correct.

Table 59.1 Statistical decision theory

	Null hypothesis (H_0)	
Decision	Correct	False
Accept H_0	Correct decision	Type II error (Ignorance)
Reject H_0	Type I error (Superstition)	Correct decision

3.2 Assuming Data Are Not Normally Distributed: Permutation Tests and the Bootstrap

Most classical statistical tests assume normal distribution of the data (more accurately, errors or residuals that remain after a model has been fitted have to be normally distributed; in many cases, normal data imply normal errors and vice versa). If this is not the case, data can be transformed to make them approximately normal, or we can use nonparametric or distribution-free tests. The vast majority of these tests are variations of permutation or randomization tests (Edginton 1987; Westfall and Young 1993). Fisher again played a crucial role in developing this approach. The major difference to parametric tests is that we make no assumptions concerning the distribution of the data. Thus:

1. Formulate a null hypothesis (H_0). For example, we propose that two groups of animals on different diets have the same final body weight.
2. Define a test statistic characterizing the difference between the two groups (e.g., difference between the two averages). Calculate the actual value of this statistic.
3. Assuming that the two groups have in fact the same final weight (i.e., H_0 is correct), assignment of the measured values to the two diets should be random. We therefore systematically establish all permutations of the data. For each permutation, we determine the value of the test statistic.
4. How likely is it that the test statistic will reach a value that is at least as extreme as the one actually measured? This value, determined from the distribution of permutated data, is called p.
5. If this probability is below a pre-established critical value α (frequently 0.05 or 0.01), we reject the null hypothesis. We label the two values as significantly different.

Even with small data collections, an exhaustive listing and evaluating of all permutations can be extremely labor-intensive. Before the advent of powerful computers, actual data were therefore first converted to ranks, which were then permutated. This generally results in a loss of statistical power (the ability to correctly reject a false H_0). With today's powerful microcomputers, actual data can be used. An extremely useful program, which allows reproducing almost all parametric and nonparametric tests, and the definition and evaluation of nonconventional test statistics, is Resampling Stats (www.resample.com). It is no longer available as stand-alone program, but can be purchased as add-in for Excel for Windows. An essentially identical giftware program, Statistics101, can be downloaded at http://statistics101.net.A brief introduction is given in Sect. 59.7 of this chapter.

Permutation tests are based on sampling without replacement, i.e., each collected value is used only once in a new "pseudosample" or "resample." Bootstrapping techniques use sampling with replacement. This means that collected values can occur more than once (Efron and Tibshirani 1993). Estimating confidence intervals and p-values by bootstrapping has become widespread in DNA sequence analyses (Chap. 62).

4 Modifications by Neyman and Pearson

Fisher's approach was expanded and modified by Neyman and Pearson (1933). In addition to H_0, at least one alternative hypothesis, H_A, is formulated. We proceed as follows:

1. Formulate a null hypothesis (H_0). For example, we propose that the final body weights of two groups of animals on different diets differ by no more than 5 mg.
2. Formulate at least one alternative hypothesis (H_A). For example, we propose that the final weights differ by at least 5 mg.
3. Define a test statistic characterizing the difference between the two groups. Measure the actual value of this statistic.
4. Generate the probability distribution of this test statistic, assuming H_0 is correct and data are normally distributed.
5. Define a critical value of α. This again defines the probability of falsely rejecting H_0 (Table 59.1).
6. In addition, define a probability β of committing a Type II error. This is the probability of falsely accepting H_0. The term $(1 - \beta)$ defines the probability of correctly accepting H_A; it is also called the power of the test (Cohen 1988).
7. Calculate p of the observed test statistic. If $p \leq \alpha$, reject H_0 and accept H_A. If $p \geq \alpha$, accept H_0 and reject H_A.

The Neyman-Pearson approach forces us to make a decision between two specified hypotheses. Depending on the costs or benefits of Type I and Type II errors, we will adjust the critical difference between H_0 and H_A, and α and β levels. For example, the commercial success of new drugs will depend on manufacturing costs and improved effectiveness (customers may be willing to pay double the price, if the drug is twice as effective, but not if the improvement is only 5%). In medical diagnosis, we must strike a balance between Type I errors (diagnosing a disease where none exists, false positive) and Type II errors (missing an existing disease, false negative). In law, we have to balance the strength of the evidence (effect size) against the potential harm of letting the guilty walk free (Type II) or wrongfully convicting the innocent (Type I).

Some textbooks define H_A simply as the opposite of H_0, i.e., H_0: difference is 0, and H_A, difference is not 0. This is not very useful, since it does not allow us to estimate β and $(1 - \beta)$. In general terms, the power of a test increases with sample size, with effect size (difference between competing hypotheses), and decreases with the variability of data. Given sufficiently large samples, p will almost always fall below any specified α. A significance test by itself, without information on the effect size and its confidence limits, is therefore considered to be meaningless by many statisticians.

The free program G∗power is useful for planning experiments (Faul et al. 2007). For example, given α, β, and effect size, it estimates the number of samples necessary to detect a significant difference. Or, it estimates the power of an experiment, provided the other parameters are known. G∗power can be downloaded at http://www.gpower.hhu.de/en.html.

5 Bayesian Statistics

As discussed above, classical hypothesis testing gives a p value that describes the probability of a value of the test statistic that is at least as extreme as the one found, provided H_0 is correct. But intuitively, we are more interested in the probability that H_0 is correct. It is a common mistake to assume that the two probabilities are identical. A simple example can illustrate this fallacy: scarlet fever is due to beta-hemolytic streptococcal bacteria. It usually causes a red rash, particularly on neck and chest. Suppose we observe a woman with a red rash. Applying statistical reasoning, we propose the null hypothesis that she does not have scarlet fever. If this is true, how likely is it she will have a red rash? Let us assume that a red rash occurs in 3 out of 100 of randomly selected women ($p = 0.03$). We might be tempted to reject H_0 and conclude that the woman has indeed been infected with scarlet fever. But the more relevant question surely is: if we observe a woman with a red rash, how likely is it that this rash is due to scarlet fever? To answer this, we would have to know all potential causes of a rash, which may include eczemas, pregnancy, allergies, measles, scarlet fever, and their relative contributions to rashes in the population. The contribution of scarlet fever to all rashes may be as low as 1%. The probability that the woman suffers from scarlet fever is therefore 0.01, and we should not reject H_0.

The Bayesian approach to statistics allows such direct probability statements. For example, we can estimate the probability that a new treatment is more effective than a control. We do this by modifying an initial estimate of this probability, which we base on prior experience or our intuition (for easy-to-follow introductions to this topic, see Berry 1996; Hilborn and Mangel 1997).

Bayes formula (Bayes 1763) allows us to determine conditional probabilities: What is the probability of event A occurring given that event B has occurred, or, in mathematical notation $p(A|B)$? A common application is an estimate of the accuracy of medical diagnoses. Assume that a virus has infected 1 out of every 1000 people. This is the a priori probability: if we randomly chose a person, the probability that he or she is infected is 1 in 1000, or 0.001.

A diagnostic test has been developed, which correctly identifies 99 out of 100 patients that have the infection. One out of 100 is incorrectly classified as noninfected. This is called a "false negative." The same test, applied to noninfected persons, correctly identifies 98 as healthy but gives a wrong result in 2 cases out of 100; this person, though healthy, is diagnosed as being infected. This is a "false positive."

Now assume that we give a test to a randomly chosen individual in the population, and it turns out to be positive. What is the probability that the person is actually infected? This is the a posteriori probability.

Bayes (1763) developed the following formula:

$$p(A \mid B) = \frac{p(A) \cdot p(B \mid A)}{p(A) \cdot p(B \mid A) + p(\text{not}A) \cdot p(B \mid \text{not}A)} \tag{59.1}$$

where

$p(A)$ = probability of A
$p(A|B)$ = probability of A given that B has occurred
$p(B|A)$ = probability of B given that A has occurred
$p(\text{not}A)$ = probability that A has not occurred
$p(B|\text{not}A)$ = probability of B, given that A has not occurred

In our example, $p(A|B)$ is the probability we are looking for: How likely is it that the person is infected, given the test was positive?

$p(A)$ stands for the probability of infections. In the population, it is 0.001.
$p(B|A)$ stands for the probability of a positive test given an infection. It is 0.99
$p(\text{not}A)$ is the probability of not being infected. It is (1–0.001) = 0.999
$p(B|\text{not}A)$ is the probability of a positive test in the absence of an infection = 0.02

We get the following result:

$$p(A \mid B) = \frac{0.001 \cdot 0.99}{0.001 \cdot 0.99 + 0.999 \cdot 0.02} = \frac{0.00099}{0.02097} = 0.0472.$$

Surprisingly, the probability of infection is less than 5%! Nevertheless, this a posteriori probability is much higher than the a priori probability of 0.001. We have modified it by experience.

The same result can be found with the following, more intuitive approach. Assume we are dealing with a population of 1 million and test every single individual. The results are summarized in Table 59.2, which also reveals the similarity of statistical tests to medical diagnosis. Similar examples could be constructed based on criminal trials (H_0 would correspond to innocence).

Of all positive tests (20,970), 990 were true positives; therefore, $p(A|B) = 990/20970 = 0.0472$. We can apply this Bayesian perspective to interpreting statistical significance. Imagine that we are testing drugs for their ability to lower blood pressure. Based on the acceptable improvement and variability of the data, we have chosen a sample size that gives a power $(1 - \beta)$ of 0.8, and we are willing to accept an α of 0.05. We run the test, and p is indeed <0.05. How likely it is that rejection of H_0 is the correct decision? The short answer is: it depends on how

Table 59.2 Bayesian analysis applied to medical diagnosis

	Infection		Total
Diagnosis	Absent	Present	
No infection	979,020 True negatives	10 False negatives	979,030
Infection	19,980 False positives	990 True positives	20,970
	999,000	1000	1000,000

False negatives and false positives correspond to Type II and Type I errors, respectively much we already know about the drug. How likely do we think a positive result is going to be? Then we apply the same reasoning outlined above. The answer can be found at: https://www.graphpad.com/guides/prism/7/statistics/index.htm?a_bayesian_perspective_on_interpreting_statistical_significance.htm. Motulsky (1995), Bärlocher (2008) and Brophy and Joseph (1995) provide more examples of how prior knowledge can modify *p* values determined from an experiment. For a more thorough introduction to Bayesian statistics, consult Berry (1996), Hilborn and Mangel (1997), and Howson and Urbach (1993). Potential applications to ecological studies have been reviewed by Ellison (2004).

6 Model Selection

Model selection replaces the traditional testing of a null hypothesis by confronting collected data with several competing models. The relative support of the data for each model is determined, allowing ranking and weighting of the models, and measuring the relative support for several competing hypotheses. Where similar levels of support are found, model averaging can provide robust parameter estimates and predictions. Model selection is based on likelihood theory and has been widely used and accepted in molecular systematics and mark-recapture analysis. Other potential applications in ecology and evolution are discussed by Hilborn and Mangel (1997) and Johnson and Omland (2004).

7 Resampling Stats

7.1 Introduction

Resampling Stats or its giftware equivalent Statistics 101 are extremely useful and powerful programs to evaluate probabilities. They allow the reproduction of almost all classical tests, which generally assume normality, as well as nonparametric or distribution-free tests based on ranks or using actual data. They can also be used for bootstrapping or Monte Carlo techniques. More information is provided at www.resamle.com and http://www.statistics101.net.

The basic idea is to generate or introduce numbers into a so-called vector, which is given a name. Whenever we manipulate these numbers (e.g., by shuffling them, determining their average, etc.), the result is placed in a new vector, which has to be given a different name.

The Print command allows to check whether our commands are doing what we want them to do. Print commands should be deleted before executing the final program.

7.2 Some Commands in Resampling Stats/Statistics101

- Data (1 2 3 4) A Places the numbers 1, 2, 3 and 4 into vector A
- Shuffle A B Shuffles numbers in A, places them in B
- Print B Prints the result on screen when you select Run

The same result can be found by writing:

- Shuffle (1 2 3 4) A
- Print A

To generate 1000 numbers of a normal distribution with an average of 2 and SD of 1, we write:

- Normal 1000 2 1 A

To prepare a histogram of the data in A, we write:

- Normal 1000 2 1 A
- Histogram A

To generate 10 numbers of a uniform distribution between 0 and 1, we write:

- Uniform 10 0 10 A

The next step calculates 2.5 and 97.5 percentiles of values in A (essentially values that enclose the central 95%; values outside this range correspond to the familiar $p = 0.05$).

- Normal 1000 2 1 A
- Histogram A
- Percentile A (2.5 97.5) B
- Print B

To take two random samples with replacement from 1, 3, 4, 5, 7, we write:

- Data (1 3 4 5 7) A places data in vector A
- Sample 2 A B takes two random samples from A and places them in B
- Print B shows which two numbers were chosen

In addition, there are numerous statistical commands, such as Boxplot, Corr, Exp, Mode, Min, Regress, etc.

7.3 List of Commands in Resampling Stats/Statistics101

Basic Commands

ADD	Adds the elements of two vectors together
CONCAT	Combines two or more vectors into one long one
COPY	Records data or copies vectors; synonym for DATA and NUMBERS

COUNT	Determines frequency of a particular number or range of numbers
DATA	Enter data; synonym for COPY and NUMBERS
DEDUP	Eliminates duplicate values
DIVIDE	Divides the contents of one vector by another
END	Ends a loop, sends you back to a "REPEAT" statement
GENERATE	Produces the desired quantity of random integers within a defined range
HISTOGRAM	Produces a histogram of trial results
IF	Succeeding commands execute only when IF condition holds
MEAN	Calculates the mean of a vector
MULTIPLES	Determines the frequency of multiplicates of a number
MULTIPLY	Multiplies the elements in one vector by those in another
NUMBERS	Enter data or define a variable; synonym for DATA and COPY
PERCENTILE	Calculates the xth percentile of a frequency distribution
PRINT	Specifies output to be shown on screen
RANDOM	Produces the desired quantity of random integers within the desired range
REMARK	Allows user to insert a remark. Precede remarks with an apostrophe.
REPEAT	Allows user to repeat a simulation up to 15,000 times
RUNS	Calculates the number of runs of a given length
SAMPLE	Samples with replacement
SCORE	Allows user to keep track of the result of each simulation
SHUFFLE	Randomizes the elements in a vector
SORT	Takes a specified number of elements from a vector and creates a new vector
URN	Creates a vector of specified quantities of specified numbers
WEED	Removes specified numbers or a specified range

Additional Mathematical and Statistical Commands

ABS	Computes absolute value of each element in a vector
BOXPLOT	Produces a Box plot
CORR	Calculates a correlation coefficient
EXP	Raises Euler's number, *e*, to the power of each vector element
EXPONENTIAL	Generates numbers from an exponential distribution
LET	Allows arithmetic expressions of the form LET x = a + b
LOGNORMAL	Generates numbers randomly from a lognormal distribution
MAX	Identifies the maximum value in a vector
MEDIAN	Calculates the median of a vector
MIN	Identifies the minimum value in a vector
MODE	Identifies the most frequent value in a vector
NORMAL	Generates numbers randomly from a normal distribution

PARETO	Generates numbers randomly from a Pareto distribution
POISSON	Generates numbers randomly from a Poisson distribution
POWER	Raises each element in the first of two vectors to the power of the corresponding element in the second
REGRESS	Runs a multiple linear regression
RANKS	Computes the ranks of elements in a vector
ROUND	Rounds each element to an integer
SQRT	Determines the square root of each element in a vector
SQUARE	Squares each element in a vector
STDEV	Calculates the standard deviation of the numbers in a vector
SUMABSDEV	Sums the absolute deviations of one vector from another
SUMSQRDEV	Sums the squared deviations of one vector from another
TIMEPLOT	Plots a vector sequentially along the x-axis of a biplot
UNIFORM	Produces random values from a continuous uniform distribution
VARIANCE	Finds the variance of the elements in a vector
WEIBULL	Generates numbers randomly from a Weibull distribution

Additional Resampling and Housekeeping Commands

CLEAR	Erases contents of a vector
FUZZ	Sets precision for comparisons
MAXSIZE	Sets maximum vector size (no longer needed in Statistics101)
PAUSE	Causes program execution to pause
PROGINFO	Prints program variables, constants, and status information to the output window
READ	Reads a file into one or more result variables (vectors)
RECODE	Changes certain elements of a vector to specified value
SEED	Sets the random number generator seed
SET	To fill a vector with one value
SIZE	Counts number of elements in input vector
WHILE	Conditional repeat
WRITE	Exports data to an ASCII file

7.4 Three Simple Examples (in Statistics101, Minor Modifications Are Needed in Resampling Stats)

Bootstrap Estimates of Confidence Intervals

Assume we wish to estimate the average height of a population of school children. We take a random sample of 15, and find the following values: 152, 140, 148, 134, 131, 156, 162, 150, 138, 153, 145, 153, 167, 143, and 130. The traditional method to estimate 95% confidence limits (CL) assumes normally distributed values; the mean and variance are estimated from the data. We get an average of 146.8, and the following confidence limits: 140.7 and 152.9.

With the bootstrap approach, we assume that the collected data are a true representation of the entire population. To reproduce this population, we simply multiply each measured value with a huge number (to approximate sampling with replacement); we then take a random sample from this pseudopopulation. We determine the average of this bootstrap sample. We do this many times, keeping track of all averages. At the end, we determine the values that enclose the central 95%; these represent bootstrapped confidence limits. Instead of multiplying the collected data with a large number, we can simply sample with replacement. Thus:

```
REPEAT 100000
    Sample 15 (152140148134131156162150138153145153167143130) A
    Mean A B
    SCORE B C
END
Mean C Aver
Print Aver
Percentile C (2.5 97.5) CL
Print CL
```

A sample run gave an average of 146.8, and confidence limits of 141.3 and 152.3, remarkably close to the theoretical values. One important difference is that bootstrapped values can never go beyond the values that have actually been measured; the lowest possible average of any pseudosample would therefore be 130. No such restriction exists when we assume that data are normally distributed; theoretically, some children would be 1 cm, others 500 cm, tall.

Comparison of Two Groups

Assume we test the effect of a fertilizer on plant growth. We have 10 replicates each; the first column gives control data (no fertilizer), and the second column gives data with fertilizer:

44	55
56	47
58	63
34	62
49	49
61	63
56	73
43	68
53	59
49	48

With an unpaired *t*-test, we find a p value of 0.04 ($t = 2.205$, 18 degrees of freedom, two-tailed test). To test our null hypothesis with Resampling Stats, we combine all 20 data (H_0 postulates that they belong to the same population). We randomly subdivide the 20 data into two groups of 10, and determine the difference between the two group averages. We do this many times. This gives us the distribution of all possible differences. The original difference of the two groups was 8.4. How likely is it that by random redistribution of the original 20 values among two groups, we find a difference that is at least as extreme as 8.4 (if we are looking at a two-tailed test, this means ≥8.4 or ≤ −8.4)?

```
REPEAT 10000

    Shuffle (44 55 56 47 58 63 34 62 49 49 61 63 56 73 43 68 53 59 49 48) A
    Take A 1, 10 B
    Mean B C
    Take A 11, 20 D
    Mean D E
    Subtract C E Dif
    Score Dif Difs

END
Count Difs > = 8.4 high
Count Difs <= −8.4 low
Add high low Tot
Print Tot
```

A sample run gave a value of 414, i.e., 414 out of 10,000 runs reached or exceeded the difference of the original data. This corresponds to a p value of 0.04 (414/10000), we would again reject H_0.

Analysis of Feeding Preferences

We wish to compare consumption of 3 food items using the experimental design presented in Chap. 52. Table 59.3 lists the daily consumption values of items A, B, and C in 10 replicates.

The grand average is $(3.46 + 1.61 + 0.31)/3 = 1.79$. As test statistic S, we choose the sum of the squared deviations of the three measured consumption rates from the grand mean, i.e., $(3.46 - 1.79)^2 + (1.61 - 1.79)^2 + (0.31 - 1.79)^2$. For the original data, this gives a value of 5.033. Next, we estimate the probability of finding a value of S at least as extreme as 5.033 (≥5.033) if the assignment of the measured consumption rates to the three food items in each container were random, i.e., if there were no consistent preferences for one food item over the other. For each shuffled data set, we calculate the value of S. We do this 10,000 times, which gives a reasonable approximation of the distribution of all possible S values. What proportion of this distribution has a value of ≥5.033? This proportion is equivalent to the classical definition of the p value. In the current example, this happens approximately 1–4

Table 59.3 Consumption (mg) of food items A, B, and C in 10 replicate containers

Replicate	A	B	C
1	3.57	2.35	0.48
2	0.35	1.87	0.40
3	3.14	0.31	0.67
4	7.17	2.29	0.28
5	3.24	3.46	0.35
6	3.07	1.11	0.17
7	5.69	0.55	0.03
8	4.45	0.85	0.34
9	1.48	1.47	0.12
10	2.48	1.85	0.22
Average	3.46	1.61	0.31

times in 10,000 trials; therefore, $p \approx 0.0002$. With Resampling Stats, the test can be run as follows:

```
REPEAT 10000 'random arrangement of data is run 10,000 times
SHUFFLE (3.57 2.35 0.48) R1 'introduce and shuffle data from first replicate
SHUFFLE (0.35 1.87 0.40) R2 'introduce and shuffle data from second replicate
SHUFFLE (3.14 0.31 0.67) R3
SHUFFLE (7.17 2.29 0.28) R4
SHUFFLE (3.24 3.46 0.35) R5
SHUFFLE (3.07 1.11 0.17) R6
SHUFFLE (5.69 0.55 0.03) R7
SHUFFLE (4.45 0.85 0.34) R8
SHUFFLE (1.48 1.47 0.12) R9
SHUFFLE (2.48 1.85 0.22) R10

TAKE R1 1 A1 'take first number from first replicate, put it in A1'
TAKE R2 1 A2 'take first number from second replicate, put it in A2'
TAKE R3 1 A3
TAKE R4 1 A4
TAKE R5 1 A5
TAKE R6 1 A6
TAKE R7 1 A7
TAKE R8 1 A8
TAKE R9 1 A9
TAKE R10 1 A10

CONCAT A1 A2 A3 A4 A5 A6 A7 A8 A9 A10 A 'put all first numbers in A'
MEAN A AVA 'calculate average consumption of A in10 containers'

TAKE R1 2 B1 'take second number from first replicate, put it in B1'
TAKE R2 2 B2 'take second number from second replicate, put it in B2'
TAKE R3 2 B3
```

```
TAKE R4 2 B4
TAKE R5 2 B5
TAKE R6 2 B6
TAKE R7 2 B7
TAKE R8 2 B8
TAKE R9 2 B9
TAKE R10 2 B10

CONCAT B1 B2 B3 B4 B5 B6 B7 B8 B9 B10 B 'put all second numbers in B'
MEAN B AVB 'calculate average consumption of B in 10 containers'

TAKE R1 3 C1 'take third number from first replicate, put it in C1'
TAKE R2 3 C2 'take third number from first replicate, put it in C2'
TAKE R3 3 C3
TAKE R4 3 C4
TAKE R5 3 C5
TAKE R6 3 C6
TAKE R7 3 C7
TAKE R8 3 C8
TAKE R9 3 C9
TAKE R10 3 C10
CONCAT C1 C2 C3 C4 C5 C6 C7 C8 C9 C10 C 'put all third numbers in C'
MEAN C AVC 'calculate average consumption of C in 10 containers'

CONCAT AVA AVB AVC AVS 'combines average consumption values'
SUBTRACT AVS 1.79 DEVS 'determines deviations from grand mean'
SQUARE DEVS SDEVS 'squares deviations'
SUM SDEVS SS 'sums squared deviations'

SCORE SS SSS 'store all values of test statistic S in SSS
END
COUNT SSS > = 5.033 RESULT 'counts all S values ≥5.033, puts it in RESULT
PRINT RESULT 'prints how often S ≥ 5.033; this number/10000 is equivalent to p.
```

In individual runs, this number generally varies between 0 and 4. An average could be determined by introducing another Repeat loop, storing the values for Result in a new vector and determining its average.

References

Bärlocher, F. (2008). *Biostatistik* (2nd ed.). Stuttgart: Thieme-Verlag.

Bayes, T. (1763). An essay towards solving a problem in the doctrine of chances. *Philosophical Transactions of the Royal Society of London, 66*, 423–437.

Berry, D. A. (1996). *Statistics: A Bayesian perspective*. Belmont: Duxbury Press.

Brophy, J. M., & Joseph, L. (1995). Placing trials in context using Bayesian analysis. *Journal of the American Medical Association, 273*, 871–875.

Cohen, J. (1988). *Statistical power analysis for the behavioral sciences*. Hillsdale: Lawrence Erlbaum.
Edginton, E. S. (1987). *Randomization tests*. New York/Basel: Marcel Dekker.
Efron, B., & Tibshirani, R. (1993). *Introduction to the bootstrap*. New York: Chapman & Hall.
Ellison, A. M. (2004). Bayesian inference in ecology. *Ecology Letters, 7*, 509–520.
Faul, F., Erdfelder, E., Lang, A.-G., & Buchner, A. (2007). G*power 3: A flexible statistical power analysis program for the social, behavioral, and biomedical sciences. *Behavior Research Methods, 39*, 175–191.
Friedman, M. (1966). *Essays in positive economics*. Chicago: University of Chicago Press.
Gigerenzer, G., Swijtink, Z., Porter, T., Daston, L., Beatty, J., & Krüger, L. (1989). *The empire of chance*. Cambridge: Cambridge University Press.
Good, P. I. (1994). *Permutation tests: A practical guide to resampling methods for testing hypotheses*. New York: Springer.
Good, P. I. (1999). *Resampling methods*. Basel: Birkhäuser.
Hilborn, R., & Mangel, M. (1997). *The ecological detective*. Princeton: Princeton University Press.
Howson, C., & Urbach, P. (1993). *Scientific reasoning: The Bayesian approach*. Peru: Open Court Publishing.
Johnson, J. B., & Omland, K. S. (2004). Model selection in ecology and evolution. *Trends in Ecology & Evolution, 19*, 101–108.
Kahnemann, D., Slovic, A., & Tversky, A. (1982). *Judgment under uncertainty: Heuristics and biases*. Cambridge: Cambridge University Press.
Manly, B. F. J. (1997). *Randomization, bootstrap and Monte Carlo methods in biology*. London: Chapman & Hall.
Mayer-Schönberger, V., & Cukier, K. (2013). *Big data: A revolution that will transform how we live, work, and think*. Boston/New York: Houghton Mifflin Harcourt.
Meehl, P. E. (1978). Theoretical risks and tabular risks: Sir Karl, Sir Ronald, and the slow progress of soft psychology. *Journal of Consulting and Clinical Psychology, 46*, 806–834.
Montello, D. (2015). To the editors. *American Scientist, 103*, 244.
Motulsky, H. (1995). *Intuitive biostatistics*. Oxford: Oxford University Press.
Neyman, J., & Pearson, E. S. (1933). On the problem of the most efficient tests of statistical hypotheses. *Philosophical Transactions of the Royal Society of London, Series A, 231*, 289–337.
Paulos, J. A. (1995). *A mathematician reads the newspaper*. New York: Doubleday.
Pinker, S. (2002). *The blank slate*. New York: Viking-Penguin.
Popper, K. R. (1935). *Logik der Forschung*. Vienna: Julius Springer.
Shipley, B. (2000). *Cause and correlation in biology*. Cambridge: Cambridge University Press.
Tukey, J. W. (1977). *Exploratory data analysis*. Reading: Addison-Wesley.
Westfall, P. H., & Young, S. S. (1993). *Resampling-based multiple testing*. New York: Wiley.
Zar, J. H. (1996). *Biostatistical analysis*. Upper Saddle River: Prentice-Hall.

Chapter 60
Determining Temperature-Normalized Decomposition Rates

Mark O. Gessner and Frank Peeters

Keywords Litter mass loss · Decay rate · Temperature correction · Temperature sensitivity · Degree-day model · Q_{10} · Arrhenius model · Metabolic theory of ecology · Negative exponential decay model · Regression analysis

1 Introduction

Temperature is the most universal factor governing rates of processes ranging from chemical reactions (Arrhenius 1889) and metabolic activities (Robinson et al. 1983; Allison et al. 2010) to transformations of organic matter in ecosystems (Davidson and Janssens 2006; Graça et al. 2015). Attention to accounting for temperature effects on physiological and ecological processes has continuously grown, prompted in part by theoretical interest (Allen et al. 2005), comparative considerations of latitudinal patterns (Irons et al. 1994; Boyero et al. 2011; Follstad Shah et al. 2017), and the challenge of quantifying and forecasting the ecological consequences of climate warming on carbon and nutrient cycling (Davidson and Janssens 2006; Allison et al. 2010). This includes efforts devoted to assessing the influence of temperature on the decomposition of plant litter in both terrestrial (e.g., Fierer et al. 2005; Prescott 2010) and aquatic ecosystems (e.g., Petersen and Cummins 1974; Hietz 1992; Ferreira and Canhoto 2015).

Electronic Supplementary Material The online version of this chapter (https://doi.org/10.1007/978-3-030-30515-4_60) contains supplementary material, which is available to authorized users.

M. O. Gessner (✉)
Department of Experimental Limnology, Leibniz Institute of Freshwater Ecology and Inland Fisheries (IGB), Stechlin, Germany

Department of Ecology, Berlin Institute of Technology (TU Berlin), Berlin, Germany
e-mail: gessner@igb-berlin.de

F. Peeters
Limnological Institute, University of Constance, Konstanz, Germany
e-mail: frank.peeters@uni-konstanz.de

F. Bärlocher et al. (eds.), *Methods to Study Litter Decomposition*,
https://doi.org/10.1007/978-3-030-30515-4_60

The process of plant litter decomposition has been typically described by first-order kinetics, where litter mass loss at any given point in time is assumed to be proportional to the litter mass present, independent of temperature (Chap. 6):

$$\frac{dm}{dt} = -k \cdot m \tag{60.1}$$

where m is the litter mass, t is time, and k is the decay rate coefficient. Dividing by m and integrating both sides of the equation over time t (i.e., from the initial time $t_0 = 0$ till t_n = elapsed time) results in the familiar negative exponential decay model:

$$\int_{tn}^{t_0} \frac{1}{m}\frac{dm}{dt}dt = \int_{t_n}^{t_0} -k \cdot dt \tag{60.2}$$

$$\Rightarrow \ln\left(m(t)\right) - \ln\left(m_0\right) = -k \cdot \left(t_n - t_0\right) \tag{60.3}$$

$$\Rightarrow m_t = m_0 \cdot e^{-k \cdot t} \tag{60.4}$$

where $m_t = m(t)$ and $m_0 = m(t = 0)$. Although not entirely realistic, this model has proved extremely useful in practice, one of the key advantages being that decomposition can be assessed based on a single parameter, the exponential decay rate coefficient k.

In aquatic environments, temperature-normalized decay rate coefficients have been reported almost invariably by assuming a linear temperature dependency of the overall decay rate (Table 63.1). Thus, the standard first-order decay model expands to

$$\frac{dm}{dt} = -k_T \cdot \frac{T(t)}{T_R} \cdot m, \tag{60.5}$$

where temperature $T(t)$ varies with time and k_T is the temperature-normalized decay rate coefficient. The reference temperature of normalization (T_R) ensures that k_T has the same dimension as k in Eq. 60.4 without temperature normalization. Setting $T_R = 1$ °C results in the same numerical value of k_T as in the degree-day models commonly used in the literature. However, other reference temperatures could be used, with 10 °C being particularly convenient, not least for ease of comparison and consistency with other models of temperature dependencies (see below). A requirement to employ this model is that $T \geq 0$ °C to ensure positive overall decomposition rates. Equation 60.5 can be solved exactly as above:

$$\int_{t_n}^{t_0} \frac{1}{m}\frac{dm}{dt}dt = \int_{t_n}^{t_0} -k_T \cdot \frac{T(t)}{T_R} \cdot dt \tag{60.6}$$

$$\Rightarrow \ln(m(t)) - \ln(m_0) = -k_T \cdot \frac{1}{T_R} \cdot \int_{t_n}^{t_0} T(t) \cdot dt \tag{60.7}$$

$$\Rightarrow m_t = m_0 \cdot e^{-k_T \cdot \frac{1}{T_R} \cdot \int_{t_n}^{t_0} T(t) \cdot dt} \tag{60.8}$$

The integral remaining in the exponent is the thermal sum commonly referred to as degree days (e.g., Irons et al. 1994; Woodward et al. 2012). It can also be viewed as the average temperature ($\bar{T}$) during the considered period of decomposition times the duration of this period ($t_n - t_0$):

$$\int_{t_n}^{t_0} T(t) \cdot dt = \bar{T} \cdot (t_n - t_0) = \bar{T} \cdot t_n \tag{60.9}$$

when $t_0 = 0$. In cases where decomposition is very slow, a more convenient unit for thermal sums can be degree years (Hietz 1992; Prescott 2010).

The same basic rationale can be used when an exponential instead of a linear dependency between temperature and decomposition rate is assumed. Accordingly,

$$\frac{dm}{dt} = -k_{\exp} \cdot e^{c \cdot (T(t) - T_R)} \cdot m \tag{60.10}$$

where $k_{\exp}$ is the temperature-normalized decay rate coefficient at the reference temperature T_R (e.g., 10 °C as above) and c is a parameter that describes the sensitivity of the temperature response. If temperature during decomposition equals the reference temperature (i.e., $T(t) = T_R$), the exponent becomes 0, and since $e^0 = 1$, Eq. 60.10 becomes Eq. 60.5 with $k_{\exp} = k$. In other words, $k_{\exp}$ is the decay rate coefficient at the reference temperature. Note that the overall decomposition rate (dm/dt) decreases with decreasing temperature but does not cease at the freezing point, in contrast to the model assuming a linear temperature dependence (Eq. 60.5). Solving Eq. 63.10 results in the same kind of relationship as in Eqs. 60.4 and 60.8, the only difference being the way in which the influence of temperature is accounted for:

$$m_t = m_0 \cdot e^{-k_{\exp} \cdot \int_{t_0}^{t_n} e^{c(T(t) - T_R)} dt} \tag{60.11}$$

A common approach when considering such exponential temperature relationships with decay rate is to apply Q_{10} values, where Q_{10} indicates by which factor a process rate coefficient is increased when temperature is raised by 10 °C. This approach is standard in modeling soil organic matter turnover (e.g., Davidson et al. 2006) but has rarely been used in litter decomposition studies in aquatic environments (but see Hietz 1992). Q_{10} is often found or assumed to be about 2, meaning that process rates double when temperature increases by 10 °C. However, values of up to 3 are not uncommon and values <2 and >3 have also been observed. Just like in Eqs. 60.10

and 60.11, the Q_{10} approach assumes an exponential relationship between temperature and decomposition rate. As a result, Q_{10} values can be readily converted to the constant c, and vice versa, according to

$$Q_{10} = e^{c \cdot 10\,^{\circ}\mathrm{C}} \tag{60.12}$$

$$\Rightarrow c = \frac{\ln(Q_{10})}{10\,^{\circ}\mathrm{C}} \tag{60.13}$$

A Q_{10} of 2.0 thus corresponds to a c of 0.069 °C^{-1}, a Q_{10} of 2.5 corresponds to a c of 0.092 °C^{-1}, and a Q_{10} of 3 is equivalent to a c of 0.110 °C^{-1}. Consequently, Eq. 60.11 can be rewritten as:

$$m_t = m_0 \cdot e^{-k_{\exp} \cdot \int_{t0}^{tn} Q_{10}^{\frac{(T(t)-T_R)}{10\,^{\circ}\mathrm{C}}} dt} \tag{60.14}$$

This chapter describes procedures to analyze temperature-dependent litter decomposition, assuming both a linear and an exponential relationship between temperature and decomposition rate. Selected rate coefficients (k and k_T) from assessments made in various streams are presented in Table 60.1. All are based on the assumption that the temperature effect is linear. Similar examples based on exponential relationships have not been published for streams and rarely for other freshwater environments (Hietz 1992).

Table 60.1 Range of standard (k) and temperature-normalized decay rates (k_T) in streams

Leaf material	Environment	k (day^{-1})	k_T (day^{-1})[a]	Reference
10 species	Costa Rica, latitude 10°N	0.0200–0.5586	0.008–0.280	1
10 species	Michigan, USA, latitude 46°N	0.0036–0.0684	0.004–0.040	1
10 species	Alaska, USA, latitude 65°N	0.0013–0.0259	0.052–0.852	1
Quercus robur	95 European streams, latitude 40–60°N, coarse-mesh bag	0.0019–0.0687	0.005–0.141	2
Quercus robur	95 European streams, latitude 40–60°N, fine-mesh bag	0.0016–0.0588	0.004–0.116	2
Alnus glutinosa	98 European streams, latitude 40–60°N, coarse-mesh bag	0.0045–0.2137	0.012–0.259	2
Alnus glutinosa	98 European streams, latitude 40–60°N, fine-mesh bag	0.0043–0.0468	0.007–0.146	2
Alnus glutinosa	4 alpine floodplain streams, coarse-mesh bag	0.0039–0.0305	0.024–0.181	3
Alnus glutinosa	4 alpine floodplain streams, fine-mesh bag	0.0029–0.0036	0.014–0.108	3

[a]Reference temperature T_R = 10 °C
1 = Irons et al. (1994), 2 = Woodward et al. (2012), 3 = Gessner et al. (1998)

2 Equipment, Materials, and Software

- Leaf litter, litter bags, drying oven, etc. to run decomposition experiment (see Chap. 6).
- Temperature data logger, calibrated and programmed to record data at hourly intervals during the expected duration of the decomposition study.
- Spreadsheet such as Excel and optionally statistical or mathematical software such as R or Matlab.

3 Procedures

3.1 Data Acquisition

1. Run a litter decomposition experiment in the field or laboratory or in mesocosms to determine litter mass remaining after at least five different time periods, spaced sufficiently to ensure that between 60% and 80% of the initial litter mass has been lost at the last collection date.
2. Before starting the experiment, program a calibrated temperature logger to record temperature at hourly intervals, deploy the data logger at the experimental site, ensuring that it is protected from direct sunlight or other heat sources.
3. Periodically collect replicate litter samples and determine the percent litter mass remaining.
4. At the final sampling date, also retrieve the logger and import the temperature data into a spreadsheet.

3.2 Linear Temperature Relationship

1. Calculate the thermal sums (degree days) reached by each of the sampling dates.
2. Construct a data sheet for regression analysis showing in separate columns the sample identifiers, elapsed time, thermal sums, percent litter mass remaining, and the natural logarithm of percent litter mass remaining (Online Appendix 1).
3. Run an ordinary least-squares regression analysis or, often superior, a nonlinear regression analysis using thermal sums divided by $T_R = 10$ °C as independent variable and, as dependent variable, the natural logarithm of percent litter mass remaining (least-squares regression; as illustrated in Online Appendix 1) or percent mass remaining (nonlinear curve-fitting, using statistical or mathematical software).
4. Ensure that the estimated initial litter mass is reasonably close to the theoretical value of 100%; otherwise consider fixing the intercept at 100% with the data

points for $t_n = 0$ omitted from the regression analysis in that case, or refrain from fitting the observational data to the model.

3.3 Exponential Temperature Relationship

1. Set an *a priori* value for the constant c in Eqs. 60.10 and 60.11 (e.g., $c = 0.069$ °C^{-1} for $Q_{10} = 2$) as well as a reference temperature T_R (e.g., 1 or 10 °C).
2. Calculate daily mean values of the temperature terms $e^{c \cdot (T - T_R))}$.
3. Compute the thermal integral reached by each of the sampling dates by summing the daily mean temperature terms from the start of the experiment to each of the sampling dates; these sums need to be multiplied by 1 day to be the formally correct integrals, although this does not change the numerical value.
4. As for the linear temperature relationship above, construct a data sheet for regression analysis showing in separate columns the sample identifiers, elapsed time, thermal integrals, percent litter mass remaining, and the natural logarithm of percent litter mass remaining (Online Appendix 1).
5. Run an ordinary least-squares regression analysis or, often preferable, a nonlinear regression analysis, using the thermal integrals as independent variable and, as dependent variable, the natural logarithm of percent litter mass remaining (least-squares regression; as illustrated in Online Appendix 1) or percent mass remaining (nonlinear curve-fitting; using statistical or mathematical software)
6. Repeat the calculations and regression analysis by testing different values for c, ranging, for example, from 0.0405 °C^{-1} ($Q_{10} = 1.5$) to 0.1099 °C^{-1} ($Q_{10} = 3.0$) to determine, by iteration, the best fitting parameter set m_0, k_{exp}, and c, as indicated by the highest coefficient of determination (r^2).
7. Ensure that the estimated initial litter mass is reasonably close to the theoretical value of 100%; otherwise, consider fixing the intercept at 100% with the data points for $t_n = 0$ omitted from the regression analysis in that case, or refrain from fitting the observational data to the model.
8. Alternatively, run regression analyses directly based on the relationship of Eq. 60.11 to fit m_0, k_{exp}, and c simultaneously, which requires mathematical or statistical software with a fitting function capable of integration or summation (Online Appendix 2).

4 Final Remarks

The linear or exponential temperature dependencies are only valid within a narrow range. This is particularly important in streams (and possibly other cold environments) where temperature optima of aquatic hyphomycetes, the main microbial decomposers of leaf litter, are low (Suberkropp 1984; Chauvet and Suberkropp 1998; Dang et al. 2009). Therefore, caution is needed when applying either linear or exponential temperature relationships to litter decomposition rates when ambient temperatures notably exceed 20 °C.

Types of temperature relationships other than linear or exponential can be modelled in the same way as outlined above. However, with more complex models (e.g., Dang et al. 2009), it may not be convenient to solve the equations and thus calculate decay rate coefficients, *k*, in a spreadsheet. Instead, the use of mathematical or statistical software such as R, Matlab, or others is recommended.

Several studies on litter decomposition have resorted to the Metabolic Theory of Ecology (Allen et al. 2005) to assess the influence of temperature on litter decomposition rates (Boyero et al. 2011; Follstad Shah et al. 2017), where the temperature dependence follows the Arrhenius law. Accordingly, the decay rate coefficient is expressed as a function of the inverse of absolute temperature (T_{abs}) in Kelvin and the Boltzmann constant, $B = 8.617 \cdot 10^{-5}$ eV K^{-1}, with the activation energy, E_A, being the fitted proportionality constant with a predicted value of 0.65 eV. Thus:

$$k = k_{\mathrm{Arr}} \cdot e^{-\frac{E_A}{B} \cdot \left(\frac{1}{T_{\mathrm{abs}}} - \frac{1}{T_{R,\mathrm{abs}}} \right)} \tag{60.15}$$

where k_{Arr} is the temperature-normalized decay rate coefficient for Arrhenius' law at the reference temperature $T_{R,\mathrm{abs}}$. The term $1/T_{R,\mathrm{abs}}$ is introduced to ensure that, similar to Eq. 60.11, k_{Arr} becomes *k* when decomposition occurs at the reference temperature, expressed here in degrees Kelvin. In essence, this approach uses yet another function for the temperature dependence of the decay rate coefficient. However, in the temperature range relevant for litter decomposition in natural environments, the temperature dependence of the function of Eq. 60.15 with $E_A \sim 0.65$ eV is very similar to the exponential (or Q_{10}) temperature dependence described above.

References

Allen, A. P., Gillooly, J. F., & Brown, J. H. (2005). Linking the global carbon cycle to individual metabolism. *Functional Ecology, 19*, 202–213.

Allison, S. D., Wallenstein, M. D., & Bradford, M. A. (2010). Soil-carbon response to warming dependent on microbial physiology. *Nature Geoscience, 3*, 336–340.

Arrhenius, S. (1889). Über die Reaktionsgeschwindigkeit bei der Inversion von Rohrzucker durch Säuren. *Zeitschrift für Physikalische Chemie, 4*, 226–248.

Boyero, L., Pearson, R. G., Gessner, M. O., Barmuta, L. A., Ferreira, V., Graça, M. A. S., Dudgeon, D., Boulton, A. J., Callisto, M., Chauvet, E., Helson, J. E., Bruder, A., Albariño, R. J., Yule, C. M., Arunachalam, M., Davies, J. N., Figueroa, R., Flecker, A. S., Ramírez, A., Death, R. G., Iwata, T., Mathooko, J. M., Mathuriau, C., Gonçalves, J. F., Jr., Moretti, M. S., Jinggut, T., Lamothe, S., M'Erimba, C., Ratnarajah, L., Schindler, M. H., Castela, J., Buria, L. M., Cornejo, A., Villanueva, V. D., & West, D. C. (2011). A global experiment suggests climate warming will not accelerate litter decomposition in streams but might reduce carbon sequestration. *Ecology Letters, 14*, 289–294.

Chauvet, E., & Suberkropp, K. (1998). Temperature and sporulation of aquatic hyphomycetes. *Applied and Environmental Microbiology, 64*, 1522–1525.

Dang, C. K., Schindler, M., Chauvet, E., & Gessner, M. O. (2009). Temperature oscillation coupled with fungal community shifts can modulate warming effects on litter decomposition. *Ecology, 90*, 122–131.

Davidson, E. A., & Janssens, I. A. (2006). Temperature sensitivity of soil carbon decomposition and feedbacks to climate change. *Nature, 440*, 165–173.

Davidson, E. A., Janssens, I. A., & Luo, Y. (2006). On the variability of respiration in terrestrial ecosystems: Moving beyond Q_{10}. *Global Change Biology, 12*, 154–164.

Ferreira, V., & Canhoto, C. (2015). Future increase in temperature may stimulate litter decomposition in temperate mountain streams: Evidence from a stream manipulation experiment. *Freshwater Biology, 60*, 881–892.

Fierer, N., Craine, J. M., McLaughlan, K., & Schimel, J. P. (2005). Litter quality and the temperature sensitivity of decomposition. *Ecology, 86*, 320–326.

Follstad Shah, J. J., Kominoski, J. S., Ardón, M., Dodds, W. K., Gessner, M. O., Griffiths, N. A., Hawkins, C. P., Johnson, S. L., Lecerf, A., LeRoy, C. J., Manning, D. W. P., Rosemond, A. D., Sinsabaugh, R. L., Swan, C. M., Webster, J. R., & Zeglin, L. H. (2017). Global synthesis of the temperature sensitivity of leaf litter breakdown in streams and rivers. *Global Change Biology, 23*, 3064–3075.

Gessner, M. O., Robinson, C. T., & Ward, J. V. (1998). Leaf breakdown in streams of an alpine glacial floodplain: Dynamics of fungi and nutrients. *Journal of the North American Benthological Society, 17*, 403–419.

Graça, M. A. S., Ferreira, V., Canhoto, C., Encalada, A. C., Guerrero-Bolaño, F., Wantzen, K. M., & Boyero, L. (2015). A conceptual model of litter breakdown in low order streams. *International Review of Hydrobiology, 100*, 1–12.

Hietz, P. (1992). Decomposition and nutrient dynamics of reed (*Phragmites australis* (Cav.) Trin. ex Steud.) litter in Lake Neusiedl, Austria. *Aquatic Botany, 43*, 211–230.

Irons, J. G., III, Oswood, M. W., Stout, R. J., & Pringle, C. M. (1994). Latitudinal patterns in leaf litter breakdown: Is temperature really important? *Freshwater Biology, 32*, 401–411.

Petersen, R. C., & Cummins, K. W. (1974). Leaf processing in a woodland stream. *Freshwater Biology, 4*, 343–368.

Prescott, C. E. (2010). Litter decomposition: What controls it and how can we alter it to sequester more carbon in forest soils? *Biogeochemistry, 101*, 133–149.

Robinson, W. R., Peters, R. H., & Zimmermann, J. (1983). The effects of body size and temperature on metabolic rate of organisms. *Canadian Journal of Zoology, 61*, 281–288.

Suberkropp, K. (1984). Effect of temperature on seasonal occurrence of aquatic hyphomycetes. *Transactions of the British Mycological Society, 82*, 53–62.

Woodward, G., Gessner, M. O., Giller, P. S., Gulis, V., Hladyz, S., Lecerf, A., Malmqvist, B., McKie, B. G., Tiegs, S. D., Cariss, H., Dobson, M., Elosegi, A., Ferreira, V., Graça, M. A. S., Fleituch, T., Lacoursière, J. O., Nistorescu, M., Pozo, J., Risnoveanu, G., Schindler, M., Vadineanu, A., Vought, L. B.-M., & Chauvet, E. (2012). Continental-scale effects of nutrient pollution on stream ecosystem functioning. *Science, 336*, 1438–1440.

Chapter 61
Biodiversity Analyses

Felix Bärlocher

Keywords Ecological functions · Rarefaction · Species evenness · Species extinction · Species heterogeneity · Species richness · Species-area curve

1 Introduction

There is great concern about the ongoing permanent loss of species. One important question is: How will this affect important aspects of ecosystem functioning? Ehrlich and Ehrlich (1981) wrote: "Ecosystems, like well-made airplanes, tend to have redundant subsystems and other 'design' features that permit them to continue functioning after absorbing a certain amount of abuse. A dozen rivets, or a dozen species, might never be missed. On the other hand, a thirteenth rivet popped from a wing flap, or the extinction of a key species involved in the cycling of nitrogen could lead to a serious accident." In recent years, a great number of studies have explored potential relationships between biodiversity and ecological functions, and tried to fit them into one of several models (Hooper et al. 2005; Cardinale et al. 2012). Most investigations dealt with plant species and terrestrial primary production (for reviews, see Hooper et al. 2005; Cardinale et al. 2012) but some also with decomposition (Wardle 2002; Srivastava et al. 2009; Gessner et al. 2010), with a focus on leaves (Swan and Palmer 2004; Frainer et al. 2015), stream invertebrates (Jonsson and Malmquist 2000), aquatic hyphomycetes (Bärlocher and Corkum 2003; Pascoal and Cássio 2008), or several trophic levels (Covich et al. 2004; Jabiol et al. 2013).

At its most basic level, diversity may simply be expressed by the number of distinct species it contains (Sect. 3.1). But clearly, a community where each species is represented by equal numbers is more diverse (more heterogeneous) than a

F. Bärlocher (✉)
Department of Biology, Mount Allison University, Sackville, NB, Canada
e-mail: fbaerloc@mta.ca

F. Bärlocher et al. (eds.), *Methods to Study Litter Decomposition*,
https://doi.org/10.1007/978-3-030-30515-4_61

community where one or a few species dominate. Both aspects are considered when quantifying diversity (Sect. 3.2).

Estimated biodiversity is most commonly based on number and proportions of species, but the accepted definition includes genetic variability within species and between populations, as well as diversity of ecosystems. An underlying objective is often an estimate of functional diversity, i.e., how many different traits are present in a habitat or ecosystem (Sect. 4).

One of the most popular definitions of a species is that of a group of actually or potentially interbreeding natural populations that are reproductively isolated from other groups of organisms. This criterion is difficult or impossible to apply in many microorganisms. Instead, an evolutionary or phylogenetic definition is often used. It focuses on monophyly: monophyletic groups are lineages that contain all known descendants of a single common ancestor. "Under the phylogenetic species concept, species are identified by estimating the phylogeny of closely related populations and finding the smallest monophyletic groups" (Herron and Freeman 2014). Phylogenies are increasingly based on DNA sequences, many of which are only known from environmental samples and have not been connected to pure cultures. These sequences are clustered according to their similarities to one another. Operational taxonomic units (OTUs) are defined based on a similarity threshold set by the researcher (generally 97–98%) and used as proxies for species. Applying DNA analyses to phylogenetics has greatly increased the estimates of extant species; Locey and Lennon (2016) predict that there may be as many as 1 trillion (10^{12}) microbial species, although others consider this number a vast overestimate (Amann and Rosselló-Móra 2016).

In laboratory studies, the number of species or OTUs can be controlled. This is generally not the case in field studies, where the number of species is unknown and has to be estimated. Sutherland (1996) and Krebs (2014) provide excellent advice and guidelines on how to conduct an unbiased census.

An important point to consider is the spatial component of diversity (Whittaker 1972). Landscapes typically contain a number of separate sites and habitats. Alpha-diversity refers to the local species pool (diversity of an individual site), generally interpreted as a mean across all relevant subunits. Gamma-diversity encompasses the diversity of the entire landscape (regional species pool), while beta-diversity represents differences in species composition among sites. Several formulations have been suggested for beta-diversity, e.g., as ratio between regional and local species diversity. Others include absolute or proportional species turnover. An important distinction is that beta-diversity can refer to directional variation along an environmental gradient or, to nondirectional variation among sample units in an area (Anderson et al. 2011; Socolar et al. 2016).

This chapter provides an introduction to some of the important concepts when measuring and comparing biological diversity.

2 Estimating Species Richness

2.1 Rarefaction

The larger a sample, the greater will be the expected number of species and the lower the evenness (Rosenzweig 1999). If we observe 88 species in a collection of 1500 individuals (community A) and 55 species in a collection of 855 individuals (community B), we do not readily know which community has more species. For a meaningful comparison, we have to standardize the sample size. We do this by a method called (individual-based) rarefaction, which was introduced by Sanders (1968). It answers the following question: If a sample had consisted of n instead of N individuals ($n < N$), how many species (s) would have been found? The largest sample in a collection is assumed to have S species distributed among N individuals; all rarefied samples have $n < N$ individuals and $s < S$ species. We can use the following formula to estimate the number of species in a sample of any size:

$$E\left(\hat{S}_n\right) = \sum_{i=1}^{S} \left[1 - \frac{\binom{N - N_i}{n}}{\binom{N}{n}} \right] \tag{61.1}$$

where:

$E(\hat{S}_n)$ = expected number of species in a random sample of n individuals
S = total number of species in entire collection
N_i = number of individuals belonging to species i
N = total number of individuals in collection
n = sample size chosen for standardization.

Alternatively, we can determine the expected number of species empirically by repeatedly taking subsamples of the size chosen for standardization. We can simulate this process by using a computer program, such as Resampling Stats or Statistics101 (Chap. 59). For example, if a sporulation experiment (Chap. 26) results in a filter with 1064 aquatic hyphomycete conidia and 8 species (Table 61.1), how many species would be expected in a sample of 250 individuals?

The formula 61.1 gives a value of 7.1. The same value can be estimated with the simple Statistics101 program listed below, which is giftware and can be downloaded at http://statistics101.net. It defines 8 species by assigning them numbers 1–8. The number of individuals belonging to each species is defined by urns: URN 550#1 25#2… implies 550 individuals of species 1, 25 of species 2, etc. The numbers are shuffled, and a sample of 250 is taken. All duplicates (i.e., identical numbers = individuals belonging to the same species) are removed, and the remaining numbers, corresponding to different species, are counted. This is repeated 10,000 times. The

Table 61.1 Fictitious result of a sporulation experiment

Species	Number of conidia
Anguillospora filiformis	550
Articulospora tetracladia	25
Clavariopsis aquatica	123
Flagellospora curvula	17
Heliscus lugdunensis	222
Lemonniera aquatica	120
Tetracladium marchalianum	5
Tumularia aquatica	2
Total of 8 species	1064

average gives the expected number of different species when a sample of 250 is taken. The commands to simulate rarefaction in Statistics101 are as follows:

```
URN 550#1 25#2 123#3 17#4 222#5 120#6 5#7 2#8 A
REPEAT 10000
SHUFFLE A B
TAKE B 1250 C
DEDUP C D
COUNT D>0 E
SCORE E F
END
MEAN F average
PRINT average
```

Individual-based rarefaction equalizes size bias by increasing all biases to match the worst bias, i.e., that of the smallest sample (Rosenzweig 1999). By contrast, sample-based rarefaction computes the expected number of species when m samples are drawn at random from a set of samples that are representative of an assemblage (Gotelli and Colwell 2011; Colwell et al. 2012). This preserves the spatial structure of the data.

An important goal may be an estimate of the total number of species in a region or ecosystem. This is often accomplished by assuming a specific underlying distribution of species and their abundances and generally relies on one or more samples. If a single sample is available, the total can be estimated by a number of estimators that are often based on the number of species represented by one or two individuals. Of these, Chao's estimator often performs reasonably well (Chao 1984), though numerous modifications have been suggested (Gotelli and Colwell 2011).

2.2 Species-Area Curve Estimates

Another way to estimate species richness is to extrapolate the species-area curve for the community. Since the number of species tends to rise with the area sampled, one can fit a regression line and use it to predict the number of species of any size. The species-area relationship generally has the following form:

$$S = c \cdot A^z \tag{61.2}$$

where:

S = number of species
c, z = constants
A = area

Provided we have several samples with known area and species number, we can estimate c and z by nonlinear curve-fitting with statistical (e.g., SYSTAT, SAS, R) or mathematical (e.g., MatLab, Mathematica) software. The samples can then be grouped based on a factor of interest (e.g., fungal conidia in streams bordered by different forest cover), to test whether the values of one group are consistently above or below the estimated species-area curve. Species-area curves have been applied to aquatic hyphomycetes by Gönczöl et al. (2001) and by Duarte et al. (2017). This method does not give reliable results for sparsely sampled sites.

2.3 Assuming an Upper Limit

Each habitat supports a limited number of species. We can estimate this upper limit by plotting the number of different species as function of the number of examined individuals or number of samples. The resulting curve often resembles a rectangular hyperbola or saturation-binding curve (also known as Monod or Michaelis-Menten type curve). Figure 61.1 shows the number of aquatic hyphomycete species found in a stream as a function of the number of monthly samples. In this particular example, the data closely resemble a hyperbola until Month 52, when the number of new species started to rise again. The estimated maximum number of 76 is therefore clearly too low in this case. Alternative methods to extrapolate to true richness in a habitat from a limited number of samples are discussed in Krebs (2014), Gotelli and Colwell (2011) and Colwell et al. (2012). They are based on various assumptions on how the "true" community is organized. Compared to conventional rarefaction approaches, they use all of the information contained in the samples.

Foggo et al. (2003) compared the performance of six techniques when estimating diversity of sandy beach macroinvertebrates; application of these techniques to aquatic hyphomycetes is discussed in Bärlocher (2005).

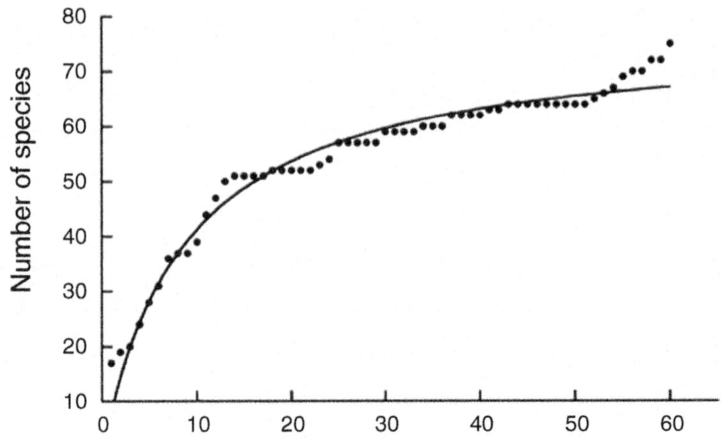

Fig. 61.1 Number of identified species with increasing number of samples. (Data from Bärlocher (2000), with nonlinear curve-fit to rectangular hyperbola)

3 Components of Species Diversity: Richness, Heterogeneity, and Evenness

3.1 Species Richness

In a community with 10 equally common species, two randomly collected individuals are unlikely to belong to the same species. In another community with 10 species, where 99% of all individuals belong to the same species, two random samples will likely recover the same species. Both communities have the same species richness, which is generally taken to be synonymous with number of species, but the first community is more heterogeneous.

3.2 Heterogeneity

Heterogeneity of a population contains two separate aspects: species richness and evenness. Simpson's index (1949) was the first attempt to combine the two in a single number; it is also known as the "repeat rate," since the index expresses the probability that two organisms selected at random from a population will "repeat" their classification, that is, that they belong to the same species. The repeat rate measure was first used in a text on cryptography (the science of analyzing and deciphering codes, ciphers, and cryptograms) in 1879 (Krebs 2014). For an infinite population, the repeat probability is given by:

$$D = \sum p_i^2 \tag{61.3}$$

where:

D = Simpson's index
p_i = proportion of species i in community

To convert this probability into a measure of diversity, usually, the complement of Simpson's index (1 – D) is taken:

Thus,

$$1 - D = 1 - \sum p_1^2 \tag{61.4}$$

Strictly speaking, this formula can only be used for infinite populations (Pielou 1969). For a finite population, the correct estimator is:

$$1 - D = 1 - \sum_{i-1}^{S} \left[\frac{n_i(n_i - 1)}{N(N-1)} \right] \tag{61.5}$$

where:

n_i = number of individuals of species i in sample
N = total number of individuals in sample
S = number of species in sample

Applying the formula to the data in Table 61.1 gives a Simpson diversity (1 – D) of 0.662.

The most popular measures of species diversity are based on information theory. The objective is to measure the amount of order (or disorder) present in a system. The underlying question is: How difficult would it be to predict correctly the species of the next individual collected? This informatics problem is the same as correctly predicting the next letter in a message. The uncertainty can be measured by the Shannon-Wiener function:

$$H' = \sum_{i=1}^{n} p_i \cdot \log_2 p_i \tag{61.6}$$

where:

H' = index of species diversity is the information content of sample (bits/individual)
S = number of species
p_i = proportion of total sample belong to the ith species

Information content measures uncertainty: The greater H', the greater is the uncertainty. A message such as BBBBB (or a community with only one species) has no uncertainty, and $H' = 0$. The Shannon-Wiener index should only be used on random samples from a large community in which the total number of species is known (Pielou 1969). If this is not the case, the Brillouin index is more appropriate (Krebs

2014). However, for large samples, the two indices give nearly identical results in practice. For example, for the data in Table 61.1, the Shannon-Wiener index is 1.955 and the Brillouin index is 1.930.

3.3 Evenness Measures

The literature on how to measure evenness (or equitability) is vast. Generally, one of the heterogeneity indices is scaled relative to the maximal value that is reached when each species is equally common. Two formulations are commonly used:

$$V = \frac{D}{D_{max}} \tag{61.7}$$

and

$$V' = \frac{D - D_{min}}{D_{max} - D_{min}} \tag{61.8}$$

where:

V, V' = evenness
D = observed index of diversity
D_{max} = maximum possible value of index, given S species and N individuals
D_{min} = minimum possible value of index, given S species and N individuals

The first expression is more commonly used, but the two converge for large samples.

A wide range of evenness indices has been proposed. Smith and Wilson (1996) prefer Simpson's, Camargo's, Smith and Wilson's, and Modified Nee's index; for an update, see Tuomisto (2012). All assume that the total number of species is known, which is almost never true. The evenness ratio is therefore generally an overestimate. Only Simpson's index is briefly introduced here. Simpson's measure of heterogeneity reaches a maximum when all species are equally abundant ($p = 1/s$); therefore:

$$D_{max} = \frac{1}{S} \tag{61.9}$$

It follows that the maximum possible value of the reciprocal of Simpson's index is always equal to the number of species observed in the sample. Simpson's index of evenness is therefore defined as:

$$E_{1/D} = \frac{1/D}{S} \quad (61.10)$$

where:

$E_{1/D}$ = Simpson's measure of evenness
D = Simpson's index
S = number of species in sample

This index is relatively little affected by rare species. For the data in Table 61.1, $E_{1/D}$ is 0.370.

4 Functional Traits and Diversity

An important rationale for protecting biodiversity is to safeguard ecosystem functions. Species richness may serve as proxy for ecological or functional diversity (Kinzig et al. 2001). The justification is that even though there is no direct effect of species number per se on ecosystem processes, the range of ecological functions expands with higher species numbers. On average, communities with more species are more productive, more resilient in the face of disturbances, and retain nutrients more efficiently. However, functional diversity based on species diversity assumes that all species are equally different (Petchey et al. 2009), while in fact, differences between species can have very specific and complex effects on diversity-function relationships. Furthermore, a species' impact is not necessarily correlated with its abundance; some are keystone actors or ecological engineers (Hooper et al. 2005).

Functional traits shape ecosystem properties and species' responses to environmental conditions; they may thus allow predictions as to how species or ecosystems respond to environmental change (Hooper et al. 2005; Petchey et al. 2009; Frainer et al. 2015). Early attempts to characterize functional diversity were based on life-forms (plants) or, more generally, guilds (Laureto et al. 2015). Cummins (1973) assigned stream invertebrates to functional feeding groups (shredders, collectors, scrapers, predators). Frainer et al. (2015) grouped leaves decaying in a stream based on a gradient of litter chemistry, while Schindler and Gessner (2009) used three decay rate categories (rapid, medium, slow) as indicators of functional diversity.

The grouping of species according to their functional traits seems at first glance a simple concept, but the actual practice of defining them and quantifying functional diversity can be difficult (Petchey et al. 2009). This is partly due to the inherent flexibility of many species. Thus, shredder invertebrates may also act as collectors or scrapers, or even as predators. Aquatic hyphomycetes (Chap. 25) are generally considered decomposers, and they have similar CAZymes (Carbohydrate-Active enZYmes). However, even within species, these and other degradative enzymes generally exist in multiple forms, which suggests some niche differentiation and therefore differences in their ecological traits (Bärlocher 2016).

Including several traits to characterize a species will emphasize its uniqueness, while using fewer traits will increase the likelihood of detecting redundancies among species (Petchey et al. 2009). Due to their inherent variability, quantitative (continuous) traits are generally preferred over qualitative (categorical) traits (Petchey et al. 2009; Laureto et al. 2015).

After selecting functional traits of a group of species, collected data have to be evaluated and compared. Petchey et al. (2009) discuss six measures, divided among three categories: measures that work directly on trait values, measures that work on the distance matrix, and measures that work on the functional dendrogram. Which measure is most appropriate has been hotly debated (e.g., Ricotta 2005), but Petchey et al. (2009) emphasize that this may be less important than the decision as to which traits should be included. They also point out that of 262 studies applying one of the diversity indices, only 16 have used empirical data.

5 Final Remarks

Krebs (2014) provides an excellent introduction to measurements and interpretations of diversity; he has also produced a computer program that automates many of the calculations mentioned in the text (https://ecological-methodology.software.informer.com/7.2).

A useful collection of free programs, EstimateS, has been provided by Robert K. Colwell (http://viceroy.colorado.edu/estimates; accessed August 14, 2017). These programs can be applied, for example, to calculate the diversity of aquatic hyphomycete spores (Chap. 26), of OTUs based on DNA sequence or other molecular analyses (Chaps. 34, 35, 36, 37, 38, and 40), or of invertebrates associated with decomposing leaves (Chap. 49).

A common objective in ecology is determining how environmental changes (e.g., eutrophication, warming) affect a community (e.g., invertebrates, fungi) in an ecosystem. This involves comparing data-rich samples listing the presence or abundances of numerous species, generally in combination with various environmental data. Analyses of such data sets, as well as measuring functional diversity, often benefit from multivariate statistical methods such as Principal Component Analysis, Correspondence Analysis, Discriminant Analysis, and many others, but discussion of these approaches is beyond the scope of this chapter. Henderson and Seaby (2008) provide an excellent introduction and overview of such techniques.

Beta-diversity, the component of regional diversity (gamma-diversity) that accumulates due to intersite differences between local species assemblages (alpha-diversity), can potentially assist in designing conservation strategies (Anderson et al. 2011; Socolar et al. 2016).

References

Amann, R., & Rosselló-Móra, R. (2016). After all, only millions? *MBio, 7*, e00999–e00916.

Anderson, M. J., Crist, T. O., Chase, J. M., Vellend, M., Inouye, B. D., Freestone, A. L., Sanders, N. J., Cornelle, H. V., Comita, L. S., Davies, K. F., Harrison, S. P. L., Kraft, N. J. B., Stegen, J. C., & Swenson, N. G. (2011). Navigating the multiple meanings of ß diversity: A roadmap for the practicing ecologist. *Ecology Letters, 14*, 19–28.

Bärlocher, F. (2000). Water-borne conidia of aquatic hyphomycetes: Seasonal and yearly patterns in catamaran brook, New Brunswick, Canada. *Canadian Journal of Botany, 78*, 157–167.

Bärlocher, F. (2005). Freshwater fungal communities. In J. Dighton, P. Oudemans, & J. White (Eds.), *The fungal community* (3rd ed.). New York: Dekker.

Bärlocher, F. (2016). Aquatic hyphomycetes in a changing environment. *Fungal Ecology, 19*, 14–27.

Bärlocher, F., & Corkum, M. (2003). Nutrient enrichment overwhelms diversity effects in leaf decomposition by stream fungi. *Oikos, 101*, 247–252.

Cardinale, B. J., Duffy, J. E., Gonzalez, A., Hooper, D. U., Perrings, C., Venail, P., et al. (2012). Biodiversity loss and its impact on humanity. *Nature, 486*, 59–67.

Chao, A. (1984). Non-parametric estimation of the number of classes in a population. *Scandinavian Journal of Statistics, 11*, 265–270.

Colwell, R. K., Chao, A., Gotelli, N. J., Lin, S.-Y., Mao, C. X., Chazdon, R. L., & Longino, J. T. (2012). Models and estimators linking individual-based and sample-based rarefaction, extrapolation and comparison of assemblages. *Journal of Plant Ecology, 5*, 3–21.

Covich, A. P., Austen, M. C., Bärlocher, F., Chauvet, E., Cardinale, B. J., Biles, C. L., Inchausti, P., Gessner, M. O., Dangles, O., Statzner, B., Solan, M., Moss, B. R., & Asmus, H. (2004). The role of biodiversity in the functioning of freshwater and marine benthic ecosystems: Current evidence and future research needs. *Bioscience, 54*, 767–775.

Cummins, K. W. (1973). Trophic relations of aquatic insects. *Annual Review of Entomology, 18*, 183–206.

Duarte, S., Cássio, F., Pascoal, C., & Bärlocher, F. (2017). Taxa-area relationship of aquatic fungi on deciduous leaves. *PLoS One, 12*, e0181545.

Ehrlich, P. R., & Ehrlich, A. H. (1981). *Extinction: The causes and consequences of the disappearance of species*. New York: Random House.

Foggo, A., Attrill, M. J., Frost, M. T., & Rowden, A. A. (2003). Estimating marine species richness: An evaluation of six extrapolative techniques. *Marine Ecology Progress Series, 248*, 15–26.

Frainer, A., Moretti, M. S., Xu, W., & Gessner, M. O. (2015). No evidence for leaf-trait dissimilarity effects on litter decomposition, fungal decomposers, and nutrient dynamics. *Ecology, 96*, 550–561.

Gessner, M. O., Swan, C. M., Dang, C. K., McKie, B. G., Bardgett, R. D., Wall, D. H., & Hättenschwiler, S. (2010). Diversity meets decomposition. *Trends in Ecology & Evolution, 25*, 372–380.

Gönczöl, J., Révay, A., & Csontas, P. (2001). Effect of sample size on the detection of species and conidial numbers of aquatic hyphomycetes collected by membrane filtration. *Archiv für Hydrobiologie, 150*, 677–691.

Gotelli, N. J., & Colwell, R. K. (2011). Estimating species richness. In A. E. Magurran & B. J. McGill (Eds.), *Frontiers in measuring biodiversity* (pp. 39–54). New York: Oxford University Press.

Henderson, P., & Seaby, R. (2008). *A practical handbook for multivariate methods*. Pennington/Lymington/Hants: Pisces Conservation.

Herron, J. C., & Freeman, S. (2014). *Evolutionary analysis* (5th ed.). Boston: Pearson.

Hooper, D. U., Chapin, F. S., III, Ewel, J. J., Hector, A., Inchausti, P., Lavorel, S., Lawton, J. H., Lodge, D. M., Loreau, M., Naeem, S., Schmid, B., Setälä, H., Synstad, A. J., Vandermeer, J., & Wardle, D. A. (2005). Effects of biodiversity on ecosystem functioning: A consensus of current knowledge. *Ecological Monographs, 75*, 3–35.

Jabiol, J., McKie, B. G., Bruder, A., Bernadet, C., Gessner, M. O., & Chauvet, E. (2013). Trophic complexity enhances ecosystem functioning in an aquatic detritus-based model system. *Journal of Animal Ecology, 82*, 1042–1051.
Jonsson, M., & Malmqvist, B. (2000). Ecosystem process rates increases with animal species richness: Evidence from leaf-eating, aquatic insects. *Oikos, 89*, 519–523.
Kinzig, A. P., Pacala, S. W., & Tilman, D. (Eds.). (2001). *The functional consequences of biodiversity*. Princeton: Princeton University Press.
Krebs, C. J. (2014). *Ecological methodology* (3rd ed.). Menlo Park: Addison-Welsey.
Laureto, L. M. O., Cianciaruso, M. V., & Samia, D. S. M. (2015). Functional diversity: An overview of its history and applicability. *Natureza & Conservçao, 13*, 112–116.
Locey, K. J., & Lennon, J. T. (2016). Scaling laws predict global microbial diversity. *Proceedings of the National Academy of Sciences of the USA, 113*, 5970–5975.
Pascoal, C., & Cássio, F. (2008). Linking fungal diversity to the functioning of freshwater ecosystems. In K. R. Sridhar, F. Bärlocher, & K. D. Hyde (Eds.), *Novel techniques and ideas in mycology* (Fungal diversity research series 20) (pp. 1–19). Hong Kong: Fungal Diversity Press.
Petchey, O. L., O'Gorman, E. J., & Flynn, D. F. B. (2009). A functional guide to functional diversity measures. In S. Naeem, D. E. Bunker, A. Hector, M. Loreau, & C. Perrings (Eds.), *Biodiversity, ecosystem functioning, and human wellbeing* (pp. 49–59). Oxford: Oxford University Press.
Pielou, E. D. (1969). *An introduction to mathematical ecology*. New York: Wiley-Interscience.
Ricotta, C. (2005). A note on functional diversity measures. *Basic and Applied Ecology, 6*, 479–486.
Rosenzweig, M. L. (1999). Species diversity. In J. McGlade (Ed.), *Advanced ecological theory* (pp. 249–281). Oxford: Blackwell.
Sanders, H. L. (1968). Marine benthic diversity: A comparative study. *American Naturalist, 102*, 243–282.
Schindler, M., & Gessner, M. O. (2009). Functional leaf traits and biodiversity effects on litter decomposition in a stream. *Ecology, 90*, 1641–1649.
Simpson, E. H. (1949). Measurement of diversity. *Nature, 163*, 688.
Smith, B., & Wilson, J. B. (1996). A consumer's guide to evenness indices. *Oikos, 76*, 70–82.
Socolar, J. B., Gilroy, J. J., Kunin, W. E., & Edwards, D. P. (2016). How should beta-diversity inform biodiversity conservation? *Trends in Ecology & Evolution, 31*, 67–80.
Srivastava, D. S., Cardinale, B. J., Downing, A. L., Duffy, J. E., Jouseau, C., Sankaran, M., & Wright, J. P. (2009). Diversity has stronger top-down than bottom-up effects on decomposition. *Ecology, 90*, 1073–1083.
Sutherland, W. J. (1996). *Ecological census techniques. A handbook*. Cambridge: Cambridge University Press.
Swan, C. M., & Palmer, M. A. (2004). Leaf diversity alters litter breakdown in a Piedmont stream. *Journal of the North American Benthological Society, 23*, 15–28.
Tuomisto, H. (2012). An updated consumer's guide to evenness and related indices. *Oikos, 121*, 1203–1218.
Wardle, D. A. (2002). *Communities and ecosystems. Linking the aboveground and belowground components*. Princeton: Princeton University Press.
Whittaker, R. H. (1972). Evolution and measurement of species diversity. *Taxon, 21*, 213–251.

Chapter 62
A Bioinformatics Primer for the Analysis of Illumina MiSeq Data of Litter-Associated Fungi and Bacteria

Sahadevan Seena, Sofia Duarte, and Christian Wurzbacher

Keywords BLAST · FastQC · Flash · Metabarcoding · Sequence database · SINA · Software · SolexaQA · Swarm · VSEARCH

1 Introduction

In recent years, metabarcoding has experienced remarkable progress toward strategies for unveiling biodiversity, based on the development of high-throughput sequencing (HTS) relying on direct DNA extraction from the environment. HTS technologies continue to improve in both accuracy and throughput (Fig. 62.1). With HTS becoming more common and complex, researchers must cope with huge amounts of raw sequence data.

Five HTS platforms have been dominating the NGS market: 454, Illumina, SOLiD, Ion Torrent, and PacBio (Fig. 62.1). They differ mainly in the length of generated reads, throughput, and speed by which runs are generated (Liu et al. 2012;

Electronic Supplementary Material The online version of this chapter (https://doi.org/10.1007/978-3-030-30515-4_62) contains supplementary material, which is available to authorized users.

S. Seena (✉)
Department of Life Sciences, MARE – Marine and Environmental Sciences Centre, University of Coimbra, Coimbra, Portugal
e-mail: seena.sahavedan@uc.pt

S. Duarte
Department of Biology, Centre of Molecular and Environmental Biology (CBMA), University of Minho, Braga, Portugal
e-mail: sduarte@bio.uminho.pt

C. Wurzbacher
Chair of Urban Water Systems Engineering, Technical University of Munich, Garching, Germany
e-mail: c.wurzbacher@tum.de

F. Bärlocher et al. (eds.), *Methods to Study Litter Decomposition*,
https://doi.org/10.1007/978-3-030-30515-4_62

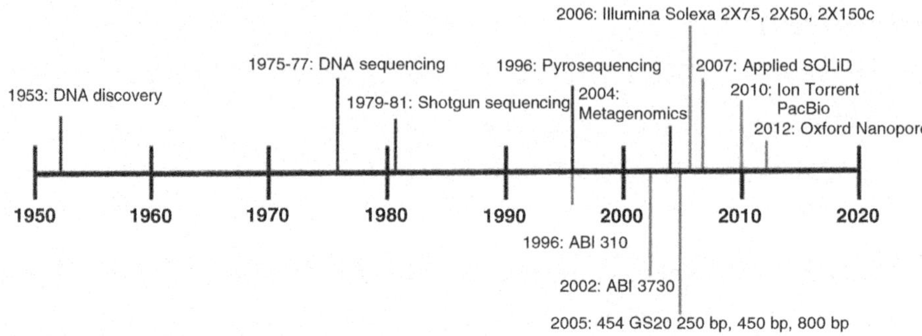

Fig. 62.1 History of sequencing and sequencers

van Dijk et al. 2014). PacBio produces the longest reads and fastest runs and has recently increased its throughput (~10 Gb). Illumina offers the highest throughput (~1800 Gb) and the lowest per-base cost and is currently the leading next-generation sequencing (NGS) platform. Fast runs are also provided by the discontinued 454 platform, but with a relatively low throughput (~700 Mb) (van Dijk et al. 2014). The history of sequencing and sequencers is summarized in Fig. 62.1.

Both 454 and Illumina platforms have proved to be highly accurate in detecting litter-associated fungi in freshwater ecosystems and in revealing new insights into freshwater fungal communities (e.g., Tolkkinen et al. 2013; Duarte et al. 2015, 2017, Wurzbacher et al. 2016; Seena et al. 2019). To our knowledge, no assessments of litter-associated fungi in freshwater have yet been conducted using SOLiD, Ion Torrent, or PacBio platforms. Very recently, the PacBio technology has been applied to metabarcoding aquatic bacteria (Singer et al. 2016), aquatic fungi (Heeger et al. 2018), and soil fungi (Tedersoo et al. 2017), with higher taxonomic classification success, but slightly inferior sequence quality and immanent errors compared to Illumina sequencing. The 454 platform appears to be the only platform that has been used to date to assess bacterial diversity on decomposing litter (Heino et al. 2014; Mykrä et al. 2017), although Illumina currently dominates analyses of aquatic prokaryotes on other substrates (e.g., Wurzbacher et al. 2017).

NGS technologies are evolving rapidly. Cheaper platforms with faster and higher throughput are likely to emerge in the next few years. However, no matter what technology is used, there are several issues that researchers face when dealing with the output of HTS. In particular, these challenges include data storage and management, and the processing of the hundreds of millions of reads that are generated by HTS sequencers. As a result, data processing can be a bottleneck in metabarcoding studies (Fig. 62.2). Another issue is the difficulty of comparing results of independent HTS studies (Divoll et al. 2018). The use of different methods for data generation, data pruning (e.g., sensitivity to detecting sequencing errors, chimeras), sequence clustering into operational taxonomic units (OTUs), and their classification may significantly affect diversity estimates (Meiser et al. 2014). Nevertheless, despite differences in the diversity and abundances found with different sequencing platforms and bioinformatics pipelines, the overall conclusions

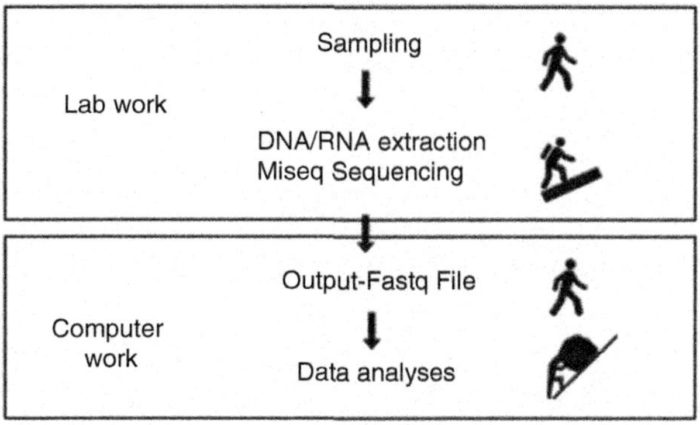

Fig. 62.2 Workflow and difficulty level in metabarcoding

regarding biology underlying the observed patterns are usually similar (e.g., Allali et al. 2017). However, since OTUs reported in independent studies may not be directly comparable, it is essential to develop standard workflows to be used for the analysis of metabarcoding data of litter-associated microbes. Consequently, one of the greatest challenges is to develop adequate software modules and bioinformatic pipelines to analyze the data generated by increasingly high-throughput NGS platforms and to establish universal pipelines that allow the comparison of OTUs reported in independent studies (Leese et al. 2016).

There are various guidelines and tutorials for pipelines to process microbial NGS data (e.g., Nilsson et al. 2011; Lindahl et al. 2013), but there is still a lack of standardized analyses of metabarcoding data. The most commonly used pipelines accompanied by a tutorial and example data are mothur (Schloss et al. 2009), QIIME (Caporaso et al. 2010), and Uparse (Edgar 2013). Their general purpose is to convert HTS data into an intuitively accessible abundance matrix in three major steps: (1) sequence trimming by eliminating low-quality data (e.g., sequences with unidentified nucleotides, short length sequences, chimeras), (2) data reduction by sequence clustering into distinct OTUs or comparable units of highly similar sequences, and (3) taxonomic classification of the OTUs by comparison with reference sequences in genetic databases and prediction of the taxa. The last step depends critically on the quality and completeness of the reference database. Therefore, the installation of a standardized pipeline may eventually be linked to the reference databases. Pioneering efforts to create reference databases have recently led to the establishment of standardized online pipelines to which users can upload their HTS data for analysis (e.g., SILVA NGS for SILVA: Glöckner et al. 2017; PipeCraft for UNITE: Anslan et al. 2017).

All these pipelines (online or local) rely on a linear sequence of open-source tools and simple text processing scripts used for each of the three steps mentioned above. This chapter presents a hands-on guide on how to use HTS data on computers. The various tools can be used in a modular way and may easily be adjusted to fit personal needs. We specifically propose a simple bioinformatic workflow to process paired MiSeq data derived from metabarcoding of litter-associated fungi and bacteria. This can be adapted or improved as soon as a universal pipeline is developed. All steps in the pipeline use a terminal on a UNIX system (Macintosh or Linux OS). Only basic computer knowledge is required to implement them.

2 Equipment and Software

2.1 Computer Hardware and Software

- Macintosh or Linux computer (at least 16 GB RAM or a server with more memory when sample size increases)
- Software to be installed on the computer:
  - FastQC: https://www.bioinformatics.babraham.ac.uk/projects/fastqc/
  - SolexaQA: http://solexaqa.sourceforge.net
  - Flash: http://ccb.jhu.edu/software/FLASH/
  - SINA: https://github.com/epruesse/SINA
  - VSEARCH: https://github.com/torognes/vsearch
  - Swarm: https://github.com/torognes/swarm
  - Mothur: https://www.mothur.org/
  - BLAST: https://www.ncbi.nlm.nih.gov/books/NBK279671/
  - R: https://www.r-project.org
  - RStudio: https://www.rstudio.com

3 Experimental Procedures

3.1 Overview (Fig. 62.3)

1. Start by quality-trimming and filtering the reads from fastq files (e.g., litter-associated fungi) with the DynamicTrim and LengthSort algorithms implemented in the SolexaQA package (Cox et al. 2010).

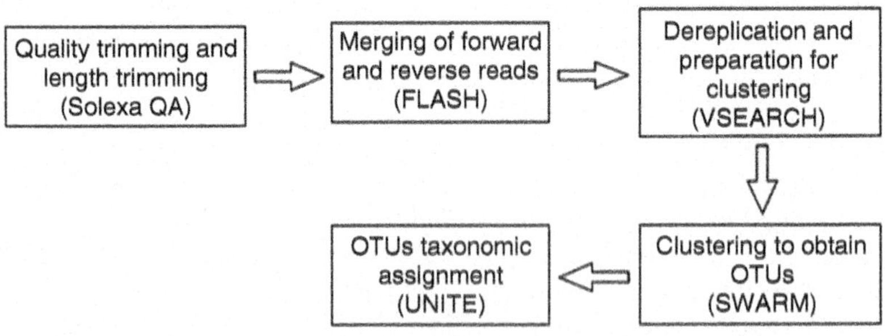

Fig. 62.3 Bioinformatics workflow chart illustrating the five key steps

2. Merge the forward and reverse reads with FLASH software (Magoč and Salzberg 2011).
3. Use dereplication to remove duplicated reads with VSEARCH software (Rognes et al. 2016).
4. Cluster the OTUs with Swarm software (Mahé et al. 2014) and then select the longest sequence from each OTU as the representative of each OTU.
5. For fungi, assign the OTU representatives to taxonomic groups with a BLAST against the UNITE data base.
6. For bacteria, use the SINA tool and Silva database to assign OTUs.

Steps 1–5 are illustrated by demonstration sequences of litter-associated fungi in streams that can be accessed at https://www.Seena_Chapter_62_Bioinformatics.ZIP.

3.2 *Quality and Length Trimming*

1. Visualize the quality of the fastq files with FastQC software.
2. Use DynamicTrim in the SolexaQA package to trim sequences dynamically by using the quality scores of bases within individual reads. In the demonstration file, the raw reads were first trimmed at a quality threshold of 25. Make a folder labelled "trimmed quality" where the generated files can be stored. The command lines for performing the appropriate tasks are given below for each step:

./SolexaQA++ dynamictrim 3927-A-MSITS3_R1.fastq 3933-A-MSITS3_R2.fastq -h 25 -d trimmed_quality

3. Use the LengthSort program of SolexaQA (length cutoff = 50 bp) to filter the reads. Make a folder labelled "trimmed length" inside the folder "trimmed quality" where the generated files can be stored. The command is:

./SolexaQA++ lengthsort 3927-A-MSITS3_R1.fastq.trimmed 3927-A-MSITS3_R2.fastq.trimmed -l 50 -d trimmed_length/

3.3 *Merging Paired-End Reads*

1. Use the script FastQ.tag.rb to rename R1 and R2 sequences. The script will be found in the folder "trimmed length."

For sequences R1 use: ./FastQ.tag.rb -i 3927-A-MSITS3_R1.fastq.trimmed.paired -p A_fungi_ -s /1 -o Tag_3927-A-MSITS3_R1.fastq.trimmed.paired

For sequences R2 use: ./FastQ.tag.rb -i 3927-A-MSITS3_R2.fastq.trimmed.paired -p A_fungi_ -s /2 -o Tag_3927-A-MSITS3_R2.fastq.trimmed.paired

2. Transfer all the files beginning with Tag (e.g., Tag_3927-A-MSITS3_R1.fastq.trimmed.paired) to a new folder named "tag."
3. Use Flash to merge paired-end reads; create inside the tag folder a new folder named "joined," where the new files can be stored after pasting the command:

```
./flash -o A_fungi -d joined/ Tag_3927-A-MSITS3_R1.fastq.trimmed.paired
    Tag_3927-A-MSITS3_R2.fastq.trimmed.paired
```

3.4 *Converting, Concatenating, and Dereplicating Sequences*

1. Use the script FastQ.toFastA.awk, which will be found in the folder tag, to convert the fastq to fasta sequences:

```
./FastQ.toFastA.awk A_fungi.extendedFrags.fastq > A_fungi.fasta
```

2. Concatenate the fasta sequences by using the command:

```
cat *.fasta > all_fungi.fasta
```

3. Dereplicate the sequences as follows:

```
vsearch \
-- derep_fulllength all_fungi.fasta \
-- sizeout \
-- fasta_width 0 \
-- output amplicons_linearized_dereplicated.fasta
```

3.5 *Clustering*

1. Prepare the sequences for OTU clustering by using the command:

```
sed '/^>/ s/$/_1/' amplicons_linearized_dereplicated.fasta > amplicons_with_total_
    abundances_fungi.fasta
```

2. Perform clustering by swarming using the command:

```
./swarm -f -t 4 -w OTU_total_fungi_representatives.fasta amplicons_with_total_
    abundances_fungi.fasta > OTU_total_fungi.swarms
```

3.6 *Assigning and Counting OTUs*

1. Assign OTUs to taxonomic groups by the BLAST tool at the US National Center for Biotechnology Information (NCBI):

```
blastn -query OTU_total_fungi_representatives.fasta -db UNITE_public_28.06.2017.fasta -out output -outfmt 6 -evalue 0.00001 -max_target_seqs=1
```

2. Count the OTUs in the "OTU_total_fungi.swarms" file that was generated after clustering.

For Macintosh use:

```
sed 's/A_fungi*/*/g' OTU_total_fungi.swarms > file1
sed 's/[^*]//g' file1 | awk '{print length}' > file2
```

For Linux use:

```
sed ’s/A_fungi*/A\t/g’ OTU.swarms > file1
sed 's/A/*/g’ file1 > file2
sed 's/[^*]//g' file2 | awk '{print length}' > file3
```

3. Copy and select the number of OTUs from file 2 and paste it in an Excel sheet named "Taxonomy," labelling the sheet in the file as "OTUs."
4. Extract the names of the representative sequences by using the command:

```
grep ">" OTU_total_fungi_representatives.fasta > names.txt
```

copy the names and paste on a new sheet (see "OTU rep+OTUs").

5. Copy the information from "output" file of BLASTs and paste this information (see sheet "OTU rep+OTUs+output").
6. Remove the duplicated taxonomic information and no hits (see OTU rep+OTUs+Output cleaned). Add one more column and name it as OTUs and start to add numbers 1, 2, and 3 beneath it, continuing until the end of the taxonomic information.

3.7 *Procedures for Bacteria*

1. For litter-associated bacteria, use fastq files of bacteria and follow up to step 5, but systematically replace "fungi" in the commands and elsewhere by "bacteria." Then use SINA to assign taxonomic groups to bacterial OTUs:

```
sina -i OTU_total_bacteria_representatives.fasta -o alignedsequences.fasta --meta-fmt csv --ptdb SSURef_NR99_128_SILVA_07_09_16_opt.arb --search --search-db SSURef_NR99_128_SILVA_07_09_16_opt.arb --lca-fields tax_slv
```

2. Follow steps 3.6.2–3.6.6 for bacteria.

3.8 Initial Statistical Analyses in R

1. Import the final table of OTUs into R.
2. Copy the OTU names and the count table (i.e., the number of occurrences per sample and individual OTUs) from the final Excel sheet into a new sheet, and save it as "csv" file (e.g., OTU_table.csv), choosing {TAB} as separator.
3. Copy the sample script detailed below (https://www.Seena_Chapter_62_Script.R) into the same folder and open it with RStudio.
4. Execute the program by pressing the Run or Ctrl+Return key. Follow the instructions of the script provided at https://www.Seena_Chapter_62_Script.R

4 Final Remarks

Mothur software can be used to check for chimeras after dereplication. To this end, paste the following commands:

chimera.uchime (fasta = input.fasta, reference = UNITE.fungi.fasta, processors=4)
input = fasta file after dereplication

Reference = UNITE public database for fungi and SILVA public database for bacteria.

Several files will be created. One of them ending ".uchime.accnos" contains the chimeric sequences. Then paste the following command to delete the chimeras from the input fasta file:

remove.seqs (accnos = input.uchimeaccnos, fasta = input.fasta)

The file "pick.fasta" will contain the sequences free from chimeras, so that they can be used for clustering.

References

Allali, I., Arnold, J. W., Roach, J., Cadenas, M. B., Butz, N., Hassan, H. M., Koci, M., Ballou, A., Mendoza, M., Ali, R., & Azcarate-Peril, M. A. (2017). A comparison of sequencing platforms and bioinformatics pipelines for compositional analysis of the gut microbiome. *BMC Microbiology, 17*, 194.

Anslan, S., Bahram, M., Hiiesalu, I., & Tedersoo, L. (2017). PipeCraft: Flexible open-source toolkit for bioinformatics analysis of custom high-throughput amplicon sequencing data. *Molecular Ecology Resources, 17*, e234–e240.

Caporaso, J. G., Kuczynski, J., Stombaugh, J., Bittinger, K., Bushman, F. D., Costello, E. K., et al. (2010). QIIME allows analysis of high-throughput community sequencing data. *Nature Methods, 7*, 335–336.

Cox, M. P., Peterson, D. A., & Biggs, P. J. (2010). SolexaQA: At-a-glance quality assessment of Illumina second-generation sequencing data. *BMC Bioinformatics, 11*, 485.

Divoll, T. J., Brown, V. A., Kinne, J., Mccracken, G. F., & O'Keefe, J. M. (2018). Disparities in second-generation DNA metabarcoding results exposed with accessible and repeatable workflows. *Molecular Ecology Resources, 18*, 590–601.

Duarte, S., Bärlocher, F., Trabulo, J., Cássio, F., & Pascoal, C. (2015). Stream-dwelling fungal decomposer communities along a gradient of eutrophication unraveled by 454 pyrosequencing. *Fungal Diversity, 70*, 127–148.

Duarte, S., Cássio, F., Pascoal, C., & Bärlocher, F. (2017). Taxa-area relationship of aquatic fungi on deciduous leaves. *PLoS One, 12*, e0181545.

Edgar, R. C. (2013). UPARSE: Highly accurate OTU sequences from microbial amplicon reads. *Nature Methods, 10*, 996–998.

Glöckner, F. O., Yilmaz, P., Quast, C., Gerken, J., Beccati, A., Ciuprina, A., Bruns, G., Yarza, P., Peplies, J., Westram, R., & Ludwig, W. (2017). 25 years of serving the community with ribosomal RNA gene reference databases and tools. *Journal of Biotechnology, 261*, 169–176.

Heeger, F., Bourne, E. C., Baschien, C., Yurkov, A., Bunk, B., Spröer, C., Overmann, J., Mazzoni, C. J., & Monaghan, M. T. (2018). Long-read DNA metabarcoding of ribosomal RNA in the analysis of fungi from aquatic environments. *Molecular Ecology Resources, 18*, 1500–1514.

Heino, J., Tolkkinen, M., Pirttilä, A. M., Aisala, H., & Mykrä, H. (2014). Microbial diversity and community – Environment relationships in boreal streams. *Journal of Biogeography, 41*, 2234–2244.

Leese, F., Altermatt, F., Bouchez, A., Ekrem, T., Hering, D., Mergen, P., Pawlowski, J., Piggott, J., Rimet, F., Steinke, D., Taberlet, P., Weigand, A., Abarenkov, K., Beja, P., Bervoets, L., Björnsdóttir, S., Boets, P., Boggero, A., Bones, A., Borja, Á., Bruce, K., Bursić, V., Carlsson, J., Čiampor, F., Čiamporová-Zatovičová, Z., Coissac, E., Costa, F., Costache, M., Creer, S., Csabai, Z., Deiner, K., DelValls, Á., Drakare, S., Duarte, S., Eleršek, T., Fazi, S., Fišer, C., Flot, J., Fonseca, V., Fontaneto, D., Grabowski, M., Graf, W., Guðbrandsson, J., Hellström, M., Hershkovitz, Y., Hollingsworth, P., Japoshvili, B., Jones, J., Kahlert, M., Kalamujic Stroil, B., Kasapidis, P., Kelly, M., Kelly-Quinn, M., Keskin, E., Kõljalg, U., Ljubešić, Z., Maček, I., Mächler, E., Mahon, A., Marečková, M., Mejdandzic, M., Mircheva, G., Montagna, M., Moritz, C., Mulk, V., Naumoski, A., Navodaru, I., Padisák, J., Pálsson, S., Panksep, K., Penev, L., Petrusek, A., Pfannkuchen, M., Primmer, C., Rinkevich, B., Rotter, A., Schmidt-Kloiber, A., Segurado, P., Speksnijder, A., Stoev, P., Strand, M., Šulčius, S., Sundberg, P., Traugott, M., Tsigenopoulos, C., Turon, X., Valentini, A., van der Hoorn, B., Várbíró, G., Vasquez-Hadjilyra, M., Viguri, J., Vitonytė, I., Vogler, A., Vrålstad, T., Wägele, W., Wenne, R., Winding, A., Woodward, G., Zegura, B., & Zimmermann, J. (2016). DNAqua-net: Developing new genetic tools for bioassessment and monitoring of aquatic ecosystems in Europe. *Research Ideas and Outcomes, 2*, e11321.

Lindahl, B. D., Nilsson, R. H., Tedersoo, L., Abarenkov, K., Carlsen, T., Kjøller, R., Kõljalg, U., Pennanen, T., Rosendahl, S., Stenlid, J., & Kauserud, H. (2013). Fungal community analysis by high-throughput sequencing of amplified markers – A user's guide. *New Phytologist, 199*, 288–299.

Liu, L., Li, Y., Li, S., Hu, N., He, Y., Pong, R., Lin, D., Lu, L., & Law, M. (2012). Comparison of next-generation sequencing systems. *Journal of Biomedicine and Biotechnology, 2012*, 251364. 11 pp.

Magoč, T., & Salzberg, S. L. (2011). FLASH: Fast length adjustment of short reads to improve genome assemblies. *Bioinformatics, 27*, 2957–2963.

Mahé, F., Rognes, T., Quince, C., de Vargas, C., & Dunthorn, M. (2014). Swarm: Robust and fast clustering method for amplicon-based studies. *PeerJ, 2*, e593.

Meiser, A., Bálint, M., & Schmitt, I. (2014). Meta-analysis of deep-sequenced fungal communities indicates limited taxon sharing between studies and the presence of biogeographic patterns. *New Phytologist, 201*, 623–635.

Mykrä, H., Tolkkinen, M., & Heino, J. (2017). Environmental degradation results in contrasting changes in the assembly processes of stream bacterial and fungal communities. *Oikos, 126*, 1291–1298.

Nilsson, R. H., Tedersoo, L., Lindahl, B. D., Kjøller, R., Carlsen, T., Quince, C., Abarenkov, K., Pennanen, T., Stenlid, J., Bruns, T., Larsson, K.-H., Kõljalg, U., & Kauserud, H. (2011). Towards standardization of the description and publication of next-generation sequencing datasets of fungal communities. *New Phytologist, 191*, 314–318.

Rognes, T., Flouri, T., Nichols, B., Quince, C., & Mahé, F. (2016). VSEARCH: A versatile open source tool for metagenomics. *PeerJ, 4*, e2584.

Schloss, P. D., Westcott, S. L., Ryabin, T., Hall, J. R., Hartmann, M., Hollister, E. B., Lesniewski, R. A., Oakley, B. B., Parks, D. H., Robinson, C. J., Sahl, J. W., Stres, B., Thallinger, G. G., Van Horn, D. J., & Weber, C. F. (2009). Introducing mothur: Open-source, platform-independent, community-supported software for describing and comparing microbial communities. *Applied and Environmental Microbiology, 75*, 7537–7541.

Seena, S., Bärlocher, F., Sobral, O., Gessner, M. O., Dudgeon, D., McKie, B. G., Chauvet, E., Boyero, L., Ferreira, V., Frainer, A., Bruder, A., Matthaei, C. D., Fenoglio, S., Sridhar, K. R., Albariño, R. J., Douglas, M. M., Encalada, A. C., Garcia, E., Ghate, S. D., Giling, D. P., Gonçalves, V., Iwata, T., Landeira-Dabarca, A., McMaster, D., Medeiros, A. O., Naggea, J., Pozo, J., Raposeiro, P. M., Swan, C. M., Tenkiano, N. S. D., Yule, C. M., & Graça, M. A. S. (2019). Biodiversity of leaf litter fungi in streams along a latitudinal gradient. *Science of the Total Environment, 661*, 306–315.

Singer, E., Bushnell, B., Coleman-Derr, D., Bowman, B., Bowers, R. M., Levy, A., Gies, E. A., Cheng, J.-F., Copeland, A., Klenk, H.-P., Hallam, S. J., Hugenholtz, P., Tringe, S. G., & Woyke, T. (2016). High-resolution phylogenetic microbial community profiling. *The ISME Journal, 10*, 2020–2032.

Tedersoo, L., Tooming-Klunderud, A., & Anslan, S. (2017). PacBio metabarcoding of Fungi and other eukaryotes: Errors, biases and perspectives. *New Phytologist, 217*, 1370–1385.

Tolkkinen, M., Mykrä, H., Markkola, A.-M., Aisala, H., Vuori, K.-M., Lumme, J., Pirttilä, A. M., & Muotka, T. (2013). Decomposer communities in human-impacted streams: Species dominance rather than richness affects leaf decomposition. *Journal of Applied Ecology, 50*, 1142–1151.

van Dijk, E. L., Auger, H., Jaszczyszyn, J., & Thermes, C. (2014). Ten years of next-generation sequencing technology. *Trends in Genetics, 30*, 418–426.

Wurzbacher, C., Grimmett, I. J., & Bärlocher, F. (2016). Metabarcoding-based fungal diversity on coarse and fine particulate organic matter in a first-order stream in Nova Scotia, Canada. *F1000Research, 4*, 1378.

Wurzbacher, C., Nilsson, R. H., Rautio, M., & Peura, S. (2017). Poorly known microbial taxa dominate the microbiome of permafrost thaw ponds. *The ISME Journal, 11*, 1938–1941.

Chapter 63
A Primer for Meta-Analysis

Verónica Ferreira and Felix Bärlocher

Keywords Effect size · Litter decomposition · Meta-regression · Publication bias · Sensitivity analysis · Software · Sub-group analysis · Systematic review

1 Introduction

Meta-analysis is a statistical tool that combines the results of several conceptually similar studies or experiments by providing a weighted average of the results of the individual studies. This pooled estimate is assumed to be close to the unknown true value. Meta-analysis allows answering the following questions: (1) Is the true effect significantly different from zero? (2) What are the magnitude and direction of the global effect? (3) Are magnitude and direction of the global effect influenced by any characteristics of individual studies or groups of studies? Additionally, it allows identifying knowledge gaps in the research field.

Meta-analysis is especially useful when sample sizes of individual studies are low or their effect sizes small or non-significant. It increases statistical power and tests consistency among individual studies. It also allows the testing of hypotheses that may be difficult to consider in individual studies (e.g. comparisons across biomes).

For meta-analysis to be useful, there needs to be a reasonably large number of empirical studies (although there is no minimum), and the collection of studies needs to be free of publication bias. All studies need to report quantitative measures

V. Ferreira (✉)
MARE – Marine and Environmental Sciences Centre & Department of Life Sciences, University of Coimbra, Coimbra, Portugal
e-mail: veronica@ci.uc.pt

F. Bärlocher
Department of Biology, Mount Allison University, Sackville, NB, Canada
e-mail: fbaerloc@mta.ca

F. Bärlocher et al. (eds.), *Methods to Study Litter Decomposition*,
https://doi.org/10.1007/978-3-030-30515-4_63

of the variables; purely descriptive reports such as case studies cannot be synthetized by meta-analysis.

Meta-analysis has been used to test ecological hypotheses and theories and the effect of covariates that are difficult to examine within a single primary study, assess the impacts of major environmental drivers and the effectiveness of management and conservation strategies, inform environmental risk assessment, and identify research gaps (Koricheva and Gurevitch 2014). Table 63.1 shows examples of meta-analyses in litter decomposition.

Meta-analysis is usually performed in the context of a systematic literature review, which includes *systematically* locating, selecting, and appraising sources (preferably peer-reviewed studies) and then synthetizing data from the selected sources. Each step is clearly documented to allow reproduction (Table 63.2).

Table 63.1 Examples of meta-analyses in litter decomposition

References	System	Factor under study	No. of studies	No. of effect sizes	Effect size	Global effect size (95% CL)
1	Soil	Nutrient enrichment	24	500	Response ratio (*R*)	0.981
2	Soil	Microarthropod presence	30	101	Hedges' *g*	1.482
3	Streams	Litter diversity	11	510	Signed deviation from additivity	–0.01[c]
4	Streams	Macroconsumer presence	17	36	Ln response ratio (ln*R*)	–0.016 (–0.023 to –0.009)[d]
5	Streams	Nutrient enrichment	99	840	Response ratio (*R*)	1.49(1.41 – 1.58)[a]
6	Streams	Heavy metal contamination	38	133	Hedges' *g*	–0.81 (–1.02 to –0.61)[b]
7	Streams	Forest change (4 types)	24	156	Response ratio (*R*)	0.82 (0.76 – 0.89)[a]
8	Streams	Acidification	17	67	Response ratio (*R*)	0.37 (0.30 – 0.46)[a]
9	Streams	Water stress	9	41	Response ratio (*R*)	0.69 (0.59 – 0.82)[a]
10	Streams	Water temperature	34	148	Hedges' *g*	1.20 (0.96 – 1.43)[b]

References: (1) Knorr et al. (2005); (2) Kampichler & Bruckner (2009); (3) Lecerf et al. (2011); (4) Mancinelli et al. (2013); (5) Ferreira et al. (2015); (6) Ferreira et al. (2016a); (7) Ferreira et al. (2016b); (8) Ferreira and Guérold (2017); (9) Sabater et al. (2018); (10) Amani et al. (2019)

[a] $R > 1$, stimulation of decomposition; $R < 1$, inhibition of decomposition

[b] $g > 0$, stimulation of decomposition; $g < 0$, inhibition of decomposition

[c] Negative values indicate synergistic response of litter decomposition to litter diversity

[d] lnR < 0, inhibition of decomposition

2 Preparation of the Database

2.1 *Definition of the Question*

The first step in a systematic literature review is to clearly define the question (or hypothesis) to address. This will determine the scope of the review and literature search (Sect. 2.2), inclusion/exclusion criteria (Sect. 2.3), and the type of information to extract (Sect. 2.5). The question should be broad enough to capture a sufficient number of empirical studies (e.g. 'Does nutrient enrichment affect litter decomposition in running waters?') but not so broad that it will become unmanageable (e.g. 'Does environmental change affect ecosystem functioning?', which would include the effect of any environmental change on any ecosystem function in any ecosystem). A useful strategy to define the question is the PICO method, where the question clearly identifies the population (P), the intervention (I), the control (C; can be implicit), and the outcome (O). In the question 'Does nutrient enrichment affect litter decomposition in running waters?', 'running waters' is the population, 'nutrient enrichment' is the intervention, it is implicit that a non-nutrient enriched condition is the control, and 'litter decomposition' is the outcome.

2.2 *Intensive and Extensive Literature Search*

The literature search should be intensive and extensive to ideally locate *all* studies that have ever addressed the question of interest. This is generally impossible because some studies may not be published (recent studies or studies not submitted or rejected due to publication bias); may belong to the 'grey literature' (e.g. theses, reports, conference abstracts), which is generally difficult to locate and retrieve; or may be inaccessible for other reasons (e.g. language bias, when published in languages unknown to the meta-analysist) (Sect. 3.6). Clearly, there needs to be a considerable effort to ensure that the studies located and retrieved are a random sample of the studies performed. The literature search protocol will always have an element of subjectivity (e.g. time frame, languages, key words, search paths), but needs to be transparent and may need to be revised repeatedly to address potential biases (Sect. 3.6).

An intensive and extensive literature search may include studies published in the mainstream literature and 'grey literature', in several languages and over a large time frame (but note that methods to determine the outcome of interest may have changed over time, which has to be coded). The literature search should be done via multiple paths, including personal literature databases, reference lists in relevant primary studies and reviews, scientific journal indices, and online databases (e.g. Google Scholar, Web of Science, Scopus). Different search paths generally retrieve different sets of studies and should be used to complement each other. The set of key search words should be clearly defined to allow the search to be reproduced. It may

Table 63.2 Steps to carry out a meta-analysis in the context of a systematic review

Systematic review	Meta-analysis
Definition of the question	Estimation of effect sizes and associated variances
Intensive and extensive literature search	Selection of the model
Definition of the inclusion/exclusion criteria	Estimation of the global effect size and precision
Critical appraisal of studies	Evaluation of publication bias
Data extraction	Test of heterogeneity and effect of moderators
Statistical synthesis (e.g., **meta-analysis**)	Sensitivity analysis
Discussion and conclusion based on the most precise studies	Interpretation and presentation of results

be useful to include a search in libraries and conference abstract books when targeting 'grey literature'. The use of mailing lists (e.g. ECOLOG-L) or direct contact with researchers known to work on the topic of interest may locate additional studies. It is essential to keep track of all steps in the literature search, annotating when and how each study was located and retrieved.

Most of the studies that seem relevant based on title and abstract can generally be retrieved without delay (e.g. by downloading from journal web pages or online databases). An additional effort may be required to retrieve older studies or data from 'grey literature'. This may include contacting the author of the study, the library, or the author of a recent study where the study of interest has been cited.

2.3 *Definition of Inclusion/Exclusion Criteria*

A literature search generally retrieves many studies that are not relevant or useful. The studies to be included in the analysis are selected based on clearly defined inclusion/exclusion criteria. Studies need to report information that will allow estimating effect sizes (and associated variability), which are interpreted as 'dependent variable' in a meta-analysis (Sect. 3.2). For instance, to address the question 'Does nutrient enrichment affect litter decomposition in running waters?', we should only consider empirical studies that report (1) litter decomposition in at least one nutrient-enriched condition and one control (non-enriched) condition, (2) a measure of variation of decomposition values (variance, SD, SE, 95% CL), and (3) sample size. If some of this information is missing, it may be available through a request to the author. If a few studies lack information on variation, they can still be included as the missing values may be imputed (Koricheva et al. 2013; Ferreira et al. 2015).

Additional inclusion/exclusion criteria may be defined. To address the question above, possible inclusion/exclusion criteria could be that the primary studies (1) report decomposition of natural litter rather than artificial substrates and (2) rely on allochthonous rather than autochthonous litter (Ferreira et al. 2015). Inclusion/exclusion criteria may also refer to research methods used (e.g. studies that use a specific method) and may need to be revised to address publication bias issues (Sect. 3.6).

2.4 Critical Appraisal of Studies

The selected studies need to be critically appraised, especially concerning methodological quality and multiple publications. The methodological quality of studies can be coded as a moderator (Sect. 3.5) or used to assess its impact on the analysis (sensitivity analysis; Sect. 3.7). Special care is needed to detect information that has been published multiple times to avoid overweighting these data. The number of studies used in the analysis may thus have to be reduced to avoid counting identical information more than once.

2.5 Data Extraction

Basic Data to Estimate Effect Size and Associated Variance

Data in ecological studies can be reported in several formats, which will determine the type of effect size (and its variance) that can be estimated (Table 63.3; Borenstein et al. 2009). If data are based on a comparison of two groups of continuous variables, then information on the variable of interest (outcome), measure of variability of the outcome (variance, SD, SE, 95% CL), and sample sizes of control and treatment conditions need to be extracted. If data are reported as a comparison of two groups in terms of categorical variables, then information on sample size and number of cases in the event and non-event situation, in the control and treatment conditions, need to be extracted. If data are reported as the relationship between two continuous variables, then Pearson's r and sample size need to be extracted.

Table 63.3 Types of data in ecology, types of effect size, and examples of common effect sizes

Type of data	Type of effect size	Effect size
Comparison of 2 groups in terms of continuous variables (e.g. control × treatment)	Effect size based on means	Raw mean difference (*D*)
		Standardized mean difference (Cohens' *d* or Hedges' *g*)
		Response ratio (*R*)
Comparison of 2 groups in terms of categorical variables (2 × 2 contigency table)	Effect size based on binary data	Risk difference (*RD*)
		Risk ratio (*RR*)
		Odds ratio (*OR*)
Relationship between 2 continuous variables	Effect size based on correlations	Pearson correlation (*r*)

For details on the estimation of each effect size and associated variance, see Borenstein et al. (**2009**); these are estimated automatically in any software for meta- analysis

Explanatory Information to Assess the Effect of Study/ Environmental Conditions

Several experimental and environmental explanatory variables – termed moderators in meta-analysis – may affect the response of the dependent variable and explain differences in effect sizes among studies. Meta-analysis allows testing the significance and strength of these moderators ('independent variables') (Sect. 3.5). Only moderators backed up by a hypothesis should be coded. In addition, research procedures likely to affect the response of the variable of interest (e.g. type of study, specific methodologies) should be coded to be later used in sensitivity analyses (Sect. 3.7).

Information on selected moderators (continuous or categorical variables) needs to be extracted from the studies and additional sources (e.g. websites for climatic information) or by contacting the author. For instance, to address the question 'Does nutrient enrichment affect litter decomposition in running waters?', categorical moderators may include type of study, scale of nutrient enrichment and identity of the nutrient used in field manipulative studies, type of aquatic decomposers, type and identity of litter, and climate. Continuous moderators may include the mean dissolved nutrient concentration in the control and the magnitude of the increase in nutrient concentrations compared to the control condition (Ferreira et al. 2015). Additionally, the type of report (i.e. published in the mainstream or 'grey literature'), the type of data (i.e. reported in the study or imputed/estimated), and methodological specifications may be coded (Ferreira et al. 2015).

3 Meta-Analysis

3.1 Software for Performing Meta-Analysis

Once the database is ready (i.e. study information, basic data for the estimation of effect sizes and associated variance, and information on moderators, generally one case by line; see, e.g. Table S1 in Ferreira et al. 2015), it can be entered into a software spreadsheet. Many options are available (Koricheva et al. 2013), including:

OpenMEE, an open-source, user-friendly software designed by ecologists for data in ecology and evolution (Wallace et al. 2017). It offers diverse and advanced statistical options based on packages developed for R without requiring programming skills. The software, as well as the user guide, can be downloaded at http://www.cebm.brown.edu/openmee/

Comprehensive Meta-Analysis (CMA), a commercial, user-friendly software developed by specialists in social sciences and medicine (Borenstein et al. 2017). It allows entering data for estimating effect sizes and associated variances in 100 different formats. A trial version, as well as user guides and a large bibliography in meta-analysis, can be downloaded at https://www.meta-analysis.com/

The *metafor package for R,* an open-source, highly versatile package that requires familiarity with R. Codes for performing a meta-analysis are provided in Viechtbauer (2010)

3.2 Effect Size and Precision

The effect size reflects the magnitude of the effect of a treatment or the strength of the relationship between two variables; it is estimated for each (case) study and used to estimate the global effect size. It may be necessary to estimate different effect sizes for different studies, depending on the format used to report data (Table 63.3). However, the various types of effect sizes are interconvertible (Table 63.3) so that the analysis is based on a single effect size.

Differences in sample size will affect the precision of the estimated effect size. Provided there is no systematic bias, the precision defines the interval containing the true effect size and indicates how much we can trust the estimated effect size. Effect sizes associated with a larger variance are given less weight in the meta-analysis. Variance estimates are specific to each effect size (Borenstein et al. 2009).

One of the most common effect sizes in ecology is the response ratio R (Koricheva and Gurevitch 2014), the ratio of the variable of interest in the treatment condition to the variable in the control condition (e.g. $R = k_{\text{nutrient-enriched}}/k_{\text{control}}$, k = decomposition rate) (Hedges et al. 1999). Being a ratio, it can only be used if outcomes differ from zero. It is very easy to interpret ($R = 1$ indicates no treatment effect, while $R > 1$ and $R < 1$ indicate higher and lower values in the treatment than in the control condition, respectively). The analyses are performed with the natural logarithm of *R* (ln*R*), which is normally distributed even in small samples, but the results can be back transformed to *R* to facilitate interpretation.

3.3 Global Effect Size

If all effect sizes had the same precision, they could be simply averaged to estimate the global effect size. However, effect sizes generally differ in their precision; the global effect size is therefore estimated by a weighted average of the individual effect sizes, with a larger weight given to the more precise effect sizes (i.e. associated with smaller variances).

There are two models to estimate the global effect size in standard weighed meta-analysis:

The *fixed-effect model* assumes a unique and true (fixed!) effect size shared by all studies and that the observed differences among individual effect sizes are due to sampling error. In this case, the weight attributed to each individual effect size is based on the inverse of the within-study variance (or sampling error). This model should be chosen only when the variation among studies is negligible; the goal is to estimate the global effect size for the studies considered and not to extrapolate beyond the analysed studies. These conditions are rarely met in ecological studies, and thus, this is not a model commonly used in ecology.

The *random-effects model* assumes that the true effect size varies among studies and that the analysed studies provide a random sample of the distribution of effect sizes, with the global effect size being the mean of that distribution. In this case, the weight attributed to each individual effect size is based on the inverse of the total variance, which is the sum of the within-study variance (or sampling error) and the between-study variance. This model should be used when the effect sizes are expected to vary among studies (e.g. due to different experimental conditions, ecosystems, species, etc.); the goal is to generalize the global effect size, which is considered to be the mean of the true effect sizes. This is the most common model in meta-analysis in ecology.

In both models, the global effect size is estimated as the sum of individual effect sizes weighted by the inverse of the corresponding variance, corrected by the sum of the weights. The variance of the global effect size is estimated as the inverse of the sum of the weights and can be converted into 95% CL. Significant effect sizes occur when 95% CL do not include 1 for effect sizes based on ratios (e.g. *R*) or 0 for effect sizes based on the natural logarithm of ratios (e.g. ln*R*), differences, or Pearson's r (Borenstein et al. 2009).

The meta-analysis result is presented in a forest plot, which shows the effect sizes of the individual studies considered as well as the global effect size, and their associated variability (generally, 95% CL; Fig. 63.1).

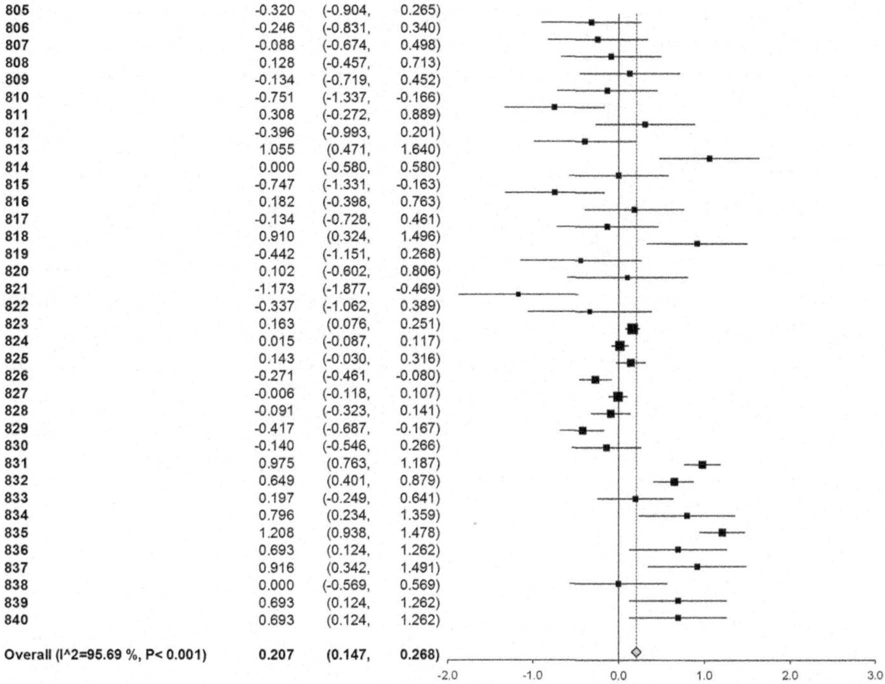

Fig. 63.1 Forest plot (detail of the last 35 effect sizes) of the response of litter decomposition to nutrient enrichment of stream water in correlative field studies ($\ln R = \ln(k_{\text{nutrient-enriched}}/k_{\text{control}})$; k = decomposition rate; $n = 521$). Individual effect sizes (squares; the size of the symbols reflects their precision) and associated 95% CL, the global effect size (blue diamond; red line) and associated 95% CL (width of the diamond), and the percentage of total variation due to true variation among effect sizes ($I^2 = 95.69\%$) are depicted. A random-effects model was used, with the restricted maximum likelihood method to determine between-study variance. The solid line ($\ln R = 0$) indicates no effect of nutrient enrichment, and $\ln R > 0$ and $\ln R < 0$ indicates stimulation or inhibition, respectively. Significant effect sizes occur when 95% CL do not include zero. To facilitate interpretation, the result is back transformed into R: $\ln R = 0.207$ (0.147–0.268) → $R = 1.23$ (1.16–1.32), which indicates that litter decomposition rates increase significantly by 23% in nutrient-enriched streams (output from OpenMEE, data from Ferreira et al. 2015)

3.4 Heterogeneity

In addition to estimating the global effect size, it is useful to identify and quantify heterogeneity in effect sizes. This addresses the sources of differences among studies rather than combining all the effect sizes into a global value.

The heterogeneity, or observed variation among effect sizes, may result from true variation in effect sizes (between-study variance) and from sampling error (within-study variance). The true variation is of primary interest. The observed variation among effect sizes is given by the Q-statistic and is estimated as the sum of squared differences between individual effect sizes and the global effect size weighted by the inverse of the variance associated with the individual effect sizes (Borenstein et al. 2009). The Q-statistic has a χ^2 distribution (H_0: $Q = df$, $df = n - 1$, n = number of effect sizes), with significant p-values ($Q > df$) suggesting that the variation among effect sizes is not all due to chance. The causes for heterogeneity are then explored (Sect. 3.5). Non-significant p-values ($Q = df$) suggest that the H_0 cannot be rejected, which suggests that effect sizes are identical (i.e. there is no between-study variance) and variation among them is due to change (i.e. sampling error); however, a non-significant result may also be due to low statistical power. Thus, it may still be useful to test for moderators when no significant variation is observed among effect sizes (Sect. 3.5). The percentage of observed variation that is due to real differences in effect sizes (i.e. between-study variance) is given by the I^2 statistics: $((Q - df)/Q)) \times 100$. I^2 values vary between 0% and 100%; I^2 values ~25% indicate low heterogeneity, I^2 values ~50% indicate moderate heterogeneity, and I^2 values ~75% indicate high heterogeneity (Fig. 63.1).

3.5 *Test of Moderators*

Variation in effect sizes may be due to differences in experimental or environmental characteristics, categorized as moderators. We can assess if they are systematically associated with variation in effect sizes. Moderators are coded, and their effects are assessed in subgroup analysis (categorical moderators) or meta-regression (continuous moderators).

Subgroup Analysis

In subgroup analysis, effect sizes are grouped by common features (i.e. levels within a given moderator), and the global effect size for each subgroup (levels) is estimated and tested for heterogeneity. A hierarchical approach to test moderators is often useful; effect sizes are stratified, and comparisons of effect sizes among levels of one moderator are made within a level of another moderator (e.g. comparison between the levels 'Coarse mesh' and 'Fine mesh' of the moderator 'Mesh size' within the level 'Leaves' of the moderator 'Litter type'; Table 63.4; Fig. 63.2).

There are multiple models and methods for performing subgroup analysis (Borenstein et al. 2009), which cannot be covered here. In ecology, the most commonly used approaches are the random-effects model with between-study variance pooled for the estimation of the effect size for each subgroup and global effect size and the random-effects model with between-study variance estimated for each subgroup (i.e. not pooled) for the estimation of the effect size for each subgroup and global effect size.

Table 63.4 Hierarchical diagram for categorical moderators (Ferreira et al. 2015)

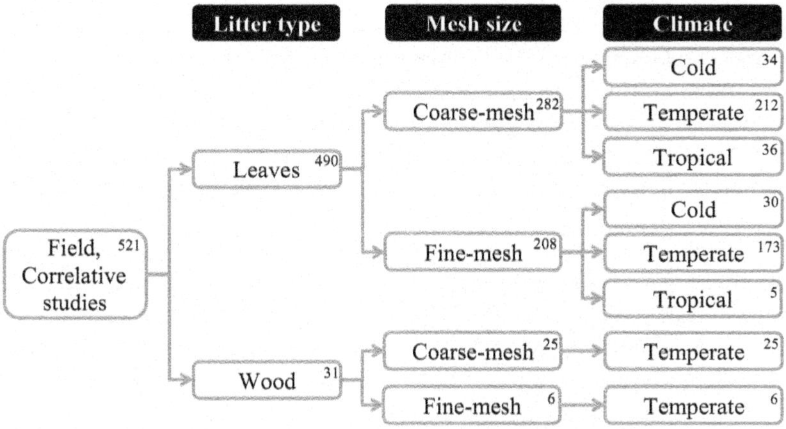

The moderator 'Litter type' has two levels ('Leaves' and 'Wood'), the moderator 'Mesh size' has two levels ('Coarse mesh' and 'Fine mesh'), and the moderator 'Climate' has three levels ('Cold', 'Temperate', and 'Tropical'). Values indicate sample size (i.e. number of effect sizes)

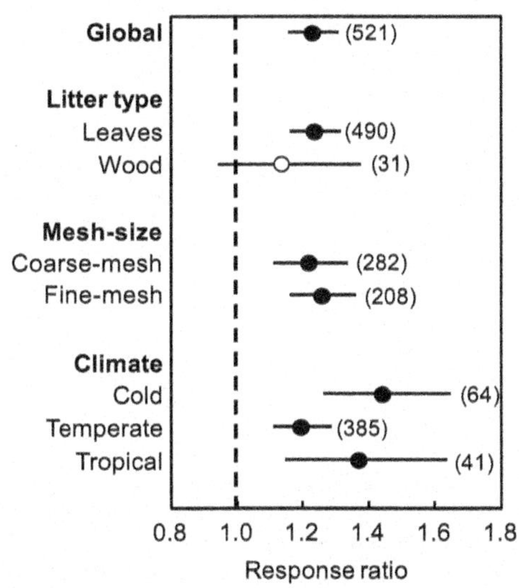

Fig. 63.2 Subgroup analyses of the response of litter decomposition to nutrient enrichment (response ratio *R*, 95% CL); same data as in Table 63.4 (random-effects model; restricted maximum likelihood method for between-study variance). Global effect size and effect size as a function of litter type (two levels), mesh size (two levels; considering only Leaves since non-significant *R* was found for Wood), and climate (three levels; considering only Leaves since non-significant *R* was found for Wood, but considering both mesh sizes since significant *R* was found for both, with no significant difference between them). The dashed line ($R = 1$) indicates no effect of nutrient enrichment on litter decomposition, and $R > 1$ indicates a stimulation of litter decomposition. The effect of nutrient enrichment is significant when 95% CL do not include 1 (black symbols). Within each moderator (in bold), levels with overlapping 95% CL do not significantly differ. Values in parenthesis indicate sample size (output from OpenMEE, data from Ferreira et al. 2015)

Meta-Regression

In meta-regression, effect sizes (weighted by their precision) are regressed against continuous moderators to explain observed variation. There are multiple procedures for meta-regression (Borenstein et al. 2009). One is based on the random-effects model. The meta-regression will produce values for the intercept, the slope, and associated statistics. The significance of the slope is assessed by the *Z*-test given by the ratio between the slope and its SE; if several covariates are used simultaneously, then the *Q*-test is used.

The model is tested using the *Q*-statistic (weighted sum of squares; Sect. 3.4). A significant Q_{model} (i.e. variation explained by the moderator) indicates that the relationship between effect sizes and the moderator is stronger than expected by chance. The goodness of fit test (Q_{resid}) assesses if there is heterogeneity that is not explained by the moderators and can be used to estimate the variance of this unexplained heterogeneity; a significant Q_{resid} indicates that some between-study variance remains unexplained. The proportion of variation in effect sizes that is explained by the model (R^2) is given by the ratio between true variance explained and total true variance (Fig. 63.3).

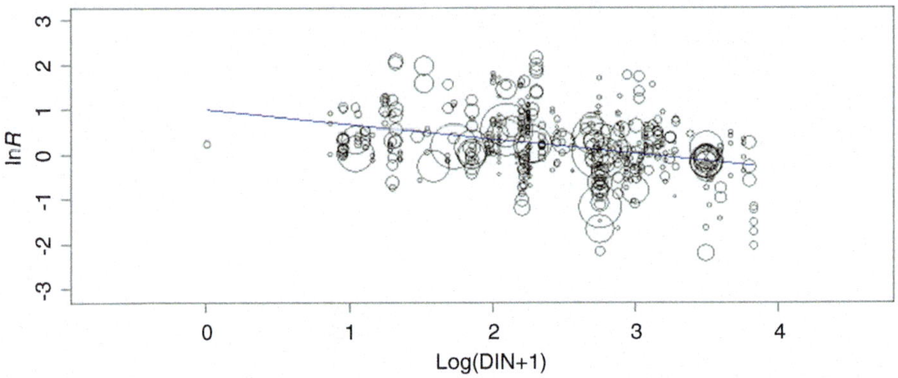

Fig. 63.3 Meta-regression to assess the effect of dissolved inorganic nitrogen (DIN) concentration in control conditions (Log (x + 1)-transformed) on the response of litter decomposition to nutrient enrichment (ln*R*) ($n = 511$). The size of the symbols reflects their precision, with effect sizes with larger symbols being more precise and thus given greater weight in the analysis. The response of litter decomposition to nutrient enrichment decreases by 0.326 (slope; $p < 0.001$) for each unit increase in DIN concentration in control conditions suggesting that the response for litter decomposition to nutrient enrichment is stronger in systems with naturally low nutrient concentration. The model explains 13% of the variation (output from OpenMEE, data from Ferreira et al. 2015)

3.6 Publication Bias

A meta-analysis allows a precise synthesis of a selected set of studies, but if the selected studies do not fairly represent all conducted studies, then the estimated global effect size will be biased. Bias may be the result of publication bias, when the results of the primary studies affect their probability of being published, the so-called file drawer problem (e.g. non-significant results or results contrary to current theory are less likely to be submitted to journals). Bias may also occur as a result of dissemination bias, when the presentation of the primary studies affects their probability of being found (e.g. language bias, English vs. other languages; citation bias, highly cited vs. less cited studies; publication site bias, high- vs. low-impact journals). Both publication bias and dissemination bias are known jointly as publication bias.

An intensive and extensive literature search (Sect. 2.2) can overcome many of these potential biases. However, there is always the need to evaluate if the database or subsets of the database (and thus the results of the analyses) are affected by publication bias. There are several methods to assess the potential impact of publication bias in the meta-analysis, which assume that bias is negatively correlated with sample size in primary studies. These methods include:

Comparing global effect size of published studies vs. *'grey literature'*. In the absence of bias, the results should not differ significantly. If they differ (and there are no important differences in methodological/environmental characteristics between the two groups), this indicates publication bias, which might be corrected by including the 'grey literature'. The database should still be tested using the methods below.

Relating effect size and precision (meta-regression). Without bias, the relationship should be non-significant; if low precision studies have larger effect sizes than more precise studies, the analysis should focus on more precise studies.

Fail safe numbers (*Nfs*; e.g. Rosenthal, Orwin, Rosenberg) estimate the number of studies with non-significant results (which may have been missed in the literature search) that are needed to nullify the global effect size (Rosenthal, Rosenberg) or to reduce it to an ecologically non-relevant value (Orwin) (Borenstein et al. 2009). It does not consider missing studies that may report results in the opposite direction. If $Nfs > 5 \times n + 10$ (n = number of effect sizes), the global effect size is robust to publication bias; if the *Nfs* are lower, the literature search and/or inclusion/exclusion criteria may need to be revised.

The *funnel plot* is a scatter plot of effect sizes vs. precision or sample sizes that, in the absence of bias, is symmetrical around the global effect size with a wider distribution of effect sizes for less precise studies. This gives the plot a funnel shape. It is less efficient when the number of effect sizes is low and it does not consider that asymmetry may have other causes (e.g. differences in experimental approaches among studies).

Duval and Tweedie's trim and fill method, which imputes the 'missing' effect sizes to the funnel plot and estimates a new global effect size (Borenstein et al. 2009). The comparison of the original global effect size and the new estimate allows assessing the degree to which the original global effect size is affected by publication bias. If publication bias has a strong effect on the results, the literature search and/or inclusion/exclusion criteria may need to be revised.

3.7 Sensitivity Analyses

Several decisions may affect the outcome of a meta-analysis. Sensitivity analyses allow assessing if and to what extent these decisions affect the results, i.e. they assess the robustness of the results. Technically, sensitivity analyses imply repeating some analyses using different criteria and performing subgroup analyses or meta-regressions on different data sets or using 'decision' moderators (Ferreira et al. 2015). Sensitivity analyses may include assessing the effect of:

Considering multiple effect sizes per study. This is typical of ecological studies but violates the assumption of independence of effect sizes. Thus, it is necessary to show that violating this assumption does not strongly affect the results. To that end, a single effect size per study is computed (using study as moderator in a subgroup analysis), and a new global effect size is estimated and compared with the global effect size estimated from all effect sizes.

Study quality by comparing effect sizes in different classes of study quality (subgroup analysis) or assessing the relationship between effect size and study quality (meta-regression).

Including 'grey literature', which may be considered of substandard quality since it did not go through peer review, by comparing effect sizes from published studies vs. 'grey literature' (subgroup analysis).

Including effect sizes with imputed or recalculated data, which may be less accurate than reported data, by comparing effect sizes from reported data vs. effect sizes from imputed or recalculated data (subgroup analysis).

Including particular studies (e.g. studies that contribute with an exceptionally large number of effect sizes, studies with unusual characteristics), by comparing results with and without these studies.

4 Quality in Meta-Analysis

The number of systematic reviews using meta-analysis is increasing in ecology, but not all reports are of high quality (Koricheva and Gurevitch 2014). Although meta-analysis is generally performed in the context of a (systematic) review, its procedures and reporting should follow closely those of a primary study. A traditional literature review generally makes no attempt to locate *all* studies, does not describe why some studies are included and others excluded, and has limited capability to deal with a large number of studies with variable outcomes. This gives it a high degree of subjectivity and low degree of reproducibility.

In a *systematic* literature review, the reviewer needs to keep record of all steps and decisions and report these with sufficient detail to allow reproduction of the results (as in a primary study!). A systematic literature review needs to start with preparing detailed protocols (as in a primary study!) for literature search, inclusion/exclusion criteria, study appraisal, and coding (these protocols can be revised during the process, with all changes being annotated). The Collaboration for Environmental Evidence website (http://www.environmentalevidence.org/) lists examples of protocols for systematic literature reviews.

The number of studies located and retained at each step of the literature search, application of inclusion/exclusion criteria, and study appraisal can be presented in a PRISMA (Preferred Reporting Items for Systematic reviews and Meta-Analyses; Liberati et al. 2009) flow diagram, which can be generated online (PRISMA flow diagram generator: http://prisma.thetacollaborative.ca/).

In order to carry out and report a high-quality meta-analysis, the quality criteria compiled by Koricheva and Gurevitch (2014) should be checked.

References

Amani, M., Graça, M. A. S., & Ferreira, V. (2019). Effects of elevated atmospheric CO2 concentration and temperature on litter decomposition in streams: A meta-analysis. *International Review of Hydrobiology, 104*, 14–25.

Borenstein, M., Hedges, L. V., Higgins, J., & Rothstein, H. R. (2009). *Introduction to meta-analysis*. Chichester: Wiley.

Borenstein, M., Hedges, L. V., Higgins, J., & Rothstein, H. R. (2017). *Comprehensive meta-analysis, version 3*. Englewood: Biostat.

Ferreira, V., & Guérold, F. (2017). Leaf litter decomposition as a bioassessment tool of acidification effects in streams: Evidence from a field study and meta-analysis. *Ecological Indicators, 79*, 382–390.

Ferreira, V., Castagneyrol, B., Koricheva, J., Gulis, V., Chauvet, E., & Graça, M. A. S. (2015). A meta-analysis of the effects of nutrient enrichment on litter decomposition in streams. *Biological Reviews, 90*, 669–688.

Ferreira, V., Koricheva, J., Duarte, S., Niyogi, D. K., & Guérold, F. (2016a). Effects of anthropogenic heavy metal contamination on litter decomposition in streams – A meta-analysis. *Environmental Pollution, 210*, 261–270.

Ferreira, V., Koricheva, J., Pozo, J., & Graça, M. A. S. (2016b). A meta-analysis on the effects of changes in the composition of native forests on litter decomposition in streams. *Forest Ecology and Management, 364*, 27–38.

Hedges, L. V., Gurevitch, J., & Curtis, P. S. (1999). The meta-analysis of response ratios in experimental ecology. *Ecology, 80*, 1150–1156.

Kampichler, C., & Bruckner, A. (2009). The role of microarthropods in terrestrial decomposition: A meta-analysis of 40 years of litterbag studies. *Biological Reviews, 84*, 375–389.

Knorr, M., Frey, S. D., & Curtis, P. S. (2005). Nitrogen additions and litter decomposition: A meta-analysis. *Ecology, 86*, 3252–3257.

Koricheva, J., & Gurevitch, J. (2014). Uses and misuses of meta-analysis in plant ecology. *Journal of Ecology, 102*, 828–844.
Koricheva, J., Gurevitch, J., & Mengersen, K. (2013). *Handbook of meta-analysis in ecology and evolution*. Princeton/London: Princeton University Press.
Lecerf, A., Marie, G., Kominoski, J. S., LeRoy, C. J., Bernadet, C., & Swan, C. M. (2011). Incubation time, functional litter diversity, and habitat characteristics predict litter-mixing effects on decomposition. *Ecology, 92*, 160–169.
Liberati, A., Altman, D. G., Tetzlaff, J., Mulrow, C., Gøtzsche, P. C., Ioannidis, J. P. A., Clarke, M., Devereaux, P. J., Kleijnen, J., & Moher, D. (2009). The PRISMA statement for reporting systematic reviews and meta-analyses of studies that evaluate health care interventions: Explanation and elaboration. *PLoS Medicine, 6*, e1000100.
Mancinelli, G., Sangiorgio, F., & Scalzo, A. (2013). The effects of decapod crustacean macroconsumers on leaf detritus processing and colonization by invertebrates in stream habitats: A meta-analysis. *International Review of Hydrobiology, 98*, 206–216.
Sabater, S., Bregoli, F., Acuña, V., Barceló, D., Elosegi, A., Ginebreda, A., Marcé, R., Muñoz, I., Sabater-Liesa, L., & Ferreira, V. (2018). Effects of human-driven water stress on river ecosystems: A metaanalysis. *Scientific Reports, 8*, 11462.
Viechtbauer, W. (2010). Conducting meta-analyses in R with the metafor package. *Journal of Statistical Software, 36*, 1–48.
Wallace, B. C., Lajeunesse, M. J., Dietz, G., Dahabreh, I. J., Trikalinos, T. A., Schmid, C. H., & Gurevitch, J. (2017). OpenMEE: Intuitive, open-source software for meta-analysis in ecology and evolutionary biology. *Methods in Ecology and Evolution, 8*, 941–947.

Index

F. Bärlocher et al. (eds.), *Methods to Study Litter Decomposition*,
https://doi.org/10.1007/978-3-030-30515-4

D

E

F

The manufacturer's authorised representative in the EU is Springer Nature Customer Service Centre GmbH, Europaplatz 3, 69115 Heidelberg, Germany. If you have any concerns regarding our products, please contact ProductSafety@springernature.com

Printed and bound by CPI Group (UK) Ltd, Croydon, CR0 4YY
30/07/2026
02182010-0007